IONIC EQUILIBRIUM

IONIC EQUILIBRIUM
Solubility and pH Calculations

James N. Butler
Division of Applied Sciences
Harvard University, Cambridge MA 02138

With a chapter by
David R. Cogley
Sudbury, MA 01776

A WILEY-INTERSCIENCE PUBLICATION
JOHN WILEY & SONS, INC

New York / Chichester / Weinheim / Brisbane / Singapore / Toronto

This book is printed on acid-free paper. ∞

Copyright © 1998 by John Wiley & Sons, Inc. All rights reserved.

Published simultaneously in Canada.

No part of this publication may be reproduced, stored in a retrieval system or transmitted in any form or by any means, electronic, mechanical, photocopying, recording, scanning or otherwise, except as permitted under Sections 107 and 108 of the 1976 United States Copyright Act, without either the prior written permission of the Publisher, or authorization through payment of the appropriate per-copy fee to the Copyright Clearance Center, 222 Rosewood Drive, Danvers, MA 01923, (978) 750-8400, fax (978) 750-4744. Requests to the Publisher for permission should be addressed to the Permissions Department, John Wiley & Sons, Inc., 605 Third Avenue, New York, NY 10158-0012, (212) 850-6011, fax (212) 850-6008, E-Mail: PERMREQ@WILEY.COM

Library of Congress Cataloging-in-Publication Data

Butler, James Newton.
    Ionic equilibrium : solubility and pH calculations / James N. Butler with a chapter by David R. Cogley.
       p.   cm.
    "A Wiley-Interscience publication."
    Includes bibliographical references and index.
    ISBN 0-471-58526-2 (alk. paper)
    1. Ionic equilibrium.   I. Cogley, David R.   II. Title.
QD561.B985   1998   97-13435
541.3'723—dc21

CONTENTS

Preface / vii

Acknowledgments / xi

1. Basic Principles / 1
2. Activity Coefficients and pH / 41
3. Strong Acids and Bases / 65
4. Monoprotic Acids and Bases / 94
5. Polyprotic Acids and Bases / 157
6. Solubility / 202
7. Complex Formation / 238
8. Organic Complexes / 289
9. Oxidation–Reduction Equilibria / 318
10. Carbon Dioxide / 365
11. pH in Brines / 461
12. Automated Computation Methods (by David R. Cogley) / 485

Index / 543

PREFACE

Almost 40 years ago, when I was just out of graduate school and beginning my career at the University of British Columbia, I was assigned to teach Analytical Chemistry to undergraduate Chemistry and Engineering students. I had been dissatisfied with the way I had been taught to do equilibrium calculations, but as I looked through the many textbooks in the library and in the offices of my colleagues, I was no more satisfied than before.

I began to work on an approach using mass and charge balances without "intuitive" approximations—much to the confusion of my students – but didn't really appreciate the depth of that approach until I read a chapter by Lars Gunnar Sillén in the Kolthoff and Elving *Treatise on Analytical Chemistry*.[1] Sillén used a graphical approach, with pH as a master variable, and showed how some of the most difficult problems in the old textbooks (e.g., "Find the pH of a solution containing 0.01M $NaHSO_4$, 0.005 M NaH_2PO_4, and 0.02 M HF") could be solved approximately with a graph, and solved exactly with a little additional numerical work.

This work developed into a 547-page book, *Ionic Equilibrium*,[2] published in 1964. Since I had originally intended to write a little book for undergraduate students, I extracted some of the simpler things from *Ionic Equilibrium*, and published them as *Solubility and pH Calculations*,[3] a 104-page paperback. Although I left UBC in 1963 to join Tyco Laboratories, a recently established research and development firm, and didn't return to academic life until I came to Harvard in 1971, *Ionic Equilibrium* and *Solubility and pH Calculations* became standard works. They were translated into Spanish and Russian, remained in print in English until about 1986, and are still cited more than 20 times per year.

I hesitate to tamper with the biggest success of my career by trying to do it again. There are many books in print now, even elementary general chemistry books, which use mass and charge balances, and present Sillén's logarithmic concentration diagrams. When I wrote *Ionic Equilibrium*, I did all the calculations with a slide rule and graph paper. Now every serious student has access to a personal computer with spreadsheet software—and this became the focus of William B. Gunther's recent book.[4]

One of the reasons *Ionic Equilibrium* was useful and lasted so long was that it gave a rigorous presentation of the principles and then showed how these were applied in examples. Each chapter had dozens of problems, and some were quite challenging—but in the back of the book, a student could find more than a numerical answer, sometimes a lengthy description of how to solve these difficult ones. There is no question in my mind that the wide availability of personal computers can make these problems more accessible, and that any new presentation should take that into account.

Another factor, which surprised me, was how easily activity coefficients could be added to the calculations without changing any of the basic concepts or approaches. In fact, activity coefficients are part of the earliest discussion in *Solubility and pH*—more radical than I would have thought, since in the early 1960s there was a strong prejudice that activity coefficients were OK for graduate courses in physical chemistry, but they shouldn't be mentioned to undergraduates. Unfortunately, the rigorous approaches were adorned with hundreds of symbols and thousands of equations,[5] and I can understand how a chemical engineer, geologist, or physiologist might find a simple approach with only a few symbols and one equation very comforting, even if it wasn't as accurate as the state of the art.

For this new edition, I have combined the two titles into one, and divided the book into two parts. The first part (Chapters 1–8) presents the basic principles of acid–base, solubility, and complex formation equilibria, beginning at the most elementary level and ending with some advanced presentations on titrations and multicomponent systems. Chapter 9 discusses oxidation–reduction equilibria; Chapter 10 develops the principles specifically for carbon dioxide; then presents some case studies of how carbon dioxide equilibria are used in physiology and oceanography. Chapter 11 explores the possibility of a pH scale for brines; and Chapter 12, by David R. Cogley, describes general automated computer programs for performing equilibrium calculations on systems of many components.

This book does not address surface reactions,[6] ion exchange,[7] chemical kinetics[8] or transport processes,[9] all of which are extremely important in chemical engineering, environmental science and engineering, and medical science. However, a solid background in equilibrium models is certainly a help for someone going on to these more difficult topics.

WHO WILL USE THIS BOOK?

This book begins at a level where a student with a solid background in general chemistry, algebra, and elementary calculus can understand it. While concepts of thermodynamics and electrolyte theory will be used, it will not be necessary for the user of the book to have studied these topics in detail.

Many examples, with details of calculations, comprise a large part of the text. Problems are included with the first eight chapters as a help and inspiration to those using this book as part of an undergraduate or graduate course. Much of the more advanced material will be useful to professionals doing research or engineering, particularly if they have had little formal instruction in these topics.

Selected equilibrium constants are included, but there are far too many data for a comprehensive appendix. The most important compilations[10] are referred to frequently, and the serious student or investigator is encouraged to make their personal acquaintance.

I have used this style of presentation for 30 years in courses attended by undergraduates concentrating in chemistry, engineering science, and earth and planetary sciences. It has also proved useful as a way of bringing graduate students in environmental science, who come with diverse backgrounds, up to the same level of understanding.

JAMES N. BUTLER

Wayland, Massachusetts

NOTES

1. L. G. Sillén, "Graphic Presentation of Equilibrium Data," Chapter 8, Volume 1 of *Treatise on Analytical Chemistry*, edited by I. M. Kolthoff, P. J. Elving, and E. B. Sandell. The Interscience Encyclopedia, New York, 1959.
2. J. N. Butler, *Ionic Equilibrium, a Mathematical Approach*. Addison-Wesley, Reading, MA, 1964, 547 pp.
3. J. N. Butler, *Solubility and pH Calculations*, Reading, MA: Addison-Wesley, 1964. 104 pp.
4. W. B. Gunther, *Unified Equilibrium Calculations*, Wiley, New York, 1990.
5. See, for example, H. S. Harned and B. B. Owen, *The Physical Chemistry of Electrolytic Solutions*, 3d Ed., Reinhold, New York, 1958.
6. See W. Stumm and J. J. Morgan, *Aquatic Chemistry*, 3d Ed., Wiley, New York, 1996, Chapters 9, 13, 14; F. M. M. Morel and J. G. Hering, *Principles and Applications of Aquatic Chemistry*, Wiley, New York, 1993; W. Stumm, *Chemistry of the Solid–Water Interface*, Wiley-Interscience, New York, 1992; A. W. Adamson, *Physical Chemistry of Surfaces*, 5th Ed. Wiley-Interscience, New York, 1990; D. A. Dzombak and F. M. M. Morel, *Surface Complexation Modeling: Hydrous Feric Oxide*, Wiley, New York, 1990.
7. See Stumm and Morgan, *op. cit.* pp. 586–93; Morel and Hering, *op. cit.* pp. 556–58; G. Sposito, *Chemical Equilibria and Kinetics in Soils*, Oxford Univ. Press, Oxford, 1994; H. F. Walton and R. D. Rocklin, *Ion Exchange in Analytical Chemistry*. CRC Press, Boca Raton, FL, 1990; J. A. Marinsky, Y. Marcus, Eds., *Ion Exchange and Solvent Extraction*, Marcel Dekker, New York, Vols. 3–10 (1973–1987).
8. See Stumm and Morgan, *op. cit.* Chapters 11, 12, 13; W. Stumm, Ed., *Aquatic Chemical Kinetics: Reaction Rates of Processes in Natural Waters*, Wiley, New York, 1990.
9. See Stumm and Morgan, *op. cit.* Chapters 10 and 15; M. M. Clark, *Transport Modeling for Environmental Engineers and Scientists*, Wiley, New York, 1996; L. J. Thibodeaux, *Environmental Chemodynamics: Movement of Chemicals in Air, Water, and Soil*, 2nd ed. Wiley, New York, 1996; R. E. Ney, *Fate and Transport of Organic Chemicals in the Environment: A Practical Guide*, 2nd ed., Government Institutes, Rockville, MD, 1995; H. E. Hemond and E. J. Fechner, *Chemical Fate and Transport in the Environment*, Academic Press, San Diego, CA, 1994.
10. R. M. Smith and A. E. Martell, *NIST Critically Selected Stability Constants of Metal*

Complexes Database, version 3.0, National Institute of Standards and Technology, Gaithersburg MD, 1997. R. M. Smith and A. E. Martell, *Critical Stability Constants*, Vols. 1–6, Plenum Press, New York, 1974, 1975, 1976, 1977, 1982, 1989; L. G. Sillén and A. E. Martell, *Stability Constants of Metal Ion Complexes*, The Chemical Society, London, Special publications Nos. 17 (1964) and 25 (1971).

ACKNOWLEDGMENTS

First, I want to thank my son, Christopher J. Butler, for introducing me to his document-processing program, FrameMaker, which I used to compose most of this book. Chris also made a major contribution to the typesetting phase of this book, working closely with House of Equations, to be sure all was correct. Dr. David Cogley is acknowledged as the author of Chapter 12, but he also reviewed Chapters 9–11, which were greatly improved by his comments. Patricia Van Ness protected me from unnecessary interruptions while letting through the important ones, and cheerfully did whatever had to be done around the office.

Drs. Thomas Maren, Giles Filley, and Neal Kindig introduced me to carbon dioxide and ion-pairing equilibria in physiological systems in 1984-1985; more recently, Dr. David Wong and his colleagues at Via Medical Corp. gave me the opportunity of studying these, and also provided some data, which are quoted in Chapter 10. I want also to thank Dr. Larry Brush and his co-workers at the Sandia National Laboratories for introducing me to the "pH in brine" problem, providing some valuable data for Chapter 11, and supporting the research behind that chapter.

The entire manuscript for this book was reviewed with meticulous attention by Professor Andrew Dickson of the Scripps Institute of Oceanography and by two anonymous reviewers. Chapter 12 was also reviewed by Dr. David L. Parkhust of the U.S. Geological Survey, Dr. William Schecher of Environmental Research Software, and Dr. Thomas J. Wolery of the Lawrence Livermore National Laboratory.

Financial support was provided by Harvard University and the Zemurray Foundation.

This book is dedicated to my wife, Rosamond Hatch Butler, and to our more than 30 years together.

IONIC EQUILIBRIUM

1

BASIC PRINCIPLES

What You Need to Know
Equilibrium
   When Is a System at Equilibrium?
   Equilibrium As a Balance of Opposing Reactions
   Simultaneous Equilibria
   Concentration Scales
The Equilibrium Constant
   Forms for the Equilibrium Constant Expression
      Ion Product of Water
      Dissociation Constant of a Weak Acid
      Dissociation Constant of a Weak Base
      Stepwise Formation Constants For Complexes
      Overall Formation Constants
      Solubility Product
   Where to Find Equilibrium Constants
      Temperature Dependence, Enthalpy, and Entropy
Mass Balances
Charge Balances
   The Proton Condition
Solving an Equilibrium Problem
   Example: The Weak Acid Problem
   Some Simple but Useful Advice
   Polynomial Equations
     Quadratic Equations
     Newton's Approximation Method
     Applying Newton's Method: Finding an Approximate Root
     Some Difficulties

The Secant Method
Automatic Function Solvers: "Goal Seek"
Problems
Note on construction of Figs. 1.1 and 1.2

WHAT YOU NEED TO KNOW

This book is more or less self-contained, but depends on some knowledge of chemistry and mathematics, specifically:

- General chemistry of salts (electrolytes), acids, bases, and gases.
- Elementary knowledge of organic chemical structures.
- Familiarity with the concept of equilibrium.
- Skill in manipulating algebraic expressions
- Knowledge of elementary differential calculus
- Familiarity with a computer system for numerical calculations

Many of the calculations presented in this book can be done with a pocket calculator, provided it has the ability to do $\log x$ and 10^x functions. Others will proceed more easily and quickly if done on a computer.

The use of personal computers has become almost universal, and there are several systems you might be familiar with. One group would include sequential programming languages such as BASIC, FORTRAN, ALGOL, C, or PASCAL. In these systems, the computer reads some input data, makes a calculation according to your instructions, and produces some output data. Their advantage is flexibility in logic, but their disadvantage is that you need to know a fair body of code for instructions and for formatting the input and output. Some currently available general programs are described in Chapter 12.

Another group contains the spreadsheets, such as Lotus™ and Excel™. These were originally developed for financial analysis, but have turned out to be highly adaptable to chemical equilibrium. In a speadsheet, input and output values are continuously displayed in the cells of a large matrix. A cell can contain a numerical value, or a formula for calculating the cell's value from the values found in other cells. The advantage of the spreadsheet is its flexibility of display and simplicity of operation for ordinary calculations; its disadvantage is that calculations involving iteration and choice (DO loops, IF, and GOTO) are limited.

Other application programs (such as Mathematica™, Maple™, MathCad™) manipulate algebraic expressions; others make numerical evaluation of expressions easier (such as TK! Solver™ and the "solver" or "goal-seek" modules for Lotus™ and Excel™). If readily available, these may ease the tedium of solving the algebraic and numerical part of equilibrium problems. But be wary of trying to solve highly nonlinear sets of equations—"solver" algorithms I have seen appear to have been designed to work primarily with linear or nearly linear systems.

Except for Chapter 12, examples that require more calculation than can be done on a pocket calculator will be illustrated by calculations in Excel™.[1]

EQUILIBRIUM

When substances are mixed and undergo a chemical change, a chemical reaction is said to have occurred. Relatively few reactions proceed in such a way that the reactants are completely consumed and only a unique product remains. Those reactions that proceed nearly to completion are useful because they provide a basis for chemical synthesis or quantitative analysis, but even those reactions that are sufficiently complete to be used for analysis are not complete in any rigorous sense. The products may react with each other, the opposition of the forward and reverse reactions resulting in equilibrium, with measurable amounts of the original reactants present. Such a situation constitutes a source of error in quantitative analysis for which corrections can be made.

This book is concerned with equilibrium in reactions that occur between ions in aqueous solutions. There are several reasons for restricting this presentation to such an apparently narrow field:

- First, many substances studied in inorganic chemistry, analytical chemistry, geochemistry, and environmental chemistry exist as ions in solution.
- Second, reactions between ions are usually very rapid, and can reach a state of equilibrium almost as fast as the reactants can be mixed. Ionic reactions at equilibrium have thus been extensively studied, and a large body of experimental data is available.
- Third, simple but powerful theoretical principles governing reactions at equilibrium are known. By applying these principles, you can make quantitative theoretical predictions about the results of chemical experiments, including the numerical estimation of many quantities that would be extremely difficult to measure directly, but that may be of importance for many applications.

When Is a System at Equilibrium?

There is a very simple conceptual test for true equilibrium: Allow the reaction to proceed in the forward direction until nothing more appears to happen, and determine the chemical composition of the system.[2] Then allow the same reaction to proceed in the reverse direction until nothing more appears to happen. If the composition of the system is the same regardless of the direction from which it was approached, then a true equilibrium exists. An example of this technique in practice is determining the solubility of a mineral by exposing it first to an undersaturated solution, then to an oversaturated solution. If the solution composition obtained is the same in both tests, then it is considered to be in equilibrium with the mineral.

It may happen that two substances are mixed but fail to react under ordinary conditions. It may happen that a different composition is reached from the forward and reverse directions. Such reactions do not proceed to a true equilibrium. Many reactions studied in organic chemistry proceed to one set of products under one condition and to a different set of products under slightly different conditions. In most cases the products of such reactions do not reach true equilibrium with the reactants. Rather, the reaction proceeds to one set of products because the rate of the reaction forming these products is somewhat faster than the rate of the reaction forming some other products. There may be hundreds of possible paths that the reaction could take if all the products could be detected, and this makes life very complicated for the chemist studying such reactions. Slow reactions with many possible paths are characteristic of the chemistry of covalent bonds.

In contrast, it has been possible in many cases of reactions involving ions in solution to reduce to simple mathematical statements the laws governing the concentrations of all the various ions and molecules in an equilibrium mixture. This means that so long as you are sure that you are dealing with a true equilibrium mixture, you can calculate theoretically the concentration of all the species present in solution, and this result can be verified by experiment.

Equilibrium as a Balance of Opposing Reactions

From the preceding discussion, it is an easy step to conceive of the equilibrium state as a dynamic balance between forward and reverse reactions. In approaching equilibrium, the reactants combine to give products, but the products themselves can combine to give the original reactants. Eventually, a balance is reached at equilibrium, where the rate at which the forward reaction proceeds is just balanced by the rate at which the backward reaction proceeds. The dynamic aspect is reflected in the observation that changes in the concentration of reactants or products, or changes in temperature, will shift the equilibrium by altering the relative rates of forward and reverse reactions.

For the sake of concreteness, consider a particularly well-studied reaction, the dissociation of acetic acid into its ions. When acetic acid (CH_3COOH, abbreviated HAc) is dissolved in water, it releases a hydrogen ion, which combines with a nearby water molecule to give a hydrated hydrogen ion (H_3O^+), leaving an acetate ion (CH_3COO^-, abbreviated Ac^-). This dissociation does not occur completely, because at the same time that molecules of acetic acid are breaking apart, acetate ions and hydrogen ions are recombining. At equilibrium these two processes exactly balance each other, and the concentrations of acetic acid, acetate ion, and hydrogen ion are independent of time. Equilibrium is represented by a double arrow in the equation for the reaction:

$$\text{Acetic acid} + \text{water} \rightleftharpoons \text{acetate ion} + \text{hydrated hydrogen ion}$$

$$HAc + H_2O \rightleftharpoons Ac^- + H_3O^+$$

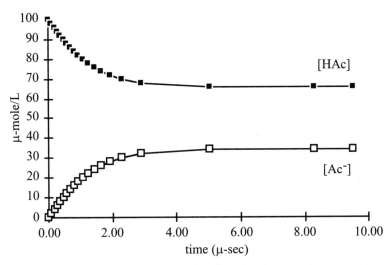

FIGURE 1.1. Variation of the concentrations of undissociated acetic acid and acetate ion with time when the initial concentration of acetic acid is 1.00×10^{-4} mole/L, and the initial concentration of acetate is zero. Note that equilibrium is reached in approximately 5 microseconds. (See the note at the end of this chapter for details of how these curves were constructed.)

The hydrated hydrogen ion actually has several more water molecules hydrogen-bonded to it (see the following). Dissociation of acetic acid takes place rapidly, being essentially at equilibrium only a few microseconds after a molecule of acetic acid is introduced into aqueous solution. Even though the equilibrium of acetic acid had been thoroughly studied before 1900, it was not until the 1960s that techniques were developed to measure the rate of dissociation of acetic acid and the rate of association of acetate ions with hydrogen ions.[3]

Consider a hypothetical experiment, the results of which can be calculated from Eigen's measurements: 100 μmole (micromoles[4]) of acetic acid are mixed instantaneously with one liter of pure water (1.00×10^{-4} mol/L). The concentrations of reactant, acetic acid, and product, acetate ion, are plotted in Fig. 1.1 as a function of time after the instantaneous mixing.

The concentration of acetic acid decreases rapidly at first, and then more slowly, until after several microseconds its concentration becomes constant at 66 μmole/L. At the same time the concentration of acetate ions rises to its maximum value of 34 μmole/L. These final values are the equilibrium concentrations present in 10^{-4} molar acetic acid in water, and they will not change unless some substance is added or the temperature is changed.

Figure 1.2 shows the approach to equilibrium from the reverse direction. Here the hypothetical experiment requires that 100 micromoles of sodium acetate and 100 micromoles of hydrochloric acid be instantaneously mixed with one liter of water. (The sodium and chloride ions do not enter into the reaction, but are there merely to keep the charges balanced in the reacting substances.) The acetate ion concentration decreases from its initial value of 100 micromoles/liter, rapidly at first, then

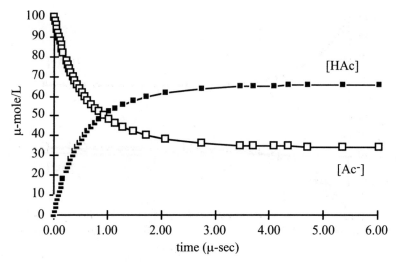

FIGURE 1.2. Variation of undissociated acetic acid and acetate ion with time when acetate ion and hydrogen ion are introduced simultaneously at a concentration of 1.00×10^{-4} mole/L, and the initial concentration of undissociated acetic acid is zero. Note that the same equilibrium values are reached as in Fig. 1.1, and that the time required to reach equilibrium is the same in both cases.

more slowly, until it reaches a stable value at 34 micromoles/liter. At the same time the acetic acid concentration increases to a maximum value of 66 micromoles/liter. Equilibrium is reached in several microseconds, and the concentrations are the same as were obtained from the forward reaction.

Note that the total concentration of acetate groups, which is equal to the sum of acetic acid and acetate ion concentrations, remains constant at 100 micromoles/liter regardless of the extent to which the reaction occurs or the direction from which equilibrium is approached. This is a direct result of the fact that the acetate groups (CH_3COO) are conserved—remain intact throughout the reaction.

Simultaneous Equilibria

The idea of equilibrium arising from two opposing reactions is a very fundamental one, but it is often incorrectly applied. Rigorously, the idea can be applied only to elementary reactions, such as the dissociation of acetic acid, which take place in a single step without the formation of any intermediate species.

Many common reactions are quite complicated when considered in detail and may involve several intermediate species. For these, considering equilibrium as resulting from the opposition of only two reactions is too simple, and to understand fully the quantitative effects of changing conditions on the equilibrium, all the intermediate species and reactions must be taken into account. For example, addition of NaOH to acetic acid may be thought of as proceeding in steps:

$$HAc \rightarrow H^+ + Ac^-$$
$$H^+ + OH^- \rightarrow H_2O$$

but if an additional base, such as NH_3, is added, an additional reaction consumes H^+

$$H^+ + NH_3 \rightarrow NH_4^+$$

and the relative rate of this reaction compared to the reaction with OH^- is important. At equilibrium the reverse reactions proceed at the same rate as the forward reactions:

$$H_2O \rightarrow H^+ + OH^-$$
$$NH_4^+ \rightarrow H^+ + NH_3$$
$$H^+ + Ac^- \rightarrow HAc$$

The concentration of Na^+ remains unaffected by these reactions, and the equilibria are summarized by

$$HAc \rightleftharpoons H^+ + Ac^-$$
$$NH_4^+ \rightleftharpoons H^+ + NH_3$$
$$H^+ + OH^- \rightleftharpoons H_2O$$

Such simultaneous acid–base reactions are discussed in Chapters 4 and 5.

Consider a more complicated example. When a solution of copper sulfate is mixed with concentrated ammonia, it changes from pale blue to a deep blue-violet color. The overall reaction is often given as

$$Cu^{2+} + 4\, NH_3 \rightleftharpoons Cu(NH_3)_4^{2+}$$

since most of the copper in the blue-violet solution is found to have four molecules of ammonia attached to it. More careful and detailed measurements of this reaction, in solutions of varying ammonia concentrations, show that four other species are present that contain copper and various numbers of molecules of ammonia.[5] The reaction can thus be represented as the five steps,

$$Cu^{2+} + NH_3 \rightleftharpoons CuNH_3^{2+}$$
$$CuNH_3^{2+} + NH_3 \rightleftharpoons Cu(NH_3)_2^{2+}$$
$$Cu(NH_3)_2^{2+} + NH_3 \rightleftharpoons Cu(NH_3)_3^{2+}$$
$$Cu(NH_3)_3^{2+} + NH_3 \rightleftharpoons Cu(NH_3)_4^{2+}$$
$$Cu(NH_3)_4^{2+} + NH_3 \rightleftharpoons Cu(NH_3)_5^{2+}$$

In salts such as $Cu(NH_3)_4SO_4 \cdot H_2O$, the tetrammine copper ion $Cu(NH_3)_4^{2+}$ is present in the crystal lattice, but in solution some of these ions lose ammonia molecules, while others may gain one, and all the possible species containing copper and ammonia are present to some extent.

X-ray diffraction measurements on crystals of hydrated copper salts such as $CuSO_4 \cdot 5H_2O$ show that in these, four water molecules are firmly bound to the copper ion. In solution, however, spectroscopic studies have indicated that two additional water molecules are bound to the copper ion at a somewhat greater distance. The reaction of copper ion with ammonia would thus be better represented as the replacement of water molecules by ammonia:

$$Cu(H_2O)_6^{2+} + NH_3 \rightleftharpoons Cu(NH_3)(H_2O)_5^{2+} + H_2O, \text{ etc.}$$

By analogy, a species with six ammonia molecules would be expected to exist, but it has not been detected in equilibrium studies. Even at the highest concentrations of ammonia that it is possible to obtain in aqueous solutions, $Cu(NH_3)_6^{2+}$ is present in concentrations too small to detect.[6]

Water molecules attached to copper ions can lose hydrogen ions in solution to form ions like $CuOH^+$; these can join together to form a dimer $Cu_2(OH)_2^{2+}$, which also exists in equilibrium with the copper ammine complexes. You will notice that extra water molecules are usually not written in the formulas for ions. For example, $Cu_2(OH)_2^{2+}$ might be written:

$$\begin{array}{ccccc} H_2O & & OH & & H_2O^{2+} \\ & \diagdown & \diagup & \diagdown & \diagup \\ & Cu & & Cu & \\ & \diagup & \diagdown & \diagup & \diagdown \\ H_2O & & OH & & H_2O \end{array}$$

Like most positively charged ions, the hydrated hydrogen ion is attached to one or more water molecules—although it is often written as H_3O^+, the unit $H_9O_4^+$ is spectroscopically identifiable even at the boiling point, and much larger units exist at lower temperatures (see Chapter 3). You may want to keep in mind this complicated hydration, but for simplicity, the hydrated hydrogen ion will be abbreviated H^+, and other hydrated ions will be written with the least number of water molecules needed to describe their composition.

Finally, in addition to its equilibria with copper ion, ammonia can take on a hydrogen ion to give an ammonium ion

$$NH_3 + H^+ \rightleftharpoons NH_4^+$$

and water dissociates to give a hydrogen ion and a hydroxyl ion

$$H_2O \rightleftharpoons H^+ + OH^-$$

In all, there may be eight or nine simultaneous equilibria necessary to quantitatively describe the species present in this solution. Such examples will be treated in Chapter 7.

Concentration Scales

Three concentration units: mole fraction (N), molarity (mole per liter of solution, M), and molality (mole per kg solvent, m), are employed in precise studies of soutions.

Molarity (mol/L) carries the practical advantage of making up solutions accurately by using volumetric glassware. It also is the scale that appears naturally during the course of applying electrostatic theory to ionic solutions, and therefore the expression of the ionic strength (see below) is normally in terms of mol/L. For example, in a solution containing 5.844 g of NaCl in 1.000 L solution, the molarity is 0.100 mol/L

Molality (mol/kg) is frequently used in precise physicochemical measurements because concentrations prepared entirely by weight can be more accurate than those dependent on volume, and even more important, are independent of temperature. For example, a solution containing 5.844 g of NaCl plus 1 kg water is 0.100 molal.

In dilute aqueous solutions, molarity and molality are approximately the same since the density of water and dilute solutions is approximately 1.00. For example,

1 L of 0.1000 molar NaCl solution weighs 1.0025 kg, contains 0.100 mole = 5.844 g of NaCl and (by difference) 0.9967 kg of water.

The molal concentration is thus 0.100 mole/0.997 kg, or 0.1003 mol/kg.

For more concentrated solutions, the density of the solution can be significantly different from 1; for example,

A solution containing 26% by weight NaCl is 260 g NaCl/kg soln, or 4.449 mol/kg soln.

Since the density of this solution is 1.1972 kg/L, 1 kg of solution occupies 0.8353 L, and the molar concentration is 4.449 mol/0.8353 L, or 5.326 mol/L.

One kg of solution contains 260 g NaCl and 1000 − 260 = 740 g water. Therefore, the molal concentration is 4.449 mol / 0.740 kg H_2O = 6.012 mol/kg H_2O

Under these circumstances, the two scales differ by 11%, and correction of one scale to the other is highly desirable.

For a single component, these scales are related by

$$N = \frac{m}{m + 1000/MW_{solvent}} \qquad (1)$$

$$M = \frac{m\rho_{soln}}{1 + 0.001m(MW_{solute})} \qquad (2)$$

where N is the mole fraction, m is the molality, and M is the molarity of the same solution. $MW_{solvent}$ is the molecular weight of water (18.0153 g/mol), MW_{solute} is the molecular weight of the solute (58.443 g/mol for NaCl in the example above), and ρ_{soln} is the density of the solution (g/mL or kg/L).[7]

Throughout this book concentration of a particular species will be denoted by square brackets around the formula of that species, for example, $[H^+]$ or $[AgCl_2^-]$. The concentration scale will be explicitly noted if it is important (mol/L in volumetric titrations or mol/kg in accurate thermodynamic measurements, or in seawater) but if not mentioned, can be assumed to be mol/L. The major exception is the activity of water, which, as mentioned above, is based on the mole fraction scale, and is usually close to 1.00.

THE EQUILIBRIUM CONSTANT

The basic principle on which the quantitative study of chemical equilibrium in solution is based is the existence of an equilibrium constant and its specific algebraic form. Guldberg and Waage proposed a "Law of Mass Action" in 1867,[8] and this family of laws was correctly formulated by Van't Hoff in 1877.[9] During the next 30 years, many notable physical chemists—Van't Hoff, Ostwald, Arrhenius, Nernst, and Gibbs—developed the quantitative understanding of equilibrium. The chemical equilibrium law can be derived from the second law of thermodynamics,[10] but for our purposes, it will be presented as an empirical result.

As an example, consider again the dissociation of acetic acid. In Table 1.1 are given concentrations of undissociated acid HAc, acetate ion Ac^-, and hydrogen ion H^+, data simplified from measurements published in the scientific literature. The table includes solutions of acetic acid in pure water, mixtures of sodium acetate and acetic acid solutions, and mixtures of hydrochloric acid with acetic acid.

The concentration equilibrium constant K_a is defined by the equation

$$K_a = \frac{[H^+][Ac^-]}{[HAc]} \tag{3}$$

which corresponds to the stoichiometric reaction

$$HAc \rightleftharpoons H^+ + Ac^-$$

Note that the concentrations of the products appear in the numerator of the expression for K_a, and the reactant concentration appears in the denominator.

To simplify typography, the equilibrium expression can be written

$$[H^+][Ac^-] = K_a [HAc] \tag{4}$$

The values of K_a computed in Table 1.1 vary from 1.74×10^{-5} to 2.1×10^{-5} as the total concentration is varied from 10^{-5} to 10^{-2} mole/L. When the total concentration is constant, as in the second group of data, K_a varies much less: from $1.89 \times$

TABLE 1.1. Equilibrium concentrations in acetic acid solutions

Total Conc.	[HAc]	[Ac⁻]	[H⁺]	K_a
\multicolumn{5}{c}{*Acetic acid in water (conc. in mol/L):*}				
1.00×10^{-5}	2.9×10^{-6}	7.1×10^{-6}	7.1×10^{-6}	1.74×10^{-5}
1.00×10^{-4}	6.6×10^{-5}	3.4×10^{-5}	3.4×10^{-5}	1.75×10^{-5}
1.00×10^{-3}	8.73×10^{-4}	1.27×10^{-4}	1.27×10^{-4}	1.85×10^{-5}
1.00×10^{-2}	9.65×10^{-3}	4.5×10^{-4}	4.5×10^{-4}	2.10×10^{-5}
\multicolumn{5}{c}{*Acetic acid–sodium acetate mixtures. Total conc. 1.00×10^{-3} mol/L*}				
	9.0×10^{-4}	1.0×10^{-4}	1.7×10^{-4}	1.89×10^{-5}
	7.0×10^{-4}	3.0×10^{-4}	4.4×10^{-5}	1.89×10^{-5}
	5.0×10^{-4}	5.0×10^{-4}	1.8×10^{-5}	1.80×10^{-5}
	4.0×10^{-4}	6.0×10^{-4}	1.3×10^{-5}	1.95×10^{-5}
\multicolumn{5}{c}{*Acetic acid–hydrochloric acid mixtures. Total conc. 1.00×10^{-3} mol/L*}				
	9.0×10^{-4}	1.7×10^{-5}	1.0×10^{-4}	1.89×10^{-5}
	7.0×10^{-4}	4.4×10^{-5}	3.0×10^{-4}	1.89×10^{-5}
	5.0×10^{-4}	1.8×10^{-5}	5.0×10^{-4}	1.80×10^{-5}
	3.0×10^{-4}	8.0×10^{-6}	7.0×10^{-4}	1.86×10^{-5}
	1.0×10^{-4}	2.0×10^{-6}	9.0×10^{-4}	1.80×10^{-5}

Note: Direct measurement of all the concentrations is not experimentally feasible. In practice the hydrogen ion concentration is measured potentiometrically (see pH) and the concentrations of the other ions are calculated from the known amounts of material put in–the "analytical" or "total" concentrations (see "mass and charge balances" below). These data were adapted from the experimental results of H. S. Harned and G. M. Murphy, *J. Am. Chem. Soc.* **53**:8–17 (1931). See Chapter 2 (p. 59) for a more rigorous analysis of these data.

10^{-5} to 1.95×10^{-5}. Much of this variability results from changes in the concentration of all ions in the solution.[11]

These deviations are expressed quantitatively by the activity coefficients γ of the various species—factors that multiply the concentration and correct for nonideal behavior[12]:

$$[H^+]\gamma_+ [Ac^-]\gamma_- = K_a^\circ [HAc]\gamma_0 \tag{5}$$

The product of a concentration and its activity coefficient is the activity, a quantity that appears in thermodynamic functions. K_a° is dependent on temperature, but not on solution composition. Sometimes K_a° is called the "thermodynamic" or "activity" equilibrium constant, in contrast to the "concentration" constant illustrated above. The concentration constant is equally valid thermodynamically, but depends on the ionic medium as well as the temperature

As implied by the subscripts on γ, the activity coefficients of different species depend on their charge as well as ionic strength; at higher solution concentrations, they also depend on the concentrations of particular oppositely charged ions. These topics are presented quantitatively in Chapter 2.

Forms for the Equilibrium Constant Expression

The general statement of the equilibrium law may be made as follows. For a reaction of the form

$$aA + bB \rightleftharpoons cC + dD$$

where A, B, C, and D are chemical species, and a, b, c, and d are the stoichiometric coefficients required to balance the equation, the following relation will be obeyed at equilibrium:

$$K° = \frac{\{C\}^c \{D\}^d}{\{A\}^a \{B\}^b} \tag{6}$$

Curly braces indicate the activity of the species within the braces. In general, the activity is interpreted as follows:

1. *Ions and molecules in dilute solutions.* Activity is approximately equal[13] to concentration in mol/L or mol/kg. The ratio of activity to concentration is the activity coefficient:

$$\{A\} = [A]\, \gamma_A \tag{7}$$

Activity coefficients can be based on any concentration scale.[14]

2. *Solvent in a dilute solution.* Activity is approximately equal to the mole fraction of solvent. The mole fraction of the solvent approaches 1.000 in dilute solutions.[15]

3. *Pure solids and liquids in equilibrium with the solution.* Activity is exactly 1.

4. *Gases in equilibrium with the solution.* Activity is partial pressure of the gas in atmospheres. At high pressures, an activity (or "fugacity") coefficient may be required to correct for nonideal behavior.[16]

5. *Mixtures of liquids.* Activity of a given component is approximately equal to its mole fraction.

Although this seems like an arbitrary set of rules, all these interpretations of activity have their roots in the same thermodynamic function, the chemical potential, and differ only by a multiplicative factor from each other. If the equation for $K°$ is accurately expressed in terms of activities, then $K°$ depends only on the temperature and total pressure, but not on the composition of the system.

Some common examples are illustrated in the following.

Ion Product of Water. A very important equilibrium in aqueous solution is the dissociation of water to give hydrogen and hydroxyl ions.

$$H_2O \rightleftharpoons H^+ + OH^-$$

THE EQUILIBRIUM CONSTANT

$$\{H^+\}\{OH^-\} = K° \{H_2O\}$$
$$[H^+]\gamma_+[OH^-]\gamma_- = K_w° \gamma_o$$
$$[H^+][OH^-] = K_w \tag{8}$$

In the sequence above, the form of the chemical reaction leads to the form of the equilibrium constant equation using activities. The activity of H_2O is then set equal to an activity coefficient γ_o (e.g., 1.000 for pure water, 0.98 for 1 M NaCl; see footnote 15) to yield the third form, which also explicitly states the activity coefficients of H^+ and OH^-. The final form is the "concentration" constant, in which the activity coefficients have been folded into the equilibrium constant (See Table 3.1 for numerical values):

$$K_w = K_w°(\gamma_o/\gamma_+\gamma_-) \tag{9}$$

The concentration constant is important since concentrations, not activities, enter the mass and charge balances (see below). In addition, if a large excess (1 to 5 mol/L) of a noncomplexing electrolyte, such as NaCl or $NaClO_4$, is added to the solution, the activity coefficients are controlled by that background electrolyte and remain constant when smaller concentrations (which may be of primary interest) are changed. Thus K_w will also remain constant—but it will be a function of the background electrolyte concentration as well as temperature and pressure.[17]

Dissociation[18] Constant of a Weak Acid. We have already discussed acetic acid as an example of a general equilibrium. Weak acids may ionize to give more than one proton: acetic acid is a monoprotic acid, yielding one proton per molecule as base is added; carbon dioxide is diprotic; and phosphoric acid is triprotic, yielding three protons. For example (see Chapter 5):

$$H_3PO_4 \rightleftharpoons H_2PO_4^- + H^+, \quad [H^+][H_2PO_4^-] = K_{a1}[H_3PO_4]$$
$$H_2PO_4^- \rightleftharpoons HPO_4^{2-} + H^+, \quad [H^+][HPO_4^{2-}] = K_{a2}[H_2PO_4^-]$$
$$HPO_4^{2-} \rightleftharpoons PO_4^{3-} + H^+, \quad [H^+][PO_4^{3-}] = K_{a3}[HPO_4^{2-}]$$

Dissociation Constant of a Weak Base. All the anions formed by dissociation of a weak acid like phosphoric acid can be considered as weak bases. Likewise, the product of ionization of a weak base such as ammonia can be considered to be a weak acid. Thus the ionization reaction

$$NH_3 + H_2O \rightleftharpoons NH_4^+ + OH^-$$

with equilibrium constant expression

$$[NH_4^+][OH^-] = K_b[NH_3]$$

could also be considered to be the dissociation of the weak acid NH_4^+:

$$NH_4^+ \rightleftharpoons NH_3 + H^+, \quad [H^+][NH_3] = K_a [NH_4^+]$$

The relation between K_b and K_a is found by substituting the ion product of water in one of the equilibria and identifying terms with the other:

$$K_a = K_w/K_b$$

This symmetry is expressed by saying, for example: "HPO_4^{2-} is the conjugate base to $H_2PO_4^-$, and HPO_4^{2-} is the conjugate acid to PO_4^{3-}."

Stepwise Formation Constants for Complexes. Similar to the polyprotic acid equilibria are the equilibria for the formation of complex ions (see Chapter 7). These are usually written in the reverse direction, as formation rather than dissociation constants. For example, the formation of four cadmium chloride complexes may be represented by the four equilibria:

$$Cd^{2+} + Cl^- \rightleftharpoons CdCl^+, \quad [CdCl^+] = K_1 [Cd^{2+}][Cl^-]$$

$$CdCl^+ + Cl^- \rightleftharpoons CdCl_2, \quad [CdCl_2] = K_2 [CdCl^+][Cl^-]$$

$$CdCl_2 + Cl^- \rightleftharpoons CdCl_3^-, \quad [CdCl_3^-] = K_3 [CdCl_2][Cl^-]$$

$$CdCl_3^- + Cl^- \rightleftharpoons CdCl_4^{2-}, \quad [CdCl_4^{2-}] = K_4 [CdCl_3^-][Cl^-]$$

The numerical index on the equilibrium constant gives the number of chlorides in the complex formed by the reaction. Activities have been replaced by concentrations for simplicity, but activity coefficients could easily be introduced:

$$[CdCl^+]\gamma_+ = K_1^\circ [Cd^{2+}][Cl^-]\gamma_{2+}\gamma_-$$

Overall Formation Constants. In calculations on complex ion systems, the equations have fewer symbols if "overall" formation constants are used. These are combinations of the stepwise formation constants. In order to describe the equilibrium completely, however, there must be just as many overall constants as there were stepwise constants. For example, the cadmium chloride system could be represented:

$$Cd^{2+} + Cl^- \rightleftharpoons CdCl^+, \quad [CdCl^+] = \beta_1 [Cd^{2+}][Cl^-], \quad \beta_1 = K_1$$

$$Cd^{2+} + 2\,Cl^- \rightleftharpoons CdCl_2, \quad [CdCl_2] = \beta_2 [Cd^{2+}][Cl^-]^2, \quad \beta_2 = K_1 K_2$$

$$Cd^{2+} + 3\,Cl^- \rightleftharpoons CdCl_3^-, \quad [CdCl_3] = \beta_3 [Cd^{2+}][Cl^-]^3, \quad \beta_3 = K_1 K_2 K_3$$

$$Cd^{2+} + 4\,Cl^- \rightleftharpoons CdCl_4^{2-}, \quad [CdCl_4^{2-}] = \beta_4 [Cd^{2+}][Cl^-]^4, \quad \beta_4 = K_1 K_2 K_3 K_4$$

As above, the numerical index on the equilibrium constant gives the number of chlorides in the complex formed. It is important to realize that whereas the stepwise formation reactions may represent elementary steps in the formation of complexes, the overall formation reactions do not represent elementary steps, but are only mathematically convenient combinations.

Occasionally, the reciprocal of the overall formation constant for the highest complex is tabulated as the "instability constant" of that complex. For the cadmium chloride system this would be

$$CdCl_4^{2-} \rightleftharpoons Cd^{2+} + 4\ Cl^-, \qquad [Cd^{2+}][Cl^-]^4 = K_{inst}\ [CdCl_4^{2-}], \qquad K_{inst} = 1/\beta_4$$

Such a constant by itself would be useful only if the concentrations of all the intermediate complexes were negligible. This is rarely the case.

Solubility Product. Another very important equilibrium constant is the solubility product. This holds when a solid ionic salt is in equilibrium with a solution containing its ions. As long as the solid is a pure substance, its activity is unity. For example, solid silver chloride dissolves to give silver and chloride ions:

$$AgCl(s) \rightleftharpoons Ag^+ + Cl^-$$

$$\{Ag^+\}\{Cl^-\} = K_{s0}^{\circ} \qquad (10)$$

In terms of concentrations:

$$[Ag^+][Cl^-] = K_{s0} \qquad (11)$$

Similarly, solids producing more than two ions have equilibrium constant expressions with more than two terms:

$$Mg(OH)_2(s) \rightleftharpoons Mg^{2+} + 2\ OH^-$$

$$[Mg^{2+}][OH^-]^2 = K_{s0} \qquad (12)$$

Although "so" could be thought of as an abbreviation for "solubility",[19] the notation K_{s0} is actually part of a larger system that indicates the degree of complexation of ions.[20] Here "s0" is "s-zero" and indicates that there are no chlorides bound to the silver ion.

When a metal forms a slightly soluble salt as well as a series of complexes with the same anion, the equilibria can be given in terms of the reactions forming the various complexes from the salt. For example, the constant K_{s2} would indicate that two chlorides were bound to the silver ion. The full series of equilibria for AgCl in excess chloride is:

$$AgCl(s) \rightleftharpoons Ag^+ + Cl^-, \qquad [Ag^+][Cl^-] = K_{s0}$$
$$AgCl(s) \rightleftharpoons AgCl(aq), \qquad [AgCl] = K_{s1}, \qquad K_{s1} = \beta_1\ K_{s0}$$
$$AgCl(s) + Cl^- \rightleftharpoons AgCl_2^-, \qquad [AgCl_2^-] = K_{s2}[Cl^-], \qquad K_{s2} = \beta_2\ K_{s0}$$
$$AgCl(s) + 2\ Cl^- \rightleftharpoons AgCl_3^{2-}, \qquad [AgCl_3^{2-}] = K_{s3}[Cl^-]^2, \qquad K_{s3} = \beta_3\ K_{s0}$$
$$AgCl(s) + 3\ Cl^- \rightleftharpoons AgCl_4^{3-}, \qquad [AgCl_4^{3-}] = K_{s4}[Cl^-]^3, \qquad K_{s4} = \beta_4\ K_{s0}$$

In the second reaction, AgCl(aq) indicates dissolved covalent[21] silver chloride molecules as distinct from solid salt or dissolved ions. All the above equilibria hold only in the presence of excess undissolved silver chloride. The indexes give the number of chlorine atoms in the complex formed by the reaction. As indicated in the last column, the constants K_{sn} can be obtained by multiplying the overall formation constant β_n by the solubility product K_{s0}.

Where to Find Equilibrium Constants

The chemical literature is vast, but well indexed. The richest and most reliable printed compilation of equilibrium constants is W. Smith and A. E. Martell, *Critical Stability Constants*.[22] These authors have selected what they consider to be the best literature values for thousands of metal–ligand equilibria, primarily at 25°C, but at a number of different ionic strengths. Enthalpy values are also given, which allow a limited amount of temperature correction near 25°C.

The predecessor volumes, by L. G. Sillén and A. E. Martell[23] did not attempt critical selection of data, but rather presented a summary of all data available at the time, with an index to all the available literature. Hence these are still good sources for data at temperatures other than 25°C and media other than water.

The most comprehensive current compilation is the NIST database.[24]

When simple compilations fail, an exhaustive search can be made, either in printed indexes or on computer systems, using *Chemical Abstracts*, *Science Citation Index*, etc.

Temperature Dependence, Enthalpy, and Entropy. From any text on thermodynamics you can find that the free energy change ΔG for a reaction is related to the equilibrium constant K, the enthalpy ΔH, and entropy ΔS by

$$\Delta G° = -RT \ln K° = \Delta H° - T \Delta S° \tag{13}$$

where R is the gas constant (1.987 cal mol^{-1} K^{-1}, or 8.314 J mole^{-1} K^{-1}), T is the absolute temperature, and the superscript ° refers to the standard state of the various components in the reaction. Values of $\Delta G°$, $\Delta H°$, and $\Delta S°$ can be found in many standard compilations of thermodynamic data.[25] An additional equation for the temperature dependence of $K°$ is

$$\frac{d \log_{10} K°}{dT} = \frac{\Delta H°}{2.303 RT^2} = (2.46 \times 10^{-3} \text{ kcal}^{-1})\Delta H° = (4.06 \times 10^{-5} \text{ J}^{-1})\Delta H° \tag{14}$$

where the numerical coefficient 2.46×10^{-3} applies at 25°C for $\Delta H°$ in kcal mole^{-1} and the coefficient 4.06×10^{-5} applies for $\Delta H°$ in J mol^{-1}. This equation is particularly useful when $K°$ and $\Delta H°$ are known at one temperature (say 25°C) and $K°$ is to be estimated at a nearby temperature (say 35°C). Since $\Delta H°$ also varies with temperature, this procedure cannot be relied on over a wide temperature range (see Fig. 1.3).

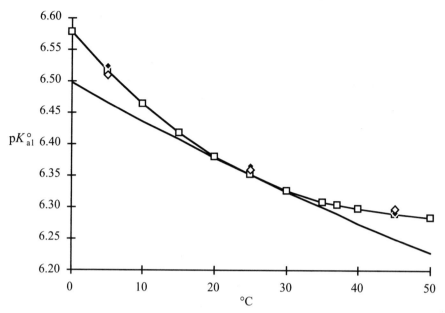

FIGURE 1.3. pK_{a1}°, the first dissociation constant of CO_2, as a function of temperature. In the region near 25°C, a constant value of $dpK_{a1}/dT = 0.0054 \times 10^{-3}$ or $\Delta H = -2.19$ kcal/mole gives a satisfactory approximation to the temperature dependence, but not at temperatures less than 15°C or above 35°C. For references to original literature, see Fig. 10.2.

MASS BALANCES

Equilibrium constants alone are not enough to determine the set of simultaneous equations to be solved. Additional relations can be obtained from mass balances on each of the components in the system. Each element or group that appears in a number of species can be the subject of a mass balance: The total amount of A put into the system (sometimes written $[A]_T$, sometimes written C_A or C_{HA} and called "analytical concentration of A or HA") must equal the sum of all the various species in which A exists:

$$[A]_T = [HA] + [A^-] \tag{15}$$

Note that concentrations, not activities, appear in mass balances.

If there are several sources of A, the total concentration is the sum of these sources. For example, if one liter of a solution contains C_1 moles of acetic acid HA, C_2 moles of sodium acetate NaA, and C_3 moles of calcium acetate CaA_2, the mass balance on acetate might be

$$[A]_T = C_1 + C_2 + 2\,C_3 = [HA] + [A^-] + [CaA^+] + \cdots \tag{16}$$

The notation "+···" means that there may be other terms in the right-hand side of the balance (such as a soluble complex NaA), but that they are small. Note that the third term in total concentration is 2 C_3 because each mole of CaA_2 introduces 2 moles of A to the solution. Once in the solution, the species come to equilibrium at concentrations different from the original ingredients. For example, [HA] is not exactly equal to C_1, and [A^-] is not exactly equal to C_2+ 2 C_3, because protons are transferred between these species (the extent depends on pH) and some acetates are taken up in the complex with calcium, CaA^+. Nevertheless, each unit of A introduced by the various ingredients ends up in one of the solution species, and hence the totals on each side of the balance are equal.

If excess of a solid phase is in equilibrium with the solution, the total amount in solution is called the *molar solubility*, denoted by S. For example, if excess barium sulfate is equilibrated with pure water, the mass balance on barium is:

$$[Ba^{2+}] = S \tag{17}$$

since there is only one dissolved species containing barium. For sulfate, there are two possible species:

$$[SO_4^{2-}] + [HSO_4^-] = S \tag{18}$$

Since S is the same in both mass balances, it can be eliminated to give

$$[Ba^{2+}] = [SO_4^{2-}] + [HSO_4^-] \tag{19}$$

Similarly, if two solutions are mixed and a precipitate is formed, the amount of precipitate can be designated P, which is included in the appropriate mass balances. For example, if a solution containing silver nitrate at total concentration [Ag]$_T$ is mixed with a solution of sodium chloride at total concentration [Cl]$_T$, the mass balances might include complexes of silver and chloride as well as the free ions and the precipitate:

$$[Ag]_T = P + [Ag^+] + [AgCl(aq)] + [AgCl_2^-] + \cdots \tag{20}$$

$$[Cl]_T = P + [Cl^-] + [AgCl(aq)] + 2\,[AgCl_2^-] + \cdots \tag{21}$$

Note that P, which represents the total number of moles of AgCl precipitate per liter of solution, enters both mass balances, as does the uncharged dissolved complex AgCl(aq). Because the complex AgCl$_2^-$ contains two chloride ions, its concentration is multiplied by 2 in the chloride mass balance. If the complex AgCl$_4^{3-}$ were included, its concentration would be multiplied by four.

A precipitate with different stoichiometry, such as Ag$_2$SO$_4$(s), would have the term P in the SO$_4$ mass balance, and the term $2P$ in the Ag mass balance. The value of P from each mass balance is the same, so that if P is not known, it can be eliminated by combining Eqs. (20) and (21) to produce a single hybrid mass balance, such as:

CHARGE BALANCES

$$[Ag]_T - [Cl]_T = [Ag^+] - [Cl^-] - [AgCl_2^-] + \cdots \quad (22)$$

A similar technique is used in acid–base equilibria to obtain the proton condition, as described below.

CHARGE BALANCES

Mass balances on hydrogen or hydroxyl ions, if done literally, can introduce large roundoff errors because they involve the solvent water (55.56 mol/L), and typical concentrations of interesting ions range from 1.0 to less than 10^{-14} mol/L. A charge balance embodies the same information, but involves only ionic species and not the solvent. For example, consider acetic acid:

$$[H^+] = [Ac^-] + [OH^-] \quad (23)$$

On the left side, all the positively charged species are listed; on the right, all the negatively charged species. Since this balance counts charges, it uses concentrations, not activities. Uncharged species (such as HA) are not included. If a species has more than one charge, its concentration is multiplied by the number of charges. For example, consider phosphoric acid:

$$[H^+] = [H_2PO_4^-] + 2\,[HPO_4^{2-}] + 3\,[PO_4^{3-}] \quad (24)$$

Note that [H_3PO_4], being an uncharged species, is not included in the charge balance.

The Proton Condition

An alternative to the charge balance is the "proton condition," which expresses the gain or loss of hydrogen ions from the "basis species," which are usually taken to be those of highest concentration. For the basis species acetic acid and water, a proton can be added to water (but not to acetic acid); a proton can be lost from water, to form OH$^-$, or from acetic acid, to give acetate ion. The gains are balanced by the losses:

$$[H^+] = [Ac^-] + [OH^-] \quad (25)$$

Note that this equation is identical to the charge balance above.

For the system sodium acetate–water, a different balance is obtained: Here the basis species are Ac$^-$ and H$_2$O. Acetate can gain a proton to form HAc, but can not lose one; water can gain a proton to form H$^+$ or lose a proton to form OH$^-$:

$$[H^+] + [HAc] = [OH^-] \quad (26)$$

This relation can also be obtained by combining the charge and mass balances for the sodium acetate–water system:

$$[H^+] + [Na^+] = [Ac^-] + [OH^-] \qquad (27)$$

$$[Na^+] = [Ac^-] + [HAc] \qquad (28)$$

Equation (28) expresses the stoichiometry of NaAc, but allows for Ac to be either in the form of the ion Ac^- or the acid HAc. Eliminating $[Na^+]$ between Eqs. (27) and (28) gives the proton condition of Eq. (26).

In general, the proton condition can always be obtained by combining mass and charge balances, and so provides no new information. Its advantage is the elimination of large terms (i.e., $[Na^+]$ and $[Ac^-]$ in the above example) and in some cases providing a simpler equation than the charge balance.

SOLVING AN EQUILIBRIUM PROBLEM

Throughout this book you will face the mathematical as well as the chemical aspects of equilibrium calculations. That is, you will be presented with a physical situation, consisting of certain substances dissolved in water, and you will desire (or be asked) to find the concentrations all the various ions and molecules that may exist in the solution. Even if only one concentration is required (as for an application to analytical chemistry), in the process of obtaining this value and checking the calculations, you obtain all the other concentrations.

Solving an equilibrium problem consists of a number of steps:

1. Establish the nature of all the species present in the solution. This involves complicated experimental and theoretical procedures, and for most chemical systems, the nature of the species present is not fully known. If you are working on a real-life problem, you will probably want to search the primary literature as well as the tables mentioned above. For the examples in this book, such information will be given as a bonus lesson in descriptive chemistry.
2. Find the equilibrium constants relating the concentrations of the various species. These are usually determined in the same experimental investigations in which the nature of the species in solution is established, and are listed in tables of stability constants. For a first approximation, concentration constants extrapolated to $I = 0$ or any other ionic strength will do, but for accurate work, all the constants should be corrected to the ionic strength and (as with sea water) to the complete electrolyte composition applicable to the problem.
3. Find enough other relations so that there are as many independent equations as unknowns. These include mass and charge balances, as well as the total (analytical) concentrations of each component introduced to the solution. Often the pH ($= -\log\{[H^+]\gamma_+\}$) is given; sometimes it is an unknown. Sometimes the solubility (an analytical concentration) is the unknown; sometimes one or more equilibrium constants are the unknowns. In any case the number of unknown quantities must be equal to the number of independent equations relating them.

SOLVING AN EQUILIBRIUM PROBLEM

4. Solve this system of n simultaneous equations for all n unknowns. To do this exactly can be tedious,[26] and part of what you will learn in this book is how to make approximations based on chemical knowledge to simplify the mathematical problem: Is this solution acidic? Basic? Is the substance slightly soluble? Very soluble? Solve the approximate equations to give a provisional answer.

5. Check the provisional answer by substitution in the full original set of exact equations. If these are not satisfied (say, to ±5%), new approximations are made and the new approximate equations solved. It may be necessary to use a more complicated approximation than before.

6. Once all the n unknowns are found, calculate the ionic strength and adjust the equilibrium constants to correspond; repeat until the answers do not change.

7. Check the answers finally by substitution in the full original set of exact equations.[27]

An example of this approach is given below.

Example: The Weak Acid Problem

Anticipating the discussion in Chapter 4, consider the problem of calculating pH in a solution consisting of a mixture of $C_{HA} = 10^{-3.0}$ mole/L of acetic acid and $C_A = 10^{-3.0}$ mole/L of sodium acetate. Following the steps given above:

1. The species to be considered are acetic acid HA, acetate ion A^-, sodium ion Na^+, hydrogen ion H^+, and hydroxyl ion OH^-. The medium is water.

2. The relevant equilibria are the dissociation of the weak acid and the ion product of water (concentration constants, which include the activity coefficients for the solution medium as mentioned above, are approximately the same as the concentration constants because the solution is dilute[28]):

$$[H^+][A^-] = K_a [HA], \quad K_a = 10^{-4.75} \tag{29}$$

$$[H^+][OH^-] = K_w, \quad K_w = 10^{-14.00} \tag{30}$$

3. There is a mass balance on acetic acid/acetate, and one on sodium:

$$C_{HA} + C_A = [HA] + [A^-] \tag{31}$$

$$C_A = [Na^+] \tag{32}$$

4. Finally there is a charge balance:

$$[H^+] + [Na^+] = [A^-] + [OH^-] \tag{33}$$

Thus there are 5 independent equations relating 9 variables, four of which are normally known: C_{HA}, C_A, K_a, and K_w, and five of which are unknowns: $[H^+]$, $[Na^+]$,

[A^-], [HA], and [OH^-]. (Alternate problems could be posed in which, for example, [H^+] is known and K_a is one of the unknowns—this is how K_a is determined experimentally.)

4. To solve these five simultaneous nonlinear equations by the "tedious but foolproof" method of direct substitution, note that [H^+] and [A^-] each occur in three equations, but [Na^+], [HA], and [OH^-] each occur in two equations. The following path is one of many: Substitute C_A = [Na^+] from Eq. (32) and [OH^-] = K_w/[H^+] from Eq. (30) in the charge balance (Eq. 33), reducing the problem to 3 equations in 3 unknowns (Eqs. 29, 31, and 34, unknowns [H^+], [A^-], [HA]):

$$[H^+] + C_A = [A^-] + K_w/[H^+] \qquad (34)$$

At this point, the variable occurring the fewest times is [HA]; it can be eliminated by solving Eq. (29) and substituting in Eq. (30):

$$C_{HA} + C_A = \frac{[H^+][A^-]}{K_a} + [A^-] \qquad (35)$$

Finally, solve Eq. (35) for [A^-] and substitute in Eq. (34):

$$[A^-] = \frac{(C_{HA} + C_A)}{([H^+]/K_a + 1)} \qquad (36)$$

$$[H^+] + C_A = \frac{(C_{HA} + C_A)}{([H^+]/K_a + 1)} + \frac{K_w}{[H^+]} \qquad (37)$$

The final equation[29] (Eq. 37) has only one unknown, [H^+], but is not trivial to solve. If you multiply it out, you will see that two terms ($C_A K_a$) cancel, and the result is a cubic in [H^+]:

$$[H^+]^3 + [H^+]^2(C_A + K_a) - [H^+](C_{HA}K_a + K_w) - K_w K_a = 0 \qquad (38)$$

This equation can be solved numerically by methods descried in the next section. The simplest is to calculate the left-hand side for a range of [H^+] values and see where it goes to zero. This is easily accomplished using a spreadsheet program, which yields a number close to zero (1% of the adjacent values) at [H^+] = 1.718 × 10^{-5} or pH = 4.765 (see Table 1.2).

Note that if you knew in advance which terms were large and which were small, an approximate answer could be obtained by combining Eqs. (32) and (33) to obtain

$$[H^+] + C_A = [A^-] + [OH^-]$$

Neglecting [H^+] and [OH^-]:

$$[A^-] = C_A + \cdots$$

SOLVING AN EQUILIBRIUM PROBLEM

TABLE 1.2. Calculation (using Excel™) of left hand side of Eq. (38) as a function of pH

Chapter 1 Eq. (38)—cubic for HA/NaA				
CHA	0.001			
CA	0.001			
Ka	1.7783E-05	=10^-4.75		
Kw	1E-14			
pH	H = 10^-pH	=H^3+H^2*(CA+Ka)-H*(CHA*Ka+Kw)-Kw*Ka		
4	0.0001	9.39955E-12		
4.5	3.1623E-05	4.87064E-13		
4.7	1.9953E-05	5.83161E-14		
4.75	1.7783E-05	1.12465E-14		
4.76	1.7378E-05	3.58367E-15		
4.765	1.7179E-05	-5.3557E-17	<--"zero"	
4.77	1.6982E-05	-3.566E-15		
4.8	1.5849E-05	-2.2202E-14		
4.9	1.2589E-05	-6.0569E-14		
5	0.00001	-7.505E-14		
5.5	3.1623E-06	-4.6025E-14		
6	0.000001	-1.6764E-14		
6.5	3.1623E-07	-5.5218E-15		
7	0.0000001	-1.7683E-15		
7.5	3.1623E-08	-5.615E-16		
8	0.00000001	-1.779E-16		

Substituting in Eq. (31) gives another approximate relation:

$$[HA] = C_{HA} + \cdots$$

and the equilibrium (Eq. 29) gives

$$[H^+] = K_a [HA]/[A^-] = K_a C_{HA}/C_A + \cdots$$

$$[H^+] = 10^{-4.75} (0.001)/(0.001) = 1.778 \times 10^{-5}$$

or pH = 4.75. Compare this with the answer obtained in Table 1.2 using all terms in the equations, which is slightly higher—pH = 4.765

5. Check the provisional answers:

$$[H^+] = 1.718 \times 10^{-5}$$

$$[OH^-] = K_w/[H^+] = 5.821 \times 10^{-10}$$

$$[Na^+] = C_A = 1.00 \times 10^{-3}$$

From Eq. (33):

$$[A^-] = [H^+] + [Na^+] - [OH^-]$$
$$= 1.718 \times 10^{-5} + 1.00 \times 10^{-3} - 5.821 \times 10^{-10} = 1.017 \times 10^{-3}$$

From Eq. (31):

$$[HA] = C_{HA} + C_A - [A^-]$$
$$= 1.00 \times 10^{-3} + 1.00 \times 10^{-3} - 1.017 \times 10^{-3} = 9.83 \times 10^{-4}$$

All the original equations have been used to obtain these five answers, except for the equilibrium, Eq. (29). Testing this equation results in:

$$[H^+][A^-] = K_a [HA]$$
$$[H^+][A^-] = (1.718 \times 10^{-5})(1.017 \times 10^{-3}) = 1.747 \times 10^{-8}$$
$$K_a [HA] = (1.778 \times 10^{-5})(9.83 \times 10^{-4}) = 1.748 \times 10^{-8}$$

The two sides agree within 0.1%.

6. The ionic strength of the solution (see Eq. 2 of Chapter 2, p. 45) is:

$$I = 0.5 \,([Na^+] + [A^-] + [H^+] + [OH^-])$$
$$= 0.5 \,(1.000 \times 10^{-3} + 9.83 \times 10^{-4} + 1.718 \times 10^{-5} + 5.821 \times 10^{-10})$$
$$= 1.000 \times 10^{-3}$$

As mentioned in Footnote 28, this ionic strength is low enough that additional adjustments, if made, would be small.[30]

Some Simple but Useful Advice

In carrying out these steps, it is important to keep a clear head. George Polya outlined four *equally important* steps in his classic book *How to Solve It*:[31]

1. **You have to understand the problem.**
   - What are the unknowns?
   - What are the knowns?
   - What condition relates them?
2. **Make a plan of solution.**
   - Do you know a related problem?
   - Look at the unknown.
   - Do you know a problem with the same unknown?
   - Is it related to the problem at hand?

SOLVING AN EQUILIBRIUM PROBLEM 25

- Go back to definitions.
- Can you solve part of the problem?
- Can you solve the problem with slightly different conditions?
- Could you change the data or unknown so that they are more closely related?
3. **Carry out the plan of solution, checking each step as you go.**
   - Is each step reasonable?
   - Can you prove that each step is logically or mathematically correct?
4. **Examine the solution.**
   - Is the result reasonable?
   - If the problem is symmetrical with respect to one of the unknowns, is the solution also symmetrical with respect to that unkown?
   - Do numerical answers fall within the limits imposed by the statement of the problem? For example, solution concentrations are always greater than zero, and species concentrations are always less than total concentrations for that material.
   - Does a complicated answer reduce to a simpler, known answer under special conditions?
   - For what other problems could you use this method or this result?

Keeping these rules in mind will make solution of the problems proposed in this book much easier and more instructive.

Polynomial Equations

The example in the previous section is typical. If no approximations are made, the final equation is a polynomial of high degree. The rest of this book will give you instruction on how to make realistic approximations using your chemical knowledge. But the ability to solve polyomial equations can be useful as a last resort.

The most general and direct method for a one-variable polynomial is "curve crawling," the technique displayed in Table 1.2. It is equivalent to plotting $f(x)$ as a function of x and noting where the curve crosses the x axis, and is ideally suited to spreadsheet calculations. It is somewhat time consuming, however, since the initial steps in x do not define the root any more accurately than the step size (0.1 in Table 1.2). Once the root is found to be between 4.7 and 4.8, a second series of smaller steps (0.01) localizes the root between 4.76 and 4.77. A third stage fixes it at 4.765.

Quadratic Equations. Early in your mathematical training, you will probably have learned the general solution to the quadratic equation $ax^2 + bx + c = 0$, which is

$$x = \frac{-b \pm \sqrt{b^2 - 4ac}}{2a}$$

To get a positive answer for x, one or two of the terms (a, b, c) must be negative. However, blind use of this formula sometimes leads to inaccurate or incorrect results. For example, if $4ac$ is very small compared to b^2, the difference under the square root is

$$b^2 - 4ac = b^2 + \cdots$$

where the dots represent the roundoff error of your calculator or computer. The result is therefore either

$$x = -b/a$$

which is negative, and hence not physically realistic, or

$$x = \frac{-b + b \pm \cdots}{2a}$$

which may be positive, but depends on roundoff errors and is not the true value of x. The remedy for this situation is: First, be aware of the size of the terms under the square root. Second, if $b^2 \gg 4ac$, solve the approximate equation with $a = 0$

$$x_0 = -c/b$$

then use this approximate answer in an iterative mode:

$$x_1 = -(c + ax_0^2)/b$$

repeating this step as many times as necessary to achieve x_1 within the desired accuracy. Here is a numerical example:

$$z^2 + 345z - 56.3 = 0$$

$$z = \frac{-345 \pm \sqrt{(345)^2 + 4(56.3)}}{2} = \frac{-345 \pm \sqrt{(1.190 + 0.00225) \times 10^5}}{2}$$

Note that the second term under the square root is only 0.2% of the first, and if the calculation is carried out to 3 or 4 significant figures, the result is

$$z = (-345 \pm 345.3)/2$$

$$z = +0.15 \quad \text{or} \quad -345.15$$

Only the positive root is meaningful for solution concentrations, and this has a greatly magnified error. If this calculation is carried out to 8 significant figures, of course, an accurate result can be obtained: $z = 0.16311$. But you do not need much imagination to see that an equation with $4ac = 10^{-32} b^2$ would be a tough proposition even for double precision!

Applying the iterative method to the problem above gives

$$z_0 = -c/b = 56.3/345 = 0.16319$$

$$z_1 = -(c + az_0^2)/b = -[-56.3 + (1)(0.16319)^2]/345 = 0.16311$$

in agreement with the accurate solution of the quadratic formula within 0.006%. In general, if b^2 is more than 10^{+3} times as large as $4ac$, an iterative method is in order, requiring less computation and producing a more reliable result.

Newton's Approximation Method. Although you will find some methods for solving higher-degree equations in mathematics books, including a general solution to the cubic equation, and the method of reducing a quartic of the form $x^4 + ax^2 + b = 0$ to a quadratic by substuting $y = x^2$, a more general method is desirable. One of the simpler general methods is Newton's approximation.

About 300 years ago, Isaac Newton (famous for developing the calculus and the laws of gravitation) proposed a simple application of differential calculus to find the root x of the function $y = f(x)$ when $y = 0$. It is important that $f(x)$ be fairly smooth[32] near the point where it crosses the x axis and $y = 0$ (see Fig. 1.4).

Let x_0 be an approximate value of the true root x_f. If a tangent is drawn to the curve $y = f(x)$ at the point x_0, it will have a slope $f'(x_0) = dy/dx$ at $x = x_0$. The tangent line will intersect the x axis at a point x_1, closer to the true root than x_0. Thus x_1 will be a better approximation to the true root. From the definition of the slope,

$$f'(x_0) = \frac{f(x_0) - 0}{x_0 - x_1} \tag{39}$$

$$x_1 = x_0 - \frac{f(x_0)}{f'(x_0)} \tag{40}$$

Substitute the approximate root x_0 in the formulas for $f(x)$ and $f'(x)$ to obtain the numbers $f(x_0)$ and $f'(x_0)$. Divide the first by the second, subtract from x_0, and ob-

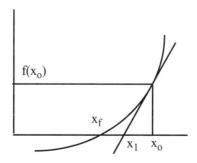

FIGURE 1.4. Diagram showing the successive approximations to the true root of the equation $f(x) = 0$.

tain a better approximation x_1. Repeat until the change $x_o - x_1$ is small enough to please you. Unlike the curve-crawling method, the closer the function $f(x)$ is to a straight line, the more rapidly will this series of approximations converge. In the limit of very small differences, the number of significant figures in the answer doubles with each iteration.[33]

Applying Newton's Method: Finding an Approximate Root. Often you will have narrowed in on a range of values in the course of testing various approximations. One strategy, if your equation is still in the form of a charge or mass balance, is to drop all but one term on each side and see what the answer is; each of these answers gives a possible value for x_o, and the one that gives the lowest value for $f(x)$ is the best starting value.

As described above under "curve crawling," you can calculate $f(x)$ for a range of x values and plot the results (or view them in a table) to see where $f(x)$ is approximately zero. Be sure to obtain both positive and negative values of $f(x)$ in order to get a good interpolation. Maxima or minima in $f(x)$ in the physically realistic range of x are not common for equilibrium problems, but steeply curving functions such as $f(x) = 10^{3x} + \cdots$ (e.g., where x is pH) frequently arise, and if x_o is too far from x_f, can lead to diverging or oscillating rather than converging answers.

Two simple checks of the equation before you begin calculating will reveal some types of error:

- At least one term (but not all terms!) of the polynomial $f(x) = 0$ must be negative, otherwise $f(x)$ could not be zero for positive real values of x.
- If only the constant term is negative, there is only one positive real root, and no maxima or minima for positive x. For such a function, $f'(x)$ is positive for all positive real x. Since maxima or minima require $f'(x) = 0$, there can be no maxima or minima for positive real x.

Consider a numerical example:

$$f(x) = x^3 + 5x^2 + 6.25x - 45 = 0$$

$$f'(x) = 3x^2 + 10x + 6.25$$

The shapes of $f(x)$ and $f'(x)$ are shown in Fig. 1.5. Note that $f(x)$ has only one negative term, the constant, and that $f'(x)$ is positive for all $x \geq 0$. This means there are no maxima or minima for $x \geq 0$. To obtain an approximate value, set each of the first 3 terms of $f(x)$ equal to the constant term:

$$x^3 = 45, \qquad x = 3.6$$
$$5x^2 = 45, \qquad x = 3.0$$
$$6.25x = 45, \qquad x = 7.2$$

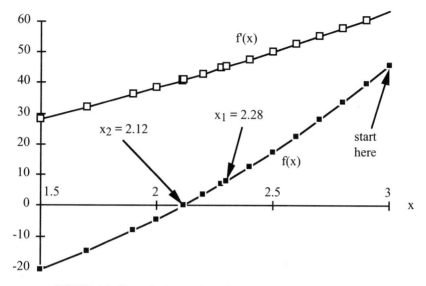

FIGURE 1.5. Numerical example of Newton's method. See text.

Since $f(x)$ increases uniformly with x, the smallest value of x gives $f(x)$ closest to zero. Choose $x_0 = 3.0$ as the starting value. Either of the others would give much the same results.

$$f(x_0) = (3.0)^3 + 5(3.0)^2 + 6.25(3.0) - 45 = 45.75$$

$$f'(x_0) = 3(3.0)^2 + 10(3.0) + 6.25 = 63.25$$

Substitute in the formula for Newton's approximation:

$$x_1 = x_0 - \frac{f(x_0)}{f'(x_0)} = 3.0 - \frac{45.75}{63.25} = 2.28$$

Repeating the process gives

$f(x_1) = 7.0$ and $f'(x_1) = 44.6$ hence $x_2 = 2.12$

$f(x_2) = 0.3$ and $f'(x_2) = 40.9$ hence $x_3 = 2.11$

Note that x_3 agrees with x_2 to within 0.5% and should be satisfactory for equilibrium calculations. One more step yields $x_4 = 2.1139$, and additional steps do not change this value. If x_0 had been taken to be 7.2, two additional iterations would have been required; If a preliminary plot had been made, x_0 might have been set at 2.12 or 2.11, and only one iteration would have been required.

Some Difficulties. Return to the weak acid example presented earlier in this chapter (p. 21). You will recall that the equilibria, mass and charge balances, when combined to eliminate all variables except $[H^+]$, resulted in the polynomial:

$$[H^+]^3 + [H^+]^2(C_A+K_a) - [H^+](C_{HA}K_a+K_w) - K_wK_a = 0 \quad (38)$$

For illustrative purposes, set[34] $C_A = 0.10$, $C_{HA} = 2.0 \times 10^{-5}$, $K_a = 10^{-4.75}$, $K_w = 10^{-14.0}$, and $[H^+] = 10^{-pH} = 10^{-x}$. This is the same set of input values as the example above, except that C_{HA} is much smaller than C_A. The equations become[35]

$$f(x) = 10^{-3x} + (0.1000)\,10^{-2x} - (3.556 \times 10^{-10})10^{-x} - 1.778 \times 10^{-19} = 0 \quad (41)$$

$$f'(x) = 2.303\{(-3)10^{-3x} + (0.1000)(-2)10^{-2x} - (3.556 \times 10^{-10})(-1)10^{-x}\} \quad (42)$$

Chemical knowledge would tell you that this is a mixture that is mostly sodium acetate (a weak base) with a trace of acetic acid; therefore, it should have a pH around 7 or 8, but you have not yet read Chapter 4. Pretend you were approaching this problem from a totally naive viewpoint. A logical mathematical starting point is $x = 0$, but this is not very helpful:

$$x_0 = 0$$
$$f(x_0) = 1.100$$
$$f'(x_0) = -7.369$$
$$x_1 = 0.1493$$

x_1 is in the right direction, but additional iterations do not seem to converge quickly: 0.1493, 0.3002, 0.4535, 0.6097, 0.9350, ... Many more steps would apparently be required to reach the final value (see below)

You could try other values for x_0 at random, but a systematic survey seems more appropriate. Here is a table of trial values:

x_0	$f(x_0)$	$f'(x_0)$	x_1
0	1.1000	−7.369	0.149
2	1.1002E−05	−5.298E−5	2.208
4	1.0011E−09	−4.614E−09	4.217
6	9.966E−14	−4.599E−13	6.217
8	6.267E−18	−3.788E−17	8.165
10	−2.124E−19	−7.730E−20	12.747
12	−1.782E−19	−8.186E−22	229.65

As x_0 increases, $f(x_0)$ becomes smaller, and eventually becomes negative beween $x = 8$ and 10 (as chemical intuition predicted!), but Newton's method does not always converge on the correct answer. Starting with $x_0 = 8$, successive approxima-

SOLVING AN EQUILIBRIUM PROBLEM

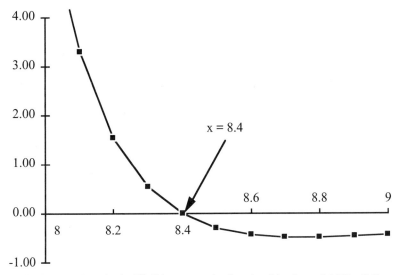

FIGURE 1.6. Sketch of $10^{18} f(x)$ versus x for "weak acid polynomial," Eq. (14).

tions are 8.165, 8.295, 8.372, 8.396, 8.398, ... That looks OK. But starting with $x_o = 10$ is a disaster. The sequence is 10.00, 12.47, 1227.3, infinity! What is to be done? The most direct approach is curve crawling: Plot $f(x)$ versus x and interpolate the value of x where $f(x) = 0$ (see Fig. 1.6).

Here is a table of the values near that point

x_o	$f(x_o)$	$f'(x_o)$	x_1
8.39	3.314E–20	–4.309E–18	8.3977
8.397	3.653E–21	–4.118E–18	8.3979
8.398	–4.524E–22	–4.093E–18	8.3979
8.399	–4.531E–21	–4.067E–18	8.3979
8.40	–8.584E–21	–4.04E–18	8.3979

$f(x_o)$ tells you that x_f is between 8.397 and 8.398. Since we are close to x_f, the x_1 value calculated by Newton's method from all points agree; the last four give $x_f = 8.3979$.

But why did Newton's method fail so badly on this problem when the starting value was 10? If you look at Fig. 1.7 and the preceding table, you will see that when $x < 8.39$, $f(x)$ is postive and $f'(x)$ is negative. Newton's method produces positive, but decreasing, increments on the previous approximation: $x_1 = x_o - f(x)/f'(x) > x_o$. When $x > 8.4$, $f(x)$ is negative but so is $f'(x)$; hence Newton's method produces decreasing negative increments $x_1 < x_o$, bringing the answer back to the true root x_f. The trouble comes when $f'(x)$ goes from negative to positive ($x = 8.75$). Then $f(x)$

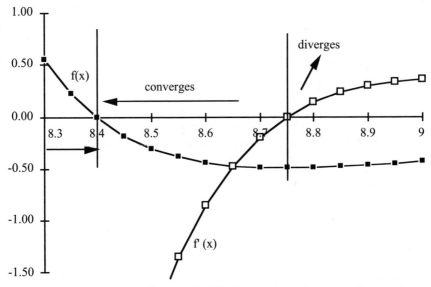

FIGURE 1.7. Comparison of $10^{18} f(x)$ and $10^{18} f'(x)$ near the points where they reach zero. Newton's method converges on the root (x_f = 8.4) provided the starting value is below 8.75. Above this, Newton's method diverges to larger and larger values. See text.

is negative and $f'(x)$ is positive; the term $-f(x)/f'(x)$ is positive, adding positive increments to the approximate value:

x_o	$f(x_o)$	$f'(x_o)$	x_1
8.9	−0.47	0.30	10.45
10.45	−0.19	0.03	17.13
17.13	−0.18	6.07E−09	2.93E+07

The result is divergence to infinity!

So long as the starting value is below 8.75, convergence is obtained. If the starting value is higher, the result diverges. But how would you know about this limit if you hadn't explored the function and its first derivative over the whole region of interest? How could you have even known that such a trap existed?

The Secant Method. A somewhat more robust technique is the secant method, which evaluates $f'(x_o)$ not by the algebraic derivative of $f(x)$ but by a "secant," a finite line segment near x_o (see footnote 33):

$$f'(x_0) \approx \frac{f(x_0) - f(x_2)}{x_0 - x_2} \qquad (43)$$

$$x_1 = x_0 - \frac{f(x_o)}{f'(x_o)} \qquad (40)$$

SOLVING AN EQUILIBRIUM PROBLEM

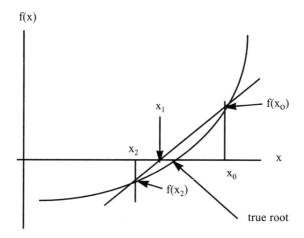

FIGURE 1.8. Sketch of the "secant method" (Eq. 43) for approximating the root of a nonlinear equations.

As sketched in Fig. 1.8, this secant is most usefully drawn between a point where $f(x_0) > 0$ and $f(x_2) < 0$, so that the root is between the two approximations. In this way, the extreme excursions which Newton's method can produce are avoided. Of course, the initial phase of problem definition requires at least a rough plot of $f(x)$ versus x.

The secant method would give a smooth result to the function in Fig. 1.5, and also in Figs. 1.6 and 1.7 provided the starting point was pH < 8.75. However, the same problems as discussed above would arise for a higher starting value.

Automatic Function Solvers: "Goal Seek." A number of software applications, including Excel™, Lotus™, and several mathematical manipulation programs, have a "goal-seek" or "solver" function that usually employs Newton's method. The default parameters are not always the best: for Excel 3.0, they are 100 iterations with an error of 0.001. This can give a quick but wrong answer if the starting value of $f(x)$ is less than 0.001. Better parameters for most iterations are 20 to 100 iterations with an error of 0.

Here are some examples of results obtained for the polynomial functions discussed above[36]:

1. The quadratic $z^2 + 345z - 56.3 = 0$ (p. 26)

	Goal Seek: Set $A$8 to 0 by adjusting $B$8		
f(z)	z		
	Maximum iterations 100, maximum change 0.001		
-1.252E-06	0.16311129		
	Max iter 20, max change 0, starting value 0		
7.1054E-15	0.16311129		
	Compare quadratic formula		
	=(-345+SQRT(345^2+4*56.3))/2		
	0.16311129		

These results are consistent and robust. The default iteration option (100 iterations, max change 0.001) gives the same result as 20 iterations with max change 0, and both agree to 8 significant figures with the direct calculation from the quadratic formula.

2. The cubic $x^3 + 5x^2 + 6.25x - 45 = 0$ (Fig. 1.5, p. 29)

	Goal Seek: Set $A$20 = 0 by adusting $B$20		
f(x)	x		
	Max iter 100, max change 0.001, starting value 0		
-8.463E-06	2.11387876		
	Max iter 20, max change 0, starting value 0		
7.1054E-15	2.11387897		
	Max iter 20, max change 0, starting value 3.00		
-2.842E-14	2.11387897		

This function is also well behaved. The default iteration option gives a value of x correct to 7 significant figures. Changing the max change from 0.001 to 0 gives a more accurate result: $f(x) = 7 \times 10^{-15}$ instead of 8×10^{-6}. A starting value higher than the root gives $f(x) = -3 \times 10^{-14}$ but the same value of x as the second choice.

3. The "weak acid" polynomial (Fig. 1.6, p. 30)

f(x) =	10^-(3*x)+0.1*10^(-2*x)-3.556e-10*10^-x-1.778e-19 = 0		
	Goal Seek: Set $A$31 = 0 by adusting $B$31		
f(x)	x		
	Max iter 100, max change 0.001, starting value 0		
0.00056351	1.22585835	"found a solution" - but it's wrong!	
	Max iter 20, max change 0, starting value 0		
2.3755E-06	2.32221968	"may not have found a solution" - and it's wrong	
	Max iter 20, max change 0, starting value 8.00		
-4.117E-33	8.39789117	"may not have found a solution" - but it's OK!	
	Max iter 100, max change 0, starting value 8.00		
-4.117E-33	8.39789117	"may not have found a solution" - but it's OK!	
	Max iter 20, max change 0, starting value 9.00		
-1.778E-19	1088.57747	"may not have found a solution" - and it's totally wrong!	

As we found in the text above, this is a rather tricky function because $f(x)$ changes steeply and goes through a minimum (Fig. 1.7). Beginning with the default option, the program says that it has "found a solution" at $x = 1.23$, because $f(x)$ is 5.6×10^{-4}, which is less than the 0.001 maximum change specified. However, this solution is quite wrong, as can be seen from a plot such as Fig. 1.6, which shows that $f(x)$ has to be less than 10^{-18}. Reducing the maximum change to 0 does not improve matters much; $f(x)$ is still relatively large (2.4×10^{-6}) and $x = 2.32$, which is still wrong.

A better starting value (8.00 instead of 0) makes a great improvement, with $f(x) = -4.2 \times 10^{-33}$ and $x = 8.3979$, in agreement with the results obtained in the manual calculation above. Increasing the number of iterations to 100 confirms the result. Finally, to emphasize how important the starting value is, a calculation starting at 9.00 produces a small $f(x) = -1.77 \times 10^{19}$ (see Fig. 1.7), but this gives a ridiculously large $x = 1089$, a value that increases rapidly as more iterations are performed.

The lesson to be learned from this example is that there may be more efficient or robust algorithms, but there are no foolproof shortcuts, and that there is no substitute for knowing your functions in some detail. So long as the equations to be solved are linear or nearly so, the answer is obtained smoothly. But highly nonlinear equations can lead to errors—some of which are described above. So be cautious, and check all your results in the original equations.

PROBLEMS

For the following reactions, set up concentration equilibrium-constant expressions (that apply in dilute solution or in a medium of constant ionic strength).

1. $HCN \rightleftharpoons H^+ + CN^-$
2. $CH_3NH_2 + H_2O \rightleftharpoons CH_3NH_3^+ + OH^-$
3. $CH_3NH_3^+ \rightleftharpoons CH_3NH_2 + H^+$
4. $BaSO_4(s) \rightleftharpoons Ba^{2+} + SO_4^{2-}$
5. $CuCl(s) + Cl^- \rightleftharpoons CuCl_2^-$
6. $CO_2(g) \rightleftharpoons CO_2(aq)$
7. $CO_2(aq) + H_2O \rightleftharpoons H^+ + HCO_3^-$
8. $CaCO_3(s) + CO_2(g) + H_2O \rightleftharpoons Ca^{2+} + 2\ HCO_3^-$
9. $Ca_3(PO_4)_2(s) + 2\ H^+ \rightleftharpoons 3\ Ca^{2+} + 2\ HPO_4^{2-}$
10. $Th^{4+} + H_3PO_4 \rightleftharpoons ThH_2PO_4^{3+} + H^+$
11. The four stepwise reactions creating $HgCl_4^{2-}$ from Hg^{2+} and Cl^-
12. The four overall reations creating the complexes $HgCl_n^{(n-2)-}$ ($n = 1, 2, 3, 4$) from Hg^{2+} and Cl^-
13. Write the instability constant of the silver thiosulfate complex $Ag(S_2O_3)_2^{3-}$ and the reaction to which it corresponds.
14. Write the three stepwise equilibria that might be expected between Ag^+ and $S_2O_3^{2-}$, and express the instability constant (problem 18) in terms of these stepwise constants.
15. In their table for formic acid HCOOH, abbreviated HL, Smith and Martell (Vol. 3, p. 1) denote an equilibrium "HL/H.L" and list "log K, 25°, $I = 0$" in their table as 3.745 ± 0.007. Interpret this by writing the equilibrium constant expression (a combination of concentrations set equal to a constant) in the form implied by this notation, and compare with the usual acid dissociation format.

Find a positive real root (accurate to ±1%) for the following equations:

16. $4.7 z^2 + 3.1 z - 17 = 0$
17. $1.8 \times 10^{-4} [NH_3]^2 + 3.6 \times 10^{-8} [NH_3] - 5.5 \times 10^{-14} = 0$
18. $S(S - 6.5 \times 10^{-4}) + 4.7 \times 10^{-8} = 0$ (find both roots)
19. $[H^+]^2 - 4.9E+4 [H^+] + 3.9E-2 = 0$, $[H^+] < 1.0$
20. In the following system of equations, all three species are of comparable concentration. Solve the system of equations exactly to obtain a quadratic. Solve this and check your results in the three original equations.

$$[HgOH^+][H^+] = 2.0 \times 10^{-4} [Hg^{2+}]$$
$$[H^+] = 10.4 [HgOH^+]$$
$$[Hg^{2+}] + [HgOH^+] = 9.76 \times 10^{-3}$$

21. $(10^{-13.62}/[OH^-]) = [OH^-] - 10^{-2.23}$
22. $10^{-14.0}[S^{2-}]^3 + 0.13 [S^{2-}]^2 = 10^{-13.05}$
23. $[H^+]^5 + 3.7 \times 10^{-20} [H^+]^3 - 2.7 \times 10^{-48} = 0$
24. Plot a graph of

$$[NH_3]^3 - 4.1 \times 10^{-7}[NH_3] + 3.7 \times 10^{-11} = 0$$

and note that there are two physically plausible roots. Find both roots.

NOTE ON CONSTRUCTION OF FIGS. 1.1 AND 1.2

These curves, showing the kinetics of approach to equilibrium, are an important pedagogical point, but their construction is much too complicated for this stage of the book. Nevertheless, for the curious, here is the derivation. The forward reaction, with rate constant k_f, is

$$HA \rightarrow H^+ + A^-$$

and the reverse reaction, with rate constant k_r, is

$$H^+ + A^- \rightarrow HA$$

The net rate of formation of A^- is equal to the rate of disappearance of HA:

$$\frac{d[A^-]}{dt} = -\frac{d[HA]}{dt} = k_f[HA] - k_r[H^+][A^-]$$

Note that at equilibrium, the rates are both zero, and $[H^+][A^-]/[HA] = k_f/k_r = K_a$. To get the time dependence, let $x = [H^+]/C = [A^-]/C$ be the degreee of dissociation; then $[HA]/C = 1 - x$, where C is the total concentration. The rate equation becomes

$$\frac{dx}{dt} = k_f(1 - x) - k_r C x^2$$

Integrating from the undissociated state ($x = 0$, $t = 0$) to an arbitrary x and t,

$$\int_0^x \frac{dx}{[k_f(1 - x) - k_r C x^2]} = \int_0^t dt$$

This is a standard integral. Let

$$Q = \sqrt{k_f^2 + 4 k_f k_r C}$$

then

$$\left(\frac{1}{Q}\right) \ln\left(\frac{2 k_r C x + k_f + Q)(k_f - Q)}{2 k_r C x + k_f - Q)(k_f + Q)}\right) = t$$

Figure 1.1 was obtained[37] using $C = 1 \times 10^{-4}$, $k_f = 10^{+5}$ sec^{-1}, $k_r = 10^{+9.75}$ Lmol^{-1} sec^{-1}, assuming values for x from 0 to 0.33, and calculating t. The corresponding equation for Fig. 1.2 is obtained by integrating from $x = 1$ to x and $t = 0$ to t. This gives

$$\left(\frac{1}{Q}\right) \ln\left(\frac{2 k_r C x + k_f + Q)(2 k_r C + k_f - Q)}{2 k_r C x + k_f - Q)(2 k_r C + k_f + Q)}\right) = t$$

which was used, assuming values of x decreasing from 1 to 0.33, to plot Fig. 2.

NOTES

1. Mention of commercially available software should not be construed as an endorsement. I mention what I know or have heard of, but to find the best current product you should keep up with expert evaluations in computer-user magazines.
2. "System" is used in physical chemistry to refer to whatever chemical and physical things are under consideration. Here the system would be the solutions comprising the two reactants; in the example just below, the system would be the piece of mineral and the solution being tested for saturation equilibrium.
3. An intense electric field applied to the solution causes the acetic acid to dissociate slightly more, and the relaxation of the system to equilibrium is measured by following on an oscilloscope the conductivity of the solution, which depends on the extent of dissociation. See M. Eigen, *Zeitschrift für Elektrochemie* **64**:115 (1960).

4. In this book, "mole" refers to Avogadro's number N_A of particles of any chemical species, and includes the terms "gram molecule," "gram ion," "gram atom," and "gram formula weight." A micromole (μmole) is 10^{-6} mole ($N_A = 6.0221367 \times 10^{23}$, *J. Research Nat. Bureau Standards* **92**, 85–95 (1987); see also http:// physics.nist.gov/ PhysRefData/codata86/physico.html).

5. The classic study of metal–ammine complexes was Jannik Bjerrum's thesis *Metal Ammine Formation in Aqueous Solutions*, P. Haase & Son, Copenhagen, 1941. His extremely careful experimental studies and critical theoretical approach provide an example of the best type of research work. More recent work is catalogued by W. Smith and A. E. Martell, *Critical Stability Constants*, Plenum, NY, 1976–1990.

6. L. G. Sillén and A. E. Martell, *Stability Constants of Metal–Ion Complexes*, Special Publication No. 17, The Chemical Society, London, 1964, pp. 152–53.

7. For a discussion of multicomponent solutions, see F. MacIntyre, *Mar. Chem.* **4**, 205 (1976)

8. C. M. Guldberg and P. Waage, *Etudes sur les affinités chimiques*, Brøgger & Christie, Christiania, 1867.

9. J. H. Van't Hoff, *Z. physik chem* **1**: 481, 1877.

10. E. A. Guggenheim, *Thermodynamics*, North Holland, Amsterdam, 3d Ed, 1957, 4th Ed, 1959, 5th (rev). 1967; G. N. Lewis and M. Randall, *Thermodynamics*, 2nd Ed, revised by K. S. Pitzer, McGraw Hill, New York, 1961; 3d (rev), 1995.

11. This will be discussed further in Chapter 4. See next footnote for definition of ionic strength.

12. As will be discussed in detail in Chapter 2, the Debye–Hückel theory of ionic solutions predicts that activity coefficients are determined by a particular combination of concentrations—the "ionic strength," computed by multiplying the concentration of each charged species by the square of its charge, adding up all such terms for the solution, and dividing by two. For a univalent electrolyte such as NaCl, each ion has charge 1, and ionic strength = $\frac{1}{2}(C_{Na} + C_{Cl}) = C_{NaCl}$.

13. For NaCl at concentration 0.1 M, $\gamma_\pm = 0.778$ (R. A. Robinson and R. H. Stokes, *Electrolyte Solutions*, 2nd Ed, Butterworths, London, 1965, p. 492); hence neglecting the difference between activity and concentration results in a 22% error. At concentration 0.001 M, $\gamma_\pm = 0.966$, a 3% error.

14. Robinson and Stokes, pp. 31–32.

15. For example, in a solution containing 0.100 moles of NaCl and 1 L (55.5 moles) of H_2O, the mole fraction of water is 55.5/55.6 = 0.9982. Similarly, for 1.00 M NaCl, $X_{H_2O} = 0.982$.

16. For example, at 177°C and 15.7 atm, the ratio of fugacity to partial pressure for CO_2 is 0.985; at 334°C and 59.2 atm, this ratio is 1.02 (A. J. Ellis and R. M. Golding, *Am. J. Sci.* **261**:47–60, 1963).

17. The combination $-\log\{a_{H^+}\} = -\log\{[H^+]\gamma_+\}$ is called "pH" and occupies a central place in acid–base chemistry, because it is easy to measure experimentally (see Chapters 2, and 11). For this reason, still another form of the above equilibrium expression has come into use, particularly in oceanography: $10^{-pH}[OH^-]_T = K'_w$ where $[OH^-]_T$ includes ion pairs such as $MgOH^+$ and $CaOH^+$ as well as free $[OH^-]$ (see Chapter 10). The hybrid constant K'_w is independent of minor species concentrations in seawater because of the large excess of sodium, chloride, magnesium, calcium, and sulfate ions. It is, however, dependent on the concentrations of these major ions, and hence on salinity.

NOTES 39

18. The terms *dissociation constant* and *ionization constant* are sometimes used interchangeably, but it is often desirable to distinguish between ionization (formation of ions) and dissociation (separation into parts).
19. In many textbooks, the solubility product is denoted K_{sp}, where "sp" is an abbreviation for solubility product. This notation does not easily extend to more complex situations.
20. This system was developed by G. Schwarzenbach and L. G. Sillén for the first *Stability Constants* tables. See footnote below for reference.
21. The same notation applies to ion pairs, which reflect weaker, primarily electrostatic, association. See Chapter 10.
22. R. M. Smith and A. E. Martell, *Critical Stability Constants*, Vols. 1–6, Plenum Press, New York, 1974, 1975, 1976, 1977, 1982, 1989.
23. L. G. Sillén and A. E. Martell, *Stability Constants of Metal Ion Complexes*, The Chemical Society, London, Special publications No. 17 (1964) and 25 (1971).
24. R. M. Smith and A.E. Martell, *NIST Critically Selected Stability Constants of Metal Complexes Database*, version 3.0, National Institute of Standards and Technology, Gaithersburg MD, 1997.
25. For example, D. Wagman et al., "The NBS tables of chemical thermodynamic properties," *J. Phys. Chem. Ref. Data* **11**: supplement No. 2, 1982; or H. C. Helgeson, J. M. Delany, H. W. Newbitt, and D. K. Bird, "Summary and Critique of the Thermodynamic Properties of Rock-Forming Minerals, *Am. J. Sci.* **278A**: 1–229, 1978.
26. Pick the unknown that occurs in the least number of equations. Call it x_1. Solve one equation for x_1 in terms of all the other unknowns. Substitute this expression for x_1 wherever x_1 occurs in the other equations. This gives $n-1$ equations in $n-1$ unknowns. Repeat this sequence until only one equation remains, and solve it for the last unknown. The intermediate results—the equations giving x_1, x_2, etc. explicitly—are then used to obtain those values. The final answers are checked by substituting in the original equations to see if they are satisfied.
27. A rigorous and historically important presentation of this approach was made by J. E. Ricci, *Hydrogen Ion Concentration: New Concepts in a Systematic Treatment*, Princeton University Press, Princeton, NJ, 1952. Ricci clarified many confusing and erroneous concepts in acid–base theory for the first time. To see the extent of confusion that existed even as recently as 1950, read the introduction to his book.
28. Note $pK_a^\circ = 4.757 \pm 0.002$ at $I = 0$ (Table 4.1), $pK_a = 4.73$ at $I = 0.001$ (Davies equation).
29. The format $[H^+] + C_A = (C_{HA} + C_A)/([H^+]/K_a + 1) + K_w/[H^+]$ for Eq. (10) is typographically simpler but not so easy to read. The format required by most computer programs is usually even less easy to read.
30. If K_a is changed to $10^{-4.73}$, the above equations give $[H^+] = 1.796 \times 10^{-5}$ and with log $\gamma_\pm = 0.015$ (Davies equation), pH = 4.761.
31. G. Polya, *How to Solve It*, 2nd Ed., New York, Doubleday, 1957.
32. That is, $f(x)$ must have a continuous, nonvanishing first derivative between the approximate root x_0 and the true root x_f, and an equal distance beyond.
33. Other approaches can also be used, including taking equal steps in x until $f(x)$ changes sign, then stepping backward in smaller steps until $f(x)$ changes sign again (HALTAFALL, by N. Ingri et al., *Tlanta* **14**:1261–86 (1967), taking the midpoint of the secant between two points on the curve of Fig. 1.4; one with a positive $f(x)$ and the other with a negative $f(x)$ (FZERO, by A. G. Dickson, Scripps Institute of Oceanography), etc.

See D. Kahaner, C. Moler and S. Nash, *Numerical Methods and Software* Prentice Hall, Englewood Cliffs, NJ, 1989.

34. In this book, small numbers occur frequently, and the notations: $\log x = -4.75$, $x = 10^{-4.75}$, $x = 1.778 \times 10^{-5}$, $x = 1.778\text{E}-05$, $x = 1.778\text{ e-}05$, and $x = 0.00001778$ may be used interchangeably.
35. Note that the derivative $d/dx(10^{ax}) = d/dx(e^{2.303ax}) = 2.303\, a\, e^{2.303ax} = 2.303\, a\, 10^{ax}$.
36. These examples were obtained with "goal seek" in Excel 3.0. Other programs may perform differently.
37. These values were chosen for illustrative purposes. The dissociation rate constant k_f could be as high as 10^{10}. See D. W. Margerum et al. (1978) in *Coordination Chemistry*, Vol. 2, A. E. Martell, Ed., Amer. Chem. Soc., Washington, DC, for experimental data on ionic reaction kinetics.

2

ACTIVITY COEFFICIENTS AND pH

Introduction
Estimating Activity Coefficients
  Ionic Interactions
    Constant Ionic Medium
  Debye–Hückel Theory
    Single-Ion Activity Coefficients
    Including an Ion Size Parameter
    The Davies Equation
    Uncharged Molecules
The pH Scale
  Definition of pH
  The Glass Electrode
  The Harned Cell
  Establishment of a pH Scale
  Equilibrium Constant Measurement
Problems

INTRODUCTION

In Chapter 1, you were briefly exposed to the concept of activity coefficient as a factor required to make the equilibrium law accurate at higher concentrations. My 1964 *Ionic Equilibrium** scrupulously avoided mention of activity coefficients until Chapter 12, where some simple methods for calculating them were presented. I no longer think that is appropriate.

---

*J. N. Butler, *Ionic Equilibrium: A Mathematical Approach,* Addison-Wesley, Reading, MA (1964).

It is possible to go through the examples in subsequent chapters using concentration equilibrium constants extrapolated to zero ionic strength or constants measured in a medium of high ionic strength, and in either case no activity coefficients need to be estimated. However, comparison with experimental results and predictions for real systems may not be satisfactory. Students who find the material in this chapter too difficult should skip to Chapter 3 and come back later when they need to know about activity coefficients for a particular application.

But I hope this will not be necessary, and so I am presenting the information from Chapter 12 of the 1964 edition as Chapter 2 of this revised book, so that I can present more rigorous discussion and more realistic examples than would otherwise be possible.

Remember, *the first approximation to an equilibrium calculation usually ignores activity coefficients* (sets them equal to 1.00), *and only when accurate results are needed are they introduced.*

ESTIMATING ACTIVITY COEFFICIENTS[1]

Ionic Interactions

The interaction between ions in solution extends over a much longer distance than the interaction between uncharged molecules. As a result, deviations from ideal laws, which assume each particle behaves independently of any other, occur at much lower concentrations in ionic solutions than in solutions of uncharged molecules. For example, in a solution of NaCl, deviations from ideal thermodynamic laws exceed 5% above about 0.002 M, whereas in a solution containing an uncharged molecule such as ethyl alcohol, deviations from ideal laws do not exceed 5% until the solute concentration is greater than about 1 M.

Fortunately, the long-range electrostatic interactions of ions in solution do not depend on the chemical nature of the ions, but only on their charge. By assuming that ions are point charges in a continuous medium of dielectric constant equal to that of water, Debye and Hückel[2] were able to derive the theoretical form that the activity coefficient should obey in dilute solutions. These long-range interactions are essentially the electrostatic attraction of an ion to distant ions of opposite charge—and the result is that in dilute solutions the activity of an ion is smaller than its concentration.

The activity coefficient bridges the gap between solution concentration (measured as mass per unit volume or per unit weight of solution) and thermodynamic activity, measured by solubility, vapor pressure, or electrochemical potential. An immense body of experimental and theoretical work stands behind the very simple Debye–Hückel theory and its extensions.

For rough calculations, or for very dilute solutions, activity coefficients can simply be taken to be unity— activities are set equal to concentrations. But the more concentrated the solution, the higher the charge on the ions, and the greater accuracy desired in the answer, the more important is an accurate estimate of the activity coefficients.

For intermediate concentrations and accuracy (say ±10% at 0.1 M) equations based on the Debye–Hückel theory (pp. 44–49) are satisfactory. In more concentrated solutions, however, other factors besides simple electrostatic attraction take over. These include:

- The ions have finite size and therefore exclude other ions from a definite volume of the solution. This factor is accounted for by including an ion-size parameter.
- Water molecules tend to exclude ions, as well as uncharged molecules, at high concentration—the "salting-out effect." This also can be accounted by another extension of the theory, using a salting-out parameter.
- Specific chemical interactions between ions of like charge as well as unlike charge become stronger as the ions are pushed closer together at high concentrations. A simple example is $MgSO_4$, which exhibits a much lower activity coefficient than would be predicted for a 2:2 electrolyte. (See Fig. 2.1. The calculations based on Debye–Hückel theory are explained later in this chapter.)
- The specific interaction between Mg^{2+} and SO_4^{2-} is often modeled as a chemical equilibrium forming an ion pair, which introduces a new parameter, the ion-pairing constant. The higher the concentration and the more components involved, the more interaction parameters are needed, and a correspondingly large body of experimental data must be behind them. Models based on this concept are developed in Chapter 10.
- When the concentration reaches the point at which essentially all the water in solution is bound in the primary hydration shell of ions, the activity coefficient can become enormous. In a solution containing 0.1 M LiBr, the mean activity coefficient of the ions is about 0.8; at a concentration of 20 M, the mean activity coefficient of LiBr is 485!

Constant Ionic Medium. One important simplification has become a basic part of the experimental approach to solution equilibria: the use of a constant ionic medium. Typically this is a salt such as NaCl or $NaClO_4$, which does not participate strongly in acid–base or complex-formation reactions. It is added to the solution at concentration (0.5 to 5 mol/L) very much higher than the ions of interest.

Thus when the acid–base or complex-formation reactions change the concentrations of the minor species, the high concentration background electrolyte keeps the total concentration of ions (and hence the ionic strength) essentially constant. As mentioned in Chapter 1, such constants depend on temperature, pressure, and ionic medium, but not on the reactant and product concentrations. Tables such as those compiled by Smith and Martell list "thermodynamic" concentration constants in media of 0.5 M, 1.0 M, 2.0 M, ..., as well as the "thermodynamic" constants obtained by extrapolation to zero ionic strength.

Thus the concentration ion product of water (equilibrium constant K_w), measured in 1.0 M NaCl, which is $10^{-13.74}$, would have a different value from that measured in 3.0 M NaCl ($10^{-14.18}$), which in turn would be different from that extrapolated to very dilute solutions ($10^{-14.00}$). (See Fig. 3.2, p. 68.)

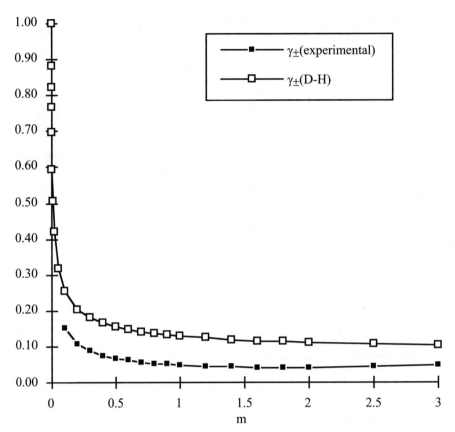

FIGURE 2.1. Activity coefficient for $MgSO_4$. Comparison of experimental data from Robinson and Stokes (1965) wih calculations from Debye–Hückel theory with $a_{Mg} = 8$ Å and $a_{SO_4} = 4$ Å (Eq. 9, p. 46).

Debye–Hückel Theory

Almost all theoretical forms for the activity coefficients of ions are derived from the Debye–Hückel theory. The essence of this theory is that ions behave like point charges in a continuous dielectric medium. From this assumption, it is straightforward, although complicated,[3] to use the laws of electrostatics and thermodynamics to obtain the "limiting law," which holds in sufficiently dilute solution for any completely dissociated electrolyte:

$$-\log \gamma_\pm = A z_+ z_- \sqrt{I} \qquad (1)$$

Here $\gamma_\pm$ is the geometric mean activity coefficient of the two ions.[4] For NaCl, $\gamma_\pm = (\gamma_+ \gamma_-)^{1/2}$. A is a constant that depends on the absolute temperature T and the dielectric constant ε of the solvent[5]:

ESTIMATING ACTIVITY COEFFICIENTS

$$A = 1.825 \times 10^6 (\varepsilon T)^{-3/2} = 0.51 \quad \text{at } 25°C \text{ in water}$$

z_+ and z_- are the charges on the two ions.

The ionic strength, defined verbally in Chapter 1, is obtained from the Debye–Hückel theory to be

$$I = \frac{1}{2} \sum_i C_i z_i^2 \qquad (2)$$

where C_i in mol/L represents the concentration of every ion, with charge z_i, that is present in the solution, not only the ions for which the activity coefficient is being calculated.[6] The fact that z_i is squared means that all terms, for positive or negative ions, are positive, and furthermore that ions of high charge count more than ions of lower charge. For example, the ionic strength of C molar NaCl is

$$I = \tfrac{1}{2}([Na^+] + [Cl^-]) = C \qquad (3)$$

but the ionic strength of C molar $MgSO_4$ is

$$I = \tfrac{1}{2}\{[Mg^{2+}](4) + [SO_4^{2-}](4)\} = 4C \qquad (4)$$

Single-Ion Activity Coefficients. It is convenient to use activity coefficients for single ions, and the theoretical equations allow you to calculate them. The Debye–Hückel theory gives an expression similar to Eq. (1) for the activity coefficient of a single ion of charge z in a sufficiently dilute solution[7]:

$$-\log \gamma_z = Az^2 \sqrt{I} \qquad (5)$$

But it is important to remember that only electrically neutral combinations of these single-ion activities can be measured experimentally. For example, measuring the vapor pressure of water over a salt solution involves the transfer of a neutral species, water. An electrochemical cell, such as Harned's cell, described in more detail later,

$$Pt \mid H_2(g) \mid H^+, Cl^- \mid AgCl(s) \mid Ag$$

can give the mean activity coefficient of HCl: $\gamma_\pm = (\gamma_+ \gamma_-)^{1/2}$, but not the individual ionic activity coefficients. It is tempting to introduce a salt bridge (as is found in commercial pH electrodes) and to assume that the potential of the H^+-responsive electrode gives the activity of H^+ alone, but that assumption neglects the liquid junction potential at the salt bridge, which cannot be rigorously measured, and can more than compensate for any changes in activity coefficient.

Even though there is a repetitive and ingeniously erroneous literature on the "measurement" of single-ion activity coefficients,[8] they are really not necessary.

Any equilibrium expression can be written using electrically neutral combinations of single-ion activity coefficients: For example[9]

$$[H^+]\gamma_+[A^-]\gamma_- = K_a^\circ[HA]\gamma_0 \tag{6}$$

$$\frac{[H^+][A^-]}{[HA]} = \frac{K_a^\circ \gamma_0}{\gamma_+ \gamma_-} = K_a \tag{7}$$

Including an Ion-Size Parameter. The limiting law predicts much smaller activity coefficients than are observed at intermediate concentrations. A slightly more complicated treatment[10] uses the same assumptions of a continuous dielectric medium but with spherical charges of finite size instead of point charges. Additional terms in the equations are included when approximations are made, which gives the extended Debye–Hückel law. For a single ion of charge z:

$$-\log \gamma_z = Az^2 \frac{\sqrt{I}}{1 + Ba\sqrt{I}} \tag{8}$$

Here A is the same constant as in Equation 1, a is an adjustable parameter, measured in Å (10^{-10} m), which corresponds roughly to the effective size of a hydrated ion, and B is a function of the temperature and dielectric constant:

$$B = 50.3 \,(\varepsilon T)^{-1/2} = 0.33 \quad \text{at } 25°C \text{ in water.}$$

For a binary electrolyte:

$$-\log \gamma_\pm = A|z_+ z_-| \frac{\sqrt{I}}{1 + Ba\sqrt{I}} \tag{9}$$

For a single 1-1 salt such as NaCl, with concentration m in mol/kg, the Debye–Hückel equation becomes:

$$-\log \gamma_\pm = A \frac{\sqrt{m}}{1 + Ba\sqrt{m}} \tag{10}$$

In 1937, Kielland[11] published a list of values for the ion-size parameter a for 130 selected ions, which is reproduced in Table 2.1. These were obtained from the measured activity coefficients of binary electrolytes, but were intended to be used to estimate activity coefficients in multicomponent systems. Thermodynamically, the use of different ion-size parameters is not consistent,[12] but this makes little difference since the results are only approximate (see Fig. 2.2).

Typical values for single-ion activity coefficients calculated using the extended Debye–Hückel equation with Kielland's ion-size parameters vary from

$$\gamma = 0.967 \text{ at } I = 0.001 \text{ with } a = 9 \text{ Å and charge 1 (e.g., H}^+\text{) to}$$

TABLE 2.1. Kielland's table: Values of the parameter *a* for 130 selected ions

Charge 1
(Abbreviations: Et = ethyl, C_2H_5; Me = methyl, CH_3; Pr = propyl, C_3H_7)

9	H^+
8	$(C_6H_5)_2CHCOO^-$, Pr_4N^+
7	$OC_6H_2(NO_3)_3^-$, $MeOC_6H_4COO^-$, Pr_3NH^+
6	$C_6H_5COO^-$, $C_6H_5CH_2COO^-$, $C_6H_4OHCOO^-$, $C_6H_4ClCOO^-$, $CH_2=CHCH_2COO^-$, $CHClCOO^-$, CCl_3COO^-, Et_4N^+, Et_3NH^+, Li^+,
5	$Me_2CCHCOO^-$, $Pr_2NH_2^+$, $PrNH_3^+$
4	$CdCl^+$, CH_3COO^-, CH_2ClCOO^-, ClO_2^-, $Co(NH_3)_4(NO_2)_2^+$, $Et_2NH_2^+$, $EtNH_3^+$ $H_2AsO_4^-$, $H_2PO_4^-$, HCO_3^-, HSO_3^-, IO_3^-, Me_4N^+, Me_3NH^+, Na^+, $^+NH_3CH_2COOH$, $NH_2CH_2COO^-$
3	Ag^+, Br^-, BrO_3^-, F^-, Cl^-, ClO_3^-, ClO_4^-, CN^-, CNO^-, CNS^-, Cs^+, $HCOO^-$, $H_2(citrate)^-$, HS^-, I^-, IO_4^-, K^+ $MeNH_3^+$, $Me_2NH_2^+$, MnO_4^-, NH_4^+, NO_2^-, NO_3^-, OH^-, Rb^+, Tl^+

Charge 2

8	Be^{2+}, Mg^{2+}
7	$(CH_2)_6(COO)_2^{2-}$, $(CH_2)_5(COO)_2^+$, (congo red)$^{2-}$
6	Ca^{2+}, $C_6H_4(COO)_2^{2-}$, $(CH_2CH_2COO)_2^{2-}$, Cu^{2+}, Co^{2+}, Fe^{2+}, $H_2C(CH_2COO)_2^{2-}$, Mn^{2+}, Ni^{2+}, Sn^{2+}, Zn^{2+}
5	Ba^{2+}, Cd^{2+}, $(CH_2COO)_2^{2-}$, $(CHOHCOO)_2^{2-}$, $(COO)_2^{2-}$, CO_3^{2-}, $Co(NH_3)_5Cl^{2+}$, $Fe(CN)_5NO^{2-}$, $H_2C(COO)_2^{2-}$, H(citrate)$^{2-}$ Hg^{2+}, MoO_4^{2-}, Pb^{2+}, Ra^{2+}, S^{2-}, $S_2O_4^{2-}$, SO_3^{2-}, Sr^{2+}, WO_4^{2-},
4	CrO_4^{2-}, Hg_2^{2+}, HPO_4^{2-}, SeO_4^{2-}, $S_2O_3^{2-}$, $S_2O_6^{2-}$, $S_2O_8^{2-}$, SO_4^{2-}

Charge 3

9	Al^{3+}, Ce^{3+}, Cr^{3+}, Fe^{3+}, In^{3+}, La^{3+}, Nd^{3+}, Pr^{3+}, Sc^{3+}, Sm^{3+}, Y^{3+}
6	Co(ethylenediamine)$_3^{3+}$
5	Citrate^{3-}
4	$Cr(NH_3)_6^{3+}$, $Co(NH_3)_6^{3+}$, $Co(NH_3)_5H_2O^{3+}$, $Fe(CN)_6^{3-}$, PO_4^{3-},

Charge 4

11	Ce^{4+}, Sn^{4+}, Th^{4+}, Zr^{4+}
6	$Co(S_2O_3)(CN)_5^{4-}$
5	$Fe(CN)_6^{4-}$

Charge 5

9	$Co(S_2O_3)_2(CN)_4^{5-}$

$\gamma = 0.755$ at $I = 0.1$ with $a = 3$ Å and charge 1 (e.g., Cl^-) to

$\gamma = 0.45$ at $I = 0.1$ with $a = 8$ Å and charge 2 (e.g., Mg^{2+}) to

$\gamma = 0.355$ at $I = 0.1$ with $a = 4$ Å and charge 2 (e.g., SO_4^{2-})

to values as low as $\gamma = 0.02$ for ions with $a = 5$ Å and charge 4 [e.g., $Fe(CN)_6^{4-}$]. For other temperatures or solvents of different dielectric constants, appropriate values of A and B must be used, but a varies only if the primary solvation shell varies in size.

How well does Eq. (9) fit real data? Figure 2.2 provides a comparison. The experimental data[13] for the mean activity coefficient of NaCl and HCl are shown as points. The closest fit to the NaCl data is $a = 5$ (higher than expected from $a = 4$ for Na^+ and $a = 3$ for Cl^- as given in Table 2.1) and the best fit at low ionic strength to the HCl data is $a = 6$ to 7 (close to the average of $a = 9$ for H^+ and $a = 3$ for Cl^-, the values given in Table 2.1). At higher ionic strengths, the experimental data are much higher than the Debye–Hückel curves, reflecting a trend resulting from salting out and repulsion of ionic hydration spheres. The NaCl data agree better than the HCl data with the theory at higher ionic strengths.

On Fig. 2.2, the Debye–Hückel limiting law is also plotted; it falls substantially below the experimental data above 0.01 M.

Guntelberg[14] proposed that for solutions where the hydrated ion size was unknown, that a be taken to be $1/B = 3.0$ Å at 25 °C, resulting in an algebraically simple formula

$$-\log \gamma_\pm = A z_+ z_- \frac{\sqrt{I}}{1 + \sqrt{I}} \quad (11)$$

but as you might guess, this formula gives $\log \gamma_\pm$ values that are too negative for most electrolytes (see Table 2.1 and Fig. 2.2). Nevertheless, this equation is impor-

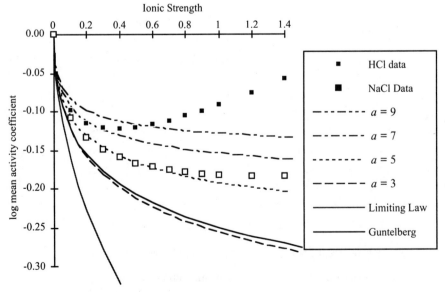

FIGURE 2.2. Extended Debye–Hückel equation for different values of a compared with experimental data for HCl and NaCl.

ESTIMATING ACTIVITY COEFFICIENTS

tant because a closely related equation is used to calculate the activity coefficient of Cl⁻ as part of the international convention for the pH scale (see Eq. 28, p. 56).

The Davies Equation. Because the extended Debye–Hückel equation does not fit experimental activity coefficient data accurately above 0.1 M for a 1:1 electrolyte, it is tempting to introduce additional parameters to produce an accurate fit to high concentrations. Guggenheim[15] suggested an unashamedly empirical approach, with sufficient polynomial terms to produce a good fit:

$$-\log \gamma_\pm = A z_+ z_- \frac{\sqrt{I}}{1+\sqrt{I}} + bI + cI^2 + dI^3 + \cdots \tag{12}$$

Such an equation is easy to use for interpolating measured data for single electrolytes, especially with the built-in curve-fitting programs of some computer applications, but has less predictive value. The empirical coefficients for multicomponent solutions depend not only on the ionic strength but on the detailed ionic composition of the solution. More recent developments have given physical meaning and hence greater predictability to empirical equations.[16]

Upon examining values of the first coefficient b (Eq. 11) for a number of 1-1 and 1-2 electrolytes, Davies[17] proposed an equation without any adjustable parameters:

$$-\log \gamma_\pm = A z_+ z_- \left(\frac{\sqrt{I}}{1+\sqrt{I}} - 0.2I \right) \tag{13}$$

or for single ions:

$$-\log \gamma_z = A z^2 \left(\frac{\sqrt{I}}{1+\sqrt{I}} - 0.2I \right) \tag{14}$$

The Davies equation is compared with experimental data for several 1-1 and 1-2 electrolytes in Fig. 2.3. The plot of Fig. 2.3 gives an idea of how accurate Davies' equation might be for 1-1 electrolytes. It is certainly more accurate than the Guntelberg approximation, but it gives a possible error of 3% at $I = 0.1$ and 10% at $I = 0.5$. Nevertheless, its convenience is attractive as an intermediate step between the assumption that all activity coefficients are unity, and the employment of more elaborate empirical equations. For interpolation of experimental data at higher ionic strength it is convenient to let the parameter b vary from 0.1 to 0.4 to fit the higher-ionic-strength data.[18]

Uncharged Molecules. The activity coefficient of most uncharged molecules (particularly CO_2, NH_3, H_2S, and other uncharged acids and bases) obey a simple salting-out model to ionic strengths as high as 5 M[19]:

$$\log \gamma_o = bI \tag{15}$$

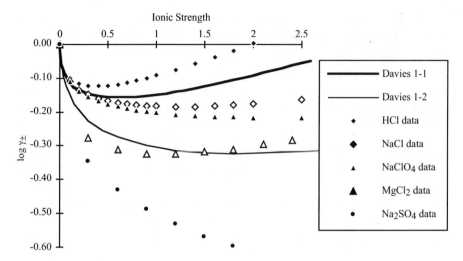

FIGURE 2.3. Comparison of experimental activity coefficients[40] for HCl, NaCl, NaClO$_4$, MgCl$_2$, and Na$_2$SO$_4$ with the Davies equation.

Typical values of b for CO_2 range from 0.11 at 10°C to 0.20 at 330°C[20,21]; for NH_3, at 30°C, $b = 0.12$.[22] b may be larger for larger organic molecules and smaller (or zero) for neutral ion pairs (see p. 59; also Chap. 12).

THE pH SCALE[23]

Definition of pH

Many acid–base scales have been suggested since Sørensen[24] proposed a "potential of hydrogen" $P_H = -\log C_H$, which was subsequently modified to pH. Some critics have objected to the way the pH scale represents higher acidity by smaller numbers; others have felt the acidity scale should have a zero point corresponding to pure water.[25] But the combination[26]

$$\text{pH} = \text{p}a_H = -\log(a_H) = -\log\{[H^+]\gamma_+\} \tag{16}$$

is now universally called "pH" and occupies a central place in acid–base chemistry. The only close competition is a concentration-based quantity similar to Sørensen's original definition:

$$\text{p}c_H = -\log[H^+] \tag{17}$$

which has found use in media of constant ionic strength, or in multicomponent solutions of higher concentration, where activity coefficients are uncertain. Examples include sea water[27] and brines[28] (See Chapters 10 and 11).

The Glass Electrode

One reason pH became so central to solution chemistry is because it is easy to measure experimentally. A glass electrode selective to hydrogen ions, a reference electrode with salt bridge to contact the test solution, and a sensitive electrometer to measure the potential between them, together with calibration standards of known pH, provide an almost universally available measurement system.

This electrochemical cell sketched in Fig. 2.4 can be represented more abstractly, for example, as

Ag | AgCl | buffer | Na$^+$ glass | *hydrated layer* | **test soln** | KCl (sat) | Hg$_2$Cl$_2$| Hg

The left part is in the "glass electrode," and the right part is in the "reference electrode"; both are often combined in a single glass body with a two-wire shielded cable to connect with the electrometer. The potential can be expressed (convention is that the right electrode's sign is the same as E) as

$$E = E°(\text{ref}) - E°(\text{glass internal}) + E_{\text{asym}} - (\text{slope}) \log\{[H^+]\gamma_+\} + E_j \quad (18)$$

The potential of the glass electrode internal elements and the potential of the reference electrode are dependent on temperature but not on the composition of the test solution.

The asymmetry potential E_{asym} of the glass electrode depends on the nature of the internal and external hydrated ion exchange layers, and thus can vary with time

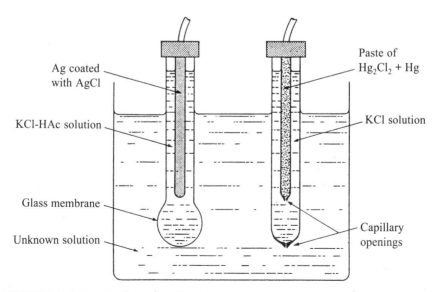

FIGURE 2.4. Schematic diagram of glass electrode system for measuring pH (*Ionic Equilibrium*, 1964, Fig. 2-6, p. 42). © James N. Butler

and temperature as well as with the presence of materials such as surface-active agents, nonaqueous solvents, polysaccharides, and proteins.

If the glass electrode were perfectly reversible to hydrogen ions, the "slope" would be the Nernst value of $\ln(10)RT/F$, or 0.05916 volts[29] at 25°C. This theoretical value is closely approximated by commercial glass electrodes provided the pH is below about 12. At high pH the ion exchange layer responds to cations such as Na^+ as well as H^+; this can be minimized by using special glass formulations or by making special calibrations that include all relevant ions.

The liquid junction potential E_j depends on the relative diffusion velocities of positive and negative ions. KCl was chosen long ago in part because the two ions have nearly identical mobilities; a high concentration (3.5 M or saturated) is used so that the effects of K^+ and Cl^- on the liquid junction potential will overwhelm any effects of ions at lower concentration in the test solution. The limitation of this technique becomes apparent when the test solution contains high concentrations of either strong acid or strong base, since the mobilities of H^+ (350) and OH^- (200) are much larger than those of K^+ (74) and Cl^- (76).[30]

Although the potential of the glass electrode system can be expressed compactly by combining terms in the equation above and employing the definition of pH:

$$E = E_{ref} + (\text{slope})\, pH \qquad (Eq.19)$$

the term E_{ref} now depends not only on the nature of the internal element and external reference electrode; it also depends on changes in the hydrated glass layer and changes in the liquid junction potential at the salt bridge. It cannot be calculated *a priori*.

The practical pH scale therefore relies on standard buffers of defined pH, which have been developed by measurements in cells *without* liquid junction, as described in the next section. For the moment, however, assume that such standards are available. Then measurement of pH consists of measuring E_s in the standard, and then E in the test solution. Most of the unknown or poorly known terms in E_{ref} cancel out, leaving

$$pH = pH_s + (E - E_s)/(\text{slope}) \qquad (20)$$

For the international pH scale, the slope is taken to be that from the Nernst equation, $(2.303\, RT/F)$.[31] However, some workers allow for a different slope (as well as compensating for liquid junction potential changes) by using two standards:

$$\frac{(pH - pH_{s1})}{(pH - pH_{s2})} = \frac{(E - E_{s1})}{(E - E_{s2})} \qquad (21)$$

This equation can be solved explicitly for pH as a function of E and the standard values. More than two standards can provide a nonlinear calibration curve, but this should be cause for suspicion if its slope deviates appreceably from the Nernst value.

THE pH SCALE

The Harned Cell

To provide standard buffers of known pH, measurements are made in cells *without* liquid junction, which employ a hydrogen electrode (a thermodynamic standard) instead of a glass electrode. Such a cell yields a potential that can be rigorously related to the mean activity of hydrogen chloride in the cell. The Harned cell (Fig. 2.5, named for a pioneer in solution thermodynamics[32]) is the most important cell of this type:

$$\text{Pt} \mid \text{H}_2(\text{gas}) \mid \text{H}^+, \text{Na}^+, \text{H}A, A^-, \text{Cl}^- \mid \text{AgCl(s)} \mid \text{Ag}$$

Here HA and A^- represent an acid–base pair that acts as a buffer in the pH region of interest (see Chapter 4). A wide variety of materials are suitable, but they must not interfere with the electrode reactions.

The chemical reaction occurring spontaneously in the cell can be represented as two electrode half-reactions

$$\tfrac{1}{2}\text{H}_2(g) \rightleftharpoons \text{H}^+ + e^-$$

$$\text{AgCl(s)} + e^- \rightleftharpoons \text{Ag(s)} + \text{Cl}^-$$

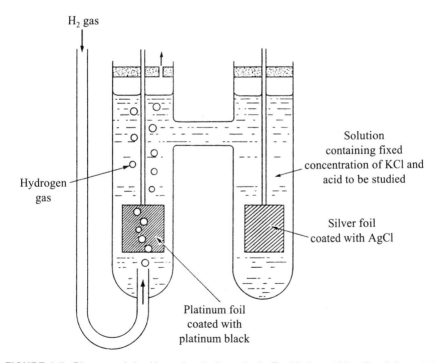

FIGURE 2.5. Diagram of the Harned cell. (from *Ionic Equilibrium*, 1964, Fig. 2-5, p. 41). © James N. Butler

or as an overall reaction:

$$\tfrac{1}{2}H_2 + AgCl(s) \rightleftharpoons H^+ + Cl^- + Ag(s)$$

The potential of this cell is given by the Nernst equation to be

$$E = E° - \frac{2.303RT}{F}\log\{[H^+][Cl^-]\gamma_+\gamma_-\} + \frac{4.605RT}{F}\log\{P_{H_2}\} \qquad (22)$$

Note that E is measured by the potentiometer, $[H^+]$ and $[Cl^-]$ can be determined by weight in moles/kg, the gas constant R is 8.3145 joule/mole K or 1.987 cal/mole K, the Faraday constant F is the charge on a mole of electrons (96485 coulombs, or 23.07 kcal/volt), and the temperature can be measured accurately (recall that $2.303RT/F = 0.05916$ volts at 25°C). The term in P_{H_2} is often omitted in the interest of simplicity. For accurate work, the partial pressure of hydrogen P_{H_2} would be measured also, and the potential corrected to 1.000 atm (typically 0.0005 volt).

This leaves three parameters to be determined for each data point: $E°$, γ_+, and γ_-. The above equation can be rearranged to put all experimentally measurable quantities on the left and the parameters on the right:

$$\begin{aligned} E°{}' &= E + \frac{2.303RT}{F}\log\{[H^+][Cl^-]\} + \frac{4.606RT}{F}\log\{P_{H_2}\} \\ &= E° - \frac{2.303RT}{R}\log\{\gamma_+\gamma_-\} \end{aligned} \qquad (23)$$

The next step is to recognize that γ_+ and γ_- approach unity as the total concentration of HCl approaches zero. (Such a standard state is consistent with the Debye–Hückel theory but in general is an arbitrary choice.) Thus the experimentally determined quantity $E°{}'$ approaches the standard potential $E°$ as total concentration decreases.

This is shown in the upper curve of Fig. 2.6.

A better extrapolation can be obtained by estimating $\gamma_\pm$ using one of the theoretical expressions discussed earlier. At this low ionic strength, it makes little difference which one is used, and for simplicity, the Debye–Hückel limiting law (Eq. 1) was used to obtain the lower curve of Fig. 2.6:

$$E°{}'' = E + \frac{2.303RT}{F}\log(m_H m_{Cl}) + \frac{2.303RT}{F}(2A)\sqrt{I} \approx E° \qquad (24)$$

For pure HCl, $m_H = m_{Cl} = I = m_{HCl}$. At 25°C, $A = 0.5109$ and $2.303\,RT/F = 0.05915$ volts. Note that the two functions do not extrapolate to exactly the same value of $E°$. This is because the upper plot is not actually linear, but curves downward in

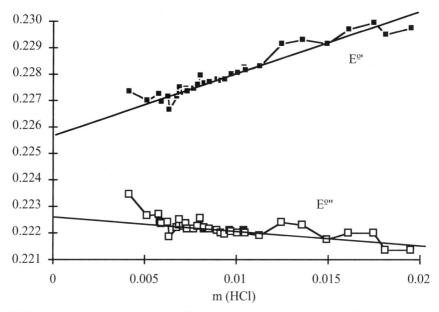

FIGURE 2.6. Extrapolation of potential[41] to obtain standard potential $E°$. $E°'$ is adjusted only for the HCl concentration; $E°'$ includes an estimate of $\gamma_\pm$ based on the Debye–Hückel limiting law.

the region below where measurements were made. The lower plot takes account of this.[33]

Once $E°$ has been obtained by extrapolation, its value depends only on temperature, and is the same for all concentrations of HCl. Since the standard potential of the hydrogen electrode is zero by definition, $E°$ is usually called the "standard potential of the silver/silver chloride electrode."

Once $E°$ is established, the Nernst equation also gives a value of $\gamma_+\gamma_-$ (or $\gamma_\pm$) for each value of E (see Fig. 2.7):

$$\log(\gamma_+\gamma_-) = 2\log(\gamma_\pm) = (E° - E)\left(\frac{F}{2.303RT}\right) - \log([H^+][Cl^-]) \qquad (25)$$

As pointed out several times above, neither ionic activity can be determined separately; only the product $\gamma_+\gamma_-$ can be determined. Thus activity coefficient measurements are often presented as a mean activity coefficient $\gamma_\pm = (\gamma_+\gamma_-)^{1/2}$ to emphasize the inseparability of single-ion activity coefficients.

Establishment of a pH Scale

Still another rearrangement of the Nernst equation for the Harned cell emphasizes that E can be taken as a measure of H^+ activity and hence pH.

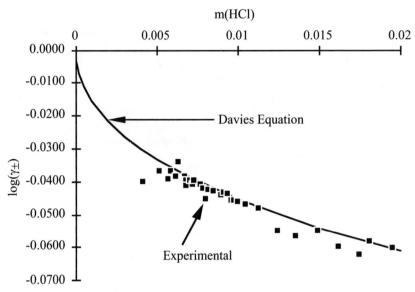

FIGURE 2.7. Activity coefficients for HCl obtained from the data of Harned and Ehlers[41] (Fig. 2.6), compared with the Davies equation. $E°$ was taken to be 0.2226.

$$\log\{[H^+][Cl^-]\gamma_+\gamma_-\} = (E° - E)(F/2.303RT)$$

$$p(a_H\gamma_{Cl}) = -\log\{[H^+]\gamma_+\gamma_-\} = (E - E°)(F/2.303RT) + \log\{[Cl^-]\} \quad (26)$$

$[Cl^-]$ can be fixed by adding a known amount of a chloride salt, $E°$ has been determined by extrapolation, E and T are measured; thus $p(a_H\gamma_{Cl})$ can be determined within experimental error, but without theoretical uncertainty. To obtain pH, γ_- or γ_{Cl} is needed.

$$pH = -\log\{[H^+]\gamma_+\} = p(a_H\gamma_{Cl}) + \log(\gamma_-) \quad (27)$$

This last step requires a nonthermodynamic assumption. Many possibilities are reasonable: γ_+ and γ_- are equal, γ_- is given by the Debye–Hückel equation with $a = 3$, γ_- is given by the Davies equation, etc. In fact, the agreement establishing the international pH scale was to use a Debye–Hückel equation with $a = 5$ to calculate the activity coefficient of chloride ion[34]:

$$-\log\gamma_{Cl} = A\frac{\sqrt{I}}{1 + 1.5\sqrt{I}} \quad (28)$$

Example 1. Consider the buffer consisting of 0.025 m $NaHCO_3$ and 0.025 m Na_2CO_3 with added KCl. Experimental data ($T = 25°C$, $E° = 0.22245$ volts) are given

THE pH SCALE

TABLE 2.2. Experimental data from the Harned cell with $NaHCO_3$-Na_2CO_3-KCl buffers[35]

m_{KCl}	I	E	$p(a_H \gamma_{Cl})$	$-\log \gamma_{Cl}$	pH
0.015	0.1150	0.92844	10.1117	0.1144	9.9973
0.010	0.1100	0.93903	10.1146	0.1127	10.0019
0.005	0.1050	0.95707	10.1186	0.1110	10.0076
0.000	0.1000	extrap.	10.121	0.1092	10.012

in Table 2.2, and the extrapolation of pH to zero added chloride is shown in Fig. 2.8.

Some standard buffer solutions that have been calibrated by this procedure are given in Table 2.3. Note that the concentrations of these standards are generally below the ionic strength of 0.1 m above which the Debye–Hückel equation might be expected to become less accurate (Fig 2.2). The pH values given in Table 2.2 have been adopted as primary standard values. Other buffers can be calibrated against these standards to provide secondary standards.

One difficulty with certifying only standards of low ionic strength is that they cannot be used reliably to calibrate pH-measuring systems in seawater, brines, biological media, etc. For such media new standards must be developed that are calibrated by a cell without liquid junction and a suitable theoretical estimate for γ_-; otherwise workers must be content with large uncertainties from liquid junction potentials as well as from poorly known activity coefficients. More details are given in Chapter 11.

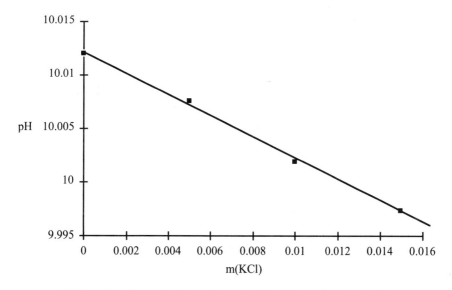

FIGURE 2.8. Extrapolation of data from Table 2.2 to zero added KCl.

TABLE 2.3. Some primary pH standards at 25°C from the U.S. National Bureau of Standards[36]

Buffer composition	pH
$KHC_4H_4O_6$–KH tartrate, sat'd at 25°C	3.557
0.05 m $KH_2C_6H_5O_7$–KH_2 citrate	3.776
0.05 m $KHC_8H_4O_4$–KH phthalate	4.008
0.025 m KH_2PO_4 + 0.025 m Na_2HPO_4	6.865
0.008695 m KH_2PO_4 + 0.03043 m Na_2HPO_4	7.413
0.01 m $Na_2B_4O_7$ (borax, sodium tetraborate)	9.180
0.025 m $NaHCO_3$ + 0.025 m Na_2CO_3	10.012

Equilibrium Constant Measurement

Recall that the equilibrium expression for a monoprotic acid–base pair is

$$[H^+][A^-]\gamma_+\gamma_{A^-} = K_a^\circ[HA]\gamma_0$$

$$pK_a^\circ = -\log(K_a^\circ) = -\log([H^+]\gamma_+) + \log([HA]\gamma_0) - \log([A^-]\gamma_{A^-})$$

The Harned cell and related experiments have been used to measure accurate values of HCl activity in buffers, and from these can be obtained accurate acid–base equilibrium constants. Indeed, such experiments are an integral part of establishing a pH standard.

Example 2. Measurements of the Harned cell containing acetic acid and either KCl or NaCl were combined with measurements of the same cell containing HCl to obtain the acidity constant of acetic acid.[37]

To obtain the concentration equilibrium constant, note that

$$-\log[H^+] = p(a_H\gamma_{Cl}) + 2\log(\gamma_\pm)$$

$\log(\gamma_\pm)$ can be evaluated from one of the theoreticial equations; but Harned and Murphy used experimental values from cells containing HCl and either NaCl or KCl. These are listed in Table 2.4. Since there are no other acids or bases in the HAc solutions, $[Ac^-] = [H^+] - [OH^-]$ and $[HAc] = m_{HAc} - [H^+] + [OH^-]$. Neglecting $[OH^-]$, the concentration acidity constant is then

$$K_a = \frac{[H^+]^2}{m_{HAc} - [H^+]} \tag{29}$$

These values are plotted in Fig. 2.9. (Compare with the data in Table 1.1, p. 11, which yield pK_a = 4.71 to 4.74)

In principle, both data sets in Fig. 2.9 should extrapolate to zero ionic strength at the same point, and yield pK_a°; but as you can see from Fig. 2.7, the activity coefficient curve is quite steep near $I = 0$. The best extrapolation method is similar to

THE pH SCALE

TABLE 2.4. Data from Harned cell containing acetic acid solutions

m_{HAc}	m_{NaCl}	m_{KCl}	$E(25°)$	$p(a_H\gamma_{Cl})$	$\log \gamma_{\pm}$(expt)	pK_a(conc)
0.2		0.05	0.46540	2.8082	−0.0932	4.5693
0.2		0.1	0.44962	2.8424	−0.1128	4.5296
0.2		0.2	0.43294	2.8615	−0.1270	4.5107
0.2		0.5	0.41170	2.9003	−0.1517	4.4895
0.2		1	0.39406	2.9031	−0.1450	4.5219
0.2		1.5	0.38394	2.9081	−0.1333	4.5792
0.2		2	0.37632	2.9042	−0.1116	4.6585
0.2		3	0.36492	2.8876	−0.0695	4.7945
0.2	0.05		0.46475	2.7972	−0.0904	4.5284
0.2	0.1		0.44860	2.8252	−0.1095	4.5078
0.2	0.2		0.43188	2.8436	−0.1250	4.4825
0.2	0.5		0.40940	2.8614	−0.1407	4.4554
0.2	1		0.39024	2.8386	−0.1213	4.4875
0.2	1.5		0.37815	2.8102	−0.0915	4.5503
0.2	2		0.36804	2.7643	−0.0514	4.6191
0.2	3		0.35222	2.6729	+0.03115	4.7670

that used to obtain $E°$—use a theoretical form for the activity coefficient to obtain approximate values of pK_a° at each ionic strength (Note that provisionally $\log \gamma_0 = 0$.)

$$pK_a^\circ = pK_a - 2 \log \gamma_{\pm}$$

and then extrapolate these values to $I = 0$. Whether the Davies equation or the experimental activity coefficients of Harned and Murphy are used, the extrapolated

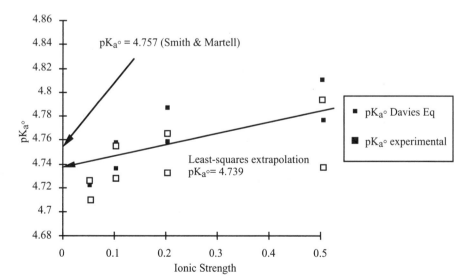

FIGURE 2.9. Extrapolation of pK_a to $I = 0$ using the Davies equation and Harned and Murphy's experimental activity coefficients yields 4.739, in agreement with others.

value is the same within experimental error ($pK_a^\circ = 4.739$) and agrees with the accepted value.[38] Interpolated values of pK_a as a function of I at high ionic strength are shown in Figure 2.10.

Example 3. See the data in Table 2.3. Although these were intended to provide a standard pH value for the bicarbonate/carbonate buffer, they can equally well be used to obtain an equilibrium constant (at ionic strength 0.100) for the acid–base reaction

$$H^+ + CO_3^{2-} \rightleftharpoons HCO_3^-$$

The simplest approach is to take the $pH = -\log\{[H^+]\gamma_+\} = 10.012$, extrapolated in Table 2.3 to $m_{KCl} = 0$, or $I = 0.100$, together with the approximate mass balances[39] $[CO_3^{2-}]_T = 0.0250$, $[HCO_3^-]_T = 0.0250$:

$$pK_a^\circ = 10.012 + \log\left(\frac{0.0250}{0.0250}\right) + \log\left(\frac{\gamma_{HCO_3}}{\gamma_{CO_3}}\right)$$

Approximate values for the activity coefficients can be obtained as described in the previous section. The acid form HA is in this case a monovalent ion HCO_3^-, and the basic form is the divalent ion CO_3^{2-}. The Davies equation yields $\log \gamma_{HCO_3} =$

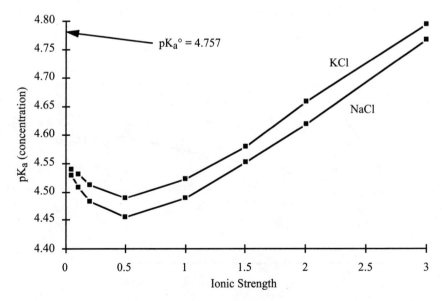

FIGURE 2.10. Concentration equilibrium constant obtained from experimental data in Table 2.4.

−0.1121 and log γ_{CO_3} = −0.4484 and hence pK_a° = 10.348. This is close to the standard value of 10.329 (see Table 10.1).

PROBLEMS

Note: Many of these problems refer to systems such as polyprotic acids, solubility products, and complex formation constants, which are not developed until later chapters. Where this applies, a cross-reference is made to that chapter.

1. Calculate the mean activity coefficient of 0.100 molar $NaNO_3$ using the Debye–Hückel limiting law, the Guntelberg equation, the Debye–Hückel equation with ion-size parameters (see Table 2.1), and the Davies equation. Compare with the experimental value of 0.762.

2. Calculate the mean activity coefficient of 0.100 molar $MgCl_2$, using the various methods given in Problem 1. Compare with the experimental value of 0.528.

3. Show that if the Davies equation is used to calculate activity coefficients, the pH of an acid HA^{z+} is given by

$$pH = pK_a^\circ + \log \frac{[HA^{z+}]}{[A^{(z-1)+}]} + (2z-1)(0.51)\left(\frac{\sqrt{I}}{1+\sqrt{I}} - 0.2I\right)$$

where pK_a° is the ionization constant of the acid at 25°C and zero ionic strength. (see Chapt. 5).

4. If the activity coefficient of ammonia in water at 25°C is given by $\log(\gamma_o) = 0.12\ I$, derive an equation for the acid ionization constant K_a for NH_4^+ (pK_a° = 9.245) (see Chapter 4).

5. Derive an equation giving the overall formation constant β_n of the complex $M(NH_3)_n^+$ as a function of ionic strength (see Chapter 7 for more information).

6. Show that the formation constant K_{s2} of the complex ion MCl_2^- from the solid salt MCl is nearly independent of the ionic strength (see Chapters 6 and 7).

7. Calculate the pH of a buffer made by mixing 50.0 mL of 0.100 molar KHK_2PO_4 with 29.1 mL of 0.100 molar NaOH and diluting to 100 mL (see Chapter 5).

8. Calculate the solubility of AgCl in 0.100 molar $NaClO_4$ (see Chapter 6).

9. Calculate the solubility of AgCl in 0.500 molar KCl (see Chapters 6 and 7).

10. Calculate the solubility of AgBr in a solution containing 3.58 moles NH_3, 0.50 mole NH_4Br, and 0.50 mole NH_4ClO_4 per liter (see Chapters 6 and 7).

11. Nair and Nancollas (*J.Chem. Soc.* 4144, 1958) measured the potential of the cell

Pt | $H_2(g)$ | HCl (m_1), $H_2SO_4(m_2)$ | AgCl | Ag

to be 0.4622 volts. Using the values $E° = 0.2224$ volt, $m_1 = 5.7655 \times 10^{-3}$ mole/kg, $m_2 = 10.100 \times 10^{-3}$ mole/kg, and the Davies equation to calculate activity coefficients, find the equilibrium constant for the reaction $HSO_4^- \rightleftharpoons H^+ + SO_4^{2-}$ at $I = 0$. Estimate the error in pK_a arising from potential measurement, concentration, and extrapolation, and estimate the overall error (see Chapter 4).

12. The dissociation equilibrium for water (See Chapter 3) is rigorously written as

$$[H^+]\gamma_+[OH^-]\gamma_- = K_w^\circ \, a_{H_2O}$$

For many applications, the activity coefficients γ_+ and γ_- as well as the activity of water are assumed to be 1.000. A less stringent approximation is that $\gamma_+ = \gamma_-$. Under this assumption, and the assumption that a_{H_2O} equals the mole fraction of H_2O in the solution, derive an expression for $pH = -\log[H^+]\gamma_+$ in a neutral salt solution containing m moles of NaCl per kilogram of water.

NOTES

1. H. S. Harned and B. B. Owen, *The Physical Chemistry of Electrolytic Solutions*, 3d Ed. Reinhold, NY, 1958; R. A. Robinson and R. H. Stokes, Electrolyte Solutions, London, Butterworths, 1959; K. S. Pitzer, Ed., *Activity Coefficients in Electrolyte Solutions*, CRC Press, Boca Raton, FL, 1991.
2. P. Debye and E. Hückel, *Physik. Z.* **24**:185–384, 1923; **25**:97, 1924.
3. See Footnotes 1 and 2 for references.
4. In general $\log \gamma_\pm = (v_+ \log \gamma_+ + v_- \log \gamma_-)/(v_+ + v_-)$, where dissolution of one mole (formula weight) of salt produces v_+ moles of cations and v_- moles of anions. Because the salt is electrically neutral, $v_+z_+ = v_-z_-$.
5. For a long time the accepted value of the dielectric constant for water was 78.54, which gives $A = 0.5085$; but the best recent data (K. S. Pitzer, Ed., *Activity Coefficients in Electrolyte Solutions*, CRC Press, Boca Raton, FL, 1991, p. 130) gives a dielectric constant of 78.3808, and hence $A = 0.5109$. The difference is not significant except for the most accurate work.
6. The ionic strength comes out of the Debye–Hückel theory in terms of molar concentration, but can also be defined in terms of molality, which is independent of temperature and pressure. See K. S. Pitzer, *Activity Coefficients in Electrolyte Solutions*, 2nd Ed., CRC Press, Boca Raton, FL, 1991. pp 83–84.
7. Even though this expression cannot be verified experimentally, the combination of values for two ions gives the correct result for the mean activity coefficient.
8. See, for example, R. G. Bates, B. R. Staples, and R. A. Robinson, *Anal. Chem.* **42**:867–70, 1970.
9. For charged acids, such as HCO_3^-, the equation corresponding to Eq. (7) appears to produce a combination of activity coefficients $(\gamma_-/\gamma_+\gamma_=)$, which are not electrically neutral:

$$K_{a2} = [\text{H}^+][\text{CO}_3^{2-}]/[\text{HCO}_3^-] = K_{a2}^\circ(\gamma_-/\gamma_+\gamma_=)$$

However, the format of the equilibrium constant hides the fact that the charge on [HCO_3^-] and [CO_3^{2-}] is compensated by an unreactive cation such as Na$^+$, and a rigorous formulation would use mean activity coefficients $\gamma_\pm$ for the salts Na$_2$CO$_3$, and H$^+$ with HCO$_3^-$ (perhaps estimated by $\gamma_\pm$ for HCl).

$$K_{a2} = [\text{H}^+\text{HCO}_3^-][(\text{Na}^+)_2\text{CO}_3^{2-}]/[\text{Na}^+\text{HCO}_3^-] = K_{a2}^\circ \gamma_\pm(\text{NaHCO}_3)/\gamma_\pm(\text{HCl})\gamma_\pm(\text{Na}_2^+\text{CO}_3)$$

10. See Footnotes 1 and 2.
11. J. Kielland, *J. Am. Chem. Soc.* **59**:1675, 1937.
12. For two electrolytes i and j, $d \log \gamma_i/dm_j = d \log \gamma_j/dm_i$ should hold. Unless $a_i = a_j$, it does not. See T. J. Wolery, *Amer. J. Sci.* **290**:296–320, 1990.
13. R. A. Robinson and R. H. Stokes, Electrolyte Solutions, Butterworths, London, 1965. App. 8.10.
14. E. Guntelberg, *Z. Phys. Chem.* **123**:199, 1926.
15. E. A. Guggenheim, *Thermodynamics*, 5th ed., North-Holland, Amsterdam, 1967.
16. K. S. Pitzer, *Activity Coefficients in Electrolyte Solutions*, 2nd Ed., CRC Press, Boca Raton, FL, 1991, Chapter 3; R. Heyrovska, *J. Electrochem. Soc.* **143**:1789–93, 1996. See also Chapter 11 of this book.
17. C. W. Davies, *Ion Association,* London, Butterworths, 1962.
18. The computer program MINEQL uses $b = 0.24$ and the program PHREEQE uses $b = 0.30$ (see Chap. 12; see also Fig. 7.10 and 9.6).
19. The form $\log \gamma = b\,I$ does not satisfy the thermodynamic consistency relation $d \log \gamma_i/dm_j = d \log \gamma_j/dm_i$ unless Σm_i is used instead of $I = \Sigma m_i z_i^2$. This caveat does not apply to the linear term in the Davies equation, since it is multiplied by z_i^2. See T. J. Wolery, *Am. J. Sci.* **290**:296–320, 1990.
20. A. J. Ellis and R. M. Golding, *Am. J. Sci.* **162**:47–60, 1963.
21. S. D. Malinin, *Geokhimiya*, 1959, No. 3, 235–45.
22. J. Bjerrum, *Metal Ammine Formation in Aqueous Solutions*, P. Haase & Son, Copenhagen, 1957.
23. R. G. Bates, *Detemination of pH*, 2nd Ed., Wiley, NY, 1973.
24. S. P. L. Sørensen, *Biochem. Z.* **21**:131 and 21–201, 1909. Bates says "In two important papers, totaling some 170 pages and published simultaneously in German and French, Sørensen compared the usefulness of the *degree of acidity* with that of the *total acidity*, proposed the hydrogen ion exponent, set up standard methods for the determination of hydrogen ion concentrations by both electrometric and colorimetric means, with a description of suitable buffers and indicators, and discussed in detail the application of pH measurements to enzymatic studies."
25. F. E. Crane, *J. Chem. Educ.* **38**:365–66, 1961.
26. The notations pa_H, p($a_\text{H}\gamma_\text{Cl}$), and p$c_\text{H}$ were introduced by R. G. Bates.
27. A. G. Dickson, *Geochim. et Cosmochim. Acta* **48**: 2299–308, 1984; C. H. Culberson, "Direct Potentiometry" in *Marine Electrochemistry*, M.Whitfield and D. Jagner, Eds., Wiley, NY, 1981, pp. 187–261.
28. K. G. Knauss, T. J. Wolery, and K. J. Jackson, *Geochim. Cosmochim. Acta* **54**:1519–23, 1990.

29. $R = 8.31451$ J/mol K, $F = 96485.3$ coul. See CODATA 1986: *J. Research Nat. Bureau Standards* **92**:85–95, 1987; also http://physics.nist.gov/PhysRefData/codata86/physico.html.
30. The values given for illustrative purposes are the limiting equivalent conductivities at 25°C in cm^2 Int. Ω^{-1} equiv^{-1}. From Robinison and Stokes, *Electrolyte Solutions*, 2nd Ed. revised, Butterworth, London, 1968, p. 463.
31. Bates, *Determination of pH*, op. cit.
32. H. S. Harned and B. B. Owen, *Physical Chemistry of Electrolytic Solutions*, 3d Ed., Reinhold, NY, 1958.
33. These data give $E° = 0.2226$ U.S. International volts (1 volt = 1.000330 V). Published values at 25°C range from $E° = 0.22234$ (R. G. Bates and V. E. Bower, *J. Res. Nat. Bur. Stds.* **53**:282, 1954) to $E° = 0.22258$. (Roy et al., *J. Phys. Chem.* **94**:7706, 1990). K. S. Pitzer, *Activity Coefficients in Electrolyte Solutions* (1991) p. 159 gives $E° = 0.22245$ or 0.22237 V. The "best value" is often taken to be 0.22240 V (A. Dickson, private communication).
34. R. G. Bates and E. A. Guggenheim, *Pure Appl. Chem.* **1**:163, 1960.
35. B. R. Staples and R. G. Bates, *J. Res. Nat. Bur. Stds.* **73A**(1):37–41, 1969. These authors cite Bates and Bower but do not give the numerical values of $E°$ that they used. $E° = 0.22245$ provides a smoother extrapolaltion (Fig. 2.7) than 0.22234.
36. R. G. Bates, *J. Res. Nat. Bur. Stds.* **66A**(2):179, 1962; B. R. Staples and R. G. Bates, *J. Res. Nat. Bur. Stds.* **73A**(1):37–41, 1969. NBS is now called National Institute for Standards and Technology (NIST).
37. H. S. Harned and G. M. Murphy, *J. Am. Chem. Soc.* **53**:8–17, 1931. These data were also used in preparing Table 1.1.
38. Smith and Martell selected $pK_a° = 4.757 \pm 0.002$. H. S.Harned and R. W. Ehlers, *J. Am. Chem. Soc.* **55**:65, 1933, obtained $pK_a° = 4.756$.
39. Combination of mass and charge balances yield: $[HCO_3^-] = C_{HCO_3} - [H^+] + [OH^-] - 2[CO_2] = 0.02512$; $[CO_3^{2-}] = C_{CO_3} + [H^+] - [OH^-] + [CO_2] = 0.02511$. These differ from the approximate values by about 0.4%. See Chapter 4 for a full development of this topic.
40. R. A. Robinson and R. H. Stokes, *Electrolyte Solutions*, Butterworths, London, 1965. App. 8.10.
41. H. S. Harned and R. W. Ehlers, *J. Am. Chem. Soc.* **54**:1350, 1932; **55**:652, 1933; **55**: 2179–93, 1933.

3

STRONG ACIDS AND BASES

The Ionization of Water
The Ion Product of Water
    Oceanographic Form of K_w
    Pure Water.
    Nonaqueous Solutions.
Strong Acids and Bases
    pH of a Strong Acid.
    pH of a Strong Base
    Mixture of Strong Acid and Strong Base
    Titration of Strong Acids and Strong Bases
        General Titration Equation
        End Point versus Equivalence Point
        Conductimetric End Points
    Titration Error
        General Equation for Titration Error
    Linearized (Gran) Titration Curves
        Nonlinear Curve Fitting
Problems

THE IONIZATION OF WATER

A number of the concepts discussed in this chapter and the next have already been briefly presented in Chapter 1 as part of the introduction to equilibria and equilibrium constants. I hope you will not take offense at a little repetition designed to make things clearer.

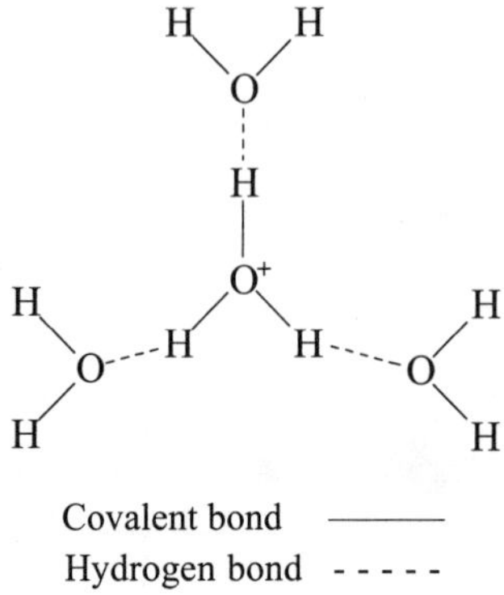

FIGURE 3.1. Structure of the primary hydration shell of the aqueous hydrogen ion.

When water ionizes, a proton is transferred from one water molecule to another, resulting in a hydrated hydrogen ion and a hydroxyl ion:

$$H_2O + H_2O \rightleftharpoons H_3O^+ + OH^-$$

This is, of course, a simplified picture of what actually happens, but it emphasizes the fact that a hydrogen ion in aqueous solution is *not* a bare proton, but is very firmly attached to a water molecule. The energy required to dissociate H_3O^+ completely into H_2O and a proton is about 258 kcal/mol[1] (three times the energy required to break most covalent bonds). H_3O^+ is further hydrated, with one water molecule attached to each of the hydrogens by hydrogen bonds. These three water molecules constitute the primary hydration shell of the hydrogen ion.[2] The hydrogen bonds (broken lines) are only about one-tenth as strong as ordinary covalent bonds, but they are stronger than the hydrogen bonds that join water molelcules into aggregates. Indeed, the smallest unit in which a hydrated proton exists in liquid water is $H_9O_4^+$ (Fig. 3.1). Many more water molecules (secondary hydration shells) are attached by weaker hydrogen bonds—at the freezing point, the size of these aggregates approaches the size of ice crystals.[3]

Since these extra hydration shells do not directly appear in acid–base reactions, the formulas of species are usually simplified by representing all hydrated hydrogen ions as "H^+" rather than "H_3O^+" or "$H_9O_4^+$."[4]

THE ION PRODUCT OF WATER

In Chapter 1, the ion product of water was introduced as an example of one format for the equilibrium constant. To recapitulate—the form of the chemical reaction:

$$H_2O \rightleftharpoons H^+ + OH^-$$

leads to the form of the equilibrium constant equation using activities:

$$a_H a_{OH} = K° a_{H_2O} \tag{1}$$

The activity of H^+ and OH^- are proportional to their concentrations, but by thermodynamic convention (Raoult's law), the activity of H_2O is proportional to the mole fraction of water in the solution. a_{H_2O} is close to 1.000 in dilute solutions (0.999 in 0.05 M NaCl), but is lower in concentrated solutions.[5] If a_{H_2O} is included in the constant, Eq. (1) yields:

$$a_H a_{OH} = [H^+]\gamma_+[OH^-]\gamma_- = K_w° \tag{2}$$

which also explicitly states the activity coefficients of H^+ and OH^-. The most common form is the "concentration" constant, in which the activity coefficients have been folded into the equilibrium constant:

$$[H^+][OH^-] = K_w \tag{3}$$

The concentration constant is related to the previous equilibrium constants by

$$K_w = K° \, a_{H_2O}/\gamma_+\gamma_- = K_w°/\gamma_+\gamma_- \tag{4}$$

and these relations also show that K_w depends not only on temperature (as does $K°$) but also on solution composition via water activity and the activity coefficients of H^+ and OH^- ions. Some selected data for the ionic strength dependence of K_w are given in Table 3.1, and the temperature dependence of $K_w°$ is given in Table 3.2.

In Fig. 3.2, data for pK_w from various sources are compared with each other and with the Davies and Debye–Hückel equations. The Davies equation represents the data from $I = 0$ to 2 with acceptable error.

The concentration constant is a valuable simplification, since concentrations, not activities, enter the mass and charge balances. In addition, if a large excess (1 to 5 mol/L) of a noncomplexing electrolyte, such as NaCl or $NaClO_4$, is added to the solution, the activity coefficients are controlled by that background electrolyte and remain constant when smaller concentrations (which may be of primary interest) are

TABLE 3.1. Concentration ion product K_w at 25°C at various ionic strengths of $NaClO_4$[6]

Ionic strength (mol/L)	pK_w
0.0	13.997 ± 0.003
0.1	13.78 ± 0.01
0.5	13.74 ± 0.02
1.0	13.79 ± 0.02
2.0	13.96 ± 0.01
3.0	14.18 ± 0.04

TABLE 3.2. Temperature dependence of pK_w° from 0 to 50°C at $I = 0^7$ (note that this table is extended to 250°C in Table 10.1)

Temp. (°C)	pK_w°
0	14.96
10	14.53
20	14.17
25	14.00
30	13.83
40	13.53
50	13.26

changed. Thus K_w will also remain constant—but it will be a function of the background electrolyte concentration (Table 3.1) as well as temperature and pressure.

Oceanographic Form of K_w

A special form is used in oceanography (see Chapter 10):

$$10^{-pH} [OH^-]_T = K_w'$$

where $[OH^-]_T$ (total OH) is the sum of free OH and ion pairs with the major cations:

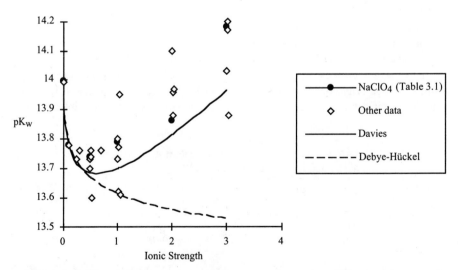

FIGURE 3.2. pK_w as a function of ionic strength. The Davies equation and Debye–Hückel equation (with $a = 9$ for H^+ and $a = 3$ for OH^-) are compared with selected data from Smith and Martell (see Table 3.1) and other data from Sillén and Martell ($NaClO_4$, NaCl, $NaNO_3$, KNO_3, KCl media).

THE ION PRODUCT OF WATER

$$[OH^-]_T = [OH^-]_f + [MgOH^+] + [CaOH^+] + [NaOH] + \cdots$$

$$= [OH^-]_f \{K_{MgOH}[Mg^{2+}]_f + K_{CaOH}[Ca^{2+}]_f + K_{NaOH}[Na^+]_f + \cdots$$

$$= [OH^-]_f \, \alpha_{OH}$$

Since the ion-pair concentrations are proportional to $[OH^-]_f$, and the major cation concentrations $[Mg^{2+}]$, $[Ca^{2+}]$, and $[Na^+]$ are in the same ratio for a wide range of salinity, α_{OH} is dependent on salinity (and hence ionic strength) but not on minor species such as $[OH^-]$. It can be included in a salinity-dependent equilibrium constant.

$$K'_w = \frac{[H^+]\gamma_+ \alpha_{OH}[OH^-]_f \gamma_-}{\gamma_-} = K^\circ_w \frac{\alpha_{OH}}{\gamma_-} = K_w \gamma_+ \alpha_{OH} \tag{5}$$

For example, the value of pK'_w at 25°C in seawater of salinity 35‰ ($I = 0.7$; see Chapter 10) is 13.60; this is significantly lower than the value for pK_w (13.76) interpolated at $I = 0.7$ from the $NaClO_4$ data in Table 3.1. The difference can be expressed in two parts: $\gamma_+ = 10^{-0.16}$ (Davies Eq.) and $\alpha_{OH} = 10^{+0.32}$ attributed to ion pairing between OH^- and cations in the seawater.[8]

Pure Water

In the very dilute solution that constitutes pure water, the activity of water is unity, as are the ionic activity coefficients. At 25°C, the value extrapolated to zero ionic strength (Table 3.1) is $\log K^\circ_w = 13.997 \pm 0.003$, or

$$[H^+][OH^-] = K^\circ_w = 1.00 \times 10^{-14} \tag{6}$$

If you think in terms of a mass balance, every molecule of water that dissociates produces one H^+ and one OH^-

$$[H^+] = [OH^-] \tag{7}$$

This equation could also be thought of as a charge balance—if the solution is to be electrically neutral, the total of all positively charged ions must balance the total of all negatively charged ions. Since the only positively charged ions are H^+, and the only negatively charged ions are OH^-, their concentrations must be equal.

Substituting $[OH^-]$ from the mass (or charge) balance into the equilibrium gives

$$[H^+]^2 = K_w = 1.00 \times 10^{-14}$$

$$[H^+] = (K_w)^{1/2} = 1.00 \times 10^{-7}$$

$$pH = -\log[H^+] = 7.00$$

This point on the acid–base scale is defined to be "neutral." Lower values of pH are acidic; higher values are basic. The exact numerical value of 7.00 is accidental—

at other temperatures, or in other aqueous solutions, the neutral point of the acid–base scale would in general be a different number. For example, at 50°C in pure water, $pK_w = 13.26$ (Table 2) and the neutral point is pH = 6.63; at 25°C in 3 mol/L $NaClO_4$, $pK_w = 14.18$ (Table 1), and the neutral point is at pH = 7.09.

Nonaqueous Solutions[9]

More general criteria of acidity and basicity are required for nonaqueus solvents. One possible criterion is in terms of the ionization of the solvent. For instance, in liquid ammonia solution, the ions are formed by transfer of a proton, just as in the ionization of water:

$$NH_3 + NH_3 \rightleftharpoons NH_4^+ + NH_2^-$$

and the equilibrium constant[10] at –60°C is

$$[NH_4^+][NH_2^-] = K = 10^{-32}$$

An acid solution in liquid ammonia would be one which has $[NH_4^+]$ larger than $[NH_2^-]$, and a basic solution would be one which has $[NH_2^-]$ larger than $[NH_4^+]$.

Because of the much smaller value of K, the pH scale in ammonia ranges from 0 to 32 instead of 0 to 14. For example, if pH is defined to be $-\log[NH_4^+]$, the pH of 1 M NH_4Cl would be approximately 0, and the pH of 1 M $NaNH_2$ would be 32.

STRONG ACIDS AND BASES

The terms *strong* and *weak* refer not to the concentration of a solution but to the extent to which a substance is dissociated to its ions in solution.[11] A strong electrolyte is one that is completely dissociated, and a weak electrolye is one that is only partly dissociated. "Strength" depends on the solvent as well as the solute. HCl is completely dissociated in water, because water is an excellent proton acceptor, but pure (glacial) acetic acid is a poor proton acceptor and HCl is a weak acid in that solvent.[12] Similarly, alkali metal hydroxides such as NaOH are fully dissociated (hence "strong") in dilute to moderate concentrations, but are somewhat associated (hence "weak") at concentrations above 1 M.[13]

Sulfuric acid H_2SO_4 is special in that one of its protons is always fully dissociated in aqueous solution, but the other is a weak acid with $pK_a = 2$. This introduces complications that will be discussed in Chapter 4.

pH of a Strong Acid

Consider the pH of C mol/L[14] HCl. In addition to the ions produced by dissociation of water, this solution contains H^+ and Cl^- from the hydrochloric acid. The known quantities are the concentration of HCl and the ion product of water. There

STRONG ACIDS AND BASES

are three concentrations to be calculated: [H$^+$], [OH$^-$], and [Cl$^-$]. Three algebraic equations connect them:

$$\text{Mass balance on Cl: } [\text{Cl}^-] = C \tag{8}$$

$$\text{Ion product: } [\text{H}^+][\text{OH}^-] = K_w = 10^{-14.00} \tag{9}$$

$$\text{Charge balance: } [\text{H}^+] = [\text{OH}^-] + [\text{Cl}^-] \tag{10}$$

Substituting [Cl$^-$] from the mass balance (Eq. 8) and [OH$^-$] from the ion product (Eq. 9) into the charge balance gives an equation in [H$^+$] alone:

$$[\text{H}^+] = K_w/[\text{H}^+] + C \tag{11}$$

Example 1. Find the pH of 0.001 molar HCl. Equation (11), with $C = 0.001$, could be solved using the quadratic formula, but it is simpler (and more in keeping with the theme of this book) to note that HCl is an acid solution, whence pH < 7 and [H$^+$] >> 10^{-7}. The first term on the right side of Eq. (11) therefore cannot be larger than $10^{-14}/10^{-7} = 10^{-7}$, which is negliglible compared to 0.001; hence

$$[\text{H}^+] = 0.001 + \cdots$$

where "$\cdots$" represents terms small compared to those explicitly displayed. Finally,

$$\text{pH} = -\log[\text{H}^+] = 3.00$$

Going back to the full Eq. (11) for a second iteration, you can see that the first term on the right is

$$(1.00 \times 10^{-14})/[\text{H}^+] = (1.00 \times 10^{-14})/(1.00 \times 10^{-3}) = 1.00 \times 10^{-11}$$

and it is indeed negligible compared to 0.001.

The general form of this calculation (Eq. 11) is plotted in Fig. 3.3. When pH << 7, as in Example 1, the approximate form of Eq. (11) is

$$\text{pH} = -\log C \tag{12}$$

At higher ionic strength ($I = C$ plus added background electrolyte) K_w changes (see Table 3.1) and pH is calculated using an appropriate activity coefficient

$$\text{pH} = -\log([\text{H}^+]\gamma_+)$$

Note that when C is large compared to 10^{-7}, the K_w term is negligible and [H$^+$] = C; when C is near 10^{-7}, both terms are important; and when $C \ll 10^{-7}$, the C term is negligible. Under this last condition, pH is the same as for pure water.

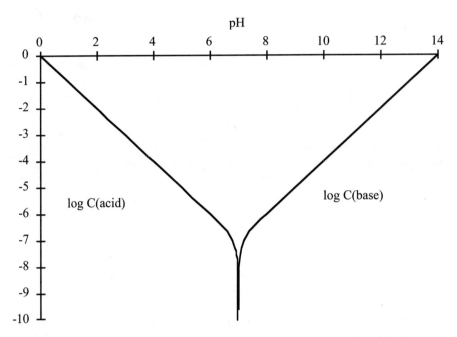

FIGURE 3.3. pH of strong acid or strong base as a function of concentration. Note that concentration approaches zero (log $C \to \infty$) as pH approaches 7 from above or from below. (compare with Fig. 4.3)

pH of a Strong Base

If you are clever, you can infer the equations for a strong base by symmetry, but for the sake of clarity, use the mass and charge balances appropriate to NaOH of concentration C:

$$\text{Mass: } [Na^+] = C \tag{13}$$

$$\text{Ion product: } [H^+][OH^-] = K_w \tag{14}$$

$$\text{Charge: } [H^+] + [Na^+] = [OH^-] \tag{15}$$

Substituting for $[Na^+]$ from the mass balance (Eq. 13) and for $[OH^-]$ from the ion product (Eq. 14) in the charge balance (Eq. 15) gives

$$[H^+] + C = K_w/[H^+] \tag{16}$$

This equation is the same as Eq. (11) for a strong acid, except that C has been replaced by $-C$. In Fig. 3.3 the base curve (right-hand branch) is the mirror image of the acid curve (left-hand branch). In basic solutions, pH > 7, and hence $[H^+] < 10^{-7}$. If $C \gg 10^{-7}$, $[H^+]$ is negligible compared to C and the equation simplifies to

STRONG ACIDS AND BASES

$$C = \frac{K_w}{[H^+]} \quad \text{or} \quad [H^+] = \frac{K_w}{C} \quad \text{or} \quad pH = pK_w + \log C \tag{17}$$

If C is near 10^{-7}, both terms on the left side of Eq. (16) are important; if $C \ll 10^{-7}$, then C is negligible compared to $[H^+]$ and the solution has the same pH as pure water.

Example 2. Find the pH of 2.0×10^{-7} m NaOH. Note that both terms in the charge balance, corresponding to $[H^+]$ and $[Na^+]$, are nearly equal. The charge balance in numerical terms becomes

$$[H^+] + 2.0 \times 10^{-7} = 1.0 \times 10^{-14}/[H^+]$$

With $x = [H^+]$, the quadratic formula can be applied:

$$x^2 + 2.0 \times 10^{-7} x - 1.0 \times 10^{-14} = 0$$

$$x = \frac{-(2.0 \times 10^{-7}) \pm \sqrt{4.0 \times 10^{-14} + 4.0 \times 10^{-14}}}{2}$$

With the positive sign, $x = 4.14 \times 10^{-8}$ (with the negative sign, $x = -2.41 \times 10^{-7}$; an unreal result). Thus

$$[H^+] = 4.14 \times 10^{-8} \quad \text{and} \quad pH = 7.38$$

Note that this is higher than the pH of pure water, and higher than the value obtained from Eq. (17): pH = $14.00 + \log(2.0 \times 10^{-7}) = 7.30$

Mixture of Strong Acid and Strong Base

If a strong acid is mixed with a strong base, hydrogen ions are neutralized by hydroxyl ions (to make water), and the resultant pH is closer to 7 than either of the initial solutions. The charge and mass balances are similar to the above examples. Consider C_a moles of strong acid such as HCl and C_b moles of strong base such as NaOH per kilogram (or liter) of solution. The equations are

$$\text{Mass: } [Cl^-] = C_a$$
$$\text{Mass: } [Na^+] = C_b$$
$$\text{Charge: } [Na^+] + [H^+] = [Cl^-] + [OH^-]$$
$$\text{Ion product: } [H^+][OH^-] = K_w$$

Combining these gives

$$C_a - C_b = [H^+] - K_w/[H^+]$$

Note that when the acid and base are perfectly neutralized ($C_a = C_b$), then $[H^+] = \sqrt{K_w}$, the same as for pure water. Of course, the water is not pure but contains Na^+ and Cl^-; if the concentration of these is high enough, the change in ionic strength will change $\gamma_\pm$ and K_w.

Titration of Strong Acids and Strong Bases

Closely related to the above discussion of mixing solutions of strong acids and bases is the process of titration. Usually this is done by adding a known volume[15] from a burette to an initial volume in a stirred beaker until an "end point" is reached. The end point can be determined in a variety of ways, as discussed below, but is always an approximation to the "equivalence point," where the strong acid and strong base exactly neutralize each other ($C_a = C_b$ in the previous section). In this section we will develop the equations for a titration curve of pH versus volume added.

The mass balances are different in form from those previously used in this chapter. Up to now, the volume of solution has been constant, and hence it was correct to balance concentrations alone. But in a titration the volume added from the burette increases from 0 to V, and at the same time the volume in the beaker increases from V_0 to $V + V_0$; so you must use moles, not moles per liter, in the mass balances. This requires that you multiply each concentration by the appropriate volume. For example, the number of moles of strong base titrant added from the burette would be $C_b V$, and the number of moles of sodium ion in the beaker would be $[Na^+](V + V_0)$.

Example 3. Titrate 50 mL (V_0) of 0.1 M hydrochloric acid with 0.2 M sodium hydroxide. Note that the equivalence point is when $V(0.2) = (50)(0.1)$, or $V = 25$ mL.

$$\text{Mass balance: } [Cl^-](V + 50.0) = (0.10)(50.0)$$

$$\text{Mass balance: } [Na^+](V + 50.0) = (0.20)V$$

$$\text{Ion product: } [H^+][OH^-] = K_w = 1.00 \times 10^{-14}$$

$$\text{Charge balance: } [H^+] + [Na^+] = [Cl^-] + [OH^-]$$

It is not too hard to solve this set of equations to obtain an equation relating V and $[H^+]$ in the region before the equivalence point. Early in the titration, when there is excess acid, $[H^+]$ is much larger than $[OH^-]$, so that $[H^+]$ is given by the excess acid that has not yet been neutralized by base: $[Cl^-] - [Na^+]$:

$$[H^+] = \frac{0.10(50.0)}{V + 50.0} - \frac{0.20V}{V + 50.0}$$

The curve of pH versus V is plotted in Fig. 4. Note that the equivalence point (at pH = 7), in principle the most interesting point on the graph, is hard to approach.

STRONG ACIDS AND BASES

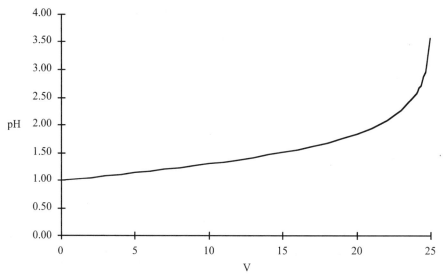

FIGURE 3.4. pH = –log[H⁺] as a function of V for the acid range of the titration example. The approximate equation gives [H⁺] = 0 (pH becomes infinite) at the equivalence point, V = 25.0.

At 24.9 mL, pH is 3.57. At 24.99, pH is 4.57. Even with the best burette technique, that is still 2.5 units away from the desired equivalence point. Furthermore, by neglecting [OH⁻], we have omitted the part of the curve with pH > 7.

General Titration Equation. Thus we are brought back to the full equations of mass, charge, and equilibrium. Consider a volume V_a of HCl, concentration C_a, titrated by a volume V_b of NaOH, concentration C_b:

$$\text{Mass balance: } [Cl^-](V_a + V_b) = (C_a)V_a \quad (18)$$

$$\text{Mass balance: } [Na^+](V_a + V_b) = (C_b)V_b \quad (19)$$

$$\text{Ion product: } [H^+][OH^-] = K_w = 1.00 \times 10^{-14} \quad (20)$$

$$\text{Charge balance: } [H^+] + [Na^+] = [Cl^-] + [OH^-] \quad (21)$$

Substituting in the charge balance (Eq. 21) for [Cl⁻] from the first mass balance (Eq. 18), for [Na⁺] from the second mass balance (Eq. 19), and for [OH⁻] from the ion product (Eq. 20), you will get:

$$[H^+] + \frac{C_b V_b}{V_a + V_b} = \frac{C_a V_a}{V_a + V_b} + \frac{K_w}{[H^+]} \quad (22)$$

This equation covers the entire pH range. At the equivalence point, where $C_b V_b =$

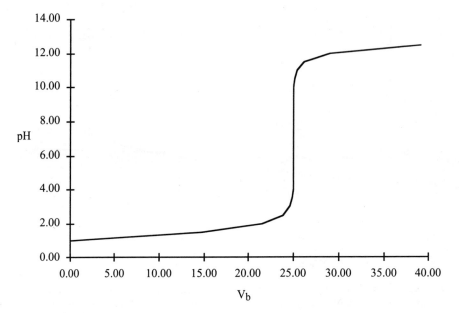

FIGURE 3.5. Titration curve for 50 mL of 0.1 M HCl with 0.2 M NaOH (Eq. 23).

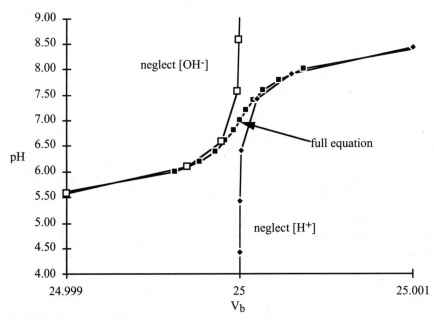

FIGURE 3.6. Enlarged section of titration curve (Eq. 23) near equivalence point. Approximate curves are obtained by neglecting [OH⁻] for the acid range (Eq. 24) and neglecting [H⁺] for the basic range (Eq. 25).

STRONG ACIDS AND BASES

TABLE 3.3. Excel spreadsheet for calculating data for Fig. 3.5 from Eq. (23)

Strong acid-strong base titration curve of Fig. 5		
Ca	0.1	
Cb	0.2	
Va	50	
Kw	1.00E-14	
H =	10^-pH	
Vb =	Va*(Ca-H+Kw/H)/(Cb+H-Kw/H)	
pH	H	Vb
1	0.1	0.000
2	0.01	21.429
3	0.001	24.627
4	0.0001	24.963
5	0.00001	24.996
6	0.000001	25.000
7	1E-07	25.000
8	1E-08	25.000
9	1E-09	25.004
10	1E-10	25.038
11	1E-11	25.377
12	1E-12	28.947
13	1E-13	100.000

$C_a V_a$, the two inner terms cancel, leaving $[H^+] = [OH^-]$, or pH = 7. A plot of this function is most easily made by solving for V_b as a function of $[H^+]$:

$$V_b = V_a \left(\frac{C_a - [H^+] + K_w/[H^+]}{C_b + [H^+] - K_w/[H^+]} \right) \tag{23}$$

Values of pH are chosen, $[H^+]$ is calculated from each, and V_b is calculated from Eq (23) (see Table 3.3). The result is plotted as pH vs V_b. Such a plot, using the numerical values ($C_a = 0.1$, $C_b = 0.2$, $V_o = 50$) from the example, is shown in Fig. 3.5. A magnified section near the equivalence point is shown in Fig. 3.6, and compared with the aproximations obtained by neglecting either $[OH^-]$ or $[H^+]$ in the charge balance (Eq. 21):

$$\text{Acid range:} \quad V_b = V_a \left(\frac{C_a + [H^+]}{C_b + [H^+]} \right) \tag{24}$$

(Compare with Example 3 and Fig. 3.4)

$$\text{Basic range:} \quad V_b = V_a \left(\frac{C_a [H^+] - K_w}{C_b [H^+] - K_w} \right) \tag{25}$$

The equations for titration of a strong base with a strong acid are identical to the above. The only change necessary is to solve for V_a instead of V_b.

$$V_a = V_b \left(\frac{C_b + [H^+] - K_w/[H^+]}{C_a - [H^+] + K_w/[H^+]} \right) \tag{26}$$

The curve is the vertical reflection of Fig. 3.5 about pH = 7, beginning with high pH values and ending with low pH.

End Point versus Equivalence Point. As mentioned above, the "equivalence point" is the theoretical point at which exactly the stoichiometric amount of titrant has been added. In the case of a strong acid titrated by strong base, if C_a mole/L[16] in volume V_a is titrated by C_b mol/L in volume V_b, the equivalence point is when $C_b V_b = C_a V_a$.

The experimental approximation to the equivalence point is the "end point," which can be determined by colored indicators, appearance of a precipitate, conductance measurement, or potentiometric (pH or ion-selective electrode) measurement. The difference between end point and equivalence point is the "titration error." Relating the end point to the equivalence point is one of the fundamental problems of volumetric analysis.

In the simplest and oldest titration procedures, the end point is determined by the color change of an indicator. Indicators are organic acids or bases, in which the color of the acidic form is different from that of the basic form (see Chapter 4). The change from one color to another is not sharp, but occurs over a range of pH; this is one cause of titration error.

Another cause is that the indicator may not change color at the same pH as the equivalence point. This is further compounded when the equivalence point is not correctly calculated because other reactions interfere. For example, the equations developed in this chapter predict that the equivalence point for a strong acid–strong base titration would be at pH = 7.00. But in normal laboratory procedure, the solutions are more or less in equilibrium with CO_2 from the atmosphere, and the titration curve is distorted by the CO_2–HCO_3^- equilibrium, so that the equivalence point occurs at pH = 4.5 to 5.0. Knowing this, one would choose an indicator such as methyl red (pH = 4.4–6.2) rather than brom thymol blue (pH = 6.2–7.6). Examples of indicators are given in Table 3.4.

The pH ranges given are the limits over which different observers are likely to choose the end point. By matching the color of the solution with that of a standard sample containing the indicator, the range can be narrowed to 0.5 pH units or less. Mixed indicators of complementary colors can change over a still narrower range. Phenol red and brom cresol green in a 2:1 ratio change from green to grey to brilliant violet in a pH range of 0.2 units near pH = 7.2.

Spectrophotometric measurements of the acidic and basic forms of the indicator, although requiring more expensive equipment than a primitive color match, can yield pH values close to those given by a glass electrode, and with comparable standard error (±0.001 units).[17] This concept will be developed in Chapter 4 (pp. 126–130).

STRONG ACIDS AND BASES

TABLE 3.4. Some common indicators for acid–base titration[18]

Indicator	pH range of end point	Color change (acidic-basic)
Methyl orange	3.1–4.5	red–orange
α–Naphthyl red	3.7–5.0	red–yellow
Brom cresol green	4.0–5.6	yellow–blue
Methyl red	4.4–6.2	red–yellow
Brom cresol purple	5.2–6.8	yellow–purple
Brom thymol blue	6.2–7.6	yellow–blue
Phenol red	6.4–8.0	yellow–red
Neutral red	6.8–8.0	red–yellow
α–Naphtholphthalein	7.3–8.7	rose–green
Phenolphthalein	8.4–10.0	colorless–red
Thymolphthalein	9.4–10.6	colorless–blue

Conductimetric End Points. The equivalent conductances of H^+ and OH^- are several times larger than most other ions, and hence as H^+ is removed by titration with NaOH, the conductance of the solution falls. It goes through a sharp minimum near the equivalence point, and increases again as excess NaOH is added. The equation governing this curve is

$$\Lambda = \lambda_H [H^+] + \lambda_{Na} [Na^+] + \lambda_{OH} [OH^-] + \lambda_{Cl} [Cl^-] \qquad (27)$$

where Λ is the overall conductance in $k\Omega^{-1}\,cm^{-1}$ and the λ factors are equivalent conductances of the various ions. Limiting conductances $\lambda°$ (at 25 °C in cm^2 Int. Ω^{-1} equiv^{-1}) are[19]:

$$H^+, \quad \lambda° = 349.8$$
$$Na^+, \quad \lambda° = 50.10$$
$$OH^-, \quad \lambda° = 199.1$$
$$Cl^-, \quad \lambda° = 76.35$$

The concentrations are given by the same equilibrium, mass balance, and charge balance equations as used in previous sections. A straightforward calculation procedure is to assume values for pH, calculate $[H^+] = 10^{-pH}/\gamma_+$, obtain $[OH^-]$ from K_w, V from the general titration curve equation, obtain $[Na^+] = C_b V/(V + V_0)$ and $[Cl^-] = C_a V_0/(V + V_0)$ from the mass balances. The results are shown in Fig. 3.7 for the same parameters as Figs. 3.5 and 3.6.

One of the advantages of conductimetric titrations is that the sharpness of the end point is not so sensitive to concentration as is the end point of a potentiometric pH titration. Figure 3.8 shows the conductimetric curve for 10^{-4} M HCl titrated with 10^{-4} M NaOH. The practical lower limit is determined by the sensitivity and repro-

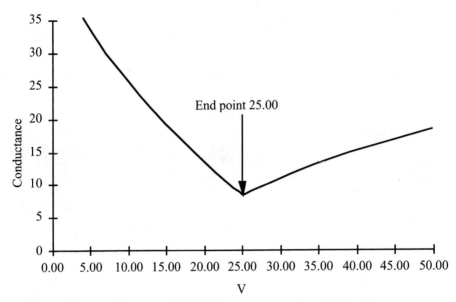

FIGURE 3.7. Conductance versus volume for 0.1 M HCl titrated with 0.2 M NaOH (same as Fig. 3.5).

ducibility of the conductivity measurement, as well as the conductivity of pure water, approximately 10^{-6} Ω^{-1} cm^{-1} (0.001 on the scale of Fig. 3.8).

The curve of Fig. 3.8 actually appears somewhat sharper than that of Fig. 3.7, but this is an artifact of the graphs. When both curves are plotted on a common logarithmic scale, and the volume scale normalized to the equivalence point, they appear nearly identical (Fig. 3.9).

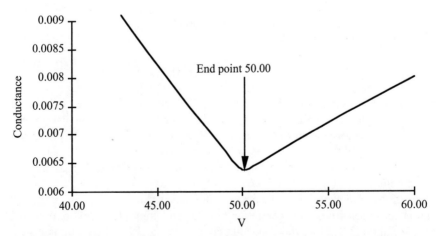

FIGURE 3.8. Conductimetric titration of 10^{-4} M HCl with 10^{-4} M NaOH. Compare with Fig. 3.7 and note that the end point is sharp in spite of dilution by a factor of 1000.

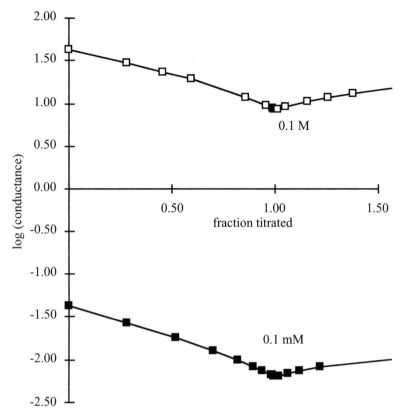

FIGURE 3.9. Comparison of conductance titration curves for 0.1 M (Fig. 3.7) and 0.1 mM (Fig. 3.8) strong acid. The equivalence point, where fraction titrated = 1, is 25 mL for Fig. 3.7 and 50 mL for Fig. 3.8.

Titration Error

Once a titration curve has been calculated, and the equivalence point V' is located, then the difference between that point and the experimental end point V_{ep} can be determined:

$$\text{Titration error} = \frac{V_{ep} - V'}{V'} \tag{28}$$

A positive value for the titration error means that the end point comes after the equivalence point; a negative value means that the end point comes before the equivalence point. This is illustrated by an example:

Example 4. Methyl red is often used as an indicator in strong acid–strong base titrations. Depending on the color standard chosen and the perception of the analytical chemist, the end point might be between 4.4 and 6.2. Consider a titration of

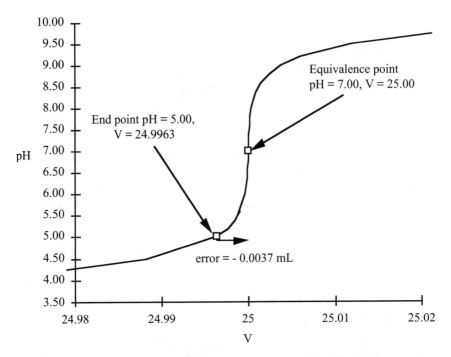

FIGURE 3.10. Enlarged section of titration curve (same data as Fig. 3.5) showing equivalence point and end point. Titration error is –0.0037 mL or –0.015%.

50 mL of 0.10 M HCl with 0.20 M NaOH (this is the same titration displayed in Figs. 3.4–3.6), and for the sake of the example, suppose the end point is at pH = 5.0. Referring to Fig. 3.10, which is an expanded section of Fig. 3.5, you can see that the end point, calculated from the titration curve equation derived above, comes slightly before the equivalence point, at 24.9963 mL instead of 25.0000 mL. The titration error is

$$(24.9963 - 25.0000) = -0.0037 \text{ mL} \quad \text{or} \quad -0.015\%$$

In spite of the large discrepancy in pH, which allows for interference from an uncertain amount of CO_2, the error in volume is well within acceptable limits.

In more dilute solutions, however, this is not the case. Figure 3.11 shows the curve of pH versus V for the titration of 1.0×10^{-4} M HCl with 1.0×10^{-4} M NaOH. Although the shape of the curve is similar to Fig. 3.10, the horizontal scale is not. The end point at pH = 5.0 occurs at V = 40 mL, a titration error of –10 mL, or –20%. This is certainly not acceptable. In order to keep the titration error smaller than 0.1%, pH must be determined to within 0.1 units. This requires an accurate pH meter and good calibration practice. Furthermore, the end point must be chosen to reflect contamination by CO_2 from the laboratory in both the standard and the

STRONG ACIDS AND BASES

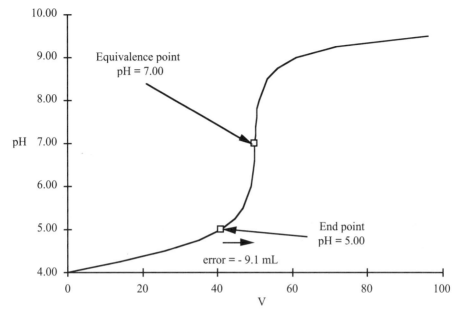

FIGURE 3.11. Titration error for 10^{-4} M HCl titrated with 10^{-4} M NaOH. Compare with Fig. 3.10.

sample. In such dilute solutions either a conductimetric titration (pp. 79–81), or a Gran titration (pp. 84–87) would be more appropriate.

General Equation for Titration Error. Define the "fraction titrated" (for the titration of a strong acid with a strong base) by

$$\phi = C_b V_b / C_a V_o \tag{29}$$

Since $\phi = 1$ at the equivalence point V', $V' = C_a V_a / C_b$, and hence the titration error is

$$\text{Titration error} = \frac{V_{ep}}{V'} - 1 = \frac{C_b V_{ep}}{C_a V_a} - 1 = \phi_{ep} - 1 \tag{30}$$

The general equation for titration of a strong acid with a strong base can be derived from Eq. (23):

$$\phi - 1 = \frac{C_a}{C_b} \left(\frac{C_a - [H^+] + K_w/[H^+]}{C_b + [H^+] - K_w/[H^+]} \right) - 1 \tag{31}$$

and this can be used directly by substituting $[H^+]_{ep}$ and ϕ_{ep} for $[H^+]$ and ϕ. However, one important simplification is that the volume at the end point is approximately the same as the volume at the equivalence point, and hence[20]

$$\frac{V_a}{V_{ep} + V_a} = \frac{V_a}{V' + V_a} = \frac{C_b}{C_b + C_a} \quad \text{(end point approximation)} \qquad (32)$$

introducing these into the charge balance (Eq. 22) gives the most useful form for calculating titration error:

$$\text{Titration error} = \phi_{ep} - 1 = \frac{C_b + C_a}{C_b C_a}\left(\frac{K_w}{[H^+]_{ep}} - [H^+]_{ep}\right) \qquad (33)$$

The examples of the previous section can be solved using this equation, by taking

$$C_a = 0.1, \; C_b = 0.2, \; [H^+]_{ep} = 10^{-5.00}, \; \text{and } K_w = 10^{-14.00}:$$

$$\phi_{ep} - 1 = (0.3/0.02)(10^{-9.00} - 10^{-5.00}) = -1.5 \times 10^{-4}, \text{ or } -0.015\%$$

Similarly, with $[H^+]_{ep} = 10^{-5.00}$, $C_a = 1.0 \times 10^{-4}$, $C_b = 1.0 \times 10^{-4}$, and $K_w = 10^{-14.00}$

$$\phi_{ep} - 1 = (2.0 \times 10^{-4}/1.0 \times 10^{-8})(10^{-9.00} - 10^{-5.00}) = -2.0 \times 10^{-1}, \text{ or } -20\%$$

Linearized (Gran) Titration Curves

Titration to a predetermined pH value, whether a pH meter or an indicator is used, is subject to errors if the solutions are not completely pure. A number of workers, especially in Scandinavia, have adopted a method of end point determination that depends on linearizing a portion of the titration curve and extrapolating it to zero.[21]

The equations derived above are a good starting point. For the titration of a strong base C_b, V_b with a strong acid C_a, V_a (Eq. 22 still applies):

$$[H^+] + \frac{C_b V_b}{V_a + V_b} = \frac{C_a V_a}{V_a + V_b} + \frac{K_w}{[H^+]} \qquad (34)$$

On the branch of the curve before the end point, pH > 7 and [H⁺] is small; after the end point pH < 7 and the term $K_w/[H^+]$ becomes negligible. In that range, the equation can be transformed to give a function linear in V:

$$(V_a + V_b)[H^+] = C_a V_a - C_b V_b + \cdots$$

or

$$f_1 = (V_a + V_b)10^{-pH} = (C_a V_a - C_b V_b)\gamma_+ + \cdots \qquad (35)$$

STRONG ACIDS AND BASES

If the experimenter determines pH as a function of V_a for the acid range, and plots f_1 as a function of V_a, he/she should obtain a straight line with slope $df_1/dV_a = C_a\gamma_+$, ideally intersecting the horizontal axis ($f_1 = 0$) at the equivalence point $V' = C_b V_b/C_a$.

Since H$^+$ is being replaced by Na$^+$, ionic strength does not change significantly in this part of the titration curve, and hence γ_+ is constant.

Some experimental data are given in Table 3.5. A plot of f_1 versus V is shown in Fig. 3.12. Extrapolation from the highest V and f_1 range (lowest pH) to the axis gives the end point at 4.68 mL. This corresponds to

$$C_b = 4.68 \times 0.02142/55.00 = 1.82 \times 10^{-3} \text{ M}$$

The accuracy of this extrapolation depends in part on your choice of which points you will draw the straight line through, in part on how carefully the line is drawn, and ultimately on how accurately the graph paper can be read.

Even if a least-squares fit is used to obtain the line, you cannot use all the points or you will get intercept 4.560, a totally wrong answer. The choice of which data points to include in the least-squares set requires some judgement. For example, points from 4.68 to 4.89 give intercept 4.672; 4.70 to 4.89 give 4.675, and 4.75 to 4.89 give 4.678. This last value is in agreement with the graphical result from Fig. 3.12.

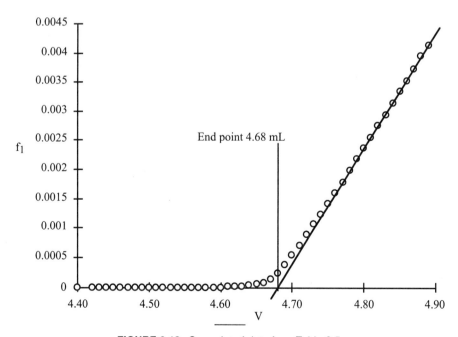

FIGURE 3.12. Gran plot of data from Table 3.5.

TABLE 3.5. Titration of dilute NaOH with HCl[22]

Volume	pH	10^{-pH}	f_1
4.40	9.719	1.91E-10	1.13E-08
4.42	9.625	2.37E-10	1.41E-08
4.44	9.574	2.67E-10	1.59E-08
4.46	9.547	2.84E-10	1.69E-08
4.48	9.485	3.27E-10	1.95E-08
4.50	9.410	3.89E-10	2.31E-08
4.52	9.338	4.59E-10	2.73E-08
4.54	9.243	5.71E-10	3.40E-08
4.56	9.076	8.39E-10	5.00E-08
4.58	8.823	1.50E-09	8.96E-08
4.60	8.028	9.38E-09	5.59E-07
4.62	7.113	7.71E-08	4.60E-06
4.64	6.416	3.84E-07	2.29E-05
4.65	6.139	7.26E-07	4.33E-05
4.66	5.892	1.28E-06	7.65E-05
4.67	5.646	2.26E-06	1.35E-04
4.68	5.400	3.98E-06	2.38E-04
4.69	5.191	6.44E-06	3.85E-04
4.70	5.044	9.04E-06	5.39E-04
4.71	4.924	1.19E-05	7.11E-04
4.72	4.828	1.49E-05	8.87E-04
4.73	4.750	1.78E-05	1.06E-03
4.74	4.685	2.07E-05	1.23E-03
4.75	4.623	2.38E-05	1.42E-03
4.76	4.572	2.68E-05	1.60E-03
4.77	4.524	2.99E-05	1.79E-03
4.78	4.479	3.32E-05	1.98E-03
4.79	4.438	3.65E-05	2.18E-03
4.80	4.404	3.94E-05	2.36E-03
4.81	4.370	4.27E-05	2.55E-03
4.82	4.336	4.61E-05	2.76E-03
4.83	4.308	4.92E-05	2.94E-03
4.84	4.281	5.24E-05	3.13E-03
4.85	4.253	5.58E-05	3.34E-03
4.86	4.229	5.90E-05	3.53E-03
4.87	4.205	6.24E-05	3.73E-03
4.88	4.181	6.59E-05	3.95E-03
4.89	4.161	6.90E-05	4.13E-03

An alternative approach is to rearrange the equation to give a value of C_b for each point:

$$C_b = \frac{C_a V_a - f_1/\gamma_+}{V_b} \tag{36}$$

STRONG ACIDS AND BASES

This requires a value for γ_+. Refer to Footnote 22 to Table 3.5: the Davies equation with $I = 0.0017$ (corresponding to 4.70 mL) gives $\gamma_+ = 10^{-0.019} = 0.95$. But this may not be accurate for several reasons: the Davies equation is only approximate; the relationship between experimental pH and [H$^+$] includes not only the activity coefficient, but also any changes in liquid junction potential between the pH standard and the sample.

Nonlinear Curve Fitting. Figures 3.13–3.15 show calculated values of C_b for three values of γ_+: 0.93, 0.88, and 0.83. In principle, C_b should be independent of pH, and indeed it nearly is, ranging from

1.824 to 1.820 mM with $\gamma_+ = 0.93$ in Fig. 3.13,

1.819 to 1.821 mM with $\gamma_+ = 0.88$ in Fig. 3.14, and

1.819 to 1.814 mM with γ_+: = 0.83 in Fig. 3.15.

The value of γ_+ obtained from the Davies equation is 0.95, much higher than 0.88, and results in a curve even steeeper than Fig. 3.13. Because γ_+ as used in Eq. (36) includes changes in liquid junction potential between the calibration buffer and the test solution, as well as the activity coefficient of H$^+$, not much weight should be given to γ_+ obtained from the Davies equation.

What is the best answer? I would say that $\gamma_+ = 0.88$ (Fig. 3.14) gives the best answer,

$$C_b = 1.820 \pm 0.001 \text{ mM}$$

because it shows the least trend with pH.

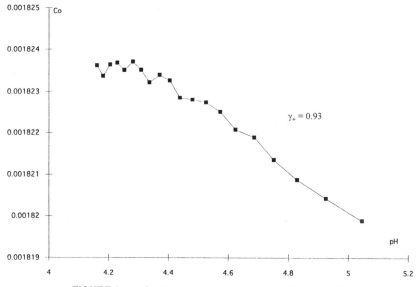

FIGURE 3.13. Calculated values of $C_o = C_b$ with $\gamma_+ = 0.93$.

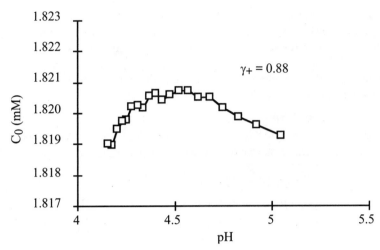

FIGURE 3.14. Calculated values of $C_o = C_b$ with $\gamma_+ = 0.88$.

A final display of these data is a direct comparison of calculated and observed values, as in Fig. 3.13. The calculated values were obtained from the full equation (Eq. 23) for the strong acid–strong base titration:

$$V_{calc} - V_{obs} = V_b \frac{C_b + [H^+] - [OH^-]}{C_a - [H^+] + [OH^-]} - V_a \tag{37}$$

with $C_a = 0.02142$, $C_b = 0.00182$, and $\gamma_+ = 0.88$. The deviations range from -0.002 to $+0.002$ mL. Although this is certainly a good precision for a burette measuring

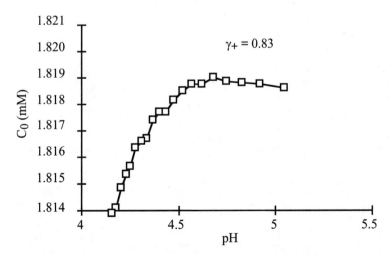

FIGURE 3.15. Calculated values of $C_o = C_b$ with $\gamma_+ = 0.83$.

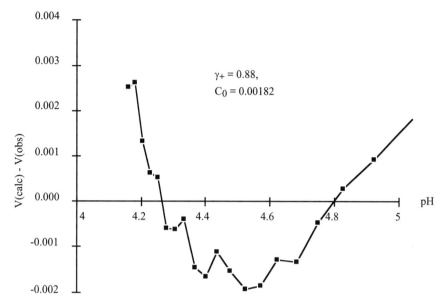

FIGURE 3.16. Deviation of calculated from observed values (Table 3.5) of V.

5 mL (0.04 %), the variation between successive points is smaller (0.01%) than the systematic variation producing the minimum, which is due to other causes, such as CO_2 absorbed from the atmosphere,[23] but probably not changes in ionic strength.[24]

PROBLEMS

Assume 25°C, complete dissociation, $I = 0$, except where noted otherwise.

1. Harned and Hamer[25] measured the ionization constant of water at various temperatures. For example, $pK_w = 14.939$ at 0°C and 13.046 at 60°C. Calculate the pH of pure water at each of these temperatures.

2. Ethanol (C_2H_5OH, abbreviated EtOH) ionizes as follows:

$$EtOH + EtOH \rightleftharpoons EtOH_2^+ + EtO^-$$

The equilibrium constant at 25°C is

$$[EtOH_2^+][EtO^-] = 8 \times 10^{-20}$$

Calculate the concentrations of the two ions in pure ethanol, and in 0.001 M Na^+EtO^-.

3. Calculate the pH of a solution made by adding 0.05 mL of 16 M (concentrated) HNO_3 to 1 L of water. Compare pH values obtained from H^+ concentration with those obained from H^+ activity.

4. Calculate the pH and pOH in a 4.5×10^{-4} M $Sr(OH)_2$ solution. At this concentration, $Sr(OH)_2$ is fully dissociated.[26]

5. Calculate the pH and pOH of 0.100 mg TlOH in 174 mL water.

6. Calculate the pH of 4.0×10^{-9} m LiOH in 0.05 m LiCl. Assume complete dissociation.

7. A strong acid with pH = 4.00 is mixed with an equal volume of strong base with pH = 9.00. Calculate the pH of the resulting mixure, recognizing the dilution that takes place.

8. 50.0 mL of 1.00×10^{-4} M HNO_3 is titrated with 0.010 M KOH. Assume values for pH and calculate V. Be sure to include the equivalence point in your range of values.

9. Calculate the pH of a solution obtained by adding 96.3 mL of 1.20 M TlOH to 87.2 mL of 0.980 M HNO_3 and diluting to 750 mL.

10. Derive the theorem that "the equivalence point in the titration of any strong acid with any strong base, or vice versa, always occurs at pH = $½pK_w$, regardless of concentration," noting what assumptions (particularly regarding activity coefficients) are required.

11. Show that the inflection point (where $d\,pH/dV$ is maximum) of the strong acid–strong base titration curve occurs at the equivalence point.

12. Methyl red is a frequently used indicator whose color changes from red to yellow as the pH changes from 4.4 to 6.2. Calculate the maximum and minimum titration errors if this indicator is used in the titration of 10 mL 0.050 M HCl with 0.010 M NaOH. If the burette can be read to ±0.02 mL, what would be the allowable range of pH to have titration error < 0.1%?

13. In the titration of 5.0×10^{-4} M HCl with 5.0×10^{-3} M NaOH, phenolphthalien is used as an indicator. This dye changes from colorless at pH = 8.4 to red at pH = 10.0. The end point is taken when the first tinge of pink is visible, pH = 8.4 to 8.8. If the burette is accurate to ±0.02 mL, what volume of sample should be taken so the titration error does not exceed the burette-reading error?

14. In a comprehensive review of the principles of titrimetric analysis, published in 1959, the following formulas for titration error in a strong acid–strong base titration was given:

$$\% \text{ error} = 200\,[H^+]_{ep}/C_o, \quad pH < 7$$
$$\% \text{ error} = 200\,[OH-]_{ep}/C_o, \quad pH > 7$$

where the notation is the same as in the text. Under what assumptions were these formulas derived? Under what conditions would they fail to apply?

15. Using the data from Table 3.5, prepare a Gran plot like Fig. 3.12 and experiment with graphical extrapolations. How precisely can you estimate the end point?

16. Perform the calculations required to produce Figs. 3.13–3.15 using five selected points between pH = 4 and 5.

17. Perform calculations like those for Fig. 3.16, using Eq. (37), trying several closely adjacent values of C_b and γ_+. Note that changes in C_b tend to make the curve translate up and down, whereas changes in γ_+ tend to change its shape.

18. In the laboratory, titrate a strong acid with a strong base to create your own data set, and analyze it using equations and graphs like those presented here.

NOTES

1. N. A. Izmailov, *Zh. Fiz. Khim.* **34**:2424, 1960; *Dokl. Akad. Nauk SSSR* **149**:884, 1963, cited by R. G. Bates, *Determination of pH*, 2nd Ed., Wiley, NY, 1973, p. 179.

2. Franks, Felix, *Water: A Comprehensive Treatise*. Vol. 1 (1972)–Vol. 7 (1982), Plenum Press, New York.

3. A recent discussion of the spatial structure in liquid water, with three-dimensional computer-generated maps, is given by P. G. Kusalik and I. M. Svischev, *Science* **265**:1219–1221, 1994. Water dimers and trimers have been studied by far infrared vibration–rotation tunneling spectroscopy (G. T. Fraser, *Int. Rev. Phys. Chem.* **10**:189, 1991; N. Pugliano and R. J. Saykally, *J. Chem. Phys.* **96**:123, 1992; N. Pugliano and R. J. Saykally, *Science* **257**:1937–40, 1992).

4. The species H_4O^{2+} has been suggested as an intermediate in superacid media, and its triarylphosphine (*L*) gold analogue $(LAu)_4O^{2+}$ has been prepared (H. Schmidbaur et al., *Nature* **377**:503, 1995).

5. For example, in a solution containing 26% by weight NaCl (26 g NaCl = 0.445 mole; 74 g H_2O = 4.107 mole) the mole fraction of water is 4.107/4.552 = 0.902.

6. W. Smith and A. E. Martell, *Critical Stability Constants*, Plenum, NY, 1976–1990. Vol. 4, p. 1; Vol. 5, p. 393; Vol. 6, p. 426. H. S. Harned and W. J. Hamer, *J. Amer. Chem. Soc.* **55**:2194; 4496, 1933 and T. Ackermann, *Z. Elektrochem.* **62**:411, 1958. These workers found $pK_w^\circ = 13.999$ at 25°C and $I = 0$; this was considered the best value for a long time. The slightly lower value $pK_w^\circ = 13.997 \pm 0.003$ was selected by Smith and Martell.

7. H. S. Harned and W. J. Hamer, *J. Amer. Chem. Soc.* **55**:2194; 4496, 1933, and T. Ackermann, *Z. Elektrochem.* **62**:411, 1958. Other sources are tabulated by Sillén and Martell (1964 and 1971). Temperature dependence of pK_w° from 0 to 250°C is given in Table 10.1.

8. α_{OH} also includes any differences between the empirical pH scale and the Davies equation. Note that $\log \gamma_\pm = -0.175$ from the Debye–Hückel equation (with $a = 9$ for H^+ and $a = 3$ for OH^- at $I = 0.7$) is in good agreement with $\log \gamma_\pm = -0.16$ from the Davies equation. See also the detailed discussion of ion pairing in seawater in Chapter 10.

9. R. G. Bates, *Determination of pH*, Wiley, NY, 1973, pp. 170–251; Pitzer, *Activity Coefficients in Aqueous Solution,* pp. 160–163.
10. M. Herlem, *Bull. Soc. Chim. France*, 1687, 1967, cited by Bates, *Determination of pH*, p. 183.
11. Thus one may have a dilute solution of a strong acid (0.0001 M HCl) or a concentrated solution of a weak acid (3 M HAc).
12. Both $HClO_4$ and HCl are completely dissociated in water and hence are "strong acids," but $HClO_4$ is said to be stronger than HCl because it is more highly dissociated in solvents like glacial acetic acid.
13. At $I = 0$, the log of the association constant $[MOH]/[M^+][OH^-]$ is +0.36 for Li, –0.2 for Na, –0.5 for K (Smith and Martell, Vol. 4, p. 1). See Chapter 4.
14. C can be in moles per liter of solution, moles per kg of solution, or moles per kg of solvent. K_w will depend slightly on the concentration scale, particularly in concentrated solutions. See "Concentration Scales" in Chapter 1.
15. If volumes are measured, the appropriate concentration unit is moles per liter of solution (M); it is also possible to do a titration by weight, in which case the appropriate unit would be moles per kilogram of solution (not a conventional unit) or moles per kilogram of water (m). Other techniques include electrochemically generated titrants, for which coulombs are measured instead of mL.
16. Since the strong acid and strong base in these examples each are monoprotic, mol/L and equivalents per liter are the same. The number of equivalents, a concept less used now than in the past, is the number of moles of one reactant, multiplied by the coefficient of the other reactant, in the stoichiometric reaction. Thus 1 mol/L of H_2SO_4 neutralizes 2 mol/L of hydroxyl:

$$H_2SO_4 + 2OH^- \rightleftharpoons 2H_2O + SO_4^{2-}$$

and so contains 2 equivalents/L. The "equivalent" concept allows a universal titration equation to be developed—otherwise different stoichiometries would introduce different multiplicative constants. However, the number of equivalents per mole is not fixed, but is determined by the particular titration reaction, and this can cause even more confusion than the introduction of multiplicative constants.

17. R. H. Byrne et al., *Deep-Sea Research* **35**:1405–10, 1988.
18. I. M. Kolthoff and V. A. Stenger, *Volumetric Analysis*, 2nd Ed., Interscience, NY, 1942, pp. 92–93.
19. R. A. Robinson and R. H. Stokes, *Electrolyte Solutions*, Butterworths, London, 1968, p. 463. Sorry for the outmoded conductance units, but these calculations are only qualitative.
20. When $\phi = 1$, $V_{ep} = V' = C_a V_a/C_b$. Substitute for V' in $V_a/(V' + V_a)$ and cancel V_a to get $C_b/(C_b + C_a)$.
21. G. Gran, *Analyst* **77**:661–71, 1952.
22. By Walt Zeltner, University of Wisconsin, 26 June 1984. $V_b = 55.00$ mL, $C_a = 0.02142$ M. Filtered deionized (milli-Q) water was used to prepare solutions. As will be seen from the analysis below, the end point is $Ve = 4.676$ mL, and the base concentration in the sample is 0.00182 M. To calculate γ_+, note that the ionic strength of the original solution was 0.00182, that of the solution at 4.7 mL, where all the NaOH is converted to

NaCl, is lower because of dilution, $(55.00)(0.00182)/(59.7) = 0.00168$, and at 4.9 mL, 0.2 mL additional HCl has been added, giving $[(55.00)(0.00182) + (0.2)(0.02142)]/59.9 = 0.00174$. γ_+ was calculated from the Davies equation to be 0.954, 0.956, and 0.955 for these three points. $[H^+] = 10^{-pH}/\gamma_+$.

23. See Chapter 10; also J. N. Butler, *Mar. Chem.* **38**:251–82, 1992.
24. See Footnote 22. At the start of the titration, only NaOH is present, and $I = 0.00182$. At 4.7 mL, all the NaOH has been converted to NaCl and the ionic strength is 0.00168. At the end of the titration 0.2 mL of excess acid has been added and $I = 0.000174$. The activity coefficients corresponding to these three conditions are 0.954, 0.956, and 0.955. Recalculating γ_+ for each point would not make a perceptible difference for this titration, but might matter for some.
25. H. S. Harned and W. J. Hamer, *J. Am. Chem. Soc.* **55**:2194, 1933.
26. $\log\{[SrOH^+]/[Sr^{2+}][OH^-]\} = 0.8 \pm 0.1$ at 25°C, $I = 0$ (Smith and Martell, Vol. 4, p. 1).

4

MONOPROTIC ACIDS AND BASES

Introduction
Dissociation (Ionization) of a Weak Acid or Base
Calculating the pH of a Weak Acid
   Flood's Diagram
   Degree of Dissociation
   Sillén's Diagram
Calculating the pH of Other Weak Acids and Bases
   Salt of a Weak Base and Strong Acid
   pH of a Weak Base
   Summary of Proton Conditions for Weak Acids and Weak Bases
   Mass Balance on H^+
Mixtures of Strong or Weak Acids and Bases
   Salt of a Weak Acid and a Weak Base
Indicators as weak acids
   pH measurement with Indicators
   Comparison of Spectrophotometric and Potentiometric pH Methods
Buffer Solutions
   The Buffer Index
Titration of a Weak Acid with Strong Base (or Weak Base with Strong Acid)
   Titration Error
   Sharpness Index
   Simple Approximations for η' (η at the Equivalence Point)
Problems

INTRODUCTION

"Monoprotic" acids are those that release one proton per molecule on dissociation; monoprotic bases are those that accept one proton per molecule. Sometimes the term *monobasic* is used for this concept, to indicate that one mole of acid is neutralized by one mole of OH⁻.

In the previous chapter, the properties of strong (completely dissociated) acids and bases were investigated. This chapter is concerned with those that are incompletely dissociated in aqueous solutions at normal concentration.

DISSOCIATION (IONIZATION) OF A WEAK ACID OR BASE

You are familiar from previous chapters with the dissociation equilibrium of a weak acid. The equations developed in Chapter 1 for acetic acid can be generalized:

$$HA \rightleftharpoons H^+ + A^-$$

$$[H^+][A^-] = K_a[HA] \quad (1)$$

K_a values for a number of inorganic and organic compounds are given in Table 4.1. At the end of Chapter 2, you saw an example of how pK_a is obtained from potentiometric data. The constants used in this chapter were obtained primarily by that method, but also by related methods such as conductance and spectrophotometric measurements.[1]

The reaction above is "dissociation," and K_a is a "dissociation constant" because HA breaks into two parts; H⁺ and A⁻; K_a is also an "ionization constant" because the reaction produces ions,[2] and K_a is sometimes called an "acidity constant" because it is a measure of acidity: The larger K_a is, the greater is [H⁺] at the same total concentration (see next section).

Ionization (dissociation) of a weak base B and its conjugate acid BH^+ are closely related. B reacts with water to produce its conjugate acid and a hydroxyl ion[3]:

$$B + H_2O \rightleftharpoons BH^+ + OH^-$$

$$[BH^+][OH^-] = K_b[B] \quad (2)$$

K_b values for a number of compounds are given in Table 4.2.

To get the relationship between K_a and K_b for a conjugate acid and base, use the format for the weak acid dissociation (Eq. 1), replacing A^- with B and HA with BH^+, to get:

$$[H^+][B] = K_a[BH^+] \quad (3)$$

Either K_a or K_b describes the acid–base equilibrium, but they are not independent.

TABLE 4.1 Dissociation constants of weak monoprotic acids at 25°C, $I = 0$[4]

Acid	Formula	pK_a
Iodic acid	HIO_3	0.8
Hydrogen sulfate ion	HSO_4^-	1.99±0.01
Chloroacetic acid	$ClCH_2COOH$	2.265±0.004
Hydrofluoric acid	HF	3.17±0.02
Nitrous acid	HNO_2	3.15
Formic acid	HCOOH	3.745±0.007
Lactic acid	$CH_3CHOHCOOH$	3.860±.002
Anilinium ion	$C_6H_5NH_3^+$	4.601±0.005
Hydrazoic acid	HN_3	4.65±0.02
Acetic acid	CH_3COOH	4.757±0.002
Propionic acid	CH_3CH_2COOH	4.874±0.001
Pyridinium ion	$C_5H_5NH^+$	5.229
Imidazolium ion	$C_4H_4NH^+$	6.993
Hypochlorous acid	HOCl	7.53±0.02
Hypobromous acid	HOBr	8.63±0.03
Boric acid	$B(OH)_3$	9.236±0.001
Ammonium ion	NH_4^+	9.244±0.005
Hydrocyanic acid	HCN	9.21±0.01
Trimethylammonium	$(CH_3)_3NH^+$	9.80±0.05
Ethylammonium	$CH_3CH_2NH_3^+$	10.636
Methylammonium	$CH_3NH_3^+$	10.64±0.02
Triethylammonium	$(C_2H_5)_3NH^+$	10.715
Dimethylammonium	$(CH_3)_2NH_2^+$	10.77±0.04

If Eq. (3) is solved for $[BH^+]$ and substituted in Eq. (2), $[B]$ cancels, leaving a general relationship between the dissociation constants.

$$[BH^+] = [H^+][B]/K_a \qquad (4)$$

$$[H^+][B][OH^-] = K_b K_a [B]$$

$$K_b K_a = K_w \qquad (5)$$

Thus in Table 4.1, $pK_a = 9.25$ is listed for NH_4^+. In Table 4.2, $pK_b = 4.75$ is listed for NH_3 and pK_b is given for some other bases. Although you will find K_b values in the literature, Table 4.2 is really not necessary, because if you know one constant, you know the other:

$$pK_b = pK_w - pK_a = 14.00 - 9.25 = 4.75 \qquad (6)$$

At other temperatures and ionic strengths, of course, the appropriate value of pK_w must be used (see Tables 3.1 and 3.2).

DISSOCIATION OF A WEAK ACID OR BASE

TABLE 4.2 Dissociation constants of weak monoprotic bases at 25°C, $I = 0$

Base	Formula	pK_b
Aniline	$C_6H_5NH_2$	9.40
Pyridine	C_5H_5N	8.77
Imidazole	C_4H_4N	7.01
Ammonia	NH_3	4.76
Trimethylamine	$(CH_3)_3N$	4.20
Ethylamine	$CH_3CH_2NH_2$	3.36
Methylamine	CH_3NH_2	3.36
Dimethylamine	$(CH_3)_2NH$	3.23

The dependence of the concentration pK_a on ionic strength is given by the usual equation:

$$K_a^\circ = \frac{[H^+][A^-]}{[HA]} \frac{\gamma_+ \gamma_-}{\gamma_o} = K_a \frac{\gamma_+ \gamma_-}{\gamma_o} \tag{7}$$

$$pK_a = pK_a^\circ + \log \gamma_+ + \log \gamma_- - \log \gamma_o \tag{8}$$

Inserting the value of pK_a° at 25°C, using the Davies equation (p. 49) to evaluate $\log \gamma_+$ and $\log \gamma_-$, and setting $\log \gamma_o = bI$ by analogy with other uncharged molecules, gives the following equation,

$$pK_a = 4.757 - (2)(0.51)\left(\frac{\sqrt{I}}{1+\sqrt{I}} - b'I\right) - bI \tag{9}$$

which is plotted in Fig. 4.1. Three curves are presented: two with $b' = 0.2$ (the usual Davies coefficient), and with $b = 0$ and 0.1 (by analogy with CO_2 and NH_3). These do not represent the data well, but an excellent fit to the experimental data is obtained with $b' = 0.3$ and $b = 0.5,6$

The temperature dependence of pK_a° over the range from 0 to 60°C is shown in Fig. 4.2. Note that pK_a° changes little in this temperature range, and that its minimum value of 4.756 occurs at 25°C. Rigorously, the function should be derived from the enthalpy and heat capacity, as described in Chapter 1:

$$\frac{d \log K}{dT} = \frac{\Delta H^\circ}{2.303RT^2}$$

$$\frac{d \Delta H^\circ}{dT} = C_p$$

C_p in general is also a function of temperature, usually represented by a polynomial. These equations can be integrated to give a function of the form (T is in K)

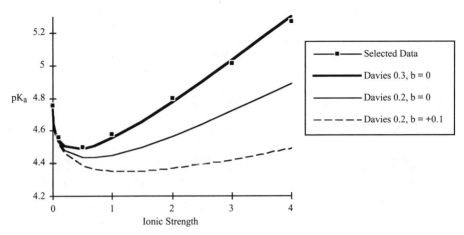

FIGURE 4.1. Ionic strength dependence of pK_a for acetic acid at 25°C (Smith and Martell Vol. 6).

$$\log K = \frac{A}{T} + B + C \log T + DT + ET^2 + \cdots$$

but this is a nonstandard least-squares fit, and the data in Fig. 4.2 can be fit (to ±0.001) by the quadratic function

$$pK_a^\circ = 4.7789 - 1.9717 \times 10^{-3} t + 4.2420 \times 10^{-5} t^2 \tag{10}$$

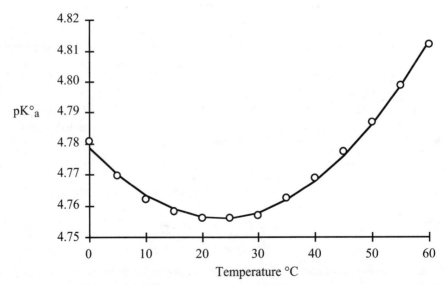

FIGURE 4.2. Temperature dependence of pK_a° for acetic acid. (Points: H. S. Harned and R. W. Ehlers, *J. Am. Chem. Soc.* **55**:65, 1933. Line: Eq. (10).

CALCULATING THE pH OF A WEAK ACID

Recall the example of "the weak acid problem" presented in Chapter 1 (pp. 21–24) to illustrate the procedure for attacking equilibrium problems. In that example, the solution was composed of C_{HA} mol/L of weak acid and C_A mol/L of its salt. In this section, we will do an even simpler problem, the solution containing C_{HA} mol/L of weak acid in pure water. Assume that C_{HA}, K_a, and K_w are known. There are four equations connecting the four unknowns [H$^+$], [A^-], [HA], and [OH$^-$][7]:

$$[H^+][A^-] = K_a[HA], \quad \text{weak acid equilibrium} \quad (11)$$

$$[H^+][OH^-] = K_w, \quad \text{ion product of water} \quad (12)$$

$$C_{HA} = [A^-] + [HA], \quad \text{mass balance} \quad (13)$$

$$[H^+] = [A^-] + [OH^-], \quad \text{charge balance} \quad (14)$$

Some combinations recur throughout acid–base calculations, and it is useful to have derived them in general.

First is the expression for [A^-] in terms of [H$^+$] and K_a. Combine the weak acid equilibrium (Eq. 11) and the mass balance (Eq. 13) and solve for [A^-]:

$$[HA] = \frac{[H^+][A^-]}{K_a}$$

$$C_{HA} = [A^-] + \frac{[H^+][A^-]}{K_a} = [A^-]\left(1 + \frac{[H^+]}{K_a}\right) = [A^-]\left(\frac{[H^+] + K_a}{K_a}\right)$$

$$[A^-] = \frac{C_{HA} K_a}{[H^+] + K_a} \quad (15)$$

Although you don't need it to calculate pH, you can also obtain [HA] from the mass balance:

$$[HA] = C_{HA} - [A^-]$$

$$[HA] = \frac{C_{HA}[H^+]}{[H^+] + K_a} \quad (16)$$

Note the symmetry: The numerator of Eq. (15) for [A^-] contains K_a; the numerator of Eq. (16) for [HA] contains [H$^+$]. C_{HA} and the denominator are the same for both. This symmetry will become even more apparent in the graphical methods described below.

From the ion product of water, get an expression for [OH⁻]:

$$[OH^-] = K_w/[H^+] \tag{17}$$

Substituting Eqs. (15) and (17) in the charge balance (Eq. 14) gives a general equation for [H⁺] in terms of C_{HA}, K_w, and K_a:

$$[H^+] = [A^-] + [OH^-] = \frac{C_{HA}K_a}{[H^+] + K_a} + \frac{K_w}{[H^+]} \tag{18}$$

which can be rearranged to a cubic polynomial (Eq. 38, p. 22, Chapter 1):

$$[H^+]^3 + [H^+]^2 (C_A + K_a) - [H^+](C_{HA}K_a + K_w) - K_w K_a = 0 \tag{19}$$

Example 1. However, rather than attempt a general algebraic solution, consider some realistic numerical values: C_{HA} = 0.010 mol/L, pK_a = 4.75 (acetic acid), pK_w = 14.00.

$$[A^-] = \frac{(0.010)(10^{-4.75})}{[H^+] + 10^{-4.75}} = \frac{10^{-6.75}}{[H^+] + 10^{-4.75}}$$

$$[OH^-] = 10^{-14.00}/[H^+]$$

Each of these expressions contains [H⁺], as does the charge balance,

$$[H^+] = [A^-] + [OH^-]$$

into which they will be substituted. But first, make a preliminary estimate for [H⁺].

"Chemical intuition" tells us that since we are adding an acid to the solution, [H⁺] > 10⁻⁷. This means that [OH⁻] < 10⁻⁷.
But what about [A⁻]? If [H⁺] < 10⁻⁵, [A⁻] ≈ 10⁻².
If [H⁺] > 10⁻⁵, [A⁻] ≈ 10⁻⁶·⁷⁵/[H⁺].
Try the simplest option: [OH⁻] ≪ [A⁻] and [H⁺] < 10⁻⁵; hence [A⁻] ≈ 10⁻². The charge balance becomes

$$[H^+] = 10^{-2} + \cdots$$

But this violates the assumption that [H⁺] < 10⁻⁵.
Try the other option:

$$[A^-] = 10^{-6.75}/[H^+] + \cdots$$

$$[H^+] = 10^{-6.75}/[H^+] + \cdots$$

$$[H^+]^2 = 10^{-6.75} + \cdots$$

$$[H^+] = 10^{-3.38} + \cdots$$

CALCULATING THE pH OF A WEAK ACID

This satisfies the assumption that $[H^+] > 10^{-5}$. If activity coefficients are close to 1, pH = 3.38.

Now check the other assumption, that $[OH^-]$ is negligible compared to $[A^-]$?

$$[H^+] = 10^{-3.385} + \cdots$$

$$[OH^-] = (10^{-14.00})(10^{+3.38}) + \cdots = 10^{-10.62}$$

$$[A^-] = \frac{(0.010)(10^{-4.75})}{10^{-3.38} + 10^{-4.75}} = \frac{10^{-6.75}}{10^{-3.36}} = 10^{-3.39}$$

and hence

$$[H^+] = [OH^-] + [A^-]$$

The assumptions are verified within roundoff errors,[8] confirming that the terms represented by "$+\cdots$" are indeed negligible.

Flood's Diagram[9]

As you may remember from Fig. 3.3, it is possible to plot pH versus C for a strong acid or strong base by assuming values of pH and calculating values of C. A similar tactic works for the more complicated equations describing the pH of a weak acid or weak base. As in the previous section, substitute the equilibrium expression in the mass balance and then in the charge balance together with the ion product of water to get Eq. (19), which can be solved for C_{HA} to give and equation which is equivalent to the cubic polynomial of Eq. (19):

$$C_{HA} = \left(\frac{[H^+] + K_a}{K_a}\right)\left([H^+] - \frac{K_w}{[H^+]}\right) \qquad (20)$$

Given values for K_a and K_w, you can assume values for $[H^+]$ and calculate C_{HA} as shown in the following table, and plot the data to obtain a curve as shown in Fig. 4.3.

pH	log C Strong acid	pH	log C pKa=4.75	pH	log C pKa=7.53	pH	log C pKa=10.72
0	0.000	0	4.750	3.8	-0.070	5.4	-0.080
1	-1.000	1	2.750	4	-0.470	5.7	-0.681
2	-2.000	2	0.751	5	-2.469	6	-1.284
2.38	-2.380	2.38	-0.008	6	-4.462	6.2	-1.691
2.7	-2.700	2.7	-0.646	6.2	-4.861	6.4	-2.108
3	-3.000	3	-1.242	6.4	-5.267	6.6	-2.555
4	-4.000	4	-3.179	6.6	-5.697	6.8	-3.102
5	-5.000	5	-4.806	6.8	-6.218	6.9	-3.518
6	-6.004	6	-5.981	6.9	-6.617	6.92	-3.638
6.2	-6.211	6.2	-6.196	6.92	-6.733	6.94	-3.787
6.4	-6.429	6.4	-6.419	6.94	-6.877	6.96	-3.989
6.6	-6.676	6.6	-6.669	6.96	-7.076	6.98	-4.328
6.8	-7.022	6.8	-7.019	6.98	-7.410		
6.9	-7.338	6.9	-7.335				

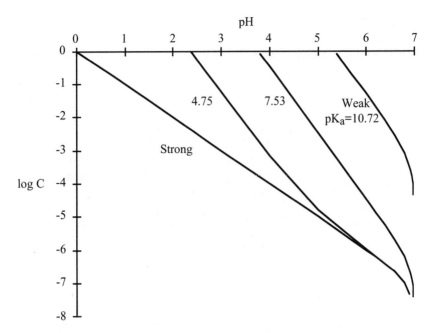

FIGURE 4.3. "Flood's Diagram" showing pH for several weak acids as a function of total concentration: acetic (pK_a = 4.75), hypochlorous (pK_a = 7.53), and methylammonium ion (pK_a = 10.72). The curve for a strong acid ($pK_a \ll 1$, see Fig. 3.1) is shown for comparison.

Note that all the curves for the weak acids fall at higher pH values than the strong acid curve. This difference is greater for higher concentrations of a given acid, and for higher pK_a at a given concentration. But regardless of concentration or pK_a, pH does not exceed 7 without addition of a base.

Degree of Dissociation

Two important parameters describing the weak acid equilibrium are the degree of dissociation and the degree of formation (association). Equations (15) and (16) can be rearranged to give these quantities. The degree of dissociation is defined to be

$$\alpha = \frac{[A^-]}{C_{HA}} = \frac{K_a}{[H^+] + K_a} \tag{21}$$

and the degree of formation is defined to be

$$1 - \alpha = \frac{[HA]}{C_{HA}} = \frac{[H^+]}{[H^+] + K_a} \tag{22}$$

These functions are shown in Fig. 4.4. Note that when pH = pK_a.

$$\alpha = 1 - \alpha = 0.50$$

Sillén's Diagram

The most useful of this suite of diagrams is the logarithmic concentration diagram introduced by the Swedish chemist Lars Gunnar Sillén[10] in the 1950s. In this diagram, the logarithms of the concentrations of the various species are plotted as a function of pH. Figure 4.5 (p. 105) is a diagram for 10^{-2} molar acetic acid (pK_a = 4.75).

For [H$^+$] a straight line of slope −1 is obtained from the definition of pH:

$$\log[H^+] = -pH \qquad (23)$$

For [OH$^-$], the ion product of water (Eq. 14) gives a straight line of slope +1:

$$\log[OH^-] = \log(K_w/[H^+]) = -pK_w + pH \qquad (24)$$

From the combination of mass balance and equilibrium, as derived above, [A^-] and [HA] can be expressed as functions of [H$^+$] (Eqs. 15 and 16):

$$[A^-] = \frac{C_{HA}K_a}{[H^+] + K_a}, \quad [HA] = \frac{C_{HA}[H^+]}{[H^+] + K_a}$$

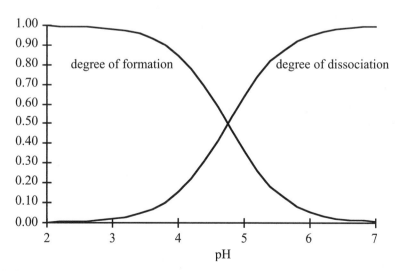

FIGURE 4.4. Degree of dissociation and degree of formation for acetic acid at I = 0, 25°C.

The following table shows some spreadsheet calculations from which Figs. 4.5 and 4.6 are derived.

Acetic acid		pKw =	14		pKa=	4.75
		Ca=	0.01		Ka=	1.78E-05
[H+]	pH	log [H+]	log [OH-]	log [HAc]	log [Ac-]	
=10^-pH	=pH	=-pH	=-pKw+pH	=LOG(Ca*H/(H+Ka))		
					=LOG(Ca*Ka/(H+Ka))	
1.00E+00	0	0	-14	-2.000	-6.750	
1.00E-01	1	-1	-13	-2.000	-5.750	
1.00E-02	2	-2	-12	-2.001	-4.751	
1.00E-03	3	-3	-11	-2.008	-3.758	
1.00E-04	4	-4	-10	-2.071	-2.821	
6.31E-05	4.2	-4.2	-9.8	-2.108	-2.658	
3.98E-05	4.4	-4.4	-9.6	-2.160	-2.510	
2.51E-05	4.6	-4.6	-9.4	-2.232	-2.382	
1.78E-05	4.75	-4.75	-9.25	-2.301	-2.301	
1.58E-05	4.8	-4.8	-9.2	-2.327	-2.277	
1.00E-05	5	-5	-9	-2.444	-2.194	
6.31E-06	5.2	-5.2	-8.8	-2.582	-2.132	
3.98E-06	5.4	-5.4	-8.6	-2.738	-2.088	
2.51E-06	5.6	-5.6	-8.4	-2.907	-2.057	
1.58E-06	5.8	-5.8	-8.2	-3.087	-2.037	
1.00E-06	6	-6	-8	-3.274	-2.024	
1.00E-06	6	-6	-8	-3.274	-2.024	
1.00E-07	7	-7	-7	-4.252	-2.002	
1.00E-08	8	-8	-6	-5.250	-2.000	
1.00E-09	9	-9	-5	-6.250	-2.000	
1.00E-10	10	-10	-4	-7.250	-2.000	
1.00E-11	11	-11	-3	-8.250	-2.000	
1.00E-12	12	-12	-2	-9.250	-2.000	
1.00E-13	13	-13	-1	-10.250	-2.000	
1.00E-14	14	-14	0	-11.250	-2.000	

In Fig. 4.5, the logarithms of these two concentrations appear as two straight line segments joined by a short curve. The "system point" at pH = pK_a and concentration C is indicated by an x and an arrow. In Fig. 4.5, the system point is at pH = 4.75 and log $C = -2$. To the left of this point, [H$^+$] is larger than K_a, and the above relations reduce to the logarithmic forms

$$\log[A^-] = \log C - pK_a + \text{pH} - \cdots \qquad (25)$$

$$\log[HA] = \log C - \cdots \qquad (26)$$

These approximations correspond to the straight line portions at the left side of Fig. 4.5. The neglected terms are shown as negative ($-\cdots$) since both curves turn concave downward as the system point is approached.

CALCULATING THE pH OF A WEAK ACID

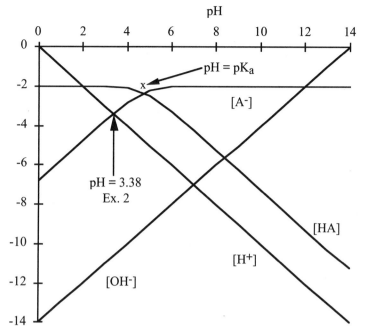

FIGURE 4.5. Logarithmic concentration diagram for 1.0×10^{-2} molar acetic acid.

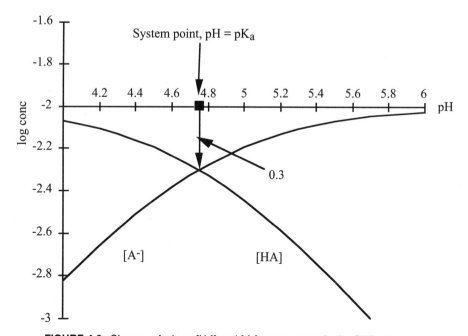

FIGURE 4.6. Closeup of where [HA] and [A^-] curves cross in the Sillén diagram.

To the right of the system point, K_a is large compared to [H$^+$], and

$$\log[A^-] = \log C - \cdots \quad (27)$$

$$\log[HA] = \log C + pK_a - pH - \cdots \quad (28)$$

As before, the neglected terms are shown as negative ($-\cdots$) since both curves turn concave downward as the system point is approached.

Near the system point (Fig. 4.6), both [H$^+$] and K_a have about the same magnitude, and both terms must be included in the denominator. The curves of Eqs. (15) and (16) cross at

$$[HA] = [A^-] = \tfrac{1}{2}C \quad (29)$$

$$\log[HA] = \log[A^-] = \log C - 0.30$$

$$pH = pK_a \quad (30)$$

This is 0.30 logarithmic units below the system point.

To construct the Sillén diagram without numerical calculations, follow these steps:

Prepare a graph with a horizontal (pH) axis scaled from 0 to 12 and a vertical (log C) axis scaled from 0 at the top to 9 at the bottom.[11]

Draw a diagonal line for [H$^+$] of slope -1 from pH = 0 and log C = 0.

Draw a diagonal line for [OH$^-$] of slope $+1$ from pH = 7 and log C = -7. This should go through the point at pH = 12 and log C = -2. [It would intersect the x axis (log C = 0) at pH = 14 if that point were on the graph.]

Locate the system point at x = pH = pK_a and y = log C.

Draw a horizontal line at log C, leaving space around the system point. The left section corresponds to [HA] and the right section to [A$^-$].

Draw a line of slope -1 (parallel to [H$^+$]) through the system point. This corresponds to [HA].

Draw a line of slope $+1$ (parallel to [OH$^-$]) through the system point. This corresponds to [A$^-$].

Join the straight sections with short curves to obtain the appearance of Figs. 4.5 and 4.6. Use a small french curve or a template cut to correspond to Fig. 4.6 or the equations derived above.

No detailed numerical calculations are required to create such a diagram. Not only is it a convenient reference when doing numerical calculations on a system, but approximate calculations can often be made by means of the diagram alone (see Examples 2–4).

Changing the concentration C shifts the [A$^-$] and [HA] curves up or down together; changing K_a shifts the [A$^-$] and [HA] curves to right (larger pK_a) or left (smaller pK_a), but neither C nor K_a shifts the [H$^+$] or [OH$^-$] lines.

Example 2. Find graphically the pH of 0.010 M acetic acid. This problem was solved numerically above, but it can also be solved using the diagram of Fig. 4.5.

CALCULATING THE pH OF A WEAK ACID

Note that the diagram contains all the information in the dissociation equilibrium, the ion product of water, and the mass balance on acetic acid species. The only remaining equation is the charge balance (Eq. 14, which is also the proton condition:

$$[H^+] = [A^-] + [OH^-]$$

Follow the $[H^+]$ line down until it meets the $[A^-]$ line. At this point $[OH^-]$ is less than 10^{-9}, which is certainly negligible compared to $[H^+]$ and $[A^-]$, both of which are between 10^{-3} and 10^{-4}. A careful interpolation of the graph yields

$$pH = 3.38$$
$$\log[A^-] = -3.38$$
$$\log[HA] = -2.02$$
$$\log[OH^-] = -10.62$$

These answers can be verified by substitution in the original equations:

$$K_a = [H^+][A^-]/[HA] = (10^{-3.38})(10^{-3.38})/(10^{-2.02}) = 10^{-4.74}$$
$$K_w = [H^+][OH^-] = (10^{-3.38})(10^{-10.62}) = 10^{-14.00}$$
$$C = [A^-] + [HA] = (10^{-3.38}) + (10^{-2.02}) = 10^{-2.00}$$
$$\text{Charge: } [H^+] - [A^-] - [OH^-] = (10^{-3.38}) - (10^{-3.38}) - (10^{-10.62}) = 10^{-10.62}$$

The only discrepancy is in K_a, which is $10^{-4.74}$ instead of $10^{-4.75}$, a result of rounding values to two decimal places. The important lesson to be learned is that this graphical method can yield an acceptable result without either algebra or numerical calculation. In addition, it gives a clear picture of how the various species concentrations vary with pH—a great advantage when dealing with polyprotic acids, amino acids, and other systems containing several acids and bases (see Chap. 5).

Example 3. If the acid concentration is close to K_a, then $[H^+]$ and $[HA]$ are close together, and the point where $[H^+]$ meets $[A^-]$ is on the curved section of that line. This is illustrated in Fig. 4.7, which is the diagram for $10^{-3.00}$ M hydrofluoric acid (HF, $pK_a = 3.17$).

The diagram embodies the dissociation equilibrium and mass balance on HF:

$$[H^+][F^-] = 10^{-3.17} [HF]$$
$$10^{-3.00} = [F^-] + [HF]$$

The line for $[H^+]$ is given by the definition $pH = -\log[H^+]$. The line for $[OH^-]$ is off the lower right of the chart. The final relationship is the charge balance (or proton condition)

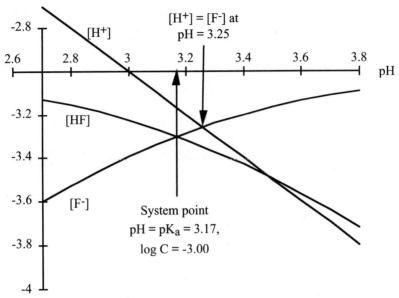

FIGURE 4.7. Sillén diagram for $10^{-3.00}$ M hydrofluoric acid ($pK_a = 3.17$).

$$[H^+] = [F^-] + [OH^-]$$

Following the $[H^+]$ line down, you can see that it meets $[F^-]$ at about pH = 3.25 or 3.26. At this point,

$$[OH^-] = \frac{K_w}{[H^+]} = \frac{10^{-14.00}}{10^{-3.25}} = 10^{-10.75}$$

which is off the chart but certainly negligible. The accuracy of the intersection depends on how accurately the curved part of the $[F^-]$ line is drawn. At pH = 3.25, you can interpolate the graph to find $[HF] = 10^{-3.34}$, and check this answer by substituting in the equilibrium.

Alternatively, you can substitute $[HF] = C - [F^-]$ from the mass balance and $[F^-] = [H^+]$ from the charge balance in the equilibrium to get a quadratic in $[H^+]$:

$$[H^+]^2 = K_a(C - [H^+])$$
$$f([H^+]) = [H^+]^2 + K_a[H^+] - K_aC = 0$$

which can be solved by the quadratic formula, with $C = 10^{-3.00}$ and $K_a = 10^{-3.17}$

$$[H^+] = \frac{-K_a + \sqrt{K_a^2 + 4K_aC}}{2} = 5.51 \times 10^{-4}$$

CALCULATING THE pH OF A WEAK ACID

which corresponds to pH = 3.259. You can also use the "goal–seek" function of your spreadsheet, adjusting H to get $f(H) = 0$;

Goal Seek 100 iter, max change 0		
f(H)	0	=H^2+Ka*H-Ka*C
H	0.000551	
pH =	3.258866	

$$K_a = \frac{[H^+][F^-]}{[HF]} = \frac{(10^{-3.23})(10^{-3.25})}{10^{-3.34}} = 10^{-3.16}$$

in good agreement with the starting value of $pK_a = 3.17$. The final check is the mass balance:

$$C = [F^-] + [HF] = (10^{-3.25}) + (10^{-3.34}) = 10^{-2.99}$$

again in good agreement with the starting value of $10^{-3.00}$.

Example 4. Here is an example where the graphical approach is best supplemented by a numerical calculation. Find the pH of 5.0×10^{-5} M HCN ($pK_a = 9.32$). Refer to the diagram of Fig. 4.8 and note that the lines for [CN$^-$] and [OH$^-$] are very

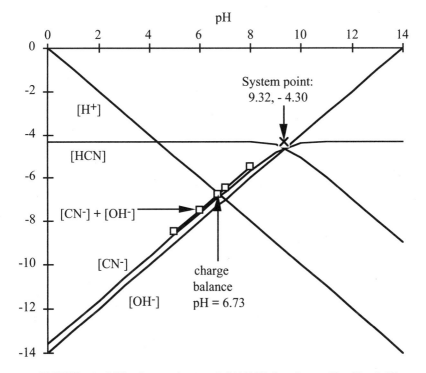

FIGURE 4.8. Sillén diagram for 5×10^{-5} M HCN (log $C = -4.30$, $pK_a = 9.32$).

close together below the system point. The equilibria and mass balance are incorporated in the diagram, and the charge balance is

$$[H^+] = [CN^-] + [OH^-]$$

Follow the $[H^+]$ line down until it meets the $[CN^-]$ and $[OH^-]$ lines. Unlike acetic acid (Ex. 2) and hydrofluoric acid (Ex. 3), both terms on the right side of this charge balance are of similar size, and neither can be neglected.

One approach at this point is to read values of $\log[CN^-]$ and $\log[OH^-]$ from the graph, calculate $[CN^-] + [OH^-] = 10^{\log[CN^-]} + 10^{\log[OH^-]}$ for a range of pH near the intersection, and find the point where the charge balance is satisfied (see points plotted on Fig. 4.8).

Another approach is to use the diagram to estimate the relative size of terms to simplify an algebraic solution. Since the intersection of the $[H^+]$ line with the $[CN^-]$ and $[OH^-]$ lines occurs approximately at pH = 7, and pK_a = 9.32, $[H^+] \gg K_a$ in the expression for $[CN^-]$:

$$[CN^-] = \frac{CK_a}{[H^+] + K_a} = \frac{(5 \times 10^{-5})(10^{-9.32})}{[H^+] + 10^{-9.32}} = \frac{10^{-13.62}}{[H^+]} - \cdots$$

The charge balance then becomes

$$[H^+] = [CN^-] + [OH^-] = \frac{10^{-13.62} + 10^{-14.00}}{[H^+]}$$

which is easily solved for $[H^+]$:

$$[H^+]^2 = 10^{-13.47}$$
$$[H^+] = 10^{-6.73}$$
$$pH = 6.73$$

Note that if the approximation $[H^+] = [CN^-]$ were made, the result would have been pH = 6.81, and if the approximation were $[H^+] = [OH^-]$, the result would have been pH = 7.00. In either case, the calculated pH would have been too high, but the worst error would have been only 0.28 pH units.

CALCULATING THE pH OF OTHER WEAK ACIDS AND BASES

Salt of a Weak Base and Strong Acid

Example 5. Find the pH of 0.01 mol/L NH_4Cl. Here the acid is ammonium ion NH_4^+, with pK_a = 9.25. As above, the equilibria, mass and charge balances are (see also Fig. 4.9):

CALCULATING THE pH OF OTHER WEAK ACIDS AND BASES

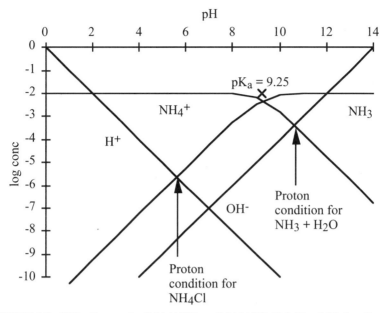

FIGURE 4.9. Sillén diagram for 0.01 M NH_3 or 0.01 M NH_4Cl (pK_a = 9.25, log C = –2).

$$[H^+][NH_3] = K_a[NH_4^+]$$
$$[H^+][OH^-] = K_w$$
$$[NH_4^+] + [NH_3] = 0.01$$
$$[Cl^-] = 0.01$$
$$[H^+] + [NH_4^+] = [OH^-] + [Cl^-]$$

Note the need to include $[Cl^-]$ in the charge balance. On Fig. 4.9, $[Cl^-]$ would be a horizontal line at –2.00, coinciding with NH_4^+ on the left and with NH_3 on the right, but it is not shown for clarity. Possible approximations (to be tested later) are $[NH_4^+]$ » $[NH_3]$, $[NH_4^+]$ » $[H^+]$, and $[Cl^-]$ » $[OH^-]$. These lead to the simplified equations

$$[H^+][NH_3] = K_a[NH_4^+]$$
$$[H^+][OH^-] = K_w$$
$$[NH_4^+] + \cdots = 0.01$$
$$[Cl^-] = 0.01$$
$$[NH_4^+] = [Cl^-] + \cdots$$

From these equations we could determine that $[NH_4^+] = [Cl^-] = 0.01 = \cdots$, but can only determine the product $[H^+][NH_3]$, not either of these alone. To break this deadlock, use the mass balances to subtract the large terms $[NH_4^+]$ and $[Cl^-]$ from the charge balance, to obtain the proton condition:

$$[NH_4^+] = 0.01 - [NH_3]$$

$$[Cl^-] = 0.01$$

$$[H^+] + 0.01 - [NH_3] = [OH^-] + 0.01$$

$$[H^+] = [NH_3] + [OH^-], \quad \text{proton condition for } NH_4Cl + H_2O$$

After that diversion, continue by assuming $[NH_3] \gg [OH^-]$ (Fig. 4.9), which gives the following simplified equations:

$$[H^+][NH_3] = K_a[NH_4^+], \quad \text{equilibrium}$$

$$[H^+][OH^-] = K_w, \quad \text{equilibrium}$$

$$[NH_4^+] + \cdots = 0.01, \quad \text{mass balance}$$

$$[Cl^-] = 0.01, \quad \text{mass balance}$$

$$[H^+] = [NH_3] + \cdots, \quad \text{proton condition}$$

Note this intersection, marked "Proton Condition for NH_4Cl" with an arrow on Fig. 4.9. Now substitute $[NH_3]$ from the proton condition into the acid dissociation equilibrium:

$$[H^+][NH_3] = [H^+]^2 = K_a[NH_4^+] = (10^{-9.25})(10^{-2.00}) = 10^{-11.25} + \cdots$$

$$[H^+] = 10^{-5.63} + \cdots$$

$$pH = -\log[H^+] = 5.63$$

The final phase is to calculate the other concentrations and to see whether the approximations were correct:

$$[NH_3] = [H^+] = 10^{-5.63} + \cdots$$

$$[NH_4^+] = 1.0 \times 10^{-2} - \cdots$$

$$[Cl^-] = 1.0 \times 10^{-2} \text{ (exactly)}$$

$$[OH^-] = K_w/[H^+] = 10^{-14.00}/10^{-5.63} = 10^{-8.38}$$

Test the approximations:

$$[NH_4^+] \gg [NH_3], \quad 10^{-4} \gg 10^{-5.63}, \quad \text{OK}$$

$$[NH_4^+] \gg [H^+], \quad 10^{-4} \gg 10^{-5.63}, \quad \text{OK}$$

CALCULATING THE pH OF OTHER WEAK ACIDS AND BASES 113

$$[Cl^-] \gg [OH^-], \quad 10^{-4} \gg 10^{-8.38}, \quad OK$$

$$[NH_3] \gg [OH^-], \quad 10^{-6.63} \gg 10^{-8.38}, \quad OK$$

Thus each of the approximations is satisfied by the final answers. It should be clear, however, that if the concentration was somewhat less or pK_a somewhat higher, some of the approximations would not hold, and more complete versions of these equations would be required.

pH of a Weak Base

Example 6. Find the pH of 0.01 M NH_3. The equilibrium and mass balance on NH_3 are the same as for Ex. 5. Only the charge balance is different—$[Cl^-]$ is absent. Since NH_3 is an uncharged base, the proton condition and the charge balance are identical.

$$[H^+][NH_3] = K_a[NH_4^+]$$

$$[H^+][OH^-] = K_w$$

$$[NH_4^+] + [NH_3] = 1.0 \times 10^{-2}$$

$$[H^+] + [NH_4^+] = [OH^-], \quad \text{proton condition for } NH_3 + H_2O$$

In this case the best approximation appears to be $[OH^-] \gg [H^+]$ and $[NH_3] \gg [NH_4^+]$, or

$$[NH_4^+] = [OH^-] + \cdots$$

This intersection is shown on Fig. 4.9 by the arrow marked "Proton Condition for $NH_3 + H_2O$." In addition,

$$[NH_3] = 1.0 \times 10^{-2} + \cdots$$

Substituting these approximations in the basic equations gives

$$[H^+](1.0 \times 10^{-2}) = K_a K_w/[H^+]$$

$$[H^+]^2 = (10^{-9.25})(10^{-14.0})/(10^{-2}) = 10^{-21.25}$$

$$pH = 10.62$$

Use this pH value to calculate the other concentrations and test the approximations:

$$[H^+] = 10^{-10.62}$$

$$[OH^-] = 10^{-3.38}, \text{ hence } [OH^-] \gg [H^+]$$

$$[NH_3] = \frac{CK_a}{K_a + [H^+]} = \frac{(10^{-2})(10^{-9.25})}{10^{-9.25} + 10^{-10.62}} = \frac{10^{-11.25}}{10^{-9.23}} = 10^{-2.02}$$

$$[NH_4^+] = \frac{[NH_3][H^+]}{K_a} = \frac{(10^{-2.02})(10^{-10.62})}{10^{-9.25}} = 10^{-3.39}$$

hence $[NH_3] \gg [NH_4^+]$.

$[H^+]$ is less than 10^{-7} as large as $[OH^-]$, truly negligible, whereas $[NH_4^+]$ is about 4% of $[NH_3]$, which might be too large an error, and the complete equations would give a more precise answer. Note also that $[OH^-] = [NH_4^+] = 10^{-3.39}$. The Sillén diagram of Fig. 4.9 shows this clearly.

Summary of Proton Conditions for Weak Acids and Weak Bases

Let us review the analogy between the calculations to obtain pH of a weak acid and and those used to obtain the pH of a weak base, the salt of a weak acid, and the salt of a weak base. There are four types of monoprotic acid–base systems, all of which can be represented by Sillén diagrams:

- Uncharged weak acid HA
- Negatively charged weak base A^-
- Positively charged weak acid BH^+
- Uncharged weak base B

All four cases employ the ion product of water.

$$[H^+][OH^-] = K_w \tag{12}$$

Recall that the dissociation constant and mass balance for the first two cases can be written in the form

$$[H^+][A^-] = K_a[HA], \quad \text{dissociation of } HA \tag{11}$$

$$[A^-] + [HA] = C, \quad \text{mass balance on } A \tag{13}$$

The final equation is the proton condition. If the acid is uncharged (HA), the charge balance produces the proton condition:

$$[H^+] = [A^-] + [OH^-], \quad \text{proton condition for } HA \tag{14}$$

Think of the components as HA and H$_2$O: $[H^+]$ is a proton excess on H$_2$O; $[A^-]$ is a proton deficiency on HA; $[OH^-]$ is a proton deficiency on H$_2$O. The proton excesses balance the proton deficiencies.

CALCULATING THE pH OF OTHER WEAK ACIDS AND BASES

For the negatively charged weak base (e.g., Na^+A^-, the salt of HA), recall the discussion of proton condition in Chapter 1. An additional mass balance on Na^+ is required, and an additional term enters the charge balance.

$$[Na^+] = C, \quad \text{mass balance on } Na^+ \quad (31)$$

$$[Na^+] + [H^+] = [A^-] + [OH^-], \quad \text{charge balance} \quad (32)$$

The two mass balances, Eqs. (13), p. 99, and (31), when substituted in the charge balance (Eq. 32), give the proton condition:

$$[HA] + [H^+] = [OH^-], \quad \text{proton condition for } A^- \quad (33)$$

If the components are A^- and H_2O, $[H^+]$ is a proton excess on H_2O; $[HA]$ is a proton excess on A^-; $[OH^-]$ is a proton deficiency on H_2O. Again, the proton excesses balance the proton deficiencies.

The acid–base equilibrium for an uncharged base B and its acid salt (e.g., BH^+Cl^-) is of precisely the same form as that for HA, and the Sillén diagram is constructed the same way.[12] The same mass balance applies to the third and fourth cases:

$$[H^+][B] = K_a [BH^+], \quad \text{dissociation equilibrium} \quad (34)$$

$$[B] + [BH^+] = C, \quad \text{mass balance on } B \quad (35)$$

As above, the proton condition for the uncharged base B is the same as the charge balance:

$$[BH^+] + [H^+] = [OH^-], \quad \text{proton condition for } B \quad (36)$$

If the components are B and H_2O, the proton excesses $[BH^+]$ and $[H^+]$ balance the proton deficiencies $[OH^-]$.

For $NH_4^+Cl^-$, the mass balance on $[Cl^-]$ (Eq. 37) and the mass balance on B (Eq. 38) are combined with the charge balance (Eq. 39):

$$C_{BH^+} = [Cl^-], \quad \text{mass balance} \quad (37)$$

$$C_{BH^+} = [B] + [BH^+], \quad \text{mass balance} \quad (38)$$

$$[H^+] + [BH^+] = [Cl^-] + [OH^-], \quad \text{charge balance} \quad (39)$$

to eliminate the large terms $[BH^+]$ and $[Cl^-]$ and obtain the proton condition:

$$[H^+] = [B] + [OH^-], \quad \text{proton condition for } BH^+ \quad (40)$$

If the components are BH^+ and H_2O, proton excesses H^+ are balanced by proton deficiencies B and OH^-.

Mass Balance on H⁺

An alternative way of looking at the proton condition is as a mass balance on H⁺; although the presence of water as a solvent obscures a true mass balance on the element H, the mass-balance approach is particularly useful in general computer programs where all components are treated by the same formalism. In general,

$$[H^+]_t = [H^+] - [OH^-] + \Sigma\,[HA_i] - \Sigma\,[B_i] \quad (41)$$

where HA_i represents acidic species, including NH_4^+ as well as HAc, and B_i represents basic species, including Ac^- as well as NH_3.

Alkalinity, a closely related concept central to the chemistry of natural waters, is developed in Chapter 10. For a monoprotic acid–strong base–water system, alkalinity is identified as the concentration of strong base cation:

$$[Alk] = [Na^+] = [A^-] + [OH^-] - [H^+] \quad (42)$$

The acidity is simply the negative of the alkalinity. The total hydrogen ion concentration equals the acidity plus the total acid concentration:

$$[Acy] = -[Alk] = [H^+] - [A^-] - [OH^-] \quad (43)$$

$$[H]_t = [H^+] - [OH^-] + [HA] + [A^-] - [A^-] = [Acy] + C_{HA} \quad (44)$$

Example 7. For HA in water alone, the mass balance on H⁺ may be thought of as two contributions: from the dissociation of HA

$$HA \rightleftharpoons H^+ + A^-$$

where each A^- formed produces one H⁺, and from the dissociation of water

$$H_2O \rightleftharpoons H^+ + OH^-$$

where each OH^- produced is matched by an equal amount of H⁺. The result is the same as the charge balance or proton condition:

$$[H^+] = [A^-] + [OH^-]$$

For this system, the total hydrogen ion concentration is given by Eq. (41):

$$[H^+]_t = [H^+] - [OH^-] + [HA]$$

$$[H^+]_t = C_{HA} = [HA] + [A^-]$$

The result is again the same as the charge balance or proton condition:

$$[H^+] = [A^-] + [OH^-]$$

Example 8. Another type of mass balance looks at all the ways that protons can manifest themselves in the solution. [OH$^-$] is negative because it consumes [H$^+$]:

$$[H^+]_t = [H^+] - [OH^-] = -C_{NaOH}, \quad \text{for a strong base}$$

$$[H^+]_t = [H^+] - [OH^-] + [HA] = C_{HA}, \quad \text{for a weak acid}$$

For a mixture of weak acid and strong base

$$[H^+]_t = [H^+] - [OH^-] + [HA] = C_{HA} - C_{NaOH}$$

Combining this with the mass balances $C_{HA} = [HA] + [A^-]$ and $C_{NaOH} = [Na^+]$ gives the familiar charge balance:

$$[Na^+] + [H^+] = [A^-] + [OH^-]$$

Example 9. Similarly, a solution containing NaOH and NH$_3$ would have a mass balance

$$[H^+]_t = [H^+] - [OH^-] - [NH_3] = -C_{NH_3} - C_{NaOH}$$

This can be converted by substituting the mass balances

$$[Na^+] = C_{NaOH} \quad \text{and} \quad [NH_3] + [NH_4^+] = C_{NH_3}$$

to obtain the familiar charge balance:

$$[Na^+] + [H^+] + [NH_4^+] = [OH^-]$$

This approach, properly applied, can be generalized to highly complex mixtures,[13] and is used in the discussion of automated computation in Chapter 12.

MIXTURES OF STRONG OR WEAK ACIDS AND BASES

The methods used in the previous sections can easily be generalized to apply to any mixture of strong or weak acids and bases. The central equation is the charge balance. Strong acids are represented by an unreactive anion[14] such as Cl$^-$ or SO$_4^{2-}$; strong bases by a cation such as Na$^+$ or K$^+$. Each separate weak acid or base is represented by an expression of the form

$$[A^-] = \frac{C_{HA} K_a}{[H^+] + K_a} \quad \text{or} \quad [BH^+] = \frac{C_{BH}[H^+]}{[H^+] + K_a} \tag{45}$$

When one term on each side of the charge balance is much larger than the others

Example 10. Find the pH of a mixture containing 0.01 M acetic acid ($pK_a = 4.75$) and 0.001 M formic acid ($pK_a = 3.75$). The charge balance is

$$[H^+] = [Ac^-] + [Fo^-] + [OH^-]$$

$$[H^+] = \frac{(0.01)(10^{-4.75})}{[H^+] + 10^{-4.75}} + \frac{(0.001)(10^{-3.75})}{[H^+] + 10^{-3.75}} + \frac{10^{-14.00}}{[H^+]}$$

This equation can be represented by superimposing the Sillén diagrams of acetic and formic acid, as in Fig. 4.10.

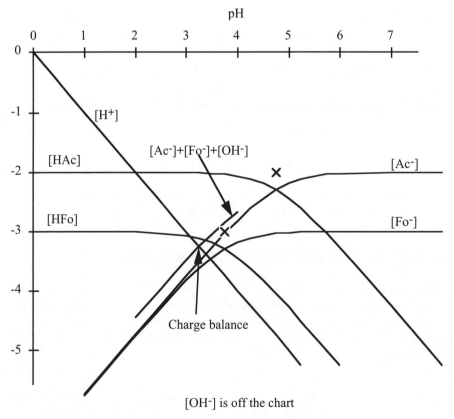

FIGURE 4.10. Sillén diagram for a mixture of 0.01 M acetic acid and 0.001 M formic acid.

MIXTURES OF STRONG OR WEAK ACIDS AND BASES **119**

Note that the largest term on the right side of the charge balance is [Ac⁻], but that [Fo⁻] is very nearly as large. At the charge balance point [HFo] and [Fo⁻] are nearly the same size, [HAc] is about ten times larger than [Ac⁻], but [OH⁻] is more than six orders of magnitude smaller and can safely be neglected. These observations lead to the approximate equation:

$$[H^+] = \frac{C_a K_a}{[H^+]} + \frac{C_f K_f}{[H^+] + K_f}$$

which is most easily solved by iteration of the equation

$$[H^+]^2 = C_a K_a + \frac{C_f K_f [H^+]}{[H^+] + K_f}$$

Start with pH = 3.5, which can be read approximately from the diagram.

$$[H^+]^2 = (0.01)(10^{-4.75}) + \frac{(0.001)(10^{-3.75})(10^{-3.5})}{10^{-3.5} + 10^{-3.75}}$$

$$[H^+]^2 = 10^{-6.75} + 10^{-6.94} = 10^{-6.54}$$

$$pH = 3.27$$

A second iteration gives pH = 3.25, and this is not changed by a third iteration.

This calculation was made more difficult by the fact that both the anion concentrations were close in magnitude, and further by the fact that pH was close to pK_a for both. More typically, one acid would have a much higher concentration than the other (as in the case of indicators), or one would have a much higher pK_a than the other. In either of these cases the charge balance reduces to $[H^+] = [A^-]$, where A is the anion with the highest concentration.

Example 11. A simple variant of the above is a mixture of a weak and a strong acid. Consider 0.001 M HCl added to 0.01 M acetic acid (Fig. 4.11). The equilibrium and mass balance for acetic acid are as above, but the charge balance contains a term for [Cl⁻], representing the strong acid.

$$[H^+] = [A^-] + [Cl^-] + [OH^-]$$

$$[H^+] = \frac{C_a K_a}{[H^+] + K_a} + [Cl^-] + [OH^-]$$

since addition of HCl will make the mixture even more acidic than pure HA (see arrow on Fig. 4.11), you can reasonably assume that $[H^+] \gg K_a$ and also that [OH⁻] is negligible. This leads to an iterative form (start with pH = 3.5):

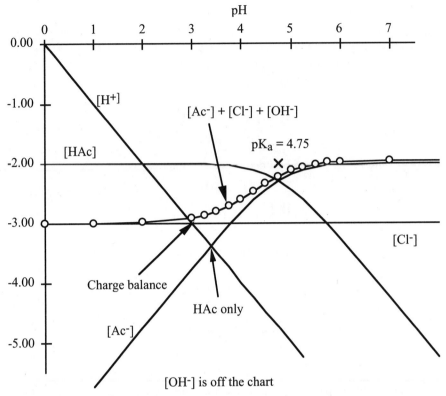

FIGURE 4.11. Sillén diagram for a mixture of 0.01 M acetic acid and 0.001 M HCl.

$$[H^+]^2 = C_a K_a + [H^+][Cl^-]$$
$$[H^+]^2 = (10^{-2.00})(10^{-4.75}) + (10^{-3.5})(10^{-3.00}) = 10^{-6.31}$$

which gives pH = 3.15. Successive iterations give 3.03, 2.98, 2.96, 2.95, and eventually converge at 2.94. This slow convergence is a result of the two terms being nearly the same size. The Sillén diagram (Fig. 4.11) shows the curve of $[Ac^-]$ + $[Cl^-]$, which crosses the $[H^+]$ line at a pH value a little to the left of where $[Cl^-]$ alone would cross.

Note that if the acid were weaker ($pK_a \gg 5$) or the HCl were more concentrated ($\gg 10^{-3}$), $[Cl^-]$ would dominate the charge balance, and the pH would be approximately $-\log(C_{HCl})$.

Salt of a Weak Acid and a Weak Base

One of the classic problems of ionic equilibrium is to find the pH of a solution containing the salt of a weak acid and a weak base. Here two independent weak

MIXTURES OF STRONG OR WEAK ACIDS AND BASES

acid systems are linked by the condition that they have the same total concentration. In general,

$$[H^+][A^-] = K_{a1} [HA] \tag{46}$$

$$[H^+][B] = K_{a2} [BH^+] \tag{47}$$

$$[H^+][OH^-] = K_w \tag{48}$$

$$C = [HA] + [A^-] = [BH^+] + [B], \quad \text{two mass balances} \tag{49}$$

$$[BH^+] + [H^+] = [A^-] + [OH^-], \quad \text{charge balance} \tag{50}$$

Normally, in this charge balance $[BH^+]$ is by far the largest term on the left side, and $[A^-]$ is by far the largest term on the right side. The critical difference between $[H^+]$ and $[OH^-]$ could get lost in the roundoff errors. Therefore, it is desirable to convert the charge balance (Eq. 50) to a proton condition by substituting from the mass balances (Eq. 49) to eliminate the large terms:

$$[HA] + [H^+] = [B] + [OH^-], \quad \text{proton condition} \tag{51}$$

If $[H^+]$ and $[OH^-]$ are small (see below and Eq. (45), p. 117),

$$\frac{C[H^+]}{[H^+] + K_{a1}} = \frac{CK_{a2}}{[H^+] + K_{a2}} \tag{52}$$

$$[H^+]^2 = K_{a1}K_{a2} \tag{53}$$

independent of C.[15]

Example 12. In Fig. 4.12, Sillén diagrams for acetic acid and ammonia, each at 0.01 M, are superimposed ($pK_{a1} = 4.75$, $pK_{a2} = 9.25$, $pK_w = 14.00$).
The balance of $[HA] = [B]$ is easily obtained at

$$pH = \tfrac{1}{2}(pK_{a1} + pK_{a2}) = \tfrac{1}{2}(4.75 + 9.25) = 7.00$$

This answer is the same as the pH of pure water because of the accidental coincidence of the pK_a values, not because of any fundamental result. Note that as the total concentration of the salt decreases, $[B]$ becomes closer to $[OH^-]$ and $[HAc]$ becomes closer to $[H^+]$. At about 10^{-5} M, all the terms in the proton condition are about equal in size. For this particular case, pH remains at 7.00 as C decreases, but in the general case where pH at high concentration is either higher or lower than 7, the pH tends toward 7 as C decreases, governed by Eq. (54) or (55) (below). Compare Fig. 4.13, where $pK_{a1} + pK_{a2} = 15.51$ and the high concentration limit is 7.755, not 7.0.

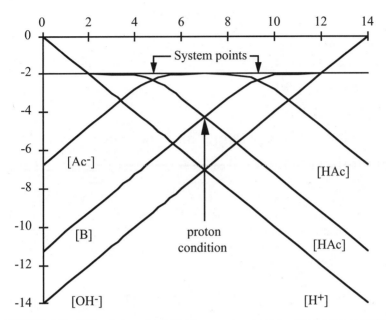

FIGURE 4.12. Sillén diagram for 0.01 M ammonium acetate. The proton condition is [HAc] = [B]. Decreasing the analytical concentration would lower all curves except [H$^+$] and [OH$^-$].

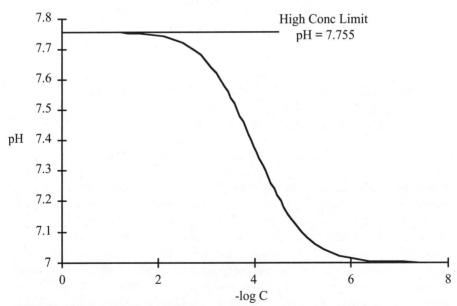

FIGURE 4.13. Concentration dependence of pH for the salt of a weak acid and weak base (dimethylammonium acetate). The high concentration limit is given by Eq. (49).

INDICATORS AS WEAK ACIDS

Algebraically, the full answer is found by using the three equilibria (Eqs. 46–48) in the proton condition (Eq. 51)

$$\frac{C[H^+]}{[H^+] + K_{a1}} + [H^+] = \frac{CK_{a2}}{[H^+] + K_{a2}} + \frac{K_w}{[H^+]} \qquad (54)$$

The direct solution of this problem involves a quartic equation in $[H^+]$, but, like the equation for pH of a weak acid, this equation is linear in C, and so can be easily rearranged to give:

$$C = \frac{K_w/[H^+] - [H^+]}{[H^+]\left(\frac{1}{[H^+] + K_{a1}} + 1\right) - \left(\frac{K_{a2}}{[H^+] + K_{a2}}\right)} \qquad (55)$$

This equation is plotted in Fig. 4.13 with $pK_{a1} = 4.75$ (acetic acid) and $pK_{a2} = 10.76$ (methylamine).

INDICATORS AS WEAK ACIDS

In Table 3.3 a number of indicators were listed together with the pH range in which they change color. Consider now the reason for this color change: the ionization of the indicator itself as a weak acid or base.

Any substance that has a light-absorption spectrum for its acid form that is different from its basic form is an indicator, and the pH range where it changes color is determined by the pK_a of the acid–base equilibrium. An example is methyl orange, normally sold as the orange sodium salt Na^+In^- (Fig. 4.14). The red acidic form HIn has a proton attached to the nitrogen on the right (Fig. 4.15).

The ionization equilibrium is the same as for any weak acid:

$$[H^+][In^-] = K_{In}[HIn] \qquad (56)$$

and for methyl orange, $pK_{In} = 3.8$. The color change occurs in the range from pH = 3.2–4.5, that is, from $[H^+] = 6.3 \times 10^{-4}$ to 3.2×10^{-5}. The ratio of the orange form In^- to the red form HIn therefore ranges from

$$[In^-]/[HIn] = K_{In}/[H^+] = 0.25 \text{ (red) to } 5.0 \text{ (orange)}$$

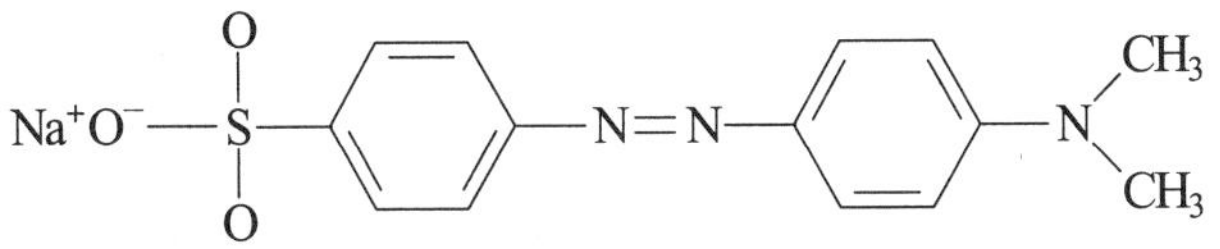

FIGURE 4.14. Structure of methyl orange sodium salt.

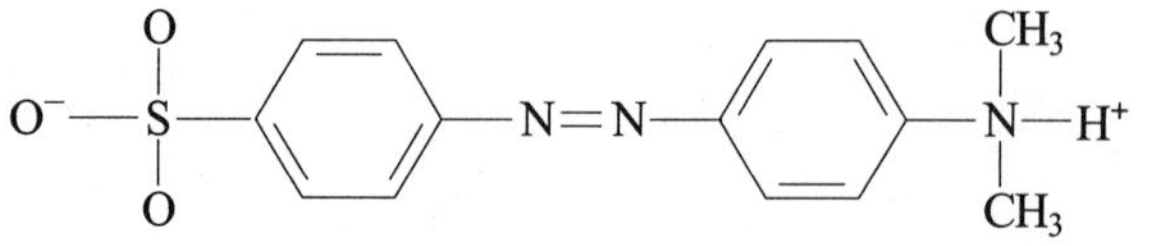

FIGURE 4.15. Structure of methyl orange acid.

Indicators are of many chemical types. The spectra of an acid and basic form of Thymol blue, a related indicator that changes from yellow to blue near pH = 8, are shown in Figs. 4.18 and 4.19 (pp. 126–127).

Another example is phenolphthalein, which at pH < 8.4 is a colorless acid H_2In (Fig. 4.16). Over the range from 8.4 to 10.0 it changes to a magenta form In^{2-} (Fig. 4.17), and in concentrated alkali it becomes $InOH^{3-}$, another colorless form. The intermediate form HIn^- is in relatively low concentration, even at its maximum, and so is usually neglected.

Because indicators are acids or bases themselves, they will modify the acidity of a solution into which they are introduced. As long as the indicator concentration is a small fraction of the main constituent concentration, the error will be small. Also, if the indicator is added in its basic form to a solution with pH » pK_{In} (or in its acidic form to a solution with pH « pK_{In}), it will not change the pH of that solution. On the other hand, as a solution is titrated and the indicator changes color, its acid–base reactions will influence the total amount of titrant required.

Example 13. A solution 10^{-4} M in NaOH is made 10^{-5} M in the sodium salt of methyl orange. Calculate the pH of the resulting solution. The charge balance is

$$[H^+] + [Na^+] = [OH^-] + [In^-]$$

The mass balances are:

$$[Na^+] = C_{NaOH} + C_{In}$$

$$[In^-] + [HIn] = C_{In}$$

And when substituted for $[Na^+]$ and $[In^-]$ in the charge balance yield the proton condition

$$[H^+] + [HIn] + C_{NaOH} = [OH^-]$$

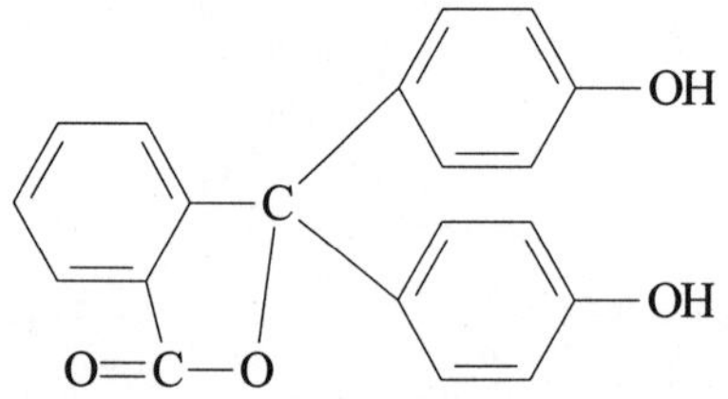

FIGURE 4.16. Structure of colorless phenolphthalein, H_2In.

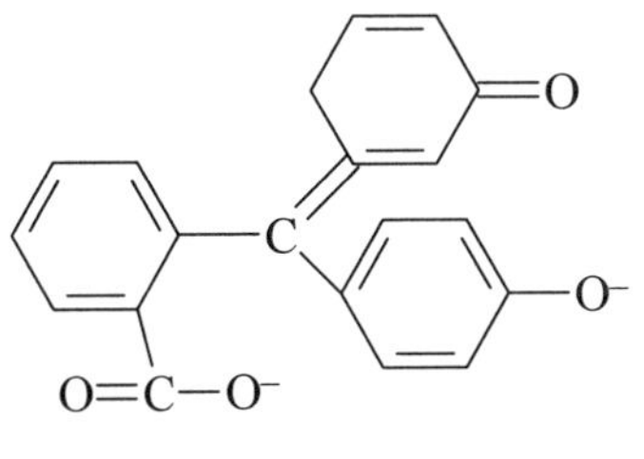

INDICATORS AS WEAK ACIDS

FIGURE 4.17. Structure of magenta phenolphthalein In^{2-}.

As a first approximation, neglect $[H^+]$ and $[In^-]$, which gives

$$[OH^-] = C_{NaOH} = 1 \times 10^{-4}$$
$$pH = pK_w + \log[OH^-] = 14.0 - 4.0 = 10.0$$

Note that $[H^+] = 10^{-10}$ and $[HIn] = C_{In}[H^+]/([H^+] + K_{In}) = (10^{-5})(10^{-10})/(10^{-10} + 10^{-3.8}) = 10^{-11.2}$, both of which are negligible compared to 10^{-4}.

Example 14. To the mixture of the previous example is added 1.2×10^{-4} M of HCl. What is the resultant pH? Note that the HCl is more than is required to react with the original NaOH, and part of the excess will be used in converting $[In^-]$ to $[HIn]$. This is all taken into account if the charge balance is modified to include $[Cl^-] = 1.2 \times 10^{-4}$:

$$[H^+] + [Na^+] = [In^-] + [OH^-] + [Cl^-]$$
$$[H^+] = [In^-] + [OH^-] + C_{HCl} - C_{NaOH} - C_{In}$$
$$[H^+] = [In^-] + [OH^-] + 1.2 \times 10^{-4} - 1.0 \times 10^{-4} - 1.0 \times 10^{-5}$$
$$[H^+] = [In^-] + [OH^-] + 1.0 \times 10^{-5}$$

Since the solution now contains excess acid, the first approximation is to neglect $[In^-]$ and $[OH^-]$, which gives $[H^+] = 1.0 \times 10^{-5}$ or $pH = 5.0$. At this pH, $[OH^-] = 10^{-9.0}$ and

$$[In^-] = C_{In}K_{In}/([H^+] + K_{In}) = 9.4 \times 10^{-6} = 10^{-5.03}$$

$[In^-]$ is not negligible compared to 1×10^{-5}, and so a better answer can be obtained by iteration of the more complete equation (still neglecting $[OH^-]$)

$$[H^+] = C_{In}K_{In}/([H^+] + K_{In}) + C_{HCl} - C_{NaOH} - C_{In}$$
$$[H^+] = (10^{-5})(10^{-3.8})/(10^{-5.0} + 10^{-3.8}) + 10^{-5} = 1.94 \times 10^{-5}$$
$$pH = 4.71$$

Successive approximations give 4.72, which is the stable answer.

What would the pH have been if the indicator were not present? Omit [In⁻] and C_{In} from the charge balance:

$$[H^+] = [OH^-] + 2 \times 10^{-5}$$

and since [OH⁻] is negligible, $[H^+] = 2 \times 10^{-5}$ or pH = 4.70, 0.02 units lower.

This type of error, due to titration of the indicator, is normally negligible because C_{In} is very much less than the concentration of material being titrated. It can be made entirely negligible, even in dilute solutions, by bringing the indicator to the pH of the end point by addition of acid or base before adding it to the titration mixture. Although the pH is initially modified by the indicator, the protons lost will be replaced by dissociation of the indicator as the titration proceeds. This approach will work even for the titration of a weak acid with a weak base or more complicated systems.

pH Measurement with Indicators

Indicators can measure pH quantitatively if used with a spectrophotometer to determine the absorbance (at the wavelengths of maximum absorption for HIn and In⁻) and calibrated as a function of pH. For example, the ratio of thymol blue indicator absorbances at 435 and 596 nm was used to calculate seawater pH.[16]

Figure 4.18 shows the absorbance of thymol blue at pH = 8.20 from 400 to 700 nm. Note the peaks at 435 and 596 nm. By adjusting the pH to a low value, say 4.0, the indicator is essentially converted to the acid form HL^-, and the absorption spectrum of that species can be obtained, as shown in Fig. 4.19. Similarly, if the pH is raised to 12.0, the spectrum obtained is that of the basic form L^{2-}. In this way

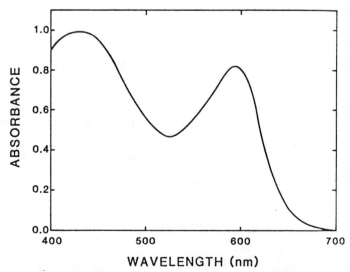

FIGURE 4.18. Relative absorbance of thymol blue in seawater at pH = 8.20. From R. H. Byrne, *Analytical Chemistry*, 1987, vol. 59, pp. 1479–1481. © American Chemical Society. Reproduced by permission.

INDICATORS AS WEAK ACIDS

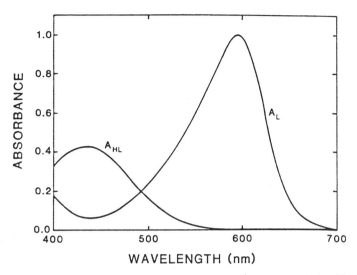

FIGURE 4.19. Relative absorbance of the acidic and basic forms of thymol blue. The indicator concentrations are different from Fig. 4.18. (See Fig. 4.18 for citation).

the relative contributions of the two indicator forms at any pH can be related to the total absorbance at that pH.

$$_1A = {_1\varepsilon_{HL}}[HL] + {_1\varepsilon_L}[L^{2-}] \tag{57}$$

$$_2A = {_2\varepsilon_{HL}}[HL] + {_2\varepsilon_L}[L^{2-}] \tag{58}$$

where $_1A$ is the total absorbance per unit cell path length at 435 nm and $_2A$ is the total absorbance at 596 nm. The coefficients $_1\varepsilon_{HL}$ and $_2\varepsilon_{HL}$ are the molar absorbances of the indicator acid at the same two wavelengths, at pH = 4; $_1\varepsilon_L$ and $_2\varepsilon_L$ are the molar absorbances of the indicator base at the same two wavelengths, at pH = 12 (Fig. 4.19).

The pH is given by the ratio of the two indicator forms according to the usual expression for a monoprotic acid:

$$\mathrm{pH} = \mathrm{p}K_2^\circ + \log\left(\frac{\gamma_=}{\gamma_-}\right) + \log\left(\frac{[L^{2-}]}{[HL]}\right) \tag{59}$$

In a constant ionic medium such as seawater, the activity coefficients can be combined with $\mathrm{p}K_2^\circ$ as $\mathrm{p}K_2$, and Eqs. (57)–(59) can be combined to give

$$\mathrm{pH} = \mathrm{p}K_2 + \log\left(\frac{\dfrac{_2A}{_1A} - \dfrac{_2\varepsilon_{HL}}{_1\varepsilon_{HL}}}{\dfrac{_2\varepsilon_L}{_1\varepsilon_{HL}} - \dfrac{_2A}{_1A}\dfrac{_1\varepsilon_L}{_1\varepsilon_{HL}}}\right) \tag{60}$$

The coefficients for thymol blue are[17] $_2\varepsilon_{HL}/_1\varepsilon_{HL} = 0.0021$, $_2\varepsilon_L/_1\varepsilon_{HL} = 2.324$, $_1\varepsilon_L/_1\varepsilon_{HL} = 0.1439$ (see Fig. 4.20), and $_2A/_1A = 0.826$ at pH = 8.20 (see Fig. 4.19). Substituting:

$$8.20 = pK_2 + \log\left(\frac{0.826 - 0.0021}{2.324 - (0.826)(0.1439)}\right)$$

which yields $pK_2 = 8.63$.

How well does the indicator pH compare with potentiometric pH standardized using NBS buffers? Byrne et al.[18] made a comparison at sea, and some of their results, obtained 20 March 1985 in the Indian Ocean at 39.37°S, 76.28°E, are shown in Fig. 4.20. The spectrophotometric pH values were obtained using phenol red at 20°C and calibrated on the "free H" scale[19] using tris(hydroxymethyl) aminomethane buffer[20]. The potentiometric values were calibrated with NBS buffers.[21] The spectrophotometric pH values are consistently about 0.02 ± 0.005 units higher than the potentiometric measurements, presumably reflecting the differences between the two calibration scales (see Chapter 11).

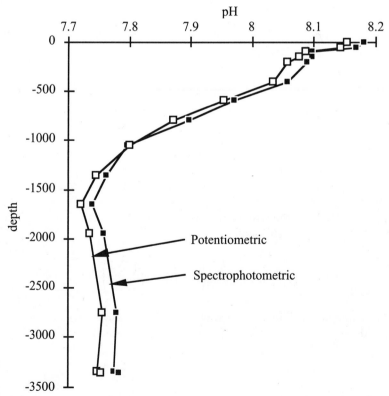

FIGURE 4.20. Comparison of spectrophotometric and potentiometric pH measurements in seawater as a function of depth.[18]

Comparison of Spectrophotometric and Potentiometric pH Methods

Colorimetric and spectrophotometric measurements of pH were widely used in the early part of this century, but have been largely supplanted by potentiometric methods because of the convenience and low cost of the combination glass/reference electrode and electrometer.

The potentiometric and spectrophotometric methods each measure hydrogen ion activity according to conventional assumptions. In the potentiometric method, standard buffers are measured using a hydrogen/silver chloride cell without liquid junction to obtain $-\log\{[H^+][Cl^-]\gamma_\pm\} + \log[Cl^-] = p(a_H\gamma_{Cl})$ and a convention adopted to calculate γ_{Cl} (see Harned cell, Chapter 2). This cell is normally calibrated with HCl, but can also be calibrated with any standard buffer for which the pK_a is known in the medium of interest. The ultimate reference state for hydrogen ion activity in the NBS system is usually taken to be infinite dilution, but it is also possible to employ a reference state based on an electrolyte of nearly constant composition (two such are seawater and 4 M $NaClO_4$).

Careful glass electrode measurements can achieve a precision of ±0.05 mV and an accuracy of ±0.5 mV (±0.001 and ±0.008 pH units, respectively). The "accuracy" quoted comes from a comparison of the activity coefficient of NaCl, measured by four different investigators, employing a sodium-selective glass electrode/silver chloride electrode cell, a sodium amalgam electrode with silver chloride electrode, and isopiestic (vapor pressure) methods.[22] Of course, the greater the difference in composition between the standard buffer and the test solution, the larger are systematic errors from the liquid junction potential.

What advantages does the spectrophotometric method provide? For samples that tend to fall in a narrow range of pH (as seawater does), and that can be conveniently brought to a spectrophotometer before their composition changes, the spectrophotometric method provides an accurate pH method with stable built-in calibration.

As seen in Eq. (56), pH is measured by the molar absorbance ratio $_2A/_1A$. Calibration of the method requires four parameters: the concentration acidity constant K_a of the indicator, and three ratios of molar absorptivity: $_2\varepsilon_{HL}/_1\varepsilon_{HL}$, $_2\varepsilon_L/_1\varepsilon_{HL}$ and $_1\varepsilon_L/_1\varepsilon_{HL}$. In general, all these parameters depend to some extent on the chemical composition of the ionic medium. For example, pK_a for cresol red varies from 7.82 in seawater to 8.55 in 1 M NaCl, 8.12 in 3 M NaCl to 8.92 in 7 M $NaClO_4$.[23]

As demonstrated by Fig. 4.20, the spectrophotometric method can give data that are apparently as accurate as the potentiometric method, although in that study[18] the spectrophotometric pH values were systematically higher. Byrne et al. have demonstrated that pH can be replicated to within 0.001 units, and that a large data set can have a precision of ±0.004. Recent results have been even better.[24]

The pH value obtained by the spectrophotometric method is $pc_H = -\log[H^+]$, not $p(a_H\gamma_{Cl})$

$$pc_H = pK_a + \log\left(\frac{[HA]}{[A^-]}\right)$$

and to relate it to the potentiometric pH requires knowledge of the mean activity coefficient coefficient $\gamma_\pm$ for HCl in the medium:

$$pc_H = -\log[H^+] = -\log\{[H^+]\gamma_H\gamma_{Cl}\} + 2\log\gamma_\pm = p(a_H\gamma_{Cl}) + 2\log\gamma_\pm$$

What are the disadvantages?

The primary disadvantage is that a spectrophotometer tends to be larger, less portable, and much more expensive than a pH meter. This often means the sample must be brought to a central laboratory, and changes in its composition during sampling, transit, and storage are more likely.

The other important disadvantage is that the range of pH that can be measured with a single indicator is narrow. The error in pH is least when the absorbances of HL and L are equal (usually near pH = pK_a), and increases as pH becomes either larger or smaller. To cover a wider pH range, a number of indicators must be calibrated independently and the pH scales so obtained reconciled where they do not agree exactly.

Absolute measurements of K_a is normally done by potentiometric titration using a standard buffer for calibration, and such buffers are normally calibrated by potentiometric measurements using the hydrogen/silver chloride cell. Thus the spectrophotometric method normally depends on a potentiometric method. One exception is calibration of the potentiometric system using HCl concentration, which yields pc_H directly. In principle, this could be used to obtain indicator pK_a values up to 3 or 4; these acid–base systems could then be used to calibrate the pK_a of indicators in a higher range, and potentiometric measurements could be avoided.

Indicator dye species, being acids or bases themselves, will enter into acid–base reactions. If the buffer capacity of the test solution is low, it is necessary to correct for reaction of the indicator species with other species, or a systematic error will result. While this is not difficult (see pp. 125–126), it is not always done. Less easy to compensate for is the possibility that indicator dyes will change due to photolysis or oxidation over the course of time. To avoid such errors, frequent calibration against secondary standards is desirable.

In summary, the spectrophotometric method, if done carefully, is as accurate as the potentiometric method for pH. Its principal advantage is the need for only one indicator–calibration experiment, and its principal disadvantage is the narrow range of pH that can be covered. Ultimately, both methods rely on the experiments used to establish the pH scale.

BUFFER SOLUTIONS

In many branches of chemistry, particularly in biochemical investigations, it is important to keep the pH of a solution relatively constant over the course of some reaction that produces or consumes hydrogen ions. To accomplish this, the reaction is carried out in a buffer solution, a fairly concentrated solution of a weak acid

BUFFER SOLUTIONS

and its conjugate base, which do not themselves enter into the reaction. If hydrogen ions are produced, they react with the base to increase the amount of conjugate acid; if hydrogen ions are consumed, the weak acid dissociates to give more hydrogen ions. The net effect is to resist any change in pH, as illustrated in the following example:

Example 15. Calculate the pH of a buffer that is 0.01 M in acetic acid and 0.01 M in sodium acetate. As in Examples 13 and 14, $pK_a = 4.75$ or $K_a = 1.75 \times 10^{-5}$, and since the solution is fairly concentrated and moderately acidic, good first approximations are $[OH^-] \ll [H^+] \ll [Ac^-]$:

$$[H^+][Ac^-] = K_a[HAc]$$

$$[Na^+] = 0.01$$

$$[HAc] + [Ac^-] = 0.02$$

$$[H^+] + [Na^+] = [Ac^-] + [OH^-] \text{ or approximately } [Na^+] = [Ac^-]$$

Combining the equilibrium and the mass balance on acetate in the usual way gives $[Ac^-]$ as a function of $[H^+]$:

$$[Na^+] = C_a K_a / ([H^+] + K_a)$$

$$0.01 = (0.02)(1.75 \times 10^{-5}) / ([H^+] + 1.75 \times 10^{-5})$$

$$[H^+] + 1.75 \times 10^{-5} = 3.50 \times 10^{-5}$$

$$[H^+] = 1.75 \times 10^{-5}$$

$$pH = 4.75$$

Now add 0.001 M NaOH to this buffer. This will increase $[Na^+]$ to 0.011 but will otherwise not change the above calculation. The result is

$$0.011 = (0.02)(1.75 \times 10^{-5}) / ([H^+] + 1.75 \times 10^{-5})$$

$$[H^+] + 1.75 \times 10^{-5} = 3.18 \times 10^{-5}$$

$$[H^+] = 1.43 \times 10^{-5}$$

$$pH = 4.84$$

This new pH is 0.09 units higher than the original buffer. Note that if 0.001 M NaOH had been added to pure water, the pH would have increased from 7 to 11, by 4.0 units. Thus the buffer does indeed effectively resist changes in pH.

This same approach can be used to obtain an equation for the pH of any buffer made from C_{HA} mol/L of a weak acid and C_A mol/L of its conjugate base. (The acid could be BH^+ and the base could be B; the equations are fully analogous.) Including $[H^+]$ and $[OH^-]$ for generality,

$$[H^+][A^-] = K_a[HA], \quad \text{equilibrium} \tag{61}$$

$$[H^+][OH^-] = K_w, \quad \text{equilibrium} \tag{62}$$

$$[H^+] + [Na^+] = [OH^-] + [A^-], \quad \text{charge bal.} \tag{63}$$

$$[A^-] + [HA] = C_A + C_{HA}, \quad \text{mass bal.} \tag{64}$$

$$[Na^+] = C_A, \quad \text{mass bal.} \tag{65}$$

It is possible to replace the charge balance by a proton condition, but since C_A and C_{HA} are usually the same order of magnitude, this maneuver does not give any special advantage.

The simplest approximation, corresponding to Example 12, is to neglect $[H^+]$ and $[OH^-]$ in the charge balance and apply the mass balances to get

$$[A^-] = C_A + \cdots \tag{66}$$

$$[HA] = C_{HA} + \cdots \tag{67}$$

$$[H^+] = K_a C_{HA}/C_A \tag{68}$$

$$pH = pK_a - \log(C_{HA}/C_A) \tag{69}$$

Equation (68) or (69) is widely used in the analytical and biochemical literature, frequently cited as the "Henderson equation" or the "Henderson–Hasselbalch equation."[25]

The more general equations are obtained by combining Eqs. (63)–(65) without approximation.

$$[A^-] = C_A + [H^+] - [OH^-] \tag{70}$$

$$[HA] = C_{HA} - [H^+] + [OH^-] \tag{71}$$

The resulting equation is third order in $[H^+]$ but linear in either K_a or C_A. Thus if C_A and C_{HA} are known, K_a can be calculated from the measured pH of the buffer solution

$$K_a = \frac{[H^+](C_A + [H^+] - [OH^-])}{(C_{HA} - [H^+] + [OH^-])} \tag{72}$$

where $[H^+] = 10^{-pH-\log(\gamma_+)}$ and $[OH^-] = K_w/[H^+] = 10^{-pK_w^\circ + pH - \log(\gamma_-)}$. This is sometimes called the "Charlot equation."[26]

Similarly the change in pH with buffer concentration can be explicitly derived from the same equations. Fix the value of C_{HA}/C_A and calculate C_A as a function of pH:

$$C_A = \frac{\left(\dfrac{K_a}{[H^+]} + 1\right)\left([H^+] - \dfrac{K_w}{[H^+]}\right)}{\left(\dfrac{K_a C_{HA}}{[H^+]C_A} - 1\right)} \tag{73}$$

BUFFER SOLUTIONS

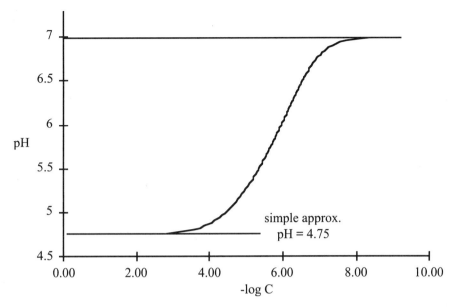

FIGURE 4.21. pH of equimolar acetic acid–sodium acetate buffer as a function of total concentration. Note that the simple approximation applies at concentrations above 10^{-3} M, and that pH increases at lower concentrations to a limit of 7 below $C = 10^{-8}$ M.

A direct solution for pH would require iteration, but if values of pH are assumed and C_A is calculated, the graph of Fig. 4.21 is easily obtained. Note that C_A is a positive number only for values of $[H^+]$ between $10^{-4.75}$ and $10^{-7.0}$ (if $[OH^-] > [H^+]$, the numerator is negative). For pH less than the simple approximation (Eq. 68), the denominator is negative.

For buffers with pH > 7, the curve of Fig. 4.21 is reflected symmetrically in the region above pH = 7, so that as a buffer of pH = 10 is diluted, its pH will decrease to a limit of 7.0. If the initial pH is 7.0, then that will hold for all concentrations. Where does the simple approximation fail? On the acid side, at a concentration about 20 times K_a; on the basic side at a concentration about 20 times $K_b = K_w/K_a$. If there is any doubt, the results should be tested by substitution in the exact equations.

The Buffer Index

In Examples 13–15, we saw that adding a small amount of base to a buffer changed the pH much less than if the same amount was added to pure water. This can be generalized as a buffer capacity, or buffer index, which is the amount of strong base required to increase the pH by a given (small) amount.[27] The larger this quantity is, the better the buffer. It is most convenient mathematically to define the buffer index to be a derivative with the units of concentration (hence "buffer capacity")

$$\beta = \frac{\partial C_b}{\partial pH} \tag{74}$$

This definition is written as a partial derivative to indicate that there may be other independent variables than pH, for example, in Eq. (74), T, P, C, K_a, etc. Adding dC_b mol of base to a liter of buffer increases pH by dpH, and raises the concentration of the basic component at the expense of the acid component. Adding C_a moles of acid produces the opposite effect, decreasing the pH, and so β can also be defined as

$$\beta = -\frac{\partial C_a}{\partial \text{pH}} \tag{75}$$

In either case, β is a positive number.[28]

The general form of the buffer index can be easily derived by differentiating the charge balance. Consider a solution containing C_b moles NaOH, C_a moles HCl, and C moles total of a weak acid and its conjugate base.

$$[\text{H}^+] + [\text{Na}^+] = [A^-] + [\text{OH}^-] + [\text{Cl}^-] \tag{76}$$

Substitute the usual mass balances and equilibria to get

$$C_b = \frac{CK_a}{[\text{H}^+] + K_a} + \frac{K_w}{[\text{H}^+]} - [\text{H}^+] + C_a \tag{77}$$

The derivative of this expression with respect to $[\text{H}^+]$ at constant C_a is

$$\frac{\partial C_b}{\partial [\text{H}^+]} = -\frac{CK_a}{(K_a + [\text{H}^+])^2} - \frac{K_w}{[\text{H}^+]^2} - 1 \tag{78}$$

At this point you need the relationship between a derivative with respect to $[\text{H}^+]$ and a derivative with respect to pH. Since

$$\text{pH} = -\log_{10}[\text{H}^+] = -\frac{1}{2.303}\ln[\text{H}^+] \tag{79}$$

$$d\text{pH} = -\frac{1}{2.303}\frac{d[\text{H}^+]}{[\text{H}^+]}, \quad \text{and hence} \quad \frac{d[\text{H}^+]}{d\text{pH}} = -2.303[\text{H}^+] \tag{80}$$

$$\beta = \frac{\partial C_b}{\partial \text{pH}} = \frac{\partial C_b}{\partial [\text{H}^+]}\frac{d[\text{H}^+]}{d\text{pH}} = -2.303[\text{H}^+]\frac{dC_b}{d[\text{H}^+]} \tag{81}$$

$$\beta = 2.303\left(\frac{CK_a[\text{H}^+]}{([\text{H}^+] + K_a)^2} + \frac{K_w}{[\text{H}^+]} + [\text{H}^+]\right) \tag{82}$$

Thus, when the pH of the buffer is known, the buffer index can be calculated by direct substitution in the above formula. The first term arises from the acid–base

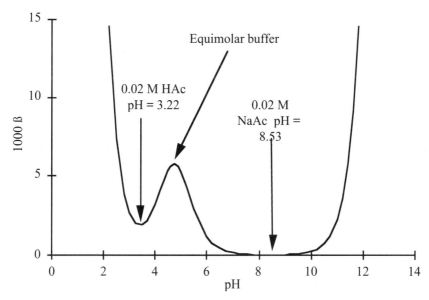

FIGURE 4.22. Buffer index of equimolar acetic acid–sodium acetate as a function of pH.

couple of the buffering agent, the second and third from the buffering effect of water. This equation can also be used to determine K_a from the measured slope of a titration curve.[29]

An identical equation can be derived for a buffer consisting of a weak base B and its conjugate acid BH^+ with acidity constant pK_a. Figure 4.22 is a plot of Eq. (82) for $C = 0.02$, $pK_a = 4.75$ and $pK_w = 14.00$. The maximum buffer index is obtained when $pH = pK_a = 4.75$. The minima occur at the pH of acetic acid alone (3.22) and sodium acetate alone (8.53).

To obtain solutions more acidic than $pH = 3.22$, it is necessary to add excess strong acid, and the large increase in buffer index below $pH = 2$ reflects the increasing concentration of strong acid required to achieve this acidity. To obtain solutions more basic than $pH = 8.53$, it is necessary to add a strong base, and the increase of buffer index above $pH = 12$ reflects that increasing concentration.

Note that a solution containing only acetic acid has an appreciable buffer capacity ($\beta = 2.7 \times 10^{-3}$) but that a solution containing only sodium acetate has a much smaller buffer capacity ($\beta = 0.015 \times 10^{-3}$). This means that while a crude measurement of the pH of acetic acid may give a nearly correct value, a similarly crude measurement of the pH of sodium acetate would be expected to be too low, unless carbon dioxide were rigorously excluded from the laboratory solutions.

Example 16. Using the buffer index, calculate the change in pH when 0.001 mole of HCl is added to one liter of equimolar acetic acid–sodium acetate buffer with total concentration $C = 0.02$ M. We found before that since the concentrations of conjugate acid and base are equal, the pH of this buffer is $pK_a = 4.75$. From Eq. (82),

$$\beta = 2.303\left(\frac{(0.02)(10^{-4.75})(10^{-4.75})}{(10^{-4.75} + 10^{-4.75})^2} + \frac{K_w}{10^{-4.75}} + 10^{-4.75}\right)$$

$$\beta = 2.303(5 \times 10^{-3} + 5.7 \times 10^{-10} + 1.75 \times 10^{-5}) = 1.15 \times 10^{-2}$$

From the definition of $\beta = -dC_a/d\text{pH}$, with finite difference $\Delta C_a = 1.0 \times 10^{-3}$,

$$\Delta\text{pH} = -\Delta C_a/\beta = (1.0 \times 10^{-3})/(1.15 \times 10^{-2}) = 0.086$$

Compare this with the increase $\Delta\text{pH} = 0.09$ calculated in Example 15 for the addition of 0.001 mole of NaOH to the same buffer.

An alternative way of expressing the buffer index equation is in terms of the concentrations of the various species. Equation (82) contains three terms. The first is almost the product of [HA] and [A⁻] except for a factor of C (see Eqs. 15 and 16):

$$\frac{[HA][A^-]}{C} = \frac{CK_a[H^+]}{([H^+] + K_a)^2} \tag{83}$$

The second and third terms are just [OH⁻] and [H⁺]. Hence Eq. (82) becomes

$$\beta = 2.303\left(\frac{[HA][A^-]}{C} + [OH^-] + [H^+]\right) \tag{84}$$

This can be envisioned as an upper envelope on the Sillén diagram of Fig. 4.5, redrawn as Fig. 4.23. Notice that since $[HA] \approx C$ at pH « pK_a and $[A^-] \approx C$ at pH » pK_a, the first term in the buffer index is $[A^-]$ at pH « pK_a and $[HA]$ at pH » pK_a. At pH = pK_a it is $C/2$, and β is approximately $0.575C$.

TITRATION OF A WEAK ACID WITH STRONG BASE (OR WEAK BASE WITH STRONG ACID)

The same equations used in the previous sections can be used to plot the change in pH as a weak acid is titrated with a strong base. The titration curve for a weak acid differs significantly from the titration curve of a strong acid (Chapter 3) but it is closely related to the dissociation curve. As with the strong acid case, the most significant point for determining titration errors is precisely where the simple approximations break down. The general form is derived as follows:

TITRATION OF A WEAK ACID WITH STRONG BASE

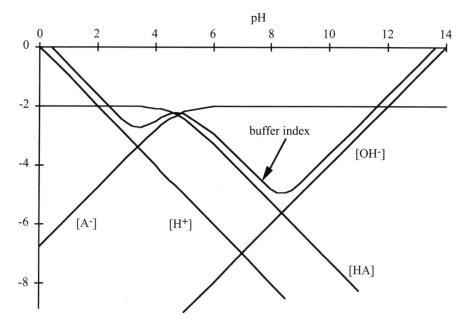

FIGURE 4.23. Sillén diagram for 0.01 M acetic acid with buffer index overlaid (top curve).

If V_o mL of weak acid HA of concentration C_o is titrated with V ml of strong base (say NaOH) of concentration C, the mass balances are:

$$[HA] + [A^-] = \frac{C_o V_o}{V + V_o} \tag{85}$$

$$[Na^+] = \frac{CV}{V + V_o} \tag{86}$$

Substituting these in the weak acid equilibrium Eq. (61) and the charge balance Eq. (63) gives

$$[H^+] + [Na^+] = [A^-] + [OH^-]$$

$$[A^-] = \frac{C_o V_o}{V + V_o} \frac{K_a}{[H^+] + K_a} \tag{87}$$

$$[H^+] + \frac{CV}{V + V_o} = \frac{C_o V_o}{V + V_o} \frac{K_a}{[H^+] + K_a} + \frac{K_w}{[H^+]} \tag{88}$$

As in Chapter 3, Eq. (29), the fraction titrated is defined to be

$$\phi = \frac{CV}{C_o V_o} \qquad (89)$$

and when this is substituted in Eq. (88) above, and rearranged, you get

$$\phi = \frac{K_a}{[H^+] + K_a} + \frac{V + V_o}{C_o V_o}\left(\frac{K_w}{[H^+]} - [H^+]\right) \qquad (90)$$

At low pH (see Fig. 4.24) the term in $[H^+]$ is the largest; at high pH, the term in K_w is largest; and in between the first term in K_a is largest. Except for values of C near 10^{-7}, at least one term is negligible at any pH.

Since V depends on ϕ, Eq. (90) is explicit only when $V \ll V_o$ (or $C \gg C_o$). However, Eq. (90) is linear in V and can be solved for ϕ in terms of the other parameters:

$$\phi = \frac{K_a}{[H^+] + K_a} + \frac{\dfrac{K_w}{[H^+]} - [H^+]}{C - \dfrac{K_w}{[H^+]} + [H^+]} \qquad (91)$$

where C is the concentration of the titrant. This equation was used to plot[30] Fig. 4.24, with $C = 0.02$ M and $pK_a = 4.75$.

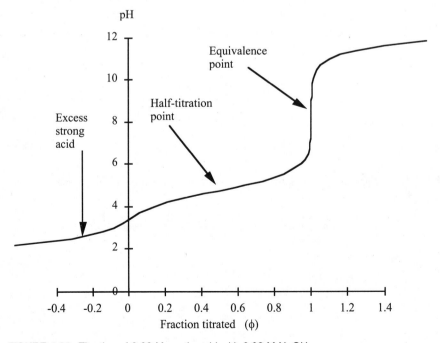

FIGURE 4.24. Titration of 0.02 M acetic acid with 0.02 M NaOH.

TITRATION OF A WEAK ACID WITH STRONG BASE

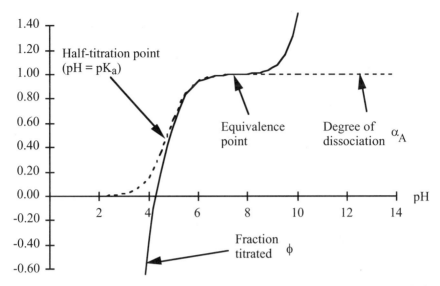

FIGURE 4.25. Comparison of the degree of dissociation α_A with the fraction titrated ϕ. In the intermediate region, $\phi = \alpha_A$, but α_A is always in the range $0 < \alpha_A < 1.0$. At low pH, ϕ becomes negative, and at high pH it exceeds 1.0. This curve of ϕ is the same as Fig. 4.24, rotated 90°.

The equation for ϕ is closely related to that for the degree of dissociation. Since $\alpha_A = K_a/([H^+] + K_a)$, Eq. (90) becomes:

$$\phi = \alpha_A + \frac{V + V_o}{C_o V_o}\left(\frac{K_w}{[H^+]} - [H^+]\right) \qquad (92)$$

and Eq. (91) becomes:

$$\phi = \alpha_A + \frac{\dfrac{K_w}{[H^+]} - [H^+]}{C - \dfrac{K_w}{[H^+]} + [H^+]} \qquad (93)$$

When the second term, dependent on both $[H^+]$ and $[OH^-]$, is small compared to α_A, the relation is simply $\phi = \alpha_A$, as shown in Fig. 4.25.

Titration Error

Titration error (see Chapter 3, p. 83) can be expressed most simply as $\phi - 1$ and so can be calculated directly from Eq. (93). One simplification that can be employed (the same as was derived for the strong acid–strong base titration) is that for $\phi \approx 1$,

$$\frac{V + V_o}{V_o} = \frac{C + C_o}{C} \qquad (94)$$

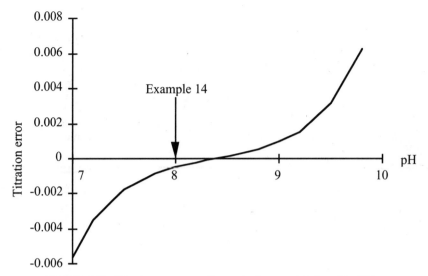

FIGURE 4.26. Titration error near the equivalence point, from Eq. (92).

and near the equivalence point pH » pK_a (See Fig. 4.24 or 4.25), or $[H^+]$ « K_a and hence

$$\alpha_A - 1 = \frac{K_a}{[H^+] + K_a} - 1 = -\frac{[H^+]}{[H^+] + K_a} \approx -\frac{[H^+]}{K_a} \quad (95)$$

When substituted in Eq. (92) to get ϕ, Eqs. (94) and (95) give

$$\phi_{ep} - 1 = \left(\frac{C + C_o}{CC_o}\right)\left(\frac{K_w}{[H^+]_{ep}} - [H^+]_{ep}\right) - \frac{[H^+]_{ep}}{K_a} \quad (96)$$

The first term is the same as that for the titration error of a strong acid–strong base titration. The second term results from the buffering action of the weak acid at the end point. This function is plotted in Fig. 4.26.

Example 17. Take $C = C_o = 0.1$, $pK_a = 4.75$, with $pH_{ep} = 8.0$ instead of the equivalence point 8.72. From Eq. (92),

$$\phi_{ep} - 1 = (2)(10^{-6} - 10^{-8}) - 10^{-8}/10^{-4.75} = 2 \times 10^{-6} - 10^{-3.25} = -5.5 \times 10^{-4}$$

Note that titration error (0.055%) is negative (the end point comes before the equivalence point; see Fig. 4.26) and the principal contribution comes from the second term, governed by the weak acid equilibrium.

The corresponding equation for the titration of a weak base with a strong acid is similar.

TITRATION OF A WEAK ACID WITH STRONG BASE

$$\phi_{ep} - 1 = \left(\frac{C + C_o}{CC_o}\right)\left([H^+]_{ep} - \frac{K_w}{[H^+]_{ep}}\right) - \frac{K_a}{[H^+]_{ep}} \qquad (97)$$

Sharpness Index

In deciding whether a titration is feasible or not, you will find the sharpness index is a useful concept. This is defined to be the magnitude of slope of the titration curve

$$\eta = \left|\frac{dpH}{d\phi}\right| \qquad (98)$$

The larger the slope of the titration curve, the smaller the titration error will be for a given error in measuring pH, and the more accurate the titration will be. The following discussion is particularly relevant to the use of an automatic titrator, which can produce an end point corresponding to the maximum slope of the titration curve.

What is a reasonable value for the sharpness index? For a titration using a classical 50 mL burette (or a microburette with a volume error of about 0.1%) and a pH indicator with an error of about 1 unit, $\eta = 10^3$. If the end point is estimated with a pH meter and a Gran-type extrapolation to within 0.01 pH units, $\eta = 10$. Thus for a titration with a sharpness index of $>10^3$ the titration accuracy will be limited by the volume measurement, but for a titration with a lower sharpness index, a more precise end point requires a more accurate measurement of pH. A sharpness index less than 10 indicates that the titration will probably not yield an acceptably accurate result.

The sharpness index can easily be derived from the buffer index. If the titrant concentration, corrected for dilution, is

$$C_b = \frac{CC_o}{C + C_o}\phi \qquad (99)$$

Since the buffer index is defined to be $\beta = dC_b/dpH$, differentiating Eq. (99) with respect to pH and using the definition of η from Eq. (98) gives

$$\eta = \frac{CC_o}{C + C_o}\frac{1}{\beta} \qquad (100)$$

Alternatively, an approximate expression for the sharpness index as a function of pH can be obtained by differentiating Eq. (96): Obtain $d(\phi - 1)/d[H^+]$ and multiply by $d[H^+]/dpH = -2.303\,[H^+]$ (Eq. 80) to get:

$$\eta = 0.434\left[\left(\frac{C + C_o}{CC_o}\right)\left(\frac{K_w}{[H^+]} + [H^+]\right) + \frac{[H^+]}{K_a}\right]^{-1} \qquad (101)$$

A second derivative, $d\eta/d\phi$, set equal to zero, gives the point of maximum slope:

$$\frac{d\eta}{d\phi} = \frac{d\eta}{d[H^+]}\frac{d[H^+]}{dpH}\frac{dpH}{d\phi} = 0.434 \frac{\left[\frac{C+C_o}{CC_o}\left([H^+] - \frac{K_w}{[H^+]}\right) - \frac{[H^+]}{K_a}\right]}{\left(\frac{C+C_o}{CC_o}\left([H^+] + \frac{K_w}{[H^+]}\right) + \frac{[H^+]}{K_a}\right)^3} \quad (102)$$

Since the denominator is always positive, $d\eta/d\phi = 0$ when

$$\left[\frac{C+C_o}{CC_o}\left([H^+] - \frac{K_w}{[H^+]}\right) - \frac{[H^+]}{K_a}\right] = 0 \quad (103)$$

Note that, within the approximation $[H^+] \ll K_a$, this is the equivalence point, where $\phi_{ep} = 1$ (Eq. 96).

In the curves of Fig. 4.27, the equivalence points [from Eq. 91 with C replaced by $CC_o/(C + C_o)$] are very close to the peak, at pH = 7.55 for C = 0.0002, 8.375 for C = 0.02, and 8.52 for C = 1.0. In general, the point of maximum slope for any combination of strong acid, strong base, weak acid, or weak base titration is nearly

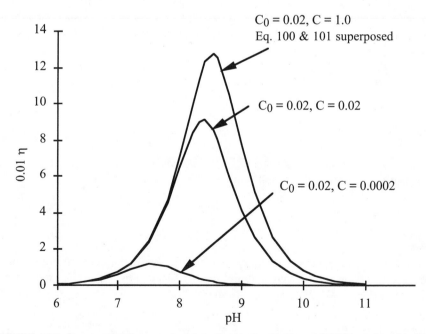

FIGURE 4.27. Sharpness index as a function of pH in the titration of 0.02 M acetic acid with 1.0 M, 0.02 M, and 0.0002 M NaOH. As C_o decreases, the peak diminishes and shifts toward pH = 7.

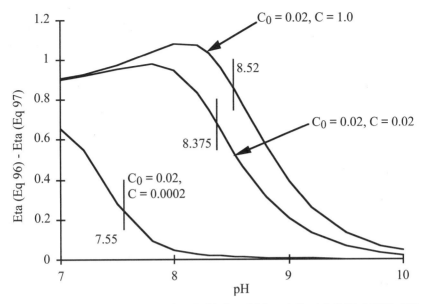

FIGURE 4.28. Sharpness index calculated with $C_o = 0.02$, and $C = 1.0, 0.02, 0.0002$. Difference between Eq. (100) (exact) and Eq. (101) (approximate). Equivalence points are indicated by vertical lines.

the same as the equivalence point. However, this derivation (Eq. 103) depends on the assumption $[H^+] \ll K_a$, and is not perfect (see Fig. 4.28).

In addition, the point of maximum slope is quite different from the equivalence point for unsymmetrical precipitation titrations (Chapter 6) or many redox titrations (Chapter 9).

Recall that β is proportional to the concentration C_o of the weak acid in the mid-pH region, at relatively high concentrations. As C_o decreases, β does not decrease proportionally because of the terms in $[H^+]$ and $[OH^-]$ (Eq. 84), and hence η decreases with decreasing C in Fig. 4.27. Note that the peak (and the equivalence point) shifts toward 7.0 as C decreases.

One method of end-point determination uses an automatic titrator, which adds a small increment of titrant and measures pH at each step, creating a data set of closely spaced points on a curve like Fig. 4.24. From these data, a curve of the slope, or sharpness index, can be obtained, and the end point determined as the point of maximum slope, as in Fig. 4.27.

The approximation used to derive Eq. (101) produces a different result from the exact Eq. (100) wherever $[H^+]$ is not small compared to K_a. On Fig. 4.27, the difference between the exact Eq. (100) and the approximate Eq. (101) (for which $[H^+] \ll K_a$ has been assumed; see Eq. (95)) is too small to be visible. However, the approximate Eq. (101) gives a smaller value of η than Eq. (100), by about 0.1 to 1%, in the region near the equivalence points at pH = 7.5 to 8.5. This is shown on a magnified scale in Fig. 4.28. The maximum difference occurs at a pH somewhat below the equivalence point.

Simple Approximations for η′ (η at the Equivalence Point)

To obtain a simple expression for η′ at the equivalence point, set $\phi = 1$ in Eq. (96):

$$\frac{[H^+]'}{K_a} = \frac{C + C_o}{CC_o}\left(\frac{K_w}{[H^+]'} + [H^+]'\right) \tag{104}$$

Since pH > 7, neglect $[H^+]'$ in the bracket on the right side of Eq. (104), solve for $[H^+]'$:

$$[H^+]' = \sqrt{\frac{C + C_o}{CC_o} K_w K_a}, \quad \text{weak acid–strong base, pH > 8} \tag{105}$$

Simplify Eq. (101) for the sharpness index by substituting $[H^+]'/K_a$ from Eq. (104)

$$\eta' = 0.434\left[\left(\frac{C + C_o}{CC_o}\right)\left(\frac{K_w}{[H^+]'} + [H^+]'\right) + \frac{[H^+]'}{K_a}\right]^{-1} \tag{106}$$

Using Eq. (104), with $[H^+]' \ll K_w/[H^+]'$, transform this to two useful expressions:

$$\eta' = 0.217\frac{K_a}{[H^+]'} = 0.217\left(\frac{CC_o}{C + C_o}\right)\frac{[H^+]'}{K_w}, \quad \text{weak acid–strong base} \tag{107}$$

Note that the last form also applies to the titration of a strong acid with a strong base, for which $[H^+]' = 10^{-7.0}$ and $K_w = 10^{-14.0}$:

$$\eta' = (2.17 \times 10^6)\left(\frac{CC_o}{C + C_o}\right), \quad \text{strong acid–strong base} \tag{108}$$

Because atmospheric carbon dioxide can shift the pH and decrease sharpness index in such a titration, a strong acid–strong base titration using indicators is usually less accurate than 0.1% if the concentrations are less than 10^{-3} M. By using a potentiometric end-point detection method and rigorously excluding carbon dioxide from the solutions,[31] concentrations as small as 10^{-5} M can be titrated with 1% accuracy. At such low concentrations, however, a conductimetric end point is more satisfactory.

Example 18. In Example 17, we calculated a titration error of −0.055% for the titration of 0.10 M HAc with 0.10 M NaOH, when the end-point pH was taken at 8.0 instead of at the equivalence point of 8.72 (calculated from Eq. 105). With these numbers, an approximate sharpness index can be calculated.

TITRATION OF A WEAK ACID WITH STRONG BASE

$$\eta = \frac{\Delta pH}{\Delta \phi} = \frac{-0.72}{-5.5 \times 10^{-4}} = 1.3 \times 10^{+3}$$

For comparison, when $C = C_o = 0.10$, $pK_a = 4.75$, and $pK_w = 14.00$ Eq. 106 or 107 yields $\eta' = 2.05 \times 10^{+3}$. This may be compared with $\eta = 900$–1300 from Fig. 4.27, for C between 0.02 and 1 and pH between 8.0 and 8.7.

For the titration of a strong or weak base with a strong acid, the analogous results are

$$\eta' = 0.217 \left(\frac{CC_o}{C + C_o} \right) \frac{1}{[H^+]}, \quad \text{strong or weak base–strong acid pH near 7} \quad (109)$$

$$\eta' = 0.217 \sqrt{\frac{CC_o}{C + C_o} \frac{1}{K_a}}, \quad \text{weak base–strong acid, pH < 6} \quad (110)$$

The relationship between the sharpness index and buffer index, as well as the various concentrations, can be seen from Sillén diagram of Fig. 4.29. As was shown

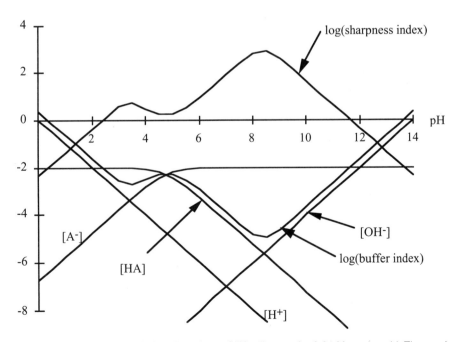

FIGURE 4.29. Sharpness index plotted on a Sillén diagram for 0.01 M acetic acid. The maximum sharpness index at 8.8×10^2 coincides with the minimum in the buffer index curve (Figs. 4.22 and 4.23).[35]

in Fig. 4.23, the buffer index closely follows the lines representing [H$^+$], [HA], [A^-], and [OH$^-$]. The sharpness index, proportional to 1/β, rises as a mirror image about the horizontal line at log $C = -1$ to a maximum value at the pH of the equivalence point.

Since the sharpness index can be expressed in terms of [H$^+$] at the equivalence point, it can also be estimated from a logarithmic concentration diagram (Fig. 4.30). If $C = C_o$, the coefficient $0.217(CC_o/(C + C_o))$ is approximately 0.1.[32] A simple rule of thumb can thus be devised:

$$\log \eta' = \log C_o + pOH' - 1, \quad \text{weak acid–strong base} \quad (111)$$
$$\log \eta' = \log C_o + pH' - 1, \quad \text{weak base–strong acid} \quad (112)$$

As a general rule, if a given intersection on a logarithmic concentration diagram corresponds to the equivalence point of a titration, then one logarithmic unit less than the distance from this point to the horizontal baseline at log C_o gives the logarithm of the sharpness index at that equivalence point.

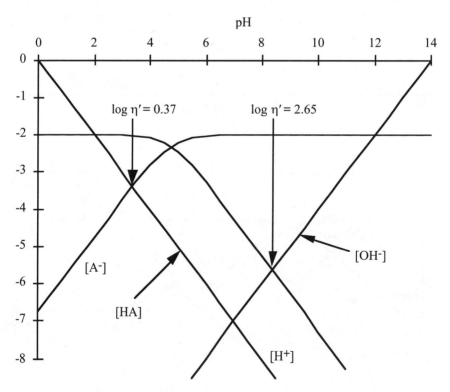

FIGURE 4.30. Logarithmic concentration diagram for 1.0×10^{-2} molar acetic acid showing how the sharpness index is estimated from the distance between baseline and equivalence point intersection.

PROBLEMS

Thus, if the intersection is four units or more below the baseline ($\eta' > 10^3$), a practical titration with an error of less than 0.1% is possible. If the intersection is less than two units below the baseline ($\eta' < 10$), even the most accurate methods of end-point determination will give a titration error greater than 1%.

Example 19. Find the sharpness index for the titration of 0.01 M HAc with 0.01 M NaOH from a diagram like Fig. 4.30. The equivalence point for titration with NaOH is where the [OH$^-$] and [HA] lines cross, at pH = 8.35 or [OH$^-$] = $10^{-5.65}$. Substituting $C_o = 10^{-2}$,

$$\log \eta' = \log C_o + \text{pOH}' - 1 = -2 + 5.65 - 1 = 2.65$$

or $\eta' = 450$.[33] For such an end point, 0.1% accuracy can be obtained if pH can be measured to within 0.5 units.

Contrast this with the titration of 0.01 M sodium acetate with HCl. The equivalence point occurs at the intersection of the [H$^+$] and [Ac$^-$] lines, at pH = 3.37

$$\log \eta' = \log C_o + \text{pH}' - 1 = -2 + 3.37 - 1 = 0.37$$

or $\eta' = 2.3$. This is a very poor end point. Even if pH can be measured to within 0.01 units, the titration error will be 5%.

PROBLEMS

1. Find several conjugate acid–base pairs in Tables 4.1 and 4.2, such as ammonium ion–ammonia and anilinium ion–aniline. Verify that $pK_a + pK_b = pK_w$.

2. Derive an equation analogous to Eq. (20), but giving the total concentration of a weak base C_B as a function of [H$^+$] and K_a.

3. Draw the symmetrical set of Flood diagram curves (like Fig. 4.3, but for the pH range 7 to 14) for NaOH and the weak bases aniline, imidazole, ammonia, and dimethylamine (pK_a is listed in Table 4.1, and pK_b in Table 4.2).

   Find the concentrations of all species and the pH of the following solutions. Check to see whether any approximations that you make are correct by substituting in the exact equations. Use logarithmic concentration diagrams to obtain approximate answers and check these by substitution in the exact equations.

4. 5.3×10^{-3} M acetic acid, $pK_a = 4.75$.

5. 0.40 M acetic acid.

6. 0.10 M HCN, $pK_a = 9.32$.

7. 1.0×10^{-6} M HF, $pK_a = 3.17$.

8. 1.0×10^{-6} M acetic acid.

9. 1.0×10^{-6} M HOCl, $pK_a = 7.53$.

10. The ionization of HF, like other weak acids, is dependent on the ionic strength of the solution. For example, the ionization constant of HF at 25°C in 0.5 M NaClO$_4$ is $pK_a = 2.91$, compared to 3.17 at $I = 0$. Calculate the pH of 1.0×10^{-3} M HF in 0.5 M NaClO$_4$, and compare with the results at $I = 0$ (see Example 3 and Problem 7). Is this shift in pK_a consistent with the Davies equation (see Chapter 2. p. 49)?

11. Although hydrochloric acid is completely dissociated to H$^+$ and Cl$^-$ in water solution, it is almost completely associated as molecular HCl in pure (glacial) acetic acid because of the low dielectric constant and high acidity of that solvent. For the reaction

$$HCl + HAc \rightleftharpoons H_2Ac^+ + Cl^-$$

the equilibrium constant is 2.8×10^{-9}. The self-dissociation of pure acetic acid is

$$HAc + HAc \rightleftharpoons H_2Ac^+ + Ac^-$$

with ion product 3.6×10^{-15} (assuming activity of the solvent is 1.0, as with the ion product of water). Calculate the concentrations of HCl, H$_2$Ac$^+$, and Cl$^-$ in 1.0×10^{-4} molar solution of HCl in glacial acetic acid.

12. At 15°C, the ionization constant of HF is $pK_a = 3.10$ and the ion product of water is $pK_w = 14.35$. Calculate the concentrations of all species and the pH of a 2.0×10^{-4} M solution of HF in water at 15°C.

13. Auerbach and Zeglin (*Z. physik. Chem.* **103**: 191, 1923) measured the degree of dissociation $\alpha_A = [A^-]/C$ of formic acid HCOOH in water solution conductimetrically. At a concentration of 1.028×10^{-3} M the degree of dissociation was calculated to be 0.348. Calculate pK_a for formic acid from these data and construct a logarithmic concentration diagram.

14. Consider a monoprotic acid HA with equilibrium constant K_a of analytical concentration C in water solution. Combine the relation for pH with the definition of degree of dissociation $\alpha_A = [A^-]/C$ to get an expression for α in terms of C and K_a alone. Find an approximate equation that applies in the range pH < 6 at 25°C (i.e., $[H^+] > 10^2[OH^-]$).

15. Sulfuric acid is completely dissociated to H$^+$ and HSO$_4^-$ in solutions more dilute than 1 M. The HSO$_4^-$ ion further dissociates to H$^+$ and SO$_4^{2-}$ with an equilibrium constant $pK_a = 1.99$. Because of the excess H$^+$ from the first dissociation, the equations derived above for a neutral HA such as acetic acid do not apply. Derive the appropriate equations and plot pH versus $-\log C$ over the range from $C = 1$ to 10^{-9} M, analogous to Flood's diagram (Fig. 4.3).

Calculate the pH of the following solutions, using logarithmic concentration diagrams if appropriate. Check your answers by substitution in the original equations.

16. 4.0×10^{-2} M trimethylamine, $pK_a = 9.91$.
17. 1.0×10^{-3} M pyridine, $pK_b = 8.82$.
18. 1.0×10^{-3} M dimethylamine, $pK_b = 2.93$.
19. 1.0×10^{-6} M ammonia, $pK_a = 9.25$.
20. Solve the four simultaneous equations governing the ionization of a weak base, to obtain a cubic equation in $[H^+]$ analogous to Eq. (19), p. 100, and plot the analogue of Flood's diagram, showing how pH varies with log C for the weak base, analogous to Fig. 4.3. Use $pK_b = 3.33$ (dimethylamine).
21. Derive approximate formulas for the pH of a weak base at high concentration ($-\log C < pK_b$), intermediate concentration ($pK_b < -\log C < 7$), and very low concentration ($-\log C \gg 7$). Relate these to straight-line segments in the diagram of Problem 17.

    Find the pH of the following solutions, using logarithmic concentration diagrams if desired. Be sure to check your answers by substitution in the exact equations.
22. 0.25 M NaCN (for HCN, $pK_a = 9.32$).
23. 5.0×10^{-2} M NaF (for HF, $pK_a = 3.17$).
24. 1.0×10^{-3} M sodium acetate (for HAc, $pK_a = 4.75$).
25. 4.0×10^{-6} M NH_4Cl (for NH_3, $pK_b = 4.75$).
26. 0.10 M $NaHSO_4$ (for HSO_4^-, $pK_a = 1.99$).
27. 0.010 mole HF and 1×10^{-4} mole HCl in one liter of water (for HF, $pK_a = 3.17$).
28. 1.0×10^{-4} mole HAc and 1.0×10^{-4} mole HCl in one liter of water (for HAc, $pK_a = 4.75$).
29. 2.5×10^{-7} mole NH_3 and 2.5×10^{-7} mole NaOH in one liter of water (for NH_3, $pK_a = 9.25$).
30. 0.10 M NH_4F (for NH_3, $pK_b = 4.75$; for HF, $pK_a = 3.17$).
31. 1.0×10^{-5} M NH_4F.
32. 0.10 mole NH_4Ac and 0.010 mole HCl per liter.
33. 0.10 mole NH_4Ac and 0.010 mole NaOH per liter.
34. 0.10 M $(NH_4)_2SO_4$ (for NH_3, $pK_b = 4.75$; for HSO_4^-, $pK_a = 1.99$).
35. 0.010 mole chloroacetic acid ($pK_a = 2.87$) and 0.10 mole dimethylamine ($pK_a = 11.07$) per liter.
36. Avery et al.[34] analyzed rainwater from North Carolina and Virginia for formic and acetic acids. The lowest average concentrations were 0.32 µM formic and 0.20 µM acetic acids; the highest were from samples stored outside, where it was suspected that photolysis of the plastic bottles contributed to the organic acid content of the water: 550 µM formic and 120 µM acetic acid.

(a) For each case, estimate the pH of the two acids plus water alone.

(b) An important factor in almost all natural waters is CO_2 (Chapter 10). In these rainwater samples, $[CO_2] = 10$ µM (equilibrium with atmospheric CO_2) and $K_a = [H^+][HCO_3^-]/[CO_2] = 10^{-6.35}$. If there is negligible alkalinity in the rain, what effect does CO_2 have on the pH of the two rainwater samples?

(c) If the measured pH of the first sample was 5.6, estimate the concentration of strong acid (i.e., H_2SO_4 or HNO_3) present in the rain.

(d) If the measured pH of the second sample was 3.5, estimate the concentration of strong acid present.

37. Derive the approximate relation for the pH of a solution of NaA: $(C[H^+] - K_w)[H^+] = K_a K_w$. Over what range of K_a and C will this apply?

38. The theorem of "isohydric solutions" originally proposed by Arrhenius (*Z. physik. Chem.* **2**: 284, 1888) is stated as follows: "If two solutions of the same pH are mixed, the pH is unchanged regardless of the composition of the solutions." Prove that this theorem applies for a mixture of a weak and strong acid; also for a mixture of a weak acid and a weak base.

39. Plot a curve of pH versus $-\log C$ for NH_4CN (for NH_3, $pK_a = 9.25$; for HCN, $pK_a = 9.32$), analogous to Fig. 4.12. At what minimum concentration does neglect of $[H^+]$ and $[OH^-]$ (i.e., use of Eq. 53) introduce an error of more than 0.1 pH units?

40. Griffith (*Trans. Faraday Soc.* **17**: 525, 1922) derived the simple formula (Eq. 53) for the salt of a weak acid and a weak base. He stated that "This formula requires slight modification in extremely dilute solution because of the failure of the assumption that the weak acid and weak base are practically completely ionized..." Express his statement in mathematical terms and criticize it.

41. Clark (*Determination of Hydrogen Ions*, Williams and Wilkins, Baltimore, 1922) derived the relation for the pH of the salt of a weak acid and weak base:

$$[H^+] = \sqrt{\frac{K_w K_a(K_b + [BH^+])}{K_b(K_a + [A^-])}}$$

He stated that this leads to the simple formula $[H^+]^2 = K_w K_a/K_b$ (Eq. 53) "if K_b and K_a are small in relation to $[BH^+]$ and $[A^-]$ and if the solution is sufficiently dilute that $[BH^+]$ and $[A^-]$ each approximate the salt concentration C..." Derive Clark's formula and note under what conditions it reduces to the simple formula. Criticize Clark's qualifying statement.

42. Compare the pH of pure 1.0×10^{-3} M HCl with that of the same solution made 10^{-5} M in the sodium salt of methyl orange ($pK_a = 3.8$). What is the ratio of the orange form In^- to the red form HIn in this solution?

PROBLEMS

43. Brom thymol blue is similar in structure to phenolphthalein. It has a yellow form H_2In and a blue form In^{-2}, which are equal in concentration at pH = 6.8. The intermediate species HIn^- is negligible. If each of the following three solutions is made 1.0×10^{-5} M in the indicator Na_2In, calculate the final pH of the solution and the ratio of the yellow to the blue form.
    (a) 10^{-3} M NaOH;
    (b) 10^{-5} M HCl;
    (c) pure water.

44. In the titration of 10^{-3} M HCl with 10^{-3} M NaOH, methyl orange is used as an indicator, the end point being taken when the concentration of the red and orange forms are equal. This end point (pH = 3.8) is at a considerably lower pH than the theoretical equivalence point (pH = 7.0), but not far from the pH (≈ 4.5) of the solution saturated with atmospheric CO_2 (see Chapter 5). If 10^{-5} M of the indicator is introduced as the basic form, what is the titration error relative to pH = 4.5?

    Calculate the pH of the following solutions and their buffer index. Check any approximations by substitution in the exact equations.

45. 1.0×10^{-2} mole HAc and 2.0×10^{-2} mole NaAc per liter ($pK_a = 4.75$).

46. 1.0×10^{-2} mole pyridine and 5.0×10^{-3} mole HCl per liter ($pK_a = 5.18$).

47. 1.0×10^{-5} mole NH_3 and 2.0×10^{-5} mole NH_4Cl per liter ($pK_a = 9.25$).

48. The imidazole group is an important constituent of many enzymes that catalyze hydrolytic reactions. Imidazole itself is a weak base

    with $pK_a = 6.91$. Find the ratio of basic to acidic form that will produce a pH of 7.00. How will this pH be affected by diluting the mixture?

49. A solution containing 2.20 g of KIO_3 and 1.80 g of HIO_3 in 100 mL of water was measured potentiometrically to have pH = 1.32. Calculate the concentration ionization constant K_a for HIO_3. Calculate activity coefficients using the Davies equation (Chapter 2) and obtain the activity constant K_a°.

50. If both $[H^+]$ and $[OH^-]$ are negligible compared to the analytical concentrations of weak acid C_{HA} and its salt C_A, the buffer index of the mixture is given by

$$\beta = 2.303 \frac{C_A C_{HA}}{C_A + C_{HA}}$$

51. Derive an equation (like Eq. 82) that gives the buffer index of a solution containing a weak base B and its conjugate acid BH^+, where K_a is the dissociation constant of BH^+.

52. Derive an expression giving the buffer index of a solution containing only strong acid or strong base. Show that the minimum buffer index occurs at the pH of pure water. What is the value of that minimum buffer index?

53. Show by differentiation (of Eq. 82) that for practical buffer solutions the buffer index is a maximum when the concentrations of conjugate acid and base are equal. Show that the buffer index is a minimum in solutions containing only the acid or its conjugate base. Under what restrictions do these theorems apply (i.e., $[H^+]$ and/or $[OH^-] \ll C$)?

54. J. W. Drenan (*J. Chem. Ed.* **32**: 36, 1955) proposed the following equation for calculating pH in buffer solutions.

$$[H^+] = K_a \frac{C_{HA} - \frac{1}{2}\sqrt{C_{HA}K_a} + \sqrt{C_A K_b}}{C_A - \frac{1}{2}\sqrt{C_A K_b} + \sqrt{C_{HA}K_a}}$$

where C_{HA} is the analytical concentration of a weak acid with ionization constant K_a, and C_A is the analytical concentration of its conjugate base. $K_b = K_w/K_a$. What approximations must be made to derive this formula? Plot it for the same series of buffers as Fig. 4.2 and compare with Eqs. (70)–(73). Does Drenan's formula give the correct result over a wider range than Henderson's equation (Eq. 68 or 69)? Where does it break down?

55. Using Drenan's formula, plot the pH of an acetic acid buffer as $C_{HA}/(C_{HA} + C_A)$ is varied between 0 and 1, holding $C_{HA} + C_A$ constant at 0.1 molar. Compare with the results of the rigorous Eq. (73) and the approximate Eq. (68).

56. Plot a titration curve pH versus ϕ for the titration of 50 mL of 0.10 M NH_3 with 0.10 M HCl. The pK_a for NH_4^+ is 9.25. Compare with the formation curve $\alpha = [NH_4^+]$ is 9.25. Compare with the formation curve $\alpha = [NH_4^+]/C$ versus pH.

57. Which of the following would be the best indicator for the titration of 0.0001 M acetic acid with 0.1 M NaOH?

Indicator	pH range	Color change
Phenolphthalein	8.4–10.0	colorless–red
Thymol blue	8.0–9.0	yellow–blue
Tetrabromophenolphthalein	8.0–9.0	colorless–violet
meta-Cresol purple	7.4–9.0	yellow–purple
α-Naphtholphthalein	7.3–8.7	rose–green
Cresol red	7.3–8.8	yellow–red

Calculate the pH at the equivalence point, and the sharpness index of the titration. Consider both the range in which the color changes and the ease of detecting the color change. Estimate the titration error range for the indicator you choose.

58. An ignorant student thinks that formic acid (pK_a = 3.75) is a strong acid. He has a sample that is approximately 0.01 M in formic acid. He fills his 50-mL burette with 0.1 M NaOH from the side shelf; fortunately the acid had been standardized that very morning and was marked 0.1045 M. He takes a 100 mL aliquot of his unknown formic acid and titrates until his pH meter reads 7.00. Since he has taken great care to standardize the meter with a standard buffer of pH = 7.00, he has probably read the end point to within < ±0.1 pH units. If he can read the burette to < ±0.05 mL, what will be the total percentage error in his value for the concentration of formic acid in the sample? How could he improve the titration procedure? What is the minimum error possible with the equipment and standards available?

59. Calculate the sharpness index in the titration of 0.10 M $NaHSO_4$ with 0.10 M NaOH (pK_a for HSO_4^- is 1.99). Within what range must pH be determined if the titration error is to be less than 0.1%?

60. Calculate the sharpness index in the titration of 0.10 M HOCl with 0.10 M NaOH (pK_a for HOCl is 7.53). Within what range must pH be determined if the titration error is to be less than 1%? Is this feasible?

61. Derive the general equation for the titration curve of a weak base with a strong acid, analogous to Eqs. (90) or (91). Show that the dissociation and titration curves are approximately the same in the region before the equivalence point. From this, derive a relation for the titration error. Compare with Eq. (96). If the pH at the end point is taken to be 5.0 ± 0.1 for the titration of 0.10 M NH_3 with 0.10 M HCl, what is the titration error?

62. Derive a relation for the sharpness index in the titration of a weak base with a strong acid. analogous to Eq. (101). Make the necessary approximations to derive Eqs. (107) and (108).

63. Show that if K_a for a weak acid is smaller than 10^{-7}, the titration of the salt of the weak acid with a strong acid is more accurate than the titration of the weak acid itself with a strong base.

64. A mixture containing 0.10 mole HCl and 0.10 mole acetic acid per liter is titrated with 0.10 M NaOH. Calculate a titration curve for this titration and note that the two equivalence points occur at two separate steps in the curve. Derive an equation giving the sharpness index as a function of pH. Be cautious about neglecting [H^+] compared to K_a. Comment on the accuracy possible in titrating a mixture of weak and strong acid.

65. For what value of K_a is the minimum titration error obtained in a mixture of

strong and weak acids if they are of the same concentration? What is this minimum titration error for the mixture of Problem 64?

66. Plot a curve of pH as a function of fraction titrated, for the titration of 0.1 M NH_3 (pK_a = 9.25) with 0.1 M chloroacetic acid (pK_a = 2.87). What titration error results if the end point is within 0.2 pH units of the equivalence point?

67. Find the titration error if 50 mL of 0.01 M chloroacetic acid is (pK_a = 2.87) titrated with 0.01 M dimethylamine (pK_a = 11.07), and the pH at the equivalence point is determined to within one unit.

68. If 50 mL of 0.2 M NH_3 is titrated with 0.2 M acetic acid and the pH at the end point is taken to be 7.25, calculate the titration error.

69. Plot a curve of pH as a function of the fraction titrated, for the titration of 0.1 M H_2SO_4 with 0.1 M NH_3. What is the approximate sharpness index at each equivalence point? What titration error results at each equivalence point if the pH is measured to within 0.1 unit?

NOTES

1. Sillén and Martell (1964 and 1971) give a note on the method for each constant in their tables, and a brief description of methods in their introduction. See also A. E. Martell and R. J. Motekaitis, *The Determination and Use of Stability Constants*, VCH, New York, 1988; M. Meloun, J. Havel, and E. Högfeldt, *Computation of Solution Equilibria: A Guide to Methods in Protentiometry, Extraction and Spectrophotometry*, Ellis Horwood, Chichester, Wiley, New York, 1988; F. J. C. Rossotti and H. Rossotti, *Determination of Stability Constants*, McGraw Hill, New York, 1961.

2. But not all dissociations produce additional ions. For example, ammonium ion NH_4^+ dissociates to produce ammonia NH_3 and a hydrated proton H^+. Although NH_4^+ breaks into two parts, no additional ions are produced. Contrariwise, not all ionizations lead to dissociation. In solvents of low dielectric constant, ionization may yield ion pairs, which may attain significant concentration, and which then dissociate in a separate step. In these circumstances the distinction between ionization and dissociation is quite important. But in solvents in high dielectric constant, such as water, the distinction between dissociation and ionization is not very important.

3. K_b is sometimes called the "basicity constant."

4. Selected from W. Smith and A. E. Martell, *Critical Stability Constants* (6 vol.), Plenum, NY, 1976–1989), and from L.G. Sillén and A. E. Martell, *Stability Constants*, The Chemical Society, Special Publications No. 17 (1964) and 25 (1971). Recall that pK_a = $-\log K_a$.

5. An equally good fit can be obtained with b' = 0.2 and b = –0.1, but I could find no literature to support this assumption.

6. C. W. Davies, *Ion Association*, Butterworths, 1962, suggested that 0.3 was as good a value as 0.2 for the linear parameter in his equation.

7. The mass and charge balances are slightly different for positively charged acids (salts of weak bases). If the acid BH^+ is introduced as the chloride to preserve electroneutrality,

an additional mass balance (on Cl) is required: C_{BH^+} = [Cl$^-$], mass balance; C_{BH^+} = [B] + [BH^+], mass balance; [H$^+$] + [BH^+] = [Cl$^-$] + [OH$^-$], charge balance.

8. Solving Eq. (19) (with $C_A = 0$) gives [H$^+$] = $10^{-3.384}$, [A^-] = $10^{-3.384}$, [HA] = $10^{-2.018}$, and [OH$^-$] = $10^{-10.616}$. Note the charge balance is exactly satisfied and C_{HA} = [A^-] + [HA] = $10^{-2.000}$.

9. H. Flood, *Z. Elektrochem.* **46**:69–675, 1940.

10. L. G. Sillén, P. W. Lange, and C. O. Gabrielson, *Problems in Physical Chemistry* (translated from the Swedish), Prentice-Hall, New York, 1952; L. G. Sillén, "Graphical presentation of equilibrium data," Chapter 8 of *Treatise on Analytical Chemistry*, edited by I. M. Kolthoff, P. J. Elving, and E. B. Sandell, The Interscience Encyclopedia, New York, 1959.

11. Why not 0 to 14? An 8½ × 11 inch sheet of millimeter-scale graph paper will conveniently provide 12.5 2-cm units on the horizontal scale and 9 2-cm units on the vertical scale.

12. The alternative form [BH^+][OH$^-$] = K_b[B] is not necessary but has been included in some discussions because of its historical importance.

13. F. M. M. Morel and J. Hering, *Principles and Applications of Aquatic Chemistry*, Wiley, N.Y., 1993, Chapter 1.

14. Note that SO_4^{2-} forms HSO_4^- at pH < 3, but at higher pH behaves as an unreactive anion.

15. Another way of stating the approximations is pK_{a1} « pH and pK_{a2} » pH (see diagram of Fig. 4.12), which also leads to [HA] ≅ [B] and hence to [H$^+$]^2 ≅ $K_{a1}K_{a2}$, independent of C.

16. R. H. Byrne, *Anal. Chem.* **59**:1479–81, 1987; R. H. Byrne, G. Robert-Baldo, S. W. Thompson, and C. T. A. Chen, *Deep-Sea Research* **35**:1405–10, 1988; R. H. Byrne and J. A. Breland, *Deep-Sea Research* **36**:803–10, 1989. Other sulfonphthalein indicators used include bromthymol blue, phenol red, and cresol red.

17. R. H. Byrne, *Anal. Chem.* **59**:1479–81, 1987.

18. R. H. Byrne, G. Robert-Baldo, S. W. Thompson, and C. T. A. Chen, *Deep-Sea Research* **35**:1405–10, 1988; R. H. Byrne and J. A. Breland, *Deep-Sea Research* **36**:803–10, 1989.

19. The "free H" scale is the "concentration pK_a," which we have been using all along: H$^+$ concentration in the expression for K_a. This is in contrast to using 10^{-pH} or total H$^+$ in seawater including HSO_4^-, HF, etc. See Chapter 10.

20. R. W. Ramette, C. H. Culberson, and R. G. Bates, *Anal. Chem.* **49**:867–70, 1977.

21. R. G. Bates, *Determination of pH, Theory and Practice*, Wiley, New York, 1973. 479 pp.

22. R. D. Lanier, *J. Phys. Chem.* **69**:2697, 1965.

23. M. Solache-Rios and G. R. Choppin, Determintion of the pcH in highly saline waters usign cresol red. Report SAND91-7068J, Sandia National Laboratories, 1991.

24. A. Dickson, Scripps Institute of Oceanography, private communication, 1997.

25. L. J. Henderson, *J. Am. Chem. Soc.* **30**:954, 1908; *Ergebnisse der physiologie* **8**:301, 1909; K. A. Hasselbalch, *Biochem. Z.* **78**, 116, 1917; J. F. McClendon, *J. Biol. Chem.* **54**: 647, 1922.

26. But I have been unable to find a primary reference—this is left as an "exercise for the reader."

27. The buffer index was first introduced by Van Slyke, *J. Biol. Chem.* **52**:525, 1922. The

differential quantity defined here is the inverse of the slope of the titration curve, which is derived in a later section. β is the usual symbol, and you should be careful not to confuse it with the overall formation constant of a complex, as defined in Chapter 7.

28. J. E. Ricci, *Hydrogen Ion Concentration*, Princeton Univ. Press, 1952, defines $\beta = |d(C_a - C_b)/d\text{pH}|$, but the absolute value symbol is not necessary, since increasing C_b always increases pH and increasing C_a always decreases pH.
29. E. Grunwald, *J. Am. Chem. Soc.* **73**:4934–38, 1951.
30. Figure 4.24 can also be obtained from Eq. (90) by setting $V = \phi\, C_o V_o/C$, where ϕ is obtained from Eq. (90) by a rapidly converging iteration.
31. Or finding the end point by a method (such as the Gran titration, Chapters 3 and 10), which can compensate in part for the CO_2 impurity.
32. If $C \gg C_o$, the coefficient $0.217(CC_o/(C + C_o)$ is 0.2 and the factor in Eqs. (105) and (106) is 0.7, not 1.0.
33. Compare with the results of Eq. (101): $\eta' = (0.217)(0.02)(0.02)(10^{-8.35+14.00})/(0.04) = 970$.
34. G. B. Avery, J. D. Willey, and C. A. Wilson, *Environ. Sci. Technol.* **25**:1875–80, 1991; J. D. Willey and G. B. Avery, *Environ. Sci. Technol.* **26**:1666–67, 1992.
35. The minimum buffer index and maximum sharpness index coincide with the inflection point of the titration curve (by definition) and agree closely with the equivalence point (Fig. 4.28) even in solutions too dilute to have good titration end points.

5

POLYPROTIC ACIDS AND BASES

Introduction
Examples of Polyprotic Acids
   Amines
   Amino Acids
   Systems Involving a Gas Phase
Ionic Strength and Temperature Dependence
Distribution Diagrams
   pH of H_3PO_4 and Its Salts
   NaH_2PO_4
   Na_2HPO_4
Overlapping Dissociation Steps
Amines
Amino Acids
Mixtures of Polyprotic Acids and Bases
Disssociation, Formation, and Titration Curves
   Titration of a Polyprotic Acid with a Strong Base
   Titration Error
   Buffer Index
Problems

INTRODUCTION

A large number of compounds yield two or more protons (or accept two or more protons) in aqueous solutions. Understanding such polyprotic acids and bases is a simple extension of what you know already. The detailed treatment of monoprotic acids in Chapter 4 provides the basis for this chapter.

The removal of each proton from a polyprotic acid constitutes a separate equilibrium, and hence there are always as many equilibrium constants for a polyprotic acid as there are hydrogens which can ionize. Some constants are given in Table 5.1. The ionization constant K_a for removing the second proton is typically much smaller than that for the first proton ($pK_{a2} \gg pK_{a1}$), but depending on the structure of the molecule, may be almost the same value.

TABLE 5.1. Ionization constants of polyprotic acids (25°C, $I = 0$)[1]

Name	Diprotic acids Formula	pK_{a1}	pK_{a2}		
Carbon dioxide	$CO_2 + H_2O$	6.36±0.01	10.33±0.01		
Hydrogen sulfide	H_2S	7.02±0.04	13.9±0.1		
Hydrogen selenide	H_2Se	3.89	15		
Thiosulfuric acid	$H_2S_2O_3$	0.60	1.6±0.1		
Oxalic acid	$(COOH)_2$	1.25	4.27±0.01		
d-Tartaric acid	HOOC–(CHOH)$_2$–COOH	3.04	4.37		
Succinic acid	HOOC–(CH$_2$)$_2$–COOH	4.21	5.64		
Glutaric acid	HOOC–(CH$_2$)$_3$–COOH	4.34±0.01	5.43±0.02		
Adipic acid	HOOC–(CH$_2$)$_4$–COOH	4.42±0.01	5.42±0.01		
o-phthalic acid	benzene-1,2-(COOH)$_2$	2.95	5.41		
Name	Triprotic acids Formula	pK_{a1}	pK_{a2}	pK_{a3}	
o-phosphoric acid	H_3PO_4	2.148±0.001	7.199±0.002	12.35±0.02	
o-arsenic acid	H_3AsO_4	2.24±0.06	6.96±0.02	11.50	
Name	Tetraprotic acids Formula	pK_{a1}	pK_{a2}	pK_{a3}	pK_{a4}
Pyrophosphoric acid (Diphosphoric acid)	$H_4P_2O_7$	0.91	2.10	6.70	9.32
Citric acid	HOOC–C(OH)(CH$_2$–COOH)–COOH with CH$_2$–COOH	3.13±0.07	4.761±0.002	6.396±0.004	16?
Ethylene diamine tetra-acetic acid (EDTA)	(HOOC–CH$_2$)$_2$N–CH$_2$–CH$_2$–N(CH$_2$–COOH)$_2$				
$I = 0$		2.2?	3.1?	6.32	11.01
$I = 0.1$		1.95	2.68	6.11	10.17

EXAMPLES OF POLYPROTIC ACIDS

For example, phosphoric acid H_3PO_4 can dissociate in three steps, corresponding to the three equilibria

$$H_3PO_4 \rightleftharpoons H_2PO_4^- + H^+, \quad [H^+][H_2PO_4^-] = K_{a1}[H_3PO_4], \quad pK_{a1} = 2.23$$

$$H_2PO_4^- \rightleftharpoons HPO_4^{2-} + H^+, \quad [H^+][HPO_4^{2-}] = K_{a2}[H_2PO_4^-], \quad pK_{a2} = 7.21$$

$$HPO_4^{2-} \rightleftharpoons PO_4^{3-} + H^+, \quad [H^+][PO_4^{3-}] = K_{a3}[HPO_4^{2-}], \quad pK_{a3} = 12.32$$

The simple tetrahedral structure of phosphoric acid

$$\begin{array}{c} OH \\ | \\ HO-P=O \\ | \\ OH \end{array}$$

is such that removing a proton gives the entire molecule a negative charge distributed over all the oxygens. This means that the removal of the second proton becomes much more difficult, and removal of the third proton even more difficult. This is reflected in the values of the ionization constants. Each successive constant is about 10^{-5} of the preceding one. For carbon dioxide (see Chap. 10) the ratio K_{a2}/K_{a1} is 10^{-4}; for hydrogen sulfide $K_{a2}/K_{a1} = 10^{-6}$; for hydrogen selenide, $K_{a2}/K_{a1} = 10^{-7}$.

At the other extreme are the first two ionization steps of pyrophosphoric acid (Table 5.1), which overlap: $pK_{a1} = 1.52$ and $pK_{a2} = 2.36$; hence $K_{a2}/K_{a1} = 10^{-0.84} = 0.14$. Even more extreme is hydrated mercuric ion $Hg(H_2O)_2^{2+}$ with $pK_{a1} = 3.7$ and $pK_{a2} = 2.1$—hence $K_{a2}/K_{a1} = 10^{+1.6}$—the second acidity constant is 40 times larger than the first. (As a result, the intermediate species $HgOH^+$ is always a small fraction of the total. See Chapter 7.)

Amines

In addition to neutral acids that dissociate into ions, polyprotic acids can also be charged (Table 5.2). An example is ethylene diamine (L stands for the neutral base $NH_2CH_2CH_2NH_2$). In its most acidic form each nitrogen accepts a proton to produced a doubly charged molecule.

$$H_2L^{2+} \rightleftharpoons H^+ + HL^+, \quad [HL^+][H^+] = K_{a1}[H_2L^{2+}], \quad pK_{a1} = 6.85$$

$$HL^+ \rightleftharpoons H^+ + L, \quad [L][H^+] = K_{a2}[HL^+], \quad pK_{a2} = 9.93$$

This system of equilibria can also be described in terms of dissociation to OH^-:

$$L + H_2O \rightleftharpoons HL^+ + OH^-, \quad [HL^+][OH^-] = K_{b1}[L], \quad pK_{b1} = 4.07$$

$$HL^+ + H_2O \rightleftharpoons H_2L^{2+} + OH^- \quad [H_2L^{2+}][OH^-] = K_{b2}[HL^+], \quad pK_{b2} = 7.15$$

Table 5.2. Ionization constants of organic molecules with several acidic or basic groups[2]

Name	Formula	pK_{a1}	pK_{a2}	pK_{a3}
Hydrazine (L, H$_2$L^{2+})	H$_2$N–NH$_2$	–0.9	7.98±0.01	
Ethylene diamine (L, H$_2$L^{2+})	H$_2$N–CH$_2$CH$_2$–NH$_2$	6.85	9.93	
1,2,3-triaminopropane (L, H$_3$L^{2+})	H$_2$N–CH$_2$CHCH$_2$–NH$_2$ | NH$_2$	3.72	7.95	9.59
8-hydroxyquinoline (L, H$_2$L$^+$)	(quinoline with OH and N)	4.92±0.04	9.82±0.01	
Glycine (HL, H$_2$L$^+$)	H$_2$N–CH$_2$–COOH	2.350±0.001	9.778±0.001	
Serine (HL, H$_2$L$^+$)	HO–CH$_2$–CH–COOH | NH$_2$	2.19	9.21	
Aspartic Acid (HL, H$_3$L^{2+})	HOOC–CH$_2$–CH–COOH | NH$_2$	1.99	3.90±0.01	10.00
Cysteine (HL, H$_3$L^{2+})	HS–CH$_2$–CH–COOH | NH$_2$	1.7?	8.36±0.03	10.74±0.04
Glutamic acid	HOOC–(CH$_2$)$_2$–CH–COOH | NH$_2$	2.23±0.07	4.42±0.01	9.95±0.01
Arginine (HL, H$_3$L^{2+})	H$_2$C⟨CH$_2$–NH–C(=NH)–NH$_2$; CH$_2$–CH(NH$_2$)–COOH⟩	1.82	8.99	12.5?

The two sets of constants are related by K_w:

$$pK_{a1} + pK_{b2} = 6.85 + 7.15 = 14.00 = pK_w$$
$$pK_{a2} + pK_{b1} = 9.93 + 4.07 = 14.00 = pK_w$$

Amino Acids

Another class of charged polyprotic acids is the amino acids (Table 5.2). These contain both carboxyl groups, which can lose a proton, and amino groups, which can gain a proton. A simple example is glycine NH$_2$CH$_2$COOH, abbreviated HG. In acidic solutions, the nitrogen accepts a proton to give $^+$NH$_3$CH$_2$COOH, abbreviated

EXAMPLES OF POLYPROTIC ACIDS

H_2G^+; in basic solution, the carboxyl group loses a proton to give $NH_2CH_2COO^-$, or G^-. Thus the glycine system can be represented as

$$H_2G^+ \rightleftharpoons HG + H^+, \quad [H^+][HG] = K_{a1}[H_2G^+], \quad pK_{a1} = 2.35$$

$$HG \rightleftharpoons G^- + H^+, \quad [H^+][G^-] = K_{a2}[HG], \quad pK_{a2} = 9.78$$

The neutral molecule HG, which dominates the mid-pH range, has a very high dipole moment, indicating that it is not NH_2CH_2COOH but the "zwitterion" $^+NH_3CH_2-COO^-$. The reason for this dipolar structure is that the COOH group tends to lose its proton at pH > 2.35, and the NH_3^+ group accepts a proton at pH < 9.77. Examples using these concepts will be given later.

Systems Involving a Gas Phase

Several environmentally important compounds involve a gas phase as well as an aqueous solution phase. The prototype is carbon dioxide, for which the equilibria are (at 25°C, I = 0—see Table 10.1)

$$CO_2(gas) \rightleftharpoons CO_2(aq), \quad [CO_2(aq)] = K_H P_{CO_2}, \quad pK_H = 1.464 \ (P \text{ in atm})$$

$$[CO_2(aq)] \rightleftharpoons H^+ + HCO_3^-, \quad [H^+][HCO_3^-] = K_{a1}[CO_2(aq)], \quad pK_{a1} = 6.363$$

$$HCO_3^- \rightleftharpoons H^+ + CO_3^{2-}, \quad [H^+][CO_3^{2-}] = K_{a2}[HCO_3^-], \quad pK_{a2} = 10.329$$

The first reaction is simple dissolution, governed by Henry's law, which states that the solution concentration of CO_2 is proportional to the partial pressure of CO_2 in the gas phase.

The second reaction is often broken into two steps. Hydration of dissolved CO_2 to form H_2CO_3 is slow (about 0.1 sec), and hence it is possible to separate this step from the much faster (10^{-6} sec) dissociation of H_2CO_3 into H^+ and HCO_3^-.

$$CO_2(aq) + H_2O \rightleftharpoons H_2CO_3, \quad [H_2CO_3] = {}^{\dagger}K[CO_2(aq)], \quad p^{\dagger}K = 2.97$$

$$H_2CO_3 \rightleftharpoons H^+ + HCO_3^-, \quad [H^+][HCO_3^-] = {}^{\dagger}K_1 H_2CO_3, \quad p^{\dagger}K_1 = 3.38$$

At equilibrium, however, $[H_2CO_3]$ is only 0.1% of $[CO_2(aq)]$. Neglecting it does not affect the results of acid–base calculations[3].

But in physiology, the rate of CO_2 hydration and dehydration is critical, and fortunately for the respiratory activities of animals, it is greatly enhanced by the enzyme carbonic anhydrase. Many applications to physiology involve both kinetics and equilibrium. Details of the carbon dioxide system and its applications are given in Chapter 10.[4]

Other important acid–base systems involving a gas phase include ammonia and hydrogen sulfide. Ammonia is more than 1000 times as soluble in water as CO_2:

$$NH_3(g) \rightleftharpoons NH_3(aq), \quad [NH_3] = K_H P_{NH_3}, \quad pK_H = -1.71$$

$$[NH_4^+] \rightleftharpoons NH_3 + H^+, \quad [NH_3][H^+] = K_a[NH_4^+], \quad pK_a = 9.245$$

Hydrogen sulfide is of intermediate solubility

$H_2S(g) \rightleftharpoons H_2S$ (aq), $\quad [H_2S] = K_H P_{H_2S}$, $\quad pK_H = 0.99$

$H_2S(aq) \rightleftharpoons H^+ + HS^-$, $\quad [H^+][HS^-] = K_{a1}[H_2S]$, $\quad pK_{a1} = 6.98$

$HS^- \rightleftharpoons H^+ + S^{2-}$, $\quad [H^+][S^{2-}] = K_{a2}[HS^-]$, $\quad pK_{a2} = 12.6$

Because solutions containing both HS^- and S^{2-} are highly alkaline, glass electrodes are subject to interference by alkali ions; perhaps the most accurate measurements would be made with a hydrogen electrode or special akaline glass electrode in conjunction with a silver sulfide ion-selective membrane electrode.[5]

IONIC STRENGTH AND TEMPERATURE DEPENDENCE

The dependence of these equilibrium constants on ionic strength follows the same principles as were presented for monoprotic acids and bases. For example, the first dissociation constant of phosphoric acid is

$$K_{a1}^\circ = \frac{[H^+]\gamma_+[H_2PO_4^-]\gamma_-}{[H_3PO_4]\gamma_o} = K_{a1}\frac{\gamma_+\gamma_-}{\gamma_o} \tag{1}$$

$$pK_{a1} = pK_{a1}^\circ + \log(\gamma_+\gamma_-) - \log(\gamma_o) \tag{2}$$

Using the Davies equation (see Chapter 2, p. 49) to evaluate γ_+ and γ_-, and setting $\log(\gamma_o) = 0.1\ I$, you get

$$pK_{a1} = pK_{a1}^\circ - (2)(0.51)\left(\frac{\sqrt{I}}{1+\sqrt{I}} - 0.2I\right) - 0.1I \tag{3}$$

This equation provides a satisfactory interpolation to the literature data (Fig. 5.1), The second ionization constant has a stronger ionic strength dependence:

$$K_{a2}^\circ = \frac{[H^+]\gamma_+[HPO_4^{2-}]\gamma_=}{[H_2PO_4^-]\gamma_-} = K_{a2}\frac{\gamma_+\gamma_=}{\gamma_-} \tag{4}$$

$$pK_{a2} = pK_{a2}^\circ + \log\left(\frac{\gamma_+\gamma_=}{\gamma_-}\right) \tag{5}$$

Using the Davies equation to evaluate γ_+ and γ_-, you get

$$pK_{a2} = pK_{a2}^\circ - (4)(0.51)\left(\frac{\sqrt{I}}{1+\sqrt{I}} - 0.2I\right) \tag{6}$$

IONIC STRENGTH AND TEMPERATURE DEPENDENCE

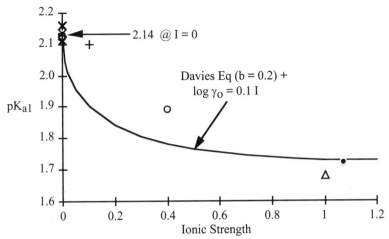

FIGURE 5.1. First ionization constant of phosphoric as a function of ionic strength at 25°C. Points represent literature data.[23]

This function is compared with literature data in Fig. 5.2. At higher ionic strengths, it gives values that are significantly higher than the experimental data, but a small adjustment of the constant factor 0.2 to 0.07 makes an excellent interpolation function:

$$pK_{a2} = pK_{a2}^\circ - (4)(0.51)\left(\frac{\sqrt{I}}{1+\sqrt{I}} - 0.07I\right) \quad (7)$$

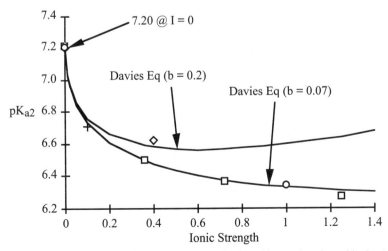

FIGURE 5.2. Second ionization constant of phosphoric acid as a function of ionic strength. Note that the normal Davies equation gives values higher than the experimental data[23], but that changing the constant term from 0.02 to 0.07 provides an excellent interpolation function.

The third ionization constant has an even steeper dependence on ionic strength.

$$K_{a3}^\circ = \frac{[H^+]\gamma_+[PO_4^{3-}]\gamma_\equiv}{[HPO_4^{2-}]\gamma_=} = K_{a3}\frac{\gamma_+\gamma_{3-}}{\gamma_=} \quad (8)$$

$$pK_{a3} = pK_{a3}^\circ + \log\left(\frac{\gamma_+\gamma_{3-}}{\gamma_=}\right) \quad (9)$$

Using the Davies equation to evaluate γ_+ and γ_-, you get

$$pK_{a3} = pK_{a3}^\circ - (6)(0.51)\left(\frac{\sqrt{I}}{1+\sqrt{I}} - 0.2I\right) \quad (10)$$

As with pK_{a2}, this function is higher than the experimental data, but adjusting the constant term is not a successful approach. To fit the higher ionic strength range, the constant must be changed from 0.2 to an unrealistic –0.2 (Fig. 5.3).

The Debye–Hückel equation usually does not give as good a fit as the Davies equation or its variants at higher ionic strength, but with ion size parameters 9 for H^+, 4 for HPO_4^{2-}, and 4 for PO_4^{3-} (see Table 2.1) it fits the data for pK_{a3} much better than does the Davies equation.

The temperature dependence of these ionization constants is varied, but good data are not available for most. For example, the temperature dependence of pK_{a1} and pK_{a2} have been accurately measured, but not pK_{a3}. Figure 5.4 shows the tempera-

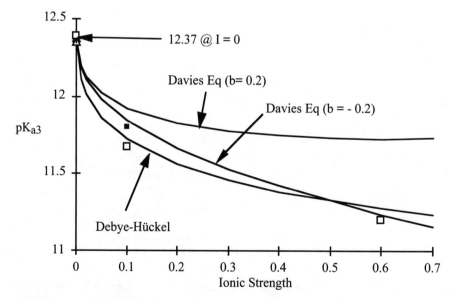

FIGURE 5.3. Third ionization constant of phosphoric acid as a function of ionic strength.

IONIC STRENGTH AND TEMPERATURE DEPENDENCE

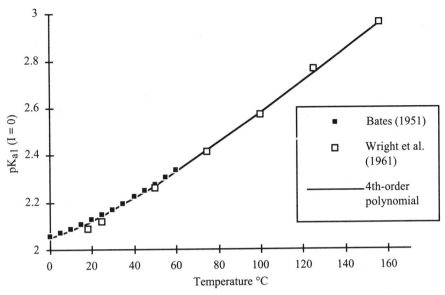

FIGURE 5.4. Temperature dependence of pK_{a1}. Literature data have been fitted with a fourth-order polynomial as described in the text.

ture dependence of pK_{a1} over the range from 0 to 160°C. The two experimental studies[6] are in agreement.

One compact way of presenting such data is by means of an empirically fitted polynomial function. For the data in Fig. 5.4, the least-squares fit to a fourth-order polynomial[7] is

$$pK_{a1}^\circ = 2.0536 + (2.7155 \times 10^{-3})t + (4.2641 \times 10^{-5})t^2 \\ - (2.1283 \times 10^{-7})t^3 + (4.3127 \times 10^{-10})t^4$$

where t is the temperature in °C. Note that pK_{a1} increases with increasing temperature; the temperature coefficient at 25°C is $d(pK_{a1})/dt = 0.0045$ deg^{-1}, corresponding to $\Delta H = +1.83$ kcal/mol.[8] This is in agreement with ΔH as calculated by the original worker,[9] and close to a calorimetric value of 1.88.[10]

The temperature dependence of pK_{a2}° is shown in Fig. 5.5. Three independent studies[11] show good agreement over the range from 0 to 60°C, and can be fit by a three-constant thermodynamic form (where $T = t + 273.16$)[12]

$$pK_{a2} = 1775.812/T - 3.9762 + 0.0175089T$$

or a fourth-order polynomial:

$$pK_{a2}^\circ = 7.3142 - (7.2925 \times 10^{-3})t + (1.2884 \times 10^{-4})t^2 \\ - (9.7655 \times 10^{-7})t^3 + (4.9911 \times 10^{-9})t^4$$

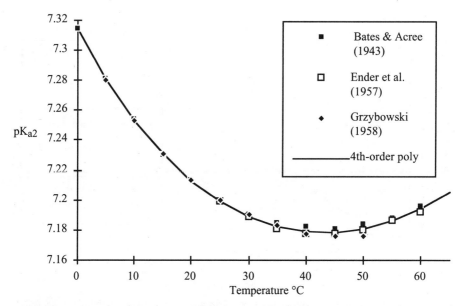

FIGURE 5.5. Temperature dependence of the second ionization constant of phosphoric acid. Literature data are compared with a fourth-order polynomial interpolation function given in the text.

For each of these functions the standard deviation of the experimental values is 0.042. Unlike pK_{a1}, pK_{a2} goes through a minimum: The temperature coefficient is negative at low temperatures, goes through zero at about 45°C, and is positive at higher temperatures. This equilibrium has been thoroughly investigated in part because $H_2PO_4^-$–HPO_4^{2-} is one of the most important standard buffer systems (see Table 2.3).

Among the almost 100 studies of pK_a for H_3PO_4 reviewed by Sillén and Martell, only one entry[13] gives an explicit temperature dependence of pK_3: 11.99 at 20°C and 11.83 at 38°C, corresponding to $d(pK_{a3})/dT = -0.0089$. However, the interpolated 25°C value (11.95) is so far below the other literature values (12.37—see Fig. 5.3) that these data must be questioned. A calorimetric study[14] gave $\Delta H = -3.50$ kcal/mole at 25°C, corresponding to $d(pK_{a3})/dT = -0.0086$ deg^{-1}, which is not too different: a decrease of roughly 0.4 log units from 0 to 50°C.

These three examples show how varied the temperature dependence of acidity constants can be: pK_{a1} increasing with temperature, pK_{a2} going through a minimum (like acetic acid—see Fig. 4.2, p. 98), and pK_{a3} decreasing with temperature. Although a value of ΔH at 25°C (as given in Smith and Martell) will tell you the temperature dependence in that region, it should not be trusted below 10°C or above 40°C (see Fig. 1.3, p. 17).

DISTRIBUTION DIAGRAMS

In Chapter 4, logarithmic concentration (Sillén) diagrams for monoprotic acids and bases were developed. A similar approach can be used for polyprotic acids. Take phosphoric acid as an example. The equations the diagram represents are the three equilibria plus the mass balance:

$$[H^+][H_2PO_4^-] = K_{a1}[H_3PO_4] \tag{11}$$

$$[H^+][HPO_4^{2-}] = K_{a2}[H_2PO_4^-] \tag{12}$$

$$[H^+][PO_4^{3-}] = K_{a3}[HPO_4^{2-}] \tag{13}$$

$$C = [H_3PO_4] + [H_2PO_4^-] + [HPO_4^{2-}] + [PO_4^{3-}] \tag{14}$$

The fraction of acid present as each species is the ratio of the concentration of that species to the analytical concentration, analogous to $\alpha_{HA} = [HA]/C$ for a monoprotic acid:

$$\alpha_3 = \frac{[H_3PO_4]}{C} \tag{15}$$

The index on α gives the number of protons attached to the molecule. Use the mass balance (Eq. 14) to calculate

$$\frac{1}{\alpha_3} = \frac{C}{[H_3PO_4]} = 1 + \frac{[H_2PO_4^-]}{[H_3PO_4]} + \frac{[HPO_4^{2-}]}{[H_3PO_4]} + \frac{[PO_4^{3-}]}{[H_3PO_4]} \tag{16}$$

and evaluate these terms using the equilibrium constants. From Eq. (11),

$$\frac{[H_2PO_4^-]}{[H_3PO_4]} = \frac{K_{a1}}{[H^+]} \tag{17}$$

From Eq. (12),

$$\frac{[HPO_4^{2-}]}{[H_3PO_4]} = \frac{K_{a1}K_{a2}}{[H^+]^2} \tag{18}$$

From Eq. (13),

$$\frac{[PO_4^{3-}]}{[H_3PO_4]} = \frac{K_{a1}K_{a2}K_{a3}}{[H^+]^3} \tag{19}$$

Substituting Eqs. (17)–(19) in Eq. (16) gives

$$\alpha_3 = \frac{[H_3PO_4]}{C} = \left(1 + \frac{K_{a1}}{[H^+]} + \frac{K_{a1}K_{a2}}{[H^+]^2} + \frac{K_{a1}K_{a2}K_{a3}}{[H^+]^3}\right)^{-1} \tag{20}$$

$$\alpha_3 = \frac{[H^+]^3}{[H^+]^3 + K_{a1}[H^+]^2 + K_{a1}K_{a2}[H^+] + K_{a1}K_{a2}K_{a3}} \tag{21}$$

Finally, evaluate the other α_n using Eqs. (17)–(19):

$$\alpha_2 = \frac{[H_2PO_4^-]}{C} = \frac{K_{a1}[H^+]^2}{[H^+]^3 + K_{a1}[H^+]^2 + K_{a1}K_{a2}[H^+] + K_{a1}K_{a2}K_{a3}} \quad (22)$$

$$\alpha_1 = \frac{[HPO_4^{2-}]}{C} = \frac{K_{a1}K_{a2}[H^+]}{[H^+]^3 + K_{a1}[H^+]^2 + K_{a1}K_{a2}[H^+] + K_{a1}K_{a2}K_{a3}} \quad (23)$$

$$\alpha_0 = \frac{[PO_4^{3-}]}{C} = \frac{K_{a1}K_{a2}K_{a3}}{[H^+]^3 + K_{a1}[H^+]^2 + K_{a1}K_{a2}[H^+] + K_{a1}K_{a2}K_{a3}} \quad (24)$$

Note that the mass balance may now be expressed, using Eqs. (14) and (21)–(24) as

$$\alpha_3 + \alpha_2 + \alpha_1 + \alpha_0 = 1 \quad (25)$$

There are a number of ways to present these functions graphically. The first is as a logarithmic concentration or Sillén diagram (Fig. 5.6). Note that the upper por-

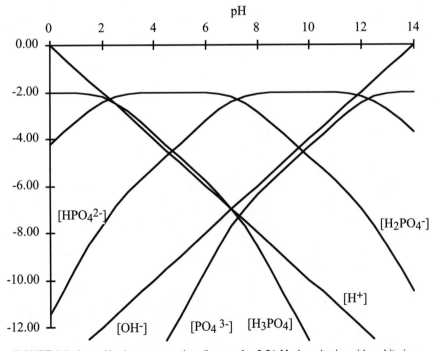

FIGURE 5.6. Logarithmic concentration diagram for 0.01 M phosphoric acid and its ions.

tion of the diagram looks exactly like a diagram for three monoprotic acids with the same total concentration. However, in the lower part of the graph, you can see that the slope of the various lines becomes steeper; this is a result of the extra terms in the denominator of Eqs. (21)–(24).

Another way of presenting these functions is as a compact distribution diagram (Fig. 5.7). The first curve is α_3; the second curve is $\alpha_3 + \alpha_2$; and the third is $\alpha_3 + \alpha_2 + \alpha_1$. The vertical axis is a linear scale from 0 to 1.0, and the horizontal axis is a pH scale. A vertical line drawn at any pH will be cut into segments by the three curves, and each segment will correspond to α for a particular species. For example, in Fig. 5.7 the line at pH = 3 is cut by the first curve into two segments

$$\alpha_3 = 0.13 \quad \text{and} \quad \alpha_2 = 0.87$$

Segments α_1 and α_0 are too small to be read from the graph at this pH. A vertical line at pH = 7.21 is divided by the second curve into two equal segments

$$\alpha_2 = 0.50 \quad \text{and} \quad \alpha_1 = 0.50$$

Segments α_3 and α_0 are too small to be read from the graph.

Each region on the graph of Fig. 5.7 is thus labeled with the formula of a particular species; the fraction of a vertical line falling in that region is the fraction present as that species. The larger the region assigned to a given species, the more "stable" is that species with respect to dissociation or association with a proton. Compact distribution diagrams like Fig. 5.7 will be used again in Chapter 7 when complex formation is discussed.

pH OF H_3PO_4 AND ITS SALTS

Phosphoric acid H_3PO_4 has a proton condition given by the charge balance

$$[H^+] = [H_2PO_4^-] + 2[HPO_4^{2-}] + 3[PO_4^{3-}] + [OH^-] \tag{26}$$

As can be seen from Fig. 5.6, the second, third, and fourth terms on the right-hand side are much smaller than the first. Hence the charge balance occurs approximately at the intersection

$$.[H^+] = [H_2PO_4^-] + \cdots \tag{27}$$

which can be read from the enlarged graph of Fig. 5.8 to be approximately pH = 2.3.

Example 1. The pH of phosphoric acid can also be obtained numerically with more accuracy but also with more effort. From Eqs. (22) and (27),

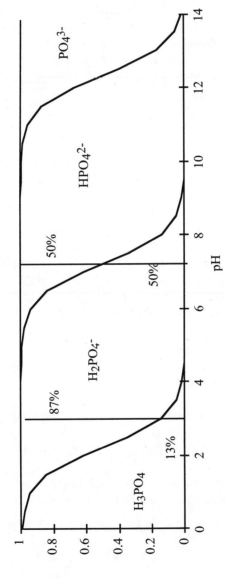

FIGURE 5.7. Compact distribution diagram a versus pH for phosphoric acid. A vertical line is divided into segments proportional to the fractions α_n.

pH OF H₃PO₄ AND ITS SALTS

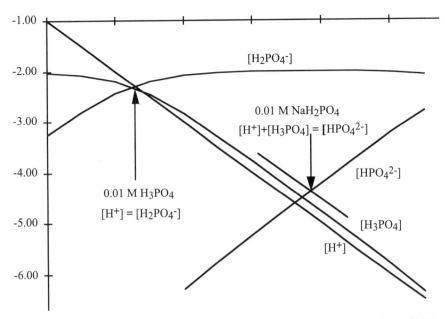

FIGURE 5.8. Enlarged section of Fig. 5.6 showing the intersections corresponding to H₃PO₄ and NaH₂PO₄.

$$[H^+] = [H_2PO_4^-] = C\alpha_2 + \cdots$$

$$[H^+] = \frac{CK_{a1}[H^+]^2}{[H^+]^3 + K_{a1}[H^+]^2 + K_{a1}K_{a2}[H^+] + K_{a1}K_{a2}K_{a3}} \tag{28}$$

Neglecting the last two terms in the denominator[15] gives

$$[H^+]^2 + K_{a1}[H^+] = CK_{a1} = (10^{-2})(10^{-2.23}) = 10^{-4.23}$$

A first approximation could be $[H^+]^2 = 10^{-4.23}$, or pH = 2.21. But note that $K_{a1}[H^+] = 10^{-4.44}$, which is only a little smaller. Iteration of the equation

$$[H^+]^2 = K_{a1}(C - [H^+])$$

gives a slowly convergent series of answers: pH = 2.21, 2.32, 2.25, 2.29, 2.27, 2.28. Alternatively the equation

$$[H^+]^2 + K_{a1}[H^+] - K_{a1}C = 0$$

can be solved using the quadratic formula

$$[H^+] = \frac{-K_{a1} + \sqrt{K_{a1}^2 + 4CK_{a1}}}{2}$$

to obtain pH = 2.278.

Thus the pH of H_3PO_4 can be calculated as if it were a monoprotic acid. The relatively small values of the other species can be seen on Fig. 5.6 as $[PO_4^{3-}]$ « $[OH^-]$ « $[HPO_4^{2-}]$ « $[H_2PO_4^-]$.

NaH_2PO_4

The equilibria and mass balance expressed by Eqs. (21)–(24) or by Fig. 5.6, describe not only H_3PO_4, but also the salts NaH_2PO_4, Na_2HPO_4, and Na_3PO_4. Each has the same equilibria and mass balance, and only the proton condition is different. For example:

For NaH_2PO_4, at concentration C the mass balances are

$$[Na^+] = C$$

$$[H_3PO_4] + [H_2PO_4^-] + [HPO_4^{2-}] + [PO_4^{3-}] = C$$

and the charge balance is

$$[H^+] + [Na^+] = [H_2PO_4^-] + 2[HPO_4^{2-}] + 3[PO_4^{3-}] + [OH^-]$$

Substituting the mass balances to eliminate the two largest terms $[Na^+]$ and $[H_2PO_4^-]$ gives the proton condition:

$$[H^+] + [H_3PO_4] = [HPO_4^{2-}] + 2[PO_4^{3-}] + [OH^-] \tag{29}$$

From Fig. 5.6, it is apparent that both terms remaining on the left, $[H^+]$ and $[H_3PO_4]$, are similar in size, and the largest term on the right is $[HPO_4^{2-}]$

$$[H^+] + [H_3PO_4] = [HPO_4^{2-}] + \cdots \tag{30}$$

As Example 1 showed, employing the full equations is cumbersome, but an approximate answer can be read from the graph. Since $[H^+]$ and $[H_3PO_4]$ are nearly equal, their sum is a line a little above the $[H^+]$ line, as shown in Fig. 5.8. Where this sum line crosses the $[HPO_4^{2-}]$ line, the proton condition is satisfied (pH = 4.8). Note that both $[PO_4^{3-}]$ and $[OH^-]$ are negligible at this pH.

Example 2. The approximate answer obtained from Fig. 5.8 can be refined numerically using the approximate proton condition (Eq. 30) and the equilibria:

$$[H^+] + \frac{[H^+][H_2PO_4^-]}{K_{a1}} = \frac{K_{a2}[H_2PO_4^-]}{[H^+]} + \cdots$$

This is easily solved for [H⁺]. From Fig. 5.6 or 5.8, note that in this pH range, $[H_2PO_4^-] \approx C$

$$[H^+]^2 \left(1 + \frac{C}{K_{a1}}\right) = K_{a2} C$$

$$[H^+]^2 = \frac{K_{a1} K_{a2} C}{K_{a1} + C}$$

With $K_{a1} = 10^{-2.23}$, $K_{a2} = 10^{-7.21}$, and $C = 10^{-2.00}$, this yields pH = 4.82, in agreement with Fig. 5.8. Calculating all the concentrations will verify the approximations.

Na₂HPO₄

For Na₂HPO₄, at concentration C the mass balances are,

$$[Na^+] = 2C$$

$$[H_3PO_4] + [H_2PO_4^-] + [HPO_4^{2-}] + [PO_4^{3-}] = C$$

and the charge balance is

$$[H^+] + [Na^+] = [H_2PO_4^-] + 2[HPO_4^{2-}] + 3[PO_4^{3-}] + [OH^-]$$

Substituting the mass balances to eliminate the two largest terms [Na⁺] and [HPO₄²⁻] gives the proton condition:

$$[H^+] + 2[H_3PO_4] + [H_2PO_4^-] = [PO_4^{3-}] + [OH^-] \tag{31}$$

From Fig. 5.6, it is apparent that in the pH range (near 10) where [HPO₄²⁻] is dominant, the largest term on the left is [H₂PO₄⁻] and the two terms on the right are of similar size. This leads to the approximate proton condition

$$[H_2PO_4^-] = [PO_4^{3-}] + [OH^-] + \cdots \tag{32}$$

Again, an approximate answer can be read from the enlarged graph of Fig. 5.9. Since [OH⁻] is a little larger than [PO₄³⁻], their sum is a line a little above the [OH⁻] line. Where this sum line crosses the [H₂PO₄⁻] line, the proton condition is satisfied (pH = 9.5). Note that both [H₃PO₄] and [H⁺] are negligible at this pH.

Example 3. As for the previous examples, a more refined calculation can be done numerically using the logarithmic diagram as a guide for approximations.

174 POLYPROTIC ACIDS AND BASES

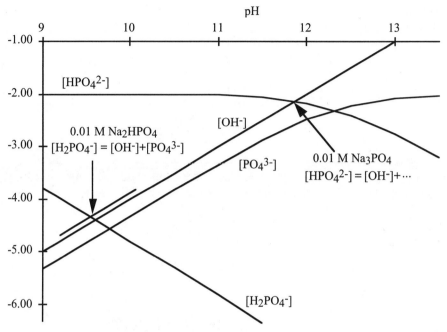

FIGURE 5.9. Enlarged section of Fig. 5.6, showing the intersection corresponding to the proton conditions for Na$_2$HPO$_4$ and Na$_3$PO$_4$.

$$[H_2PO_4^-] = [PO_4^{3-}] + [OH^-] + \cdots$$

$$[HPO_4^{2-}] + \cdots = C$$

$$\frac{[H^+][HPO_4^{2-}]}{K_{a2}} = \frac{K_{a3}[HPO_4^{2-}]}{[H^+]} + \frac{K_w}{[H^+]} \tag{33}$$

$$[H^+]^2 = K_{a2}K_{a3} + \frac{K_{a2}K_w}{C}$$

With $K_{a2} = 10^{-7.21}$, $K_{a3} = 10^{-12.32}$, $K_w = 10^{-14.00}$, $C = 10^{-2.00}$, this gives pH = 9.52, in agreement with Fig. 5.9.

Finally, consider the proton condition for Na$_3$PO$_4$. At concentration C the mass balances are,

$$[Na^+] = 3C$$

$$[H_3PO_4] + [H_2PO_4^-] + [HPO_4^{2-}] + [PO_4^{3-}] = C$$

and the charge balance is

pH OF H₃PO₄ AND ITS SALTS

$$[H^+] + [Na^+] = [H_2PO_4^-] + 2[HPO_4^{2-}] + 3[PO_4^{3-}] + [OH^-]$$

Substituting the mass balances to eliminate the two largest terms $[Na^+]$ and $[PO_4^{3-}]$ gives the proton condition:

$$[H^+] + 3[H_3PO_4] + 2[H_2PO_4^-] + [HPO_4^{2-}] = [OH^-] \tag{34}$$

From Fig. 5.9 in the pH range (near 13) where $[PO_4^{3-}]$ is dominant, the largest term on the left is $[HPO_4^{2-}]$ and there is only one term on the right. This leads to the approximate proton condition

$$[HPO_4^{2-}] = [OH^-] + \cdots \tag{35}$$

Again, an approximate answer can be read from the graph of Fig. 5.9, where the $[HPO_4^{2-}]$ line curves down to meet $[OH^-]$. There the proton condition is satisfied at pH = 11.9. Note that $[H_3PO_4]$, $[H_2PO_4]$, and $[H^+]$ are negligible at this pH. Although $[PO_4^{3-}]$ is nearly as large as $[OH^-]$, as shown in Fig. 5.9, it does not enter the proton condition of Eq. (34) or Eq. (35).

Example 4. The more refined numerical calculation for Na_3PO_4 is simplified by using the logarithmic diagram as a guide for approximations.

$$[HPO_4^{2-}] = [OH^-] + \cdots$$

From Fig. 5.9, note that this intersection occurs at a pH where $[HPO_4^{2-}]$ is larger than $[PO_4^{3-}]$; hence $[PO_4^{3-}]$ is less than C and the approximations used in Examples 2 and 3 do not apply. Instead, $[HPO_4^{2-}] + [PO_4^{3-}] + \cdots = C$ and

$$\left(1 + \frac{K_{a3}}{[H^+]}\right)[HPO_4^{2-}] = C$$

$$[HPO_4^{2-}] = \frac{C[H^+]}{[H^+] + K_{a3}} = \frac{K_w}{[H^+]} \tag{36}$$

$$[H^+]^2 = \frac{K_w}{C}(K_{a3} + [H^+])$$

With $K_{a3} = 10^{-12.32}$, $K_w = 10^{-14.00}$, $C = 10^{-2.00}$, iteration of Eq. (36) gives pH = 11.88, in agreement with Fig. 5.9. If the assumption $[PO_4^{3-}] = C$ were followed, the result would have been too high: pH = 12.16.

Overlapping Dissociation Steps

In the examples using phosphoric acid, each dissociation step was treated to a good approximation as a separate monoprotic acid–base pair.

The general form for a diprotic acid is obtained (pp. 167–168) from the equilibria and mass balance to be

$$[H_2A] = \frac{C[H^+]^2}{[H^+]^2 + K_{a1}[H^+] + K_{a1}K_{a2}} \tag{37}$$

$$[HA^-] = \frac{CK_{a1}[H^+]}{[H^+]^2 + K_{a1}[H^+] + K_{a1}K_{a2}} \tag{38}$$

$$[A^{2-}] = \frac{CK_{a1}K_{a2}}{[H^+]^2 + K_{a1}[H^+] + K_{a1}K_{a2}} \tag{39}$$

If $K_{a2} \ll K_{a1}$, a good approximation is to treat the diprotic acid as two monoprotic acids:

$$[H_2A] = \frac{C[H^+]}{[H^+] + K_{a1}} - \cdots \tag{40}$$

$$[HA^-] = \frac{CK_{a1}}{[H^+] + K_{a1}} - \cdots = \frac{C[H^+]}{[H^+] + K_{a2}} - \cdots \tag{41}$$

$$[A^{2-}] = \frac{CK_{a2}}{[H^+] + K_{a2}} + \cdots \tag{42}$$

But when two pK_a values are close together, and the intermediate species at its maximum concentration never reaches the baseline at $-\log C$, the two can no longer be treated as separate monoprotic acids.[16]

The smaller K_{a2} is compared to K_{a1}, the less reliable are these approximations. See Problem 32 for a more detailed discussion and some diagrams. These approximations will also be used in deriving an expression for the buffer index (Eqs. 72–76 and Figs. 5.22 and 5.23; see also Chap. 10).

Example 5. In Fig. 5.10, a portion of the logarithmic concentration diagram for pyrophosphoric acid $H_4P_2O_7$, abbreviated H_4A ($pK_{a1} = 0.91$, $pK_{a2} = 2.10$, $pK_{a3} = 6.70$, $pK_{a4} = 9.32$),[17] shows how the concentration of the intermediate species $[H_3A^-]$ rises as pH increases to pK_{a1}, but as soon as pH > pK_{a2}, $[H_3A^-]$ decreases to make room for $[H_2A^{2-}]$.

Calculate the pH of NaH_3A. The proton condition is obtained from the mass balance on Na and the mass balance on A, eliminating the two large terms $[Na^+]$ and $[H_3A^-]$, as was done in the previous examples:

$$[H^+] + [H_4A] = [H_2A^{2-}] + [HA^{3-}] + [A^{4-}] + [OH^-] \tag{43}$$

pH OF H_3PO_4 AND ITS SALTS

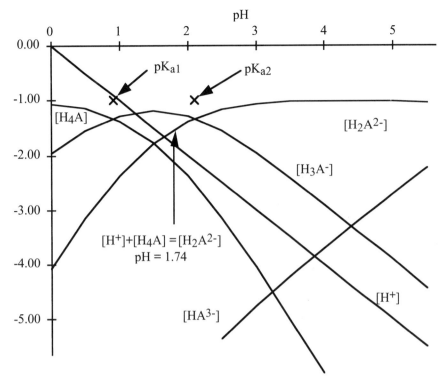

FIGURE 5.10. Logarithmic concentration diagram for 0.1 M pyrophosphoric acid H_4A illustrating that when two acidity constants are close together ($pK_{a1} = 0.91$, $pK_{a2} = 2.10$), the intermediate species H_3A^- is dominant only in a narrow pH range between the two pK values.

From Fig. 5.10, you can see that the last three terms are small or negligible, and

$$[H^+] + [H_4A] = [H_2A^{2-}] + \cdots$$

This intersection ocurs just below the maximum in $[H_3A^-]$, and all four concentrations ($[H^+]$, $[H_4A]$, $[H_3A^-]$, and $[H_2A^{2-}]$) are between 10^{-1} and 10^{-2}. The most direct way of approaching this problem is to use the full mass balance and equilibrium expressions, but to neglect $[HA^{3-}]$, $[A^{4-}]$, and $[OH^-]$. Making the same sort of manipulations as for H_3PO_4 (Eqs. 16–24) with these approximations, you get

$$[H_4A] = \frac{C[H^+]^4}{[H^+]^4 + K_{a1}[H^+]^3 + K_{a1}K_{a2}[H^+]^2 + K_{a1}K_{a2}K_{a3}[H^+] + K_{a1}K_{a2}K_{a3}K_{a4}}$$

At pH near 2, the first three terms in the denominator (10^{-8}, $10^{-7.52}$, and $10^{-7.88}$) are much larger than the last two terms ($10^{-12.48}$ and $10^{-19.73}$), which can be neglected. Canceling a factor $[H^+]^2$ between numerator and denominator simplifies the equations slightly, so that they now are identical to the corresponding Eqs. (37)–(39) for a diprotic acid H_2A.

$$[H_4A] = \frac{C[H^+]^2}{[H^+]^2 + K_{a1}[H^+] + K_{a1}K_{a2} + \cdots} \tag{44}$$

$$[H_3A^-] = \frac{CK_{a1}[H^+]}{[H^+]^2 + K_{a1}[H^+] + K_{a1}K_{a2} + \cdots} \tag{45}$$

$$[H_2A^{2-}] = \frac{CK_{a1}K_{a2}}{[H^+]^2 + K_{a1}[H^+] + K_{a1}K_{a2} + \cdots} \tag{46}$$

The pH is given by the proton condition, $[H^+] + [H_4A] = [H_2A^{2-}]$, Eq. (43):

$$[H^+] + \frac{C[H^+]^2}{[H^+]^2 + K_{a1}[H^+] + K_{a1}K_{a2} + \cdots} = \frac{CK_{a1}K_{a2}}{[H^+]^2 + K_{a1}[H^+] + K_{a1}K_{a2} + \cdots}$$

$$[H^+]^3 + (K_{a1} + C)[H^+]^2 + K_{a1}K_{a2}[H^+] = CK_{a1}K_{a2}$$

Solving this cubic equation for $[H^+]$ with a given value of C is complicated, but the reverse process is easy, since the equation is linear in C. Choose values for pH and calculate C:

$$C = \frac{[H^+]^3 + K_{a1}[H^+]^2 + K_{a1}K_{a2}[H^+]}{K_{a1}K_{a2} - [H^+]^2} \tag{47}$$

This function (with $pK_{a1} = 0.91$, $pK_{a2} = 2.10$) is plotted in Fig. 5.11. When $C = 0.10$, pH = 1.74. Note that lower values of C correspond to higher pH, but higher values of C reach an asymptote where the denominator of Eq. (47) becomes zero, at $[H^+]^2 = K_{a1}K_{a2}$. That is, no matter how high the concentration of NaH_3A, pH cannot be less than 1.51.

AMINES

Example 6. Find the pH of 0.01 M ethylene diamine (L) and its di-hydrochloride $[H_2L^{2+}(Cl^-)_2]$. Recall from the beginning of the chapter that[18]

$$H_2L^{2+} \rightleftharpoons H^+ + HL^+, \quad [HL^+][H^+] = K_{a1}[H_2L^{2+}], \quad pK_{a1} = 6.85$$
$$HL^+ \rightleftharpoons H^+ + L, \quad [L][H^+] = K_{a2}[HL^+], \quad pK_{a2} = 9.93$$

The proton condition for L is the same as the charge balance:

$$[H^+] + 2[H_2L^{2+}] + [HL^+] = [OH^-] \tag{48}$$

AMINES

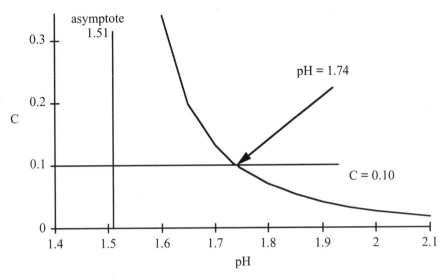

FIGURE 5.11. Plot of Eq. (47), showing that pH = 1.74 when C = 0.10.

The various concentrations are plotted as a function of pH in Fig. 5.12. When $[HL^+] = [OH^-]$, at pH near 12, $[H^+]$ and $2[H_2L^{2+}]$ are very much smaller.

At the intersection of $[HL^+]$ and $[OH^-]$ the equilibria give a numerical answer:

$$\frac{[L][H^+]}{K_{a2}} = \frac{K_w}{[H^+]}$$

$$[H^+]^2 = \frac{K_w K_{a2}}{[L]} \tag{49}$$

With $[L] = 0.01$, Eq. 49 gives pH = 10.965.

Note from Fig. 5.12, that at the intersection, $[L]$ is slightly less than $C = 0.01$. As with Na_3PO_4 (Example 4), $[L]$ can be calculated from the equilibrium and approximate mass balance

$$[HL^+] + [L] + \cdots = C$$

$$[L] = \frac{CK_{a2}}{K_{a2} + [H^+]}$$

Substituting this in Eq. (49) gives a form suitable for iteration

$$[H^+]^2 = \frac{K_w}{C}(K_{a2} + [H^+])$$

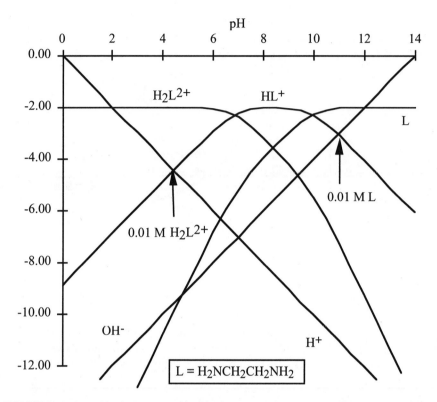

FIGURE 5.12. Logarithmic concentration diagram for ethylene diamine (L) and its protonated species. $pK_{a1} = 6.85$, $pK_{a2} = 9.93$.

Successive iterations give pH = 10.966, 10.946, converging at 10.945. This is only 0.02 pH units lower than the answer from the approximation $[L] = C$.

For the acidic form H_2L^{2+}, the approximation $[H_2L^{2+}] = C$ is much better, and the calculation is very simple. The proton condition is

$$[H^+] = [HL^+] + [OH^-]$$

$$[H^+] = [HL^+] + \cdots = \frac{[H_2L^{2+}]K_{a1}}{[H^+]} + \cdots = \frac{CK_{a1}}{[H^+]} + \cdots$$

$$[H^+]^2 = CK_{a1}$$

With $C = 0.01$ and $pK_{a1} = 6.85$, this gives pH = 4.43. The omitted terms ($+\cdots$) correspond to less than 0.01 pH units.

Just like phosphoric acid, the equilibria and mass balances of ethylene diamine and its acidic ions can be represented by a compact distribution diagram, Fig. 5.13.

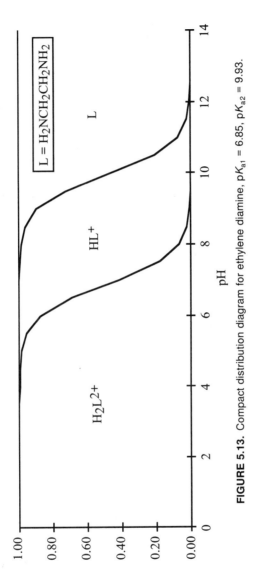

FIGURE 5.13. Compact distribution diagram for ethylene diamine, $pK_{a1} = 6.85$, $pK_{a2} = 9.93$.

AMINO ACIDS

As was mentioned in the introduction to this chapter, amino acid equilibria can be displayed in the same format as polyprotic acids and amines. Figure 5.14 is a logarithmic concentration diagram for glycine, which has $pK_{a1} = 2.35$ (for H2G$^+$ dissociation) and $pK_{a2} = 9.77$ (for HG dissociation). Note the wide range of pH over which HG is the dominant species.

A significant parameter for amino acids is the "isoelectric point," where there are equal concentrations of positively and negatively charged species. Experimentally, the isoelectric point is determined by conductance (ionic mobility). At low pH, glycine, being positively charged, migrates to the cathode; at high pH, being negatively charged, it migrates to the anode. Between these two extremes is a pH where it does not migrate. This is the isoelectric point. For glycine, this is when

$$[H_2G^+] = [G^-]$$

From the equilibria,

$$[H^+][HG] = K_{a1}[H_2G^+] \tag{50}$$

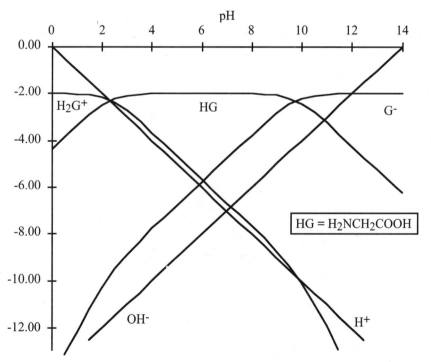

FIGURE 5.14. Logarithmic concentration diagram for 0.01 M glycine. $pK_{a1} = 2.35$, $pK_{a2} = 9.77$.

MIXTURE OF POLYPROTIC ACIDS AND BASES

$$[H^+][G^-] = K_{a2}[HG] \tag{51}$$

$$\frac{[H^+][HG]}{K_{a1}} = \frac{K_{a2}[HG]}{[H^+]} \tag{52}$$

$$[H^+]^2 = K_{a1}K_{a2}$$

$$[H^+]^2 = 10^{-2.35-9.77} = 10^{-12.12} \tag{53}$$

$$pH = 6.06$$

It is close to, but not exactly the pH of glycine in pure water, which includes $[H^+]$ and $[OH^-]$ in the charge balance:

$$[H^+] + [H_2G^+] = [G^-] + [OH^-] \tag{54}$$

$$[H^+]\left(1 + \frac{[HG]}{K_{a1}}\right) = \frac{(K_{a2}[HG] + K_w)}{[H^+]} \tag{55}$$

If $pK_{a1} \ll pH \ll pK_{a2}$ and [HG] is large compared to $[H^+]$ and $[OH^-]$, as in Fig. 5.14, Eq. (55) gives the same result as the isoelectric point (Eq. 53); but when [HG] is small compared to $[H^+]$ and $[OH^-]$, Eq. (55) gives the same pH as pure water. In between these two extremes, more than one term on each side of Eq. (55) is significant.

Like other polyprotic acids, the glycine equilibria can be displayed as a compact distribution diagram (Fig. 5.15).

MIXTURES OF POLYPROTIC ACIDS AND BASES

You can now calculate pH in a mixture of any number of strong, monoprotic, or polyprotic acids and bases. No new principles or techniques are required; the logarithmic concentration diagrams for the various components can be overlaid (as was shown for strong acid–weak acid mixtures in Chapter 4, pp. 117–123), and the critical intersection established using the proton condition.

Example 7. Find the pH of 0.1 M $(NH_4)_2HPO_4$. In Fig. 5.16, the logarithmic diagrams for 0.1 M H_3PO_4 and 0.2 M NH_3 are overlaid. The pH and concentration scales have been expanded to keep the diagram uncluttered. The proton condition can be obtained by balancing species with excess protons against those deficient in protons, using NH_4^+, HPO_4^{2-}, and H_2O as the baseline species.[19]

$$[H^+] + 2[H_3PO_4] + [H_2PO_4^-] = [NH_3] + [PO_4^{3-}] + [OH^-]$$

184 POLYPROTIC ACIDS AND BASES

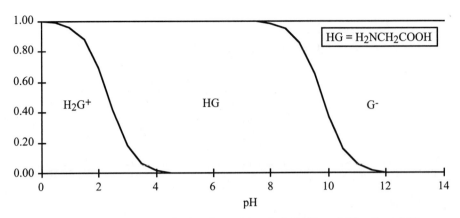

FIGURE 5.15. Compact distribution diagram for glycine, $pK_{a1} = 2.35$, $pK_{a2} = 9.77$.

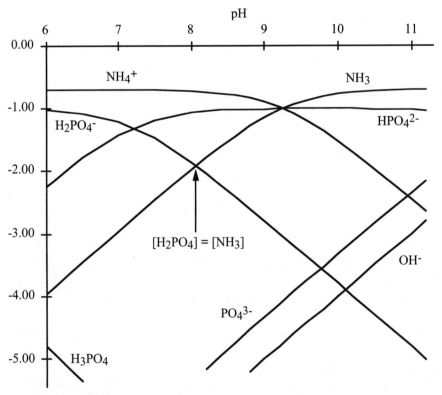

FIGURE 5.16. Logarithmic diagram for 0.1 M H_3PO_4 superposed on 0.2 M NH_3, with the intersection corresponding to the pH of 0.1 M $(NH_4)_2HPO_4$ shown by the arrow.

Selecting only the largest term on either side gives the desired intersection at pH = 8:

$$[H_2PO_4^-] = [NH_3]$$

Note that at this pH, [NH_4^+] and [HPO_4^{2-}] are large, essentially equal to their respective total concentrations; and the others are at least a factor of 10^3 smaller. [H^+] is so small, it does not even show on the graph.

DISSSOCIATION, FORMATION, AND TITRATION CURVES

In Chapter 4, the relationship between the dissociation curve and titration curve was derived for a monoprotic acid. Anticipating the discussion later in this chapter about titration of polyprotic acids, the dissociation and formation curves are presented here.

The "degree of formation," $\bar{n}$ ("n-bar") for the phosphoric acid system is defined as the average number of protons bound to a phosphate ion. Counting each phosphate ion as many times as there are protons bound to that species, and dividing by the total concentration, gives:

$$\bar{n} = \frac{[HPO_4^{2-}] + 2[H_2PO_4^-] + 3[H_3PO_4]}{C} \tag{56}$$

or in terms of the fractions α_n as defined above:

$$\bar{n} = \alpha_1 + 2\alpha_2 + 3\alpha_3 \tag{57}$$

The plot of $\bar{n}$ as a function of pH was first introduced by Niels Bjerrum[20] and called the "formation curve." The function for phosphoric acid is shown in Fig. 5.17.

The analogous degree of dissociation is $N - \bar{n}$, where N is the maximum number of protons that can be bound to the acid. For phosphoric acid, this is $3 - \bar{n}$, the average number of protons lost from an H_3PO_4 molecule. Making use of the mass balance,

$$3 - \bar{n} = 3(\alpha_0 + \alpha_1 + \alpha_2 + \alpha_3) - (\alpha_1 + 2\alpha_2 + 3\alpha_3)$$
$$3 - \bar{n} = 3\alpha_0 + 2\alpha_1 + \alpha_2 \tag{58}$$

This function, the dissociation curve, is also plotted in Fig. 5.17, and is the mirror image of $\bar{n}$.

Note that in Fig. 5.17 plateaus in $\bar{n}$ or $3 - \bar{n}$ occur at 1.0 and 2.0, corresponding to the salts NaH_2PO_4, and Na_2HPO_4. If the pK_a values were less widely separated, these plateaus would not be so flat, and even when $\bar{n}$ was an integral number, there would be significant quantities of the ions with more or fewer protons.

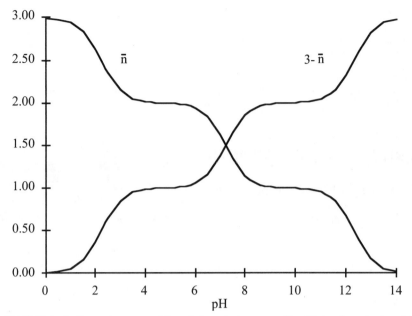

FIGURE 5.17. Formation curve ($\bar{n}$) and dissociation curve ($3 - \bar{n}$) for phosphoric acid.

Example 8. Recall the discussion of pyrophosphoric acid (p. 136, see Fig. 5.10). Fig. 5.18 shows $\bar{n}$ for pyrophosphoric acid in the pH range below 5. Note that the curve passes steeply through $\bar{n} = 3.0$ but shows a definite plateau at $\bar{n} = 2.0$. These points correspond to the salts NaH_3A and Na_2H_2A. In the example given with Fig. 5.10, a calculation of the pH of NaH_3A showed that $[H_4A]$, $[H_3A^-]$, and $[H_2A^{2-}]$ all were of comparable size, and all had to be included in the mass balance and charge balance.

Titration of a Polyprotic Acid with a Strong Base

The same principles as were developed for monoprotic acid titrations (pp. 136–147) apply to polyprotic acid titrations, especially the relationship between the titration curve and the dissociation curve.

A derivation from first principles begins with the charge and mass balances for the acid (H_3A, concentration C_o, volume V_o) and the titrant (NaOH, concentration C, volume V).

$$[H^+] + [Na^+] = [H_2A^-] + 2\,[HA^{2-}] + 3\,[A^{3-}] + [OH^-] \quad (59)$$

$$[Na^+] = \frac{CV}{(V + V_o)} \quad (60)$$

$$[H_3A] + [H_2A^-] + [HA^{2-}] + [A^{3-}] = \frac{C_o V_o}{(V + V_o)} = \sum_k [H_k A] \quad (61)$$

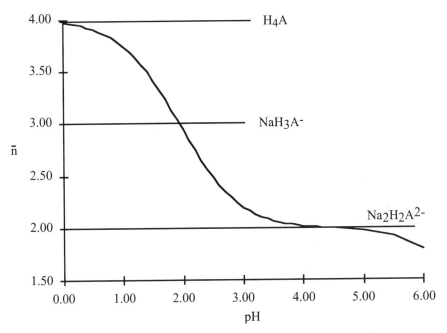

FIGURE 5.18. Plot of $\bar{n}$ for pyrophosphoric acid. Note the absence of a plateau at $\bar{n} = 3.0$, corresponding to NaH_3A^-, but the presence of a plateau at $\bar{n} = 2.0$, corresponding to $Na_2H_2A^{2-}$.

Since

$$\alpha_2 = \frac{[H_2A]}{\Sigma_k[H_kA]}, \quad \text{etc.,} \quad [H_2A] = \alpha_2\left(\frac{C_oV_o}{(V+V_o)}\right) \tag{62}$$

$$[H^+] + \left(\frac{CV}{(V+V_o)}\right) = (\alpha_2 + 2\alpha_1 + 3\alpha_0)\left(\frac{C_oV_o}{(V+V_o)}\right) + [OH^-] \tag{63}$$

$$\left(\frac{CV}{(V+V_o)}\right) = (3-\bar{n})\left(\frac{C_oV_o}{(V+V_o)}\right) - [H^+] + [OH^-] \tag{64}$$

Recalling that the fraction titrated is $\phi = CV/C_oV_o$,

$$\phi = (3-\bar{n}) + \frac{V+V_o}{C_oV_o}([OH^-] - [H^+]) \tag{65}$$

Figure 5.19 shows that the approximation $\phi = 3 - \bar{n}$ applies except at very high or very low pH.

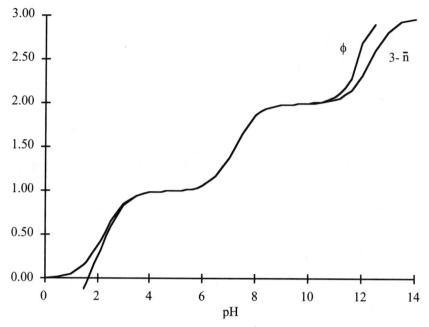

FIGURE 5.19. Titration of 0.01 M H_3PO_4 with 0.01 M NaOH. At higher concentrations, ϕ and $3 - \bar{n}$ are equal over an even wider range.

Note that V is included in ϕ as well as in the factor that multiplies $[OH^-] - [H^+]$. It is possible to calculate ϕ iteratively, setting $V = \phi\, C_o V_o / C$, an easy choice if $C \gg C_o$, since V can then be neglected compared to V_o. Alternatively, Eq. (65) can be solved directly for V, yielding a form that does not require iteration:

$$\phi = (3 - \bar{n}) + \frac{[OH^-] - [H^+]}{C - [OH^-] + [H^+]} \tag{66}$$

Note that if C is large compared to $[H^+]$ and $[OH^-]$, this also reduces to $\phi = 3 - \bar{n}$.

Titration Error

The titration error at any given end point can be easily calculated using Eq. (65) or (66). Near the equivalence point, $\phi \approx 1$ and $(V + V_o)/C_o V_o = (C + C_o)/C_o C$ (see Eq. (32), Chapter 3, p. 84 and note 20, p. 92). For the first end point of the phosphoric acid titration, $3 - \bar{n} - 1 = 2\alpha_0 + \alpha_1 - \alpha_3$, and the titration error is

$$\phi_{ep} - 1 = (2\alpha_0 - \alpha_1 - \alpha_3)_{ep} + \frac{C + C_o}{CC_o}\left(\frac{K_w}{[H^+]_{ep}} - [H^+]_{ep}\right) \tag{67}$$

Note that this end point corresponds to NaH$_2$PO$_4$, and that α_2 has been subtracted out of the equation. A plot of the error for titrating H$_3$PO$_4$ with NaOH is shown in Fig. 5.20. Note that since this is a titration of an acid with a base, the error is negative at low pH, positive at high pH, and zero at the equivalence point.

At the second (Na$_2$HPO$_4$) end point, ϕ is near 2.0 and the titration error (plotted in Fig. 5.20) is given by

$$\phi_{ep} - 2 = (\alpha_0 - \alpha_2 - 2\alpha_3)_{ep} + \left(\frac{C + 2C_o}{CC_o}\right)\left(\frac{K_w}{[H^+]_{ep}} - [H^+]_{ep}\right) \quad (68)$$

The slope $d\phi/d$pH of the curves in Fig. 5.20 is the inverse of the sharpness index $\eta = d$pH$/d\phi$. The first end point is slightly sharper than the second end point.

As with monoprotic acid titrations, the sharpness index can be estimated from the logarithmic concentration diagram (pp. 144–147). In each case, $1 + \log \eta'$ is the distance between the baseline and the intersection that determines the equivalence point. This is demonstrated by the diagram of Fig. 5.21, prepared for 0.01 M phosphoric acid. Note that the maximum sharpness index occurs at approximately the same pH as the equivalence points for NaH$_2$PO$_4$ and Na$_2$HPO$_4$. The third

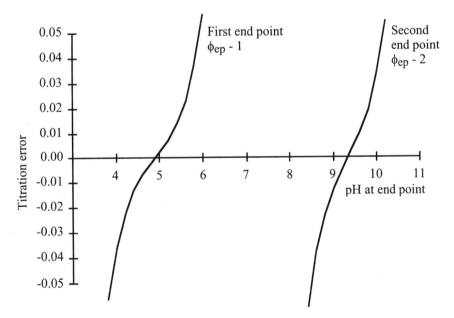

FIGURE 5.20. Titration error for first (NaH$_2$PO$_4$) and second (Na$_2$HPO$_4$) end points of the titration of 0.01 M H$_3$PO$_4$ with 0.01 M NaOH.

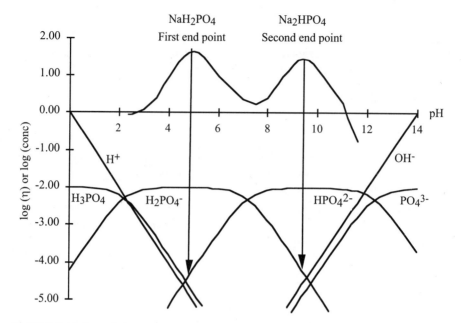

FIGURE 5.21. Log of sharpness index (upper graph) in the titration of 0.01 M H_3PO_4 with NaOH. The lower part of the figure is the same as Fig. 5.6, the logarithmic concentration diagram for H_3PO_4. Note that the maximum sharpness index occurs at the end points corresponding to the equivalence points for NaH_2PO_4 and Na_2HPO_4.

equivalence point, corresponding to Na_3PO_4, has a sharpness index less than 1 and cannot be used as an end point in a practical titration.

Buffer Index

As with monoprotic acids and bases (pp. 133–136), the buffer index is defined to be

$$\beta = \frac{\partial C_b}{\partial pH} \tag{69}$$

and it can be calculated by differentiating the charge balance with respect to [H⁺]. For a series of monoprotic acids HA_1, HA_2, etc., this gives

$$[H^+] + [Na^+] = [A_1^-] + [A_2^-] + \cdots + [OH^-] \tag{70}$$

$$C_b = \frac{C_1 K_{a1}}{K_{a1} + [H^+]} + \frac{C_2 K_{a2}}{K_{a2} + [H^+]} + \cdots + \frac{K_w}{[H^+]} - [H^+] \tag{71}$$

Making use of the relationship derived in Chapter 4 (Eqs. (79)–(80), p. 134), $d[H^+]/dpH = -2.303\,[H^+]$, Eq. (71) gives

$$\beta = 2.303\left(\frac{C_1 K_{a1}[H^+]}{(K_{a1}+[H^+])^2} + \frac{C_2 K_{a2}[H^+]}{(K_{a2}+[H^+])^2} + \cdots + \frac{K_w}{[H^+]} + [H^+]\right) \quad (72)$$

which can be rewritten in terms of the species concentrations:

$$\beta = 2.303\left(\frac{[HA_1][A_1^-]}{C_1} + \frac{[HA_2][A_2^-]}{C_2} + \cdots + [OH^-] + [H^+]\right) \quad (73)$$

Equation (73) can be expanded to include any number of monoprotic acids, but can also serve as a good approximation for many polyprotic acids. So long as the pK_a values are separated by more than one or two units, each of the dissociation steps of a polyprotic acid can be treated as an independent monoprotic acid with the same concentration.[21]

The basis for the rule of thumb in the previous paragraph is a more rigorous derivation of β for a diprotic acid H_2A. Omitting the terms in $[OH^-]$ and $[H^+]$ for simplicity, the charge balance gives

$$C_b = [HA^-] + [A^{2-}] = \frac{C(K_{a1}[H^+] + K_{a1}K_{a2})}{[H^+]^2 + K_{a1}[H^+] + K_{a1}K_{a2}} \quad (74)$$

Differentiating this and converting from $d[H^+]$ to dpH gives

$$\beta_{H_2A} = 2.303\, CK_{a1}[H^+]\frac{[H^+]^2 + 4K_{a2}[H^+] + K_{a1}K_{a2}}{([H^+]^2 + K_{a1}[H^+] + K_{a1}K_{a2})^2} \quad (75)$$

As before, the various terms can be identified with species concentrations:

$$\beta_{H_2A} = \frac{2.303}{C}([H_2A][HA^-] + 4[H_2A][A^{2-}] + [HA^-][A^{2-}]) \quad (76)$$

The first term corresponds to H_2A behaving as a monoprotic acid with $pK_a = pK_{a1}$, and the last term corresponds to HA^- behaving as a monoprotic acid with $pK_a = pK_{a2}$. The middle term is small unless the two pK_a values are close together. This is most easily seen by comparing the three terms graphically.

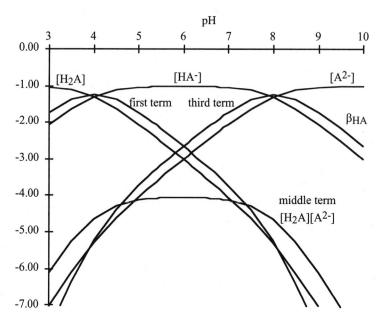

FIGURE 5.22. Logarithmic diagram for a hypothetical diprotic acid with $pK_{a1} = 4$ and $pK_{a2} = 8$. Note that the first and third terms of Eq. (76) correspond to the buffer indexes of H_2A and HA^- each behaving as monoprotic acids. The middle term is a factor of ten or more below the other two terms.

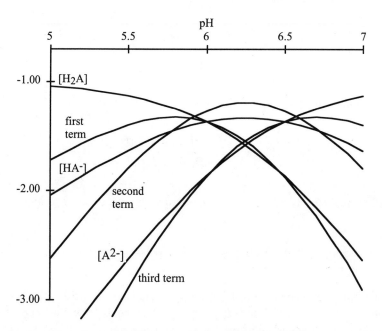

FIGURE 5.23. Diagram showing terms in Eq. (76) when $pK_{a1} = 6$ and $pK_{a2} = 6.5$. All three terms are of similar magnitude in the pH range from 5.5 to 7.

PROBLEMS

Figure 5.22 shows the various terms in Eq. (76) for a diprotic acid with $pK_{a1} = 4$ and $pK_{a2} = 8$. The first and third terms behave exactly like two independent monoprotic acids with concentration C. The middle term is always small compared to the other two. Figure 5.23 shows the situation when $pK_{a1} = 6$ and $pK_{a2} = 6.5$. All three terms are of similar magnitude and the "two monoprotic acids" simplification does not apply.

PROBLEMS

Use equilibrium constant data from Table 5.1 or Table 5.2 where necessary.

1. Derive formulas for the fraction of a diprotic acid present as each of the three species H_2A, HA^-, and A^{2-}, as functions of $[H^+]$.

2. Draw a logarithmic and compact distribution diagram for oxalic acid at total concentration 0.1 and 10^{-4} M.

3. Draw a logarithmic and compact distribution diagram for the amino acid serine [HO–CH_2–CH(NH_2)–COOH] at 0.01 M.

4. When carbon dioxide solutions are kept in a sealed system (no gas phase), they behave like a simple diprotic acid, except for the distinction between CO_2 + H_2O and H_2CO_3. Draw a logarithmic diagram for the various CO_2 species as a function of pH. If [H_2CO_3] is ignored in the mass balances, how much error is introduced?

5. Draw a distribution diagram for hydrogen sulfide in a sealed system, at total concentration 0.01 M. What fraction is present as H_2S, HS^-, and S^{2-} at the pH of seawater (8.3).

6. Derive an expression for the $[H^+]$ at which the intermediate species HA^- of a diprotic acid is a maximum. Compare with the expression for $[H^+]$ at which $[H_2A] = [A^{2-}]$.

7. As the hydrocarbon chain between acid groups of a dicarboxylic acid is made longer, the ratio of ionization constants K_{a1}/K_{a2} decreases

Name	Formula	pK_{a1}	pK_{a2}	K_{a1}/K_{a2}
Oxalic	HOOC–COOH	1.25	4.28	1070
Malonic	HOOC–CH_2–COOH	2.85	5.66	646
Succinic	HOOC–CH_2–CH_2–COOH	4.21	5.63	26.3
Glutaric	HOOC–$(CH_2)_3$–COOH	4.34	5.43	12.3
Adipic	HOOC–$(CH_2)_4$–COOH	4.42	5.42	10.0
Sebacic	HOOC–$(CH_2)_8$–COOH	4.40	5.22	6.6

Plot $\log(K_{a1}/K_{a2})$ versus number of carbons in the chain and extrapolate to an infinitely long chain. Make a statistical argument, using the concept of equilibrium as a balance of opposing reaction, for a limiting value $K_{a1}/K_{a2} = 4$ for a diprotic acid with a very long hydrocarbon chain.

8. Derive formulas for the fraction of a tetraprotic acid present as each of the species H_4A, H_3A^-, H_2A^{2-}, HA^{3-}, and A^{4-}, as functions of $[H^+]$. Plot a logarithmic diagram and a compact distribution diagram for pyrophosphoric acid ($H_4P_2O_7$) and EDTA [which stands for ethylene diamine tetra-acetic acid ($HOOC-CH_2)_2N-CH_2CH_2-N(CH_2-COOH)_2$].

9. In the early literature, one can find the ionization constants of amino acids expressed in the "classical" manner, where the acid ionization constant is given for each of the COOH groups and the basic ionization constant is given for each of the NH_2 groups. Calculate the "classical constants" in terms of the stepwise dissociation constants of the most highly protonated form (as we have done in Table 5.2) for an amino acid with:

    (a) one COOH group and one NH_2 group (e.g., glycine);

    (b) two COOH groups and one NH_2 group (e.g., aspartic acid);

    (c) one COOH group and two NH_2 groups (e.g., arginine).

10. When it was first realized that amino acids existed as zwitterions, Niels Bjerrum proposed that the ionization constants of amino acids be presented in the zwitterionic form, where acid ionization constants are given for the NH_3^+ groups on the zwitterion and basic ionization constants are given for the COO^- groups. Relate these to the "classical" constants described in Problem 9, and to the stepwise dissociation constants as given in Table 5. 2. In one table, the constants for aspartic acid are listed as $pK_{a1} = 2.08$, $pK_{a2} = 3.94$, $pK_b = 4.02$. To what equilibrium expressions do these correspond? What ambiguities might arise from a failure to specify which of the three systems was being used?

11. Draw a distribution diagram for citric acid. What fraction of the total concentration would be found as the A^{4-} species in 0.1 M NaOH?

12. Find the pH of the following (use data from Tables 5.1 and 5.2, and logarithmic concentration diagrams as appropriate)

    (a) 0.10 M H_3PO_4;

    (b) 10^{-5} M H_3PO_4;

    (c) 10^{-4} M Na_2HPO_4;

    (d) 0.01 M NaH_2AsO_4;

    (e) 0.002 M $NaHCO_3$ (a typical fresh water);

    (f) 0.1 M Na_2CO_3;

(g) 0.1 M potassium hydrogen phthalate (an international pH standard);

(h) 0.001 M disodium citrate;

(i) 0.1 M HCl, 0.1 M NaHSO$_4$, and 0.1 M acetic acid;

(j) 0.1 M oxalic acid + 10^{-3} M acetic acid;

(k) 0.1 M ammoniuim glycinate (NH$_4$G);

(l) 0.01 M ammonium carbonate [(NH$_4$)$_2$CO$_3$];

(m) 0.01 M serine; compare with the pH at the isoelectric point;

(n) 0.1 M ethylene diamine (L) + 0.1 M glycine hydrochloride (GH$^+$Cl$^-$);

(o) 0.1 M arginine (HL) + 0.2 M adipic acid (H$_2$A).

13. Derive the approximate expression for pH of the salt NaHA (where H$_2$A is a diprotic acid)

$$pH = \tfrac{1}{2}(pK_{a1} + pK_{a2})$$

What approximations are necessary? How accurate is this formula for each of the diprotic acids listed in Problem 7?

14. Derive the approximate expression for pH in a tetraprotic acid salt. when H$_2$A^{2-} is a maximum:

$$pH = \tfrac{1}{2}(pK_{a2} + pK_{a3})$$

Note that this is independent of total concentration of the salt. As the salt concentration decreases, the approximations required to derived this simple formula break down. What would be a more accurate expression, which would take account of the limiting case of zero concentration, where pH = 7.0?

15. Aspartic acid is referred to as an "acidic amino acid," glycine as a "neutral amino acid," and arginine as a "basic amino acid." Compare the pH of these three compounds in pure water at a moderate concentration (0.1 M), and at the isoelectric point.

16. Compare the pH of glycine with that of an equimolar mixture of acetic acid and methylamine, which would have the same concentration of COOH and NH$_2$ groups. What structural factors might account for the difference?

17. Compute a dissociation courve for a mixture containing 0.10 mole each of HCl, NaHSO$_4$, and acetic acid per liter. Define $\alpha = 1 + [SO_4^{2-}]/C_{NaHSO_4} + [Ac^-]/C_{Ac}$ and plot as a function of pH. Compare this curve with the titration curve of the three-acid mixture with 0.1 M NaOH. Locate end points that could allow the three components to be determined separately. Which is most accurate?

POLYPROTIC ACIDS AND BASES

18. Derive an expression for the fraction titrated as a function of $[H^+]$ for the titration of a diprotic acid with a strong base.

19. Derive an expression for the fraction titrated as a function of $[H^+]$ if neutral glycine is titrated with a strong acid, and also with a strong base.

20. The predominant acid–base system in natural waters is carbon dioxide and its ions. An important measure of this is "alkalinity," which is the amount of strong acid required to bring the water from its original pH (6 to 9) to the approximate pH of 10^{-3} M CO_2 in water (pH = 4.5, methyl orange). Derive an expression for the fraction titrated as a function of pH. Plot the titration curve for total of all carbonate species = 10^{-3} M in a sealed system if the initial pH is 8.0.

21. Strong acid–strong base titrations are often biased by the presence of traces of carbon dioxide, which shift the end point from 7.0 to a more acidic value. Calculate the titration curve for 0.001 M NaOH containing 10^{-5} M carbonate, with 0.001 M HCl.

22. Estimate the sharpness index for the two end points in a titration of Na_2CO_3 with HCl. Which is better?

23. Estimate the sharpness index for the two end points in the titration of phthalic acid with NaOH. Which is the best end point for a quantitative analysis? Why is potassium hydrogen phthalate a good candidate for an international pH standard?

24. In many analytical chemistry textbooks and handbooks, the following formulas are given for pH during the titration of a diprotic acids with strong base:

$$\text{At start:} \quad pH = \tfrac{1}{2}(pK_{a1} - \log C)$$
$$\text{First end point:} \quad pH = \tfrac{1}{2}(pK_{a1} + pK_{a2})$$
$$\text{Second end point:} \quad pH = \tfrac{1}{2}(pK_w + pK_{a2} + \log C)$$

Derive each of these formulas, noting the approximations made. Use the logarithmic concentration diagram to test how accurate they are for the following diprotic acids:

(a) oxalic acid;
(b) adipic acid;
(c) carbon dioxide;
(d) hydrogen sulfide.

25. Derive simple formulas (as in Problem 19) for pH at the end points in the titration of 10^{-4} M Na_3PO_4 with 10^{-2} M HCl. Use the logarithmic concentration diagram to test how accurate they are.

PROBLEMS

26. Find the pH and the buffer index of the following solutions:
    (a) 0.2 M $NH_4H_2PO_4$;
    (b) 0.1 M potassium hydrogen oxalate;
    (c) 0.001 M NH_4HCO_3;
    (d) a "universal buffer" consisting of 0.1 M $(NH_4)_2HPO_4$, 0.1 M Na borate, and 0.1 M NaAc;
    (e) 0.1 M N_2H_4 and 0.1 M NH_4Cl.

27. Calculate the buffer index as a function of pH over the range $2 < \text{pH} < 12$ for the "universal buffer" of Problem 26d.

28. Derive an equation analogous to Eq. (76), but for a triprotic acid.

29. Using Eq. (75) for the buffer index of a diprotic acid, show that when $K_{a1} = 4K_{a2}$ (the statistical limit for a long-chain dicarboxylic acid), β_{H_2A} has the same form as β for a monoprotic acid, but with a concentration $2C$ and an effective K_a that is the geometric mean of K_{a1} and K_{a2}.

30. Show that when $K_{a1} > K_{a2}$, the central minimum in the buffer index curve for a diprotic acid is obtained at pH = $\frac{1}{2}(pK_{a1} + pK_{a2})$.

31. Plot β/C from Eq. 75 for a diprotic acid with $K_{a1} = K_{a2} = K_a$. On the same graph plot β/C for a monoprotic acid with the same K_a. Compare the curves.

32. Equation (37), p. 176, for the concentration of the diprotic acid species H_2A can be expanded in series to give part of the neglected terms in Eq. (40)[22]:

$$[H_2A] = \frac{C[H^+]}{[H^+] + K_{a1}} - \frac{CK_{a1}K_{a2}}{([H^+] + K_{a1})^2} + \frac{C(K_{a1}K_{a2})^2}{H([H^+] + K_{a1})^3} + \cdots \quad (40a)$$

The ratio of the second term to the first term is $K_{a1}K_{a2}/[H^+]([H^+] + K_{a1})$, which decreases linearly as K_{a2} decreases when $[H^+]$ is close to K_{a1}. Likewise, Eq. (38) can be expanded to give Eq. (41) with extra terms:

$$[HA^-] = \frac{CK_{a1}}{[H^+] + K_{a1}} - \frac{CK_{a1}^2 K_{a2}}{[H^+]([H^+] + K_{a1})^2} + \cdots \quad (38a)$$

$$[HA^-] = \frac{C[H^+]}{[H^+] + K_{a2}} - \frac{C[H^+]^3}{K_{a1}([H^+] + K_{a2})^2} + \cdots \quad (38b)$$

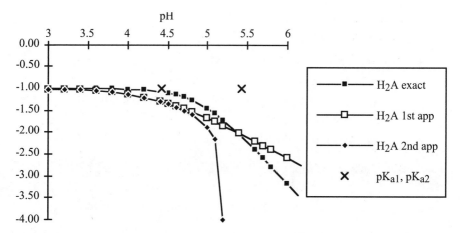

FIGURE 5.24. Approximations to [H$_2$A].

and Eq. 39 can be expanded to give Eq. 42 with extra terms:

$$[A^{2-}] = \frac{CK_{a2}}{[H^+] + K_{a2}} - \frac{CK_{a2}[H^+]^2}{K_{a1}([H^+] + K_{a2})^2} \quad (42a)$$

These various approximations are compared in Figs. 5.24–5.26, which show log concentration versus pH for the species H$_2$A, HA$^-$, and A^{2-}, calculated us-

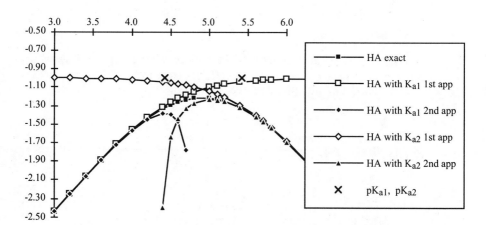

FIGURE 5.25. Approximations to [HA$^-$].

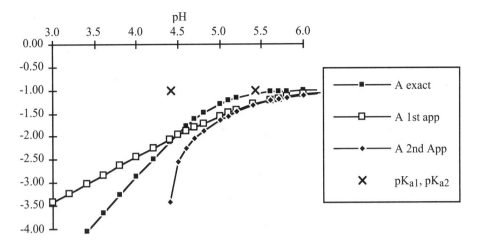

FIGURE 5.26. Approximations to [HA^{2-}].

ing the constants for adipic acid (pK_{a1} = 4.42, pK_{a2} = 5.42). Since these pKs are only one unit apart, this is a situation where the approximations would be expected to break down.

(a) Over what range does each of the approximations hold? To what precision?

(b) Does adding a third or fourth term to Eq. (38a), (38b), (42a) improve matters?

(c) Compare this case with the first two proton dissociations of pyrophosphoric acid (see text).

(d) Compare this case with phthalic acid (pK_{a1} = 2.95, pK_{a2} = 5.41).

33. Use the approximations of Problem 22 to evaluate the buffer index equation for two monoprotic acids (Eq. 73) and the exact equation for a diprotic acid (Eq. 76). Evaluate the range of C and pH over which each of the approximations holds. To what precision?

NOTES

1. Selected from W. Smith and A. E. Martell, *Critical Stability Constants*, 6 volumes. Plenum, N.Y., 1976–1989. These authors have selected the best literature values at 25°C, but do not give temperature or pressure dependence except near 25°C. See also L. G. Sillén and A. E. Martell, *Stability Constants of Metal–Ion complexes*, Special Publications 17 (1964) and 25 (1971), the Chemical Society, London. They did not attempt critical selection of data, but rather presented a summary of all data available at the time,

with an index to all the available literature. Hence these are still good sources for data at temperatures other than 25°C, pressures other than 1 atm, and a variety of nonaqueous media.

2. See Note 1. Formulas indicating the number of ionizable hydrogen atoms of the neutral and most higly protonated forms are given after the name of the compound. Protons are generally lost from COOH groups and gained by NH_2 groups. On 8-hydroxyquinoline, the proton on the OH group ionizes. On arginine, two protons can add to the NH_2 groups, and one proton can be lost from the COOH group. The NH groups do not enter into acid–base reactions in aqueous solutions.

3. A more rigorous alternative is to define a combined equilibrium species that is $[CO_2]$ + $[H_2CO_3]$ as the acid. The constants do not change significantly.

4. For further details on CO_2, see J. N. Butler, *Carbon Dioxide Equilibria and Their Applications*, CRC–Lewis, 1991.

5. S. Licht, *Anal. Chem.* **57**:514, 1985; S. Licht and J. Manassen, *J. Electrochem. Soc. Electrochem. Sci. Technol.* **134**:918, 1987.

6. R. G. Bates, *J. Res. Nat. Bur. Stand.* **47**:127, 1951; J. M. Wright, W. T. Lindsay, Jr., and T. R. Druga, TID-4500 or WAPD-TM-204, 1961, cited by Sillén and Martell, 1964, pp. 181, 182, 189.

7. At the beginning of Chapter 4 (pp. 97–98), we noted that a more thermodynamically rigorous form for temperature dependence is $\log K = A/T + B + C \log T + \cdots$. Although this form is less convenient than a simple polynomial for least–squares fitting, it has been used for pK_{a2} (p. 165).

8. $d \log K/dT = \Delta H/(2.3RT^2) = 0.00246\ \Delta H$ at 25°C for ΔH in kcal/mol.

9. R. G. Bates, *J. Res. Nat. Bur. Stand.* **47**:127, 1951.

10. K. S. Pitzer, *J. Amer. Chem. Soc.* **59**:2365, 1937.

11. R. G. Bates and S. F. Acree, *J. Res. Nat. Bur. Stand.* **30**:129, 1943; F. Ender, W. Teltschik, and K. Schäfer, *Z. Elektrochem.* **61**:775, 1957; A. K. Grzybowski, *J. Phys. Chem.* **62**:555, 1958.

12. A. K. Grzybowski, *J. Phys. Chem.* **62**:555, 1958.

13. I. N. Kugelmass, *Biochem. J.* **23**:587, 1929.

14. K. S. Pitzer, *J. Amer. Chem. Soc.* **59**:2365, 1937.

15. At pH = 2.28, the four terms are $[H^+]^3 = 10^{-6.83}$, $K_{a1}[H^+]^2 = 10^{-6.79}$, $K_{a1}K_{a2}[H^+] = 10^{-11.72}$, and $K_{a1}K_{a2}K_{a3} = 10^{-21.76}$. Term 1 $\approx$ term 2 » term 3 » term 4.

16. H. S. Simms, *J. Am. Chem. Soc.* **48**:1239, 1926, is noted for his use of paper templates in curve fitting, as well as for his analysis of diprotic acids with pK_{a1} and pK_{a2} close together. See F. J. C. Rossotti and H. Rosotti, *The Determination of Stability Constants*, McGraw-Hill, N.Y., 1961, pp. 88, 98.

17. R. P. Mitra, H. C. Malhotra and D. V. S. Jain, *Trans. Faraday Soc.* **62**:167, 1966.

18. Constants at 25°C from R. Nasanen and P. Merilainen, *Suomen Kem.* **B36**:97, 1963; ibid. *Acta Chem. Scand.* **18**:1337, 1964; R. Nasanen, M. Koskinen and K. Kajander, *Suomen Kem.* **B38**:103, 1965; R. Nasanen and M. Koskinen, *Suomen Kem.* B40:108, 1967, cited by Sillén and Martell (1971). The ionic strength dependence is also given: $pK_{a1} = 6.84 + 1.018\ \sqrt{I}/(1 + 1.381\sqrt{I}) + 0.209\ I$; $pK_{a2} = 9.93 + 0.308\ I$.

19. Alternatively, the mass and charge balances can be combined to eliminate the large terms [NH$_4^+$] and [HPO$_4^{2-}$]
20. N. Bjerrum, *Selected Papers*, E. Munksgkaard, Copenhagen, 1949.
21. H. S. Simms, *J. Am. Chem. Soc.* **48**:1239, 1926.
22. Note $C(A + B) = C/A - CB/A^2 + CB^2/A^3 + \cdots$.
23. Selected from Sillén and Martell (1964 and 1971). See note 1.

6

SOLUBILITY

The Solubility Product
  Solubility in Pure Water
  The Common-Ion Effect
  Mixing two solutions
  Logarithmic Concentration Diagrams
  Separation of Compounds by Precipitation
  Effects of Ionic Strength
Evaluating pK_{so} from Data
Precipitation Titrations
  Point of Maximum Slope for Symmetrical Titration
  Unsymmetrical Titrations
  Point of Maximum Slope for Unsymmetrical Titration
Complications in Solubility Calculations
  Acid–Base Effects on Solubility
  Salt of a Weak Acid
  Effect of Complex Formation on Solubility
Problems

THE SOLUBILITY PRODUCT

In 1899, W. Nernst showed that the equilibrium between a solid ionic salt and its solution in water was governed by the solubility-product expression.[1] For example, consider the salt AgCl, which dissolves slightly in water to give primarily the cation Ag^+ and the anion Cl^-. The reaction

$$AgCl(s) \rightleftharpoons Ag^+ + Cl^-$$

THE SOLUBILITY PRODUCT

TABLE 6.1. Solubility products at 25°C and zero ionic strength[2]

Ions of equal charge		Ions of unequal charge	
Salt	pK_{so}	Salt	pK_{so}
TlCl	3.74±0.02	Ag_2SO_4	4.83±0.03
$AgBrO_3$	4.26±0.02	$Ca(OH)_2$	5.19±0.02
$CaSO_4$	4.62±0.02	BaF_2	5.76
$(Hg_2)SO_4$	6.13±0.04[3]	$Cu(IO_3)_2$	7.13±0.01
$SrSO_4$	6.50±0.05	MgF_2	8.18
$AgIO_3$	7.51±0.01	SrF_2	8.54
$PbSO_4$	7.79±0.02	CaF_2	10.41
AgCl	9.752±0.002	$Ce(IO_3)_3$	10.86
$BaSO_4$	9.96±0.03	$La(IO_3)_3$	10.99±0.07
AgSCN	11.97±0.03	$Mg(OH)_2$(brucite)	11.15±0.2
AgBr	12.30±0.02	$Mg(OH)_2$ (active)	9.2
AgI	16.08	$Pb(IO_3)_2$	12.61
		$(Hg_2)Cl_2$	17.91±0.03

has the equilibrium expression

$$[Ag^+][Cl^-] = K_{so} \qquad (1)$$

where K_{so} is the solubility product constant of silver chloride. Because the activity of a pure solid is constant, it has been included in the equilibrium constant. A more formal way of saying this is that because the solid is pure, it has unit activity, and hence does not appear in the equilibrium expression. The solubility product applies only in a saturated solution.

A number of salts that yield only two ions in appreciable concentration when dissolved in pure water are listed in Table 6.1, together with their solubility products at 25°C and zero ionic strength

Solubility in Pure Water

If S moles of AgCl dissolve in pure water to form a liter of saturated solution, a mass balance on cation and anion give

$$[Ag^+] = S \quad \text{and} \quad [Cl^-] = S$$

Substituting these two equations in the solubility product expression gives

$$[Ag^+][Cl^-] = S^2 = K_{so} \qquad (2)$$

$$S = \sqrt{K_{so}} \qquad (3)$$

The solubility of an ionic salt in pure water thus depends on the solubility product alone, provided that only these two ions are in solution. The same form of equation holds for $BaSO_4$ or any other salt where the two ions have the same charge.

Example 1. Calculate the solubility of $BaSO_4$ in pure water at 25°C. From Table 6.1, $pK_{so} = 9.96$ and from Eq. (2) or (3), $S = \sqrt{K_{so}} = 10^{-(9.96/2)} = 10^{-4.98}$.

If the charges on the ions are not equal, the solubility product becomes slightly more complicated. For the general salt M_zX_y, which yields only the cation M^{y+} and anion X^{z-} in solution in pure water, the dissolution reaction is

$$M_zX_y(s) \rightleftharpoons zM^{y+} + yX^{z-}$$

and the solubility product is

$$[M^{y+}]^z[X^{z-}]^y = K_{so} \qquad (4)$$

In pure water, mass balances on cation and anion give

$$[M^{y+}] = zS \quad \text{and} \quad [X^{z-}] = yS$$

Substitution in the solubility product yields

$$(zS)^z(yS)^y = S^{(z+y)}z^zy^y = K_{so} \qquad (5)$$

$$S = \left(\frac{K_{so}}{z^zy^y}\right)^{1/(z+y)} \qquad (6)$$

Unfortunately, only a small fraction of all ionic salts can be treated by this simple formalism, because many ions are weak acids or bases, many salts yield a number of complex ions on solution in water, and many are so soluble that serious activity coefficient corrections are required, making K_{so} a function of concentration. These complications are discussed later in this chapter and in Chapter 7.

The Common-Ion Effect

Calculating the solubility of salts in pure water can be very simple, but does not make the most valuable use of the solubility product principle. In a saturated solution of an ionic salt, addition of another, more soluble, salt containing one of those ions will decrease the solubility of the first. This is known as the "common-ion effect."

Example 2. Calculate the solubility of $BaSO_4$ in 0.01 M $BaCl_2$. As above, the solubility product is

$$[Ba^{2+}][SO_4^{2-}] = 10^{-9.96}$$

THE SOLUBILITY PRODUCT

at 25°C. The mass balances are a little different, since $[Ba^{2+}]$ can come both from the dissolution of $BaSO_4$ and from the added $BaCl_2$:

$$[Ba^{2+}] = S + 0.01$$
$$[SO_4^{2-}] = S$$

Substituting these into the solubility product gives a quadratic equation:

$$S(S + 0.01) = 10^{-9.96}$$

From Example 1, we know that the solubility of $BaSO_4$ in pure water is about 10^{-5} M, and because of the common-ion effect, its solubility in 0.01 M $BaCl_2$ will be less than 10^{-5} M—hence negligible compared to 0.01. The quadratic thus reduces to

$$S(0.01) = 10^{-9.96}$$
$$S = 10^{-7.96}$$

which is indeed less than 10^{-5} M.

The common-ion effect for a more general salt MX in a solution containing C mole/L of a soluble salt of X^- can be formulated as

$$[M^+][X^-] = K_{so} \tag{7}$$
$$S(S + C) = K_{so} \tag{8}$$

Exactly the same equation is obtained if a soluble salt of M^+ at C mole/L is used.

If $C \gg S$, this reduces to $S = K_{so}/C$; if not, the quadratic equation must be solved. Note that in the limit where $C = 0$, $S = \sqrt{K_{so}}$.

Figure 6.1 is a plot of Eq. (8) for $BaSO_4$ with added salt (either Na_2SO_4 or $BaCl_2$).

Mixing Two Solutions

Consider an example where two solutions are mixed and a salt precipitates. (This is a simple version of a precipitation titration, some of which are discussed later.) Unless the two solutions are exactly in the stoichiometric ratio, the common-ion effect will decrease the solubility of the salt and make the precipitation more complete.

Example 3. $V = 50$ mL of a solution containing $C_1 = 30$ μmol/L (μM) $BaCl_2$ are mixed with $V_o = 100$ mL of $C_2 = 45$ μM Na_2SO_4. Calculate the fraction of barium precipitated as $BaSO_4$. Let P be the number of moles of $BaSO_4$ precipitated per liter of solution:

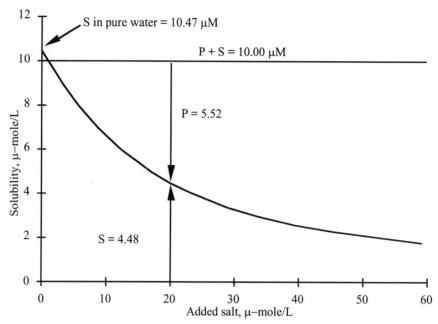

FIGURE 6.1. Solubility of BaSO$_4$ in excess Na$_2$SO$_4$ (or BaCl$_2$). With no added salt, the solubility is 10.47 µM. If 20 µM salt is added to 10 µM Ba^{2+}, P = 5.52 µmole/L of BaSO$_4$ is precipitated; S = 4.48 µmole/L is still dissolved.

$$[Ba^{2+}] + P = C_1\left(\frac{V}{V + V_o}\right) = 1.00 \times 10^{-5}$$

$$[SO_4^{2-}] + P = C_2\left(\frac{V}{V + V_o}\right) = 3.00 \times 10^{-5}$$

Note that three times the amount of sulfate theoretically required to react with all the barium has been added. The above equations, together with the solubility product, provide three equations in three unknowns. Eliminating P from the two mass balances gives

$$[Ba^{2+}] + 2 \times 10^{-5} = [SO_4^{2-}]$$

Substituting $[SO_4^{2-}]$ in the solubility product gives a quadratic in $[Ba^{2+}]$:

$$[Ba^{2+}]([Ba^{2+}] + 2 \times 10^{-5}) = K_{so}$$
$$[Ba^{2+}]^2 + 2 \times 10^{-5}\,[Ba^{2+}] - 1.096 \times 10^{-10} = 0$$

THE SOLUBILITY PRODUCT

The quadratic formula yields $[Ba^{2+}] = 4.48 \times 10^{-6}$, or 4.48 μM. From the mass balances,

$$[SO_4^{2-}] = 2 \times 10^{-5} - [Ba^{2+}] = (20 - 4.48) \times 10^{-6} = 15.52 \; \mu M$$

$$P = 1 \times 10^{-5} - [Ba^{2+}] = (10.00 - 4.48) \times 10^{-6} = 5.52 \; \mu M \; \text{or} \; 55.2\%$$

$$S = [Ba^{2+}] = 4.48 \; \mu M \; \text{or} \; 44.8\%$$

$$S + P = 10.00 \; \mu M \; \text{or} \; 100\%$$

Under the conditions specified, about half the barium is precipitated. Note that $P + S$ is 10.0 μM, which is somewhat less than the solubility of $BaSO_4$ in pure water (10.47 μM; see Fig. 6.1). Thus if exactly the stoichiometric amount of sulfate were added to a 10.0 μM barium solution, the solution would be unsaturated and P would be zero. But if a small excess of either barium or sulfate were present, the solubility would decrease and some $BaSO_4$ would precipitate. As long as the solution is saturated and sulfate is in excess, dissolved barium is governed by the solubility product $[Ba^{2+}][SO_4^{2-}] = K_{so}$, with $S = [Ba^{2+}]$, and $S + P$ = total barium in the system.[4]

Logarithmic Concentration Diagrams

The solubility product law can be represented on a logarithmic concentration diagram analogous to those used in describing acid–base equilibria. For a salt MX, the solubility product can be transformed to the logarithmic form:

$$[M][X] = K_{so} \tag{9}$$

$$-\log[M] - \log[X] = pK_{so} \tag{10}$$

$$\log[M] = -pK_{so} + pX \tag{11}$$

where $pX = -\log[X]$ by analogy with pH. These functions are plotted in Fig. 6.2.

When X is in excess, the dissolved concentration includes the contribution from the dissolved salt $[X] = C_X + S$. For relatively large amounts of added X, $C_X \gg S$ and $[X] = C_X$. When C_X is small enough to approach S, the full mass balance must be used (see previous section):

$$S(C_X + S) = K_{so}$$

Similarly, when M is in excess, $S = [X]$ and $[M] = C_M + S$:

$$S(C_M + S) = K_{so}$$

In the chart of Fig. 6.2, begin at the center, where $[M] = [X]$ and the solubility is that in pure water. Proceeding left, the line sloping upward to the left represents

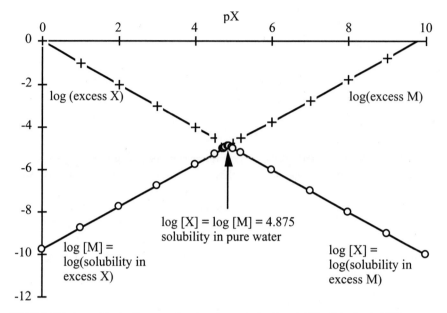

FIGURE 6.2. Logarithmic diagram showing the variation in solubility [M] of salt MX in excess X (left side of chart) and solubility [X] in excess M (right side of chart). Calculations are for AgCl, with pK_{so} = 9.75: [M] = [Ag$^+$] and [X] = [Cl$^-$].

excess X (actually $[X] = C_X + S$), and the line sloping downward to the left represents the solubility $[M]$ of MX. On the right side of the chart, added M ($[M] = C_M + S$) is given by the line sloping upward to the right, and the solubility $[X]$ is given by the line sloping downward to the right. Thus the maximum solubility occurs when exactly the stoichiometric amounts of M and X are present: $C_M = C_X$ and $[M] = [X]$. Figure 6.3 shows a magnified section of Fig. 6.2 near the point where $[M] = [X]$, illustrating the deviation between $[M]$ and C_M or $[X]$ and C_X. Figure 6.4 is a linear plot of the Fig. 6.3 data.

Separation of Compounds by Precipitation

One of the oldest methods of chemical separation is fractional precipitation. This involves the addition of a reagent that precipitates most of one metal ion and leaves another mostly in solution. Simple calculations based on equilibrium constants can tell the maximum degree of separation that can be achieved under given conditions. Example 4 is a typical calculation.

In practice, the separation will be poorer than calculated due to coprecipitation. In this phenomenon are included the mechanical enclosure of some solution by the precipitate, adsorption of foreign ions on the surface of the precipitate, solid solution formation, etc.[5]

Example 4. A solution is 0.010 M in $BaCl_2$ and 0.010 M in $SrCl_2$. Concentrated sodium sulfate solution is added. Which ion precipitates first? What is its concentration when the second ion begins to precipitate?

THE SOLUBILITY PRODUCT

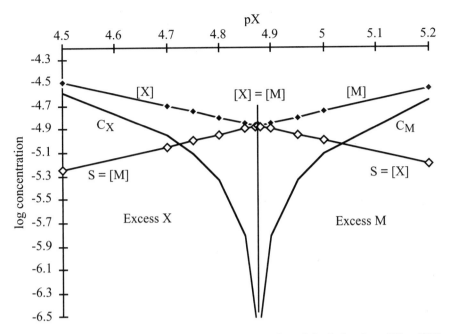

FIGURE 6.3. Magnified section of Fig. 6.2, showing how C_X and C_M deviate from [X] and [M], both approaching zero as [X] approaches [M].

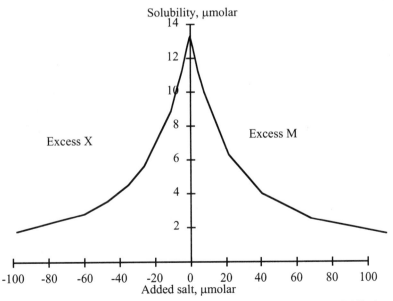

FIGURE 6.4. Linear plot of the same data as are displayed in Fig. 6.3. The solubility is maximum when $C_M = C_X = 0$. It decreases (to the left) as excess X is added, and decreases (to the right) as excess M is added. The concentration of added X is plotted as a negative number for mathematical convenience.

Two solubility products are involved, each holding *only* in the presence of the corresponding solid salt.

$$[Ba^{2+}][ISO_4^{2-}] = 10^{-9.96} = 1.1 \times 10^{-10}$$
$$[Sr^{2+}][SO_4^{2-}] = 10^{-6.17} = 2.8 \times 10^{-7}$$

In a solution saturated with both solids, elimination of $[SO_4^{2-}]$ between these two equations shows that $[Sr^{2+}]$ is much larger than $[Ba^{2+}]$:

$$\frac{[Sr^{2+}]}{[Ba^{2+}]} = 10^{-6.17+9.96} = 10^{+3.79} = 2.5 \times 10^{+3}$$

Thus when the first crystal of $SrSO_4$ precipitates, the solubility equilibria predict that $[Ba^{2+}]$ is only 0.04% as large as $[Sr^{2+}]$, the rest of the barium having precipitated as $BaSO_4$. At this point, $[Sr^{2+}]$ is still at its initial value of 0.010, and hence

$$[Ba^{2+}] = (0.010)/(2.5 \times 10^{+3}) = 4.0 \times 10^{-6} \text{ M}$$

When any sulfate in excess of the amount required to start precipitation of Sr^{2+} has been added, the solubility product of $SrSO_4$ will also hold, and the composition of the solution will remain unchanged as $SrSO_4$ and $BaSO_4$ precipitate together.

The optimum separation will be obtained if exactly the stoichiometric amount of sulfate required to precipitate all the barium is added. At this point all the Sr^{2+} is in solution (except for that which coprecipitates with the $BaSO_4$) and 99.96% of the barium has been precipitated. That is fine in theory, but in most gravimetric analyses, the precise stoichiometric amount of reagent to add is not known accurately, and so some external method of determining that point is needed.[6]

Another approach is to introduce an indicator, either an ion such as chromate that produces a colored precipitate or a dye that selectively adsorbs to one of the precipitates. Electrochemical measurements can also be used to determine the concentration of certain free ions such as Ag^+. At this point, one is progressing toward precipitation titrations, which are discussed in a later section.

Effects of Ionic Strength

Even if only two ions are formed by dissolution of a salt, their activity will be affected by the presence of other salts. For example, the solubility of AgCl in pure water is given by the condition $[Ag^+] = [Cl^-]$, but so is its solubility in 0.1 M $NaClO_4$. These two solubilities are different because of the different ionic strength—both the solubility and ionic strength are 13.3 μM in the first case, but in the second, the ionic strength is 0.1 M and the solubility is 17.5 μM (Fig. 6.5).

The formal way to account for this effect is to include activity coefficients in the solubility product (compare with Eq. 1):

$$[Ag^+]\gamma_+ [Cl^-]\gamma_- = K_{so}^{\circ} = 10^{-9.75} \tag{12}$$

THE SOLUBILITY PRODUCT

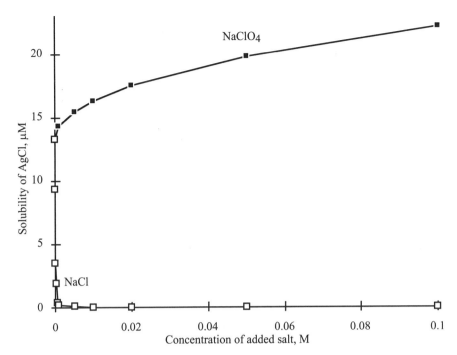

FIGURE 6.5. Addition of a nonreactive salt (NaClO$_4$) decreases the activity coefficients of silver and chloride ions, hence increases the solubility of silver chloride. This effect is small compared to the orders of magnitude decrease when a salt with a common ion (NaCl) is added.

As you will recall from Chapter 2, γ_+ and γ_- cannot be determined separately, but only as the combination $\gamma_\pm = (\gamma_+\gamma_-)^{1/2}$, so that Eq. (12) is more correctly written

$$[Ag^+][Cl^-]\gamma_\pm^2 = K_{so}^\circ \qquad (13)$$

Often the solubility product K_{so} is written in terms of concentrations at a definite ionic strength instead of activities:

$$[Ag^+][Cl^-] = K_{so} = K_{so}^\circ/\gamma_\pm^2 \qquad (14)$$

Thus, while K_{so}°, the "activity" constant, or the constant extrapolated to zero ionic strength, is a function of temperature alone, K_{so}, the concentration constant, depends on the ionic strength as well. This dependence is shown in Fig. 6.6, where $\gamma_\pm$ is calculated from the Davies equation, p. 49, (at 25°C)

$$\log \gamma_\pm = -0.5\left(\frac{\sqrt{I}}{1+\sqrt{I}} - 0.2I\right) \qquad (15)$$

and also from the extended Debye–Huckel equation

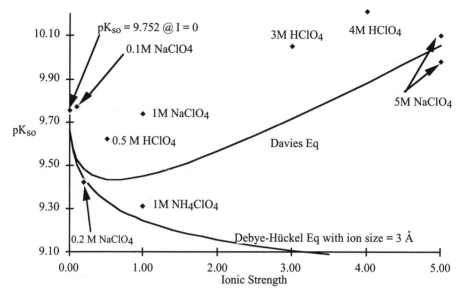

FIGURE 6.6. Concentration solubility product pK_{so} as a function of ionic strength. Experimental data are from the compilations of Sillén and Martell (1964, 1971) and Smith and Martell (1976–1986). The activity product pK_{so}° = 9.752 ± 0.002 is the value extrapolated to I = 0.

$$\log \gamma_{\pm} = -0.5\left(\frac{\sqrt{I}}{1 + 0.33a\sqrt{I}}\right) \tag{16}$$

with the ion-size parameter a = 3 Å (see Chap. 2, pp. 46–48). In Fig. 6.6, K_{so} from Eqs. (14)–(16) is compared with experimental values of K_{so} obtained in NaClO$_4$ medium at different ionic strengths. The general trend follows the Davies equation[7] better than the Debye–Hückel equation for $I > 1$, but the difference between the different experimental studies is disappointingly large.

However, the variation of K_{so} with ionic strength is small compared to the common-ion effect, as can be seen in Fig. 6.5, where the solubility of AgCl in excess NaClO$_4$—a nonreactive mediuim showing the ionic strength effect—is compared with the solubility in excess NaCl of the same concentration range—showing common-ion effect. For example, the solubility at zero ionic strength is $10^{-(\frac{1}{2}pK_{so})}$ = 1.33 × 10^{-5} M. In 0.1 M NaClO$_4$ it increases to 2.21 × 10^{-5} M. In contrast, the solubility of AgCl in 0.1 M NaCl is 4.9 × 10^{-9} M, nearly four orders of magnitude smaller, (and thus a trace of Cl$^-$ in the NaClO$_4$ could introduce serious errors in the solubility of AgCl).

Nevertheless, ionic strength can be a highly significant factor in the solubility of polyvalent materials. For example, the concentration solubility product of Fe(OH)$_3$ is

$$K_{so} = [Fe^{3+}][OH^-]^3 = K_{so}^{\circ}/[\gamma_{3+}(\gamma_-)^3]$$

EVALUATING pK_{so} FROM DATA

A simple approximation such as the Davies equation predicts that log γ varies as the square of the ionic charge; that is, $\gamma_{3+} = (\gamma_+)^9$ and $\gamma_- = \gamma_+$. Hence

$$K_{so} = K_{so}^\circ/(\gamma_+)^{12}$$

At $I = 1.0$, γ_+ is approximately 0.7, and so $K_{so}/K_{so}^\circ = (0.7)^{-12} = 72$. This means that the solubility [Fe^{3+}], at constant [OH$^-$], will be nearly two orders of magnitude larger at $I = 1$ than at $I = 0$.[8]

EVALUATING pK_{so} FROM DATA

Solubility products do not grow in tables from spores. They depend on a great deal of tedious work, assuring that solutions are saturated with the solid, and accurately analyzing the solution over a range of composition. Earlier, I noted that there is no solubility product given for silver acetate in the standard tables. However, there are several studies that have produced quantitative solubility data.[9]

Table 6.2 lists the pooled data of three investigations into the solubility of AgAc in AgNO$_3$ solutions. Three different extrapolation methods are shown. The first is to calculate the concentration solubility product

$$pK_{so}(\text{conc}) = -\log\{[\text{Ag}^+][\text{Ac}^-]\} \tag{17}$$

which gives a rather steep extrapolation curve (Fig. 6.7) and yields p$K_{so}^\circ = 2.53$ at $I = 0$.[10]

TABLE 6.2. Solubility of silver acetate in silver nitrate solutions

AgNO$_3$ (mol/L)	S (mol/L)	Ionic str.	pK_{so} (conc)	pK_{so} (Davies)	pK_{so} (DH)	Ref.[11]
0	0.0602	0.060	2.441	2.629	2.685	1
0.015	0.0539	0.069	2.430	2.628	2.691	2
0.0163	0.0595	0.076	2.346	2.551	2.619	2
0.0294	0.0491	0.079	2.413	2.621	2.692	1
0.0494	0.0503	0.100	2.605	2.830	2.917	3
0.0589	0.0419	0.101	2.374	2.599	2.688	1
0.0653	0.0444	0.110	2.312	2.544	2.639	2
0.0708	0.0458	0.117	2.489	2.725	2.826	3
0.0883	0.0383	0.127	2.314	2.556	2.664	1
0.095	0.0415	0.137	2.404	2.651	2.768	3
0.106	0.0381	0.144	2.393	2.645	2.767	3
0.1177	0.0341	0.152	2.285	2.540	2.668	1
0.1307	0.0348	0.166	2.239	2.501	2.638	2
0.1766	0.0264	0.203	2.272	2.547	2.712	1
0.1984	0.0308	0.229	2.214	2.498	2.680	3
0.2001	0.0307	0.231	2.212	2.496	2.679	3
0.2355	0.0192	0.255	2.311	2.602	2.802	1
0.3082	0.0274	0.336	2.073	2.379	2.632	3

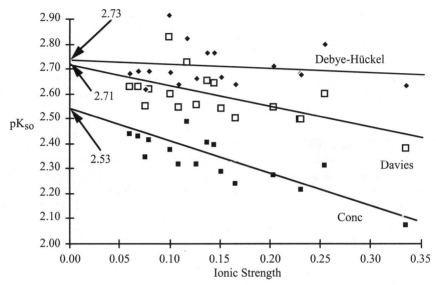

FIGURE 6.7. Three ways to extrapolate pK_{so} for AgAc in $AgNO_3$ solutions to $I = 0$. (Data from Table 6.2).

The second method is to use the Davies equation (Eq. 15) to calculate $\gamma_\pm$, and hence

$$pK_{so}(\text{conc}) = -\log\{[Ag^+][Ac^-]\gamma_\pm^2\} \qquad (18)$$

This yields a less steep curve, which extrapolates to $pK_{so}^\circ = 2.71$ at $I = 0$ (Fig. 6.7). The best method for these data appears to be to calculate $\gamma_\pm$ from the Debye–Hückel equation (Eq 16) with ion size $a = 3$, which gives results for pK_{so}° that have essentially no trend with ionic strength, and that extrapolate to 2.73 at $I = 0$ (Fig. 6.7).

Data from NaAc solutions are given in Table 6.3. The average of the $AgNO_3$ data in Table 6.2 is $pK_{so}^\circ = 2.71 \pm 0.04$[13]; the NaAc data in Table 6.3 gives $pK_{so}^\circ = 2.67 \pm 0.04$, in agreement.

TABLE 6.3. Solubility of silver acetate in sodium acetate solutions[12]

NaAc (mol/L)	S (mol/L)	Ionic str.	pK_{so} (conc)	pK_{so} (Davies)	pK_{so} (DH)
0.000	0.067	0.067	2.347	2.543	2.605
0.010	0.062	0.071	2.357	2.557	2.622
0.041	0.049	0.090	2.354	2.571	2.652
0.074	0.040	0.114	2.336	2.571	2.670
0.175	0.034	0.208	2.156	2.434	2.602
0.426	0.019	0.445	2.076	2.393	2.714
0.583	0.017	0.600	1.988	2.311	2.722
0.877	0.016	0.892	1.855	2.168	2.736

PRECIPITATION TITRATIONS

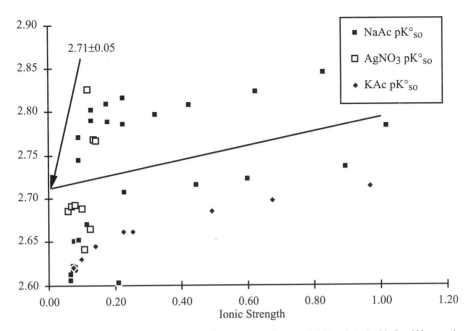

FIGURE 6.8. Extrapolation of pK°_{so} for silver acetate from solubility data in NaAc, KAc, and AgNO$_3$, corrected by the Debye–Hückel Equation (with $a = 3$) to $I = 0$.

Figure 6.8 shows the data from Tables 6.2 and 6.3, plus some additional data obtained in NaAc and KAc solutions, going out to 1 M ionic strength. This extrapolation gives an overall result $pK^\circ_{so} = 2.71 \pm 0.05$. Note that the KAc data fall slightly below the NaAc data, but that the AgNO$_3$ data span the range of both.

This exercise has shown that old, uninterpreted data can still be useful, that the extrapolation method makes a significant difference in the result, and that pK_{so} from three different electrolytes (AgNO$_3$, NaAc, and KAc) is the same within experimental error, quantitatively confirming the common-ion effect.

PRECIPITATION TITRATIONS

Many ions that form insoluble salts can be determined by titration if a suitable method of estimating the end point is available. The most convenient methods are potentiometric, but chemical indicators are also available for a number of cases. Plots of metal ion concentration as a function of volume added resemble the titration curves calculated for strong acid–strong base systems in Chapter 3, because the mathematical form of the equations is the same.

Consider the titration of an anion X (such as Cl$^-$) with a metal M (such as Ag$^+$). The two ions form a relatively insoluble salt MX with solubility product[14]

$$[M][X] = K_{so} \tag{19}$$

Let C_o be the initial concentration of X in a volume V_o, and C be the concentration of the titrant M, with volume V. At each point in the titration, P moles per liter of MX are present as precipitate. The mass balances are:

$$[M] + P = \frac{CV}{V + V_o} \tag{20}$$

$$[X] + P = \frac{C_o V_o}{V + V_o} \tag{21}$$

Eliminating P between Eqs. (20) and (21), and making use of the solubility product, (Eq. 1 or 19), one obtains an equation for the titration curve (V vs $[M]$ or vice versa):

$$V\left(C - [M] + \frac{K_{so}}{[M]}\right) = V_o\left(C_o + [M] - \frac{K_{so}}{[M]}\right) \tag{22}$$

By analogy with acid–base titrations, the fraction titrated ϕ is defined to be

$$\phi = \frac{CV}{C_o V_o}$$

and so Eq. (22) can be expressed as

$$\phi = \frac{C\left(C_o + [M] - \frac{K_{so}}{[M]}\right)}{C_o\left(C - [M] + \frac{K_{so}}{[M]}\right)} \tag{23}$$

From this equation, a titration curve can be plotted, as in Fig. 6.9. At the left of the diagram, chloride is in excess and $[Ag^+]$ is small (pAg is large); at the right, Ag^+ is in excess and its concentration approaches that of the diluted titrant. Between these extremes is a very sharp change in pAg, analogous to the sharp change in pH in a strong acid–strong base titration.

Equations (19)–(21) can also be combined to give ϕ as a function of $[M]$ (or vice versa):

$$\frac{C_o V_o}{V + V_o}(\phi - 1) = [M] - \frac{K_{so}}{[M]} \tag{24}$$

This form, mathematically equivalent to Eq. (23), has the formal disadvantage of including V both explicitly and as part of ϕ, but the advantage of simplicity in plotting the titration curve if the titrant is very concentrated ($C \gg C_o$ and $V \ll V_o$).

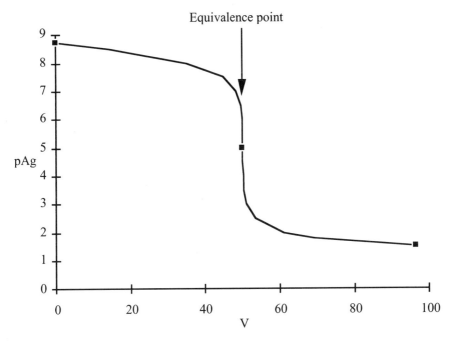

FIGURE 6.9. Plot of pAg = −log[Ag⁺] versus volume of titrant added in the titration of 0.1 M NaCl with 0.1 M AgNO$_3$. $\phi = 1$ at $V = 50$.

Equation (24) is also useful in deriving a simpler equation for the titration error. Recall from Chapter 4 that near the equivalence point, V is approximately C_oV_o/C and hence

$$\frac{C_oV_o}{V+V_o} = \frac{C_oC}{C+C_o} \tag{25}$$

Thus the approximate titration error is given by substituting Eq. (25) in Eq. (24):

$$\phi_{ep} - 1 = \left(\frac{C+C_o}{CC_o}\right)\left([M]_{ep} - \frac{K_{so}}{[M]_{ep}}\right) \tag{26}$$

where the subscript "ep" refers to the practical end point. The closer ϕ_{ep} is to the true equivalence point at $\phi = 1$, the more accurate Eq. (26) will be.

For the reverse titration of silver ion with chloride, only the signs on the right-hand side need to be changed:

$$\phi_{ep} - 1 = \left(\frac{C+C_o}{CC_o}\right)\left(\frac{K_{so}}{[M]_{ep}} - [M]_{ep}\right) \tag{27}$$

Example 5. Using a silver electrode as a detector, one can measure the activity (hence concentration) of dissolved silver ion in a solution, and the variation of this quantity can be used to determine the end point of a titration of chloride, bromide, or other ions with silver nitrate. To plot the titration curve (Fig. 6.9), use Eq. (15) with $pK_{so} = 9.75$, $V_o = 50$ mL, $C_o = 0.10$ M, and $C = 0.10$ M, and calculate V as a function of $[Ag^+]$. The graph is plotted as $-\log[Ag]$ as a function of V. What is the titration error if $[Ag^+]_{ep} = 10^{-5.0}$? Substitute in Eq. (27) to obtain

$$\phi_{ep} - 1 = 20(10^{-9.75+5.0} - 10^{-5.0}) = 1.56 \times 10^{-4} = +0.016\%$$

Example 6. One of the classic silver chloride titrations is the Mohr method, which employs a small concentration of chromate as an indicator. Consider the same titration as Example 5, but with 1.0 mL of 0.1 M K_2CrO_4 added. The end point is taken when the red precipitate of Ag_2CrO_4 appears. Since $V + V_o = 100$ mL at the end point, $[CrO_4^{2-}] = (1.0)(0.1)/(100) = 1.0 \times 10^{-3}$ M. The solubility product of Ag_2CrO_4 (Table 6.3) is $pK_{so} = 11.92$ at 25°C and zero ionic strength.

$[Ag^+]$ at the point where the first precipitate appears is therefore

$$[Ag^+]^2[CrO_4^{2-}] = 10^{-11.92}$$

$$[Ag^+]^2 = 10^{-11.92+3.0} = 10^{-8.92}$$

$$[Ag^+] = 10^{-4.46}$$

Again making use of Eq. (26), we find

$$\phi_{ep} - 1 = 20(10^{-4.46} - 10^{-9.75+4.40}) = 6.04 \times 10^{-4} = +0.06\%$$

This model of the Mohr method overestimates the equivalence point by approximately 0.06%.

If a higher $[Ag^+]$ is needed to produce a visible color, say, $10^{-4.0}$, the error would be larger, 0.2%. A lower concentration of chromate would yield a higher concentration of silver at the end point, and make matters worse. A higher concentration of chromate would use up a significant amount of titrant, but this could be corrected for since the amount added is known. The most straightforward approach to optimizing the Mohr method would be to develop titration curve equations including both chloride and chromate mass balances, and to adjust the concentration of chromate so that Ag_2CrO_4 precipitates at exactly the desired point (see problems).

Point of Maximum Slope for Symmetrical Titration

One popular method of determining an end point, particularly if an automatic titrator is used, is to find the point where the maximum change occurs in the quantity monitored. For strong acid–strong base and monoprotic acid–base titrations, this point, which is the point of minimum buffer index or maximum sharpness index, normally coincides with the equivalence point. For polyprotic acids it is also a good

PRECIPITATION TITRATIONS

approximation provided the pK_a values are not too close together (see buffer index discussion, Chapter 5).

Precipitation titrations forming a symmetrical salt MX also have a maximum slope at the equivalence point.[15] As in the discussions of buffer index in Chapters 4 and 5, the derivative of $-\log[M] = pM$ with respect to ϕ is

$$\frac{dpM}{d\phi} = -\left(\frac{0.434}{[M]}\right)\left(\frac{d[M]}{d\phi}\right) = -\left(\frac{0.434}{[M]}\right)\left(\frac{d\phi}{d[M]}\right)^{-1} \quad (28)$$

Rigorously, $d\phi/d[M]$ should be obtained by differentiating Eq. (23), but for simplicity, since the final result applies only near the equivalence point, take the derivative of Eq. (26) and substitute in Eq. (28):

$$\frac{d\phi}{d[M]} = \left(\frac{C+C_o}{CC_o}\right)\left(1 + \frac{K_{so}}{[M]^2}\right) \quad (29)$$

$$\frac{dpM}{d\phi} = -\left(\frac{0.434 CC_o}{C+C_o}\right)\left([M] + \frac{K_{so}}{[M]}\right)^{-1} \quad (30)$$

The point of maximum negative slope (Fig. 6.10) occurs where the derivative of the slope $dpM/d\phi$ is zero:

$$\frac{d^2pM}{d\phi^2} = \frac{d}{d\phi}\left(\frac{dpM}{d\phi}\right) = -0.434\left(\frac{CC_o}{C+C_o}\right)\frac{\left(1 - \frac{K_{so}}{[M]^2}\right)}{\left([M] + \frac{K_{so}}{[M]}\right)^2}\left(\frac{d[M]}{d\phi}\right) = 0 \quad (31)$$

Substituting the reciprocal of $d\phi/d[M]$ as given by Eq. (29) gives

$$\frac{d^2pM}{d\phi^2} = -0.434\left(\frac{CC_o}{C+C_o}\right)^2 \frac{\left(1 - \frac{K_{so}}{[M]^2}\right)}{\left([M] + \frac{K_{so}}{[M]}\right)^2 \left(1 + \frac{K_{so}}{[M]^2}\right)} = 0 \quad (32)$$

All the terms in this complicated equation are positive except for the bracket in the numerator, which must therefore be

$$1 - \frac{K_{so}}{[M]^2} = 0 \quad (33)$$

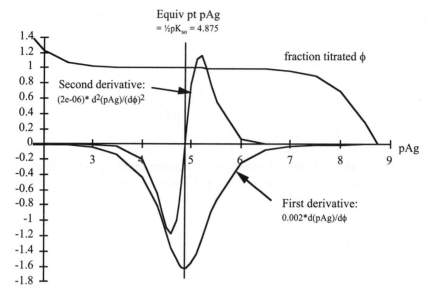

FIGURE 6.10. Titration curve for Ag^+ and Cl^-. The curve marked "fraction titrated" is the same as Fig. 6.9, but rotated by 90°. The first derivative $d(pAg)/d\phi$ reaches a maximum negative value at the equivalence point, and the second derivative $d^2(pAg)/d\phi^2$ goes through zero at the same point.

$$[M] = \sqrt{K_{so}} \tag{34}$$

This simple result is the same as the value of $[M]$ at the equivalence point. Differentiation of the complete Eq. (23) gives the same final result, but somewhat more complicated versions of Eqs. (30)–(32).

Unsymmetrical Titrations

For titrations forming unsymmetrical salts such as Ag_2SO_4 or Hg_2Cl_2 (recall Hg_2^{2+} behaves as a single divalent ion), the mass balances and transformations to obtain the titration curve are entirely analogous. For example, the titration of a divalent ion X^{2-} (concentration C_o, volume V_o) with a univalent ion M^+ (concentration C, volume V) to give the insoluble salt M_2X

$$X^{2-} + 2M^+ \rightleftharpoons M_2X \,(S)$$

yields the following expression for the titration curve of $[M]$ versus ϕ. From the mass balances on M and X

$$\frac{CV}{2C_oV_o} - 1 = \left(\frac{V+V_o}{2C_oV_o}\right)\left([M] - \frac{2K_{so}}{[M]^2}\right) \tag{35}$$

PRECIPITATION TITRATIONS

Here $K_{so} = [M]^2[X]$. If all the terms in V are collected on one side and all the terms in V_o on the other, the equation can be solved explicitly for V—and hence for the fraction titration, ϕ: Note that because two moles of M are required to react with one mole of X, ϕ carries a factor of 2 in its denominator.

$$\phi = \frac{CV}{2C_oV_o} = \left(\frac{C}{2C_o}\right)\frac{\left(2C_o + [M] - \frac{2K_{so}}{[M]^2}\right)}{\left(C - [M] + \frac{2K_{so}}{[M]^2}\right)} \quad (36)$$

An approximate form[16] in which CV is approximately $2C_oV_o$ can be obtained by expansion of either Eq. (35) or (36). It gives a simpler equation for the titration error near the equivalence point.

$$\phi_{ep} - 1 = \left(\frac{2C_o + C}{2CC_o}\right)\left([M^+]_{ep} - \frac{2K_{so}}{[M^+]_{ep}^2}\right) \quad (37)$$

In Fig. 6.11, the full equation (36) and the approximate equation (37) are compared for a sample titration of sulfate with silver to give the precipitate Ag_2SO_4. Near the

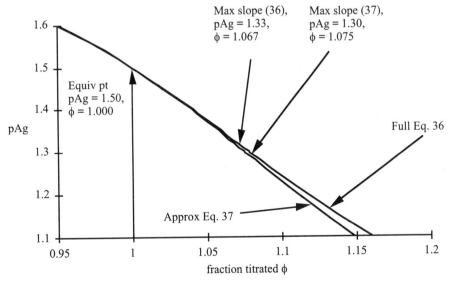

FIGURE 6.11. Titration of 0.50 M sulfate with 1.0 M silver ion, with $pK_{so} = 4.8$ for Ag_2SO_4. The equivalence point $\phi = 1$ occurs at $[Ag^+]^3 = 2K_{so}$, or $pAg = 1.50$, for both the approximate equation (37) and the full equation (36), but the slopes of the two curves are slightly different. See below for a discussion of the maximum slope.

equivalence point the two equations give approximately the same values, but the approximate equation (Eq. 37) gives lower values of pAg both before and after the equivalence point.

For the reverse titration, the signs in the right-hand brackets of Eq. (36) or (37) are changed; and for titrations where [X] is monitored instead of [M], [M]$_{ep}$ can be replaced by

$$[M]_{ep} = \sqrt{\frac{K_{so}}{[X]_{ep}}} \tag{38}$$

which gives for the titration error:

$$\phi_{ep} - 1 = \left(\frac{2C_o + C}{2CC_o}\right)\left(\sqrt{\frac{K_{so}}{[X]_{ep}}} - 2[X]_{ep}\right) \tag{39}$$

Point of Maximum Slope for Unsymmetrical Titration

For an unsymmetrical titration, given by Eqs. (35)–(37), the point of maximum slope is not the same as the equivalence point. Differentiating Eq. (36) gives the slope as a function of [M]:

$$\frac{d\phi}{d[M]} = \left(\frac{CC + 2C_o}{2C_o}\right) \frac{\left(1 + \frac{4K_{so}}{[M]^3}\right)}{\left(C - [M] + \frac{2K_{so}}{[M]^2}\right)^2} \tag{40}$$

From Eq. (40) (and Eq. 30 in the previous section), the accurate expression for the derivative $dpM/d\phi$ is (brackets on M and the subscript "ep" have been omitted for simplicity):

$$\frac{dpM}{d\phi} = -\left(\frac{0.434}{M}\right)\left(\frac{d\phi}{dM}\right)^{-1} = -\left(\frac{0.434}{M}\right)\left(\frac{2C_o}{C(C+2C_o)}\right) \frac{\left(C - M + \frac{2K_{so}}{M^2}\right)^2}{\left(1 + \frac{4K_{so}}{M^3}\right)} \tag{41}$$

To avoid complicated algebra, it is tempting to differentiate Eq. (37), the approximate form that applies near the equivalence point[17]:

$$\frac{d\phi_{ep}}{d[M]_{ep}} = -\left(\frac{2C_o + C}{2CC_o}\right)\left(1 + \frac{4K_{so}}{[M^+]_{ep}^3}\right) \tag{42}$$

PRECIPITATION TITRATIONS

$$\frac{dpM}{d\phi} = \left(\frac{0.434}{M}\right)\left(\frac{d\phi}{dM}\right)^{-1} = \left(\frac{0.868CC_o}{(2C_o + C)}\right)\left(\frac{M^2}{M^3 + 4K_{so}}\right) \quad (43)$$

The point of maximum slope occurs when the derivative of Eq. (43) is zero.

$$\frac{d^2pM}{d\phi^2} = \frac{d}{dM}\left(\frac{dpM}{d\phi}\right)\left(\frac{dM}{d\phi}\right) = 0 \quad (44)$$

Differentiate Eq. (42) and substitute this and the reciprocal of Eq. (41) in Eq. (44) to obtain

$$\frac{d^2pM}{d\phi^2} = 0.434\left(\frac{CC_o}{2C_o + C}\right)^2\left(\frac{2M}{M^3 + 4K_{so}} - \frac{M^2(3M^2)}{(M^3 + 4K_{so})^2}\right)\left(\frac{M^3}{M^3 + 4K_{so}}\right) = 0 \quad (45)$$

Eliminate all positive multiplying terms; Eq. (45) reduces to

$$[M]^3 = 8 K_{so} \quad (46)$$

This result is different from the equivalence point ($\phi = 1$ occurs at $[M]^3 = 2 K_{so}$ in Eq. 36 or 37).

In Fig. 6.12 is plotted the slope of the titration curve of Fig. 6.11. The difference between the maximum slope point and the equivalence point, as well as the difference between the full Eq. (41) and the approximate Eq. (43) show clearly, whether the slope is plotted versus pAg or ϕ. For comparison, some values of $d(pAg)/d\phi$ obtained by numerical differentiation [$\Delta(pAg) = 0.01$] of the full titration curve (Eq. 36, Fig. 6.11) are also given. They agree with the algebraic derivative Eq. (41).

Note that the derivative of the approximate equation, as Eq. (46) showed, reaches a maximum where $[Ag^+]^3 = 8 K_{so}$ (pAg = 1.30, $\phi = 1.075$). However, this approximation overestimates the titration error; the full derivative given by Eq. (41) reaches its maximum at pAg = 1.33 and $\phi = 1.067$, as shown in Fig. 6.12.

For the titration of a divalent ion with a monovalent ion, such as sulfate with silver ion (Figs. 6.11 and 6.12), the maximum slope end point comes after the equivalence point ($\phi_{ep} > 1$). An approximate value for ϕ at the point of maximum slope can be obtained by substituting Eq. (46) in Eq. (38) to get

$$\phi_{ep} - 1 = \frac{3(2C_o + C)}{4CC_o} K_{so}^{1/3} \quad (47)$$

For the titration of Fig. 6.11, with $C = 1.0$, $C_o = 0.5$, and $pK_{so} = 4.8$, Eq. (46) gives pAg = 1.30 and Eq. (47) gives $\phi_{ep} = 1.075$, in agreement with the approximate curve of Fig. 6.12.

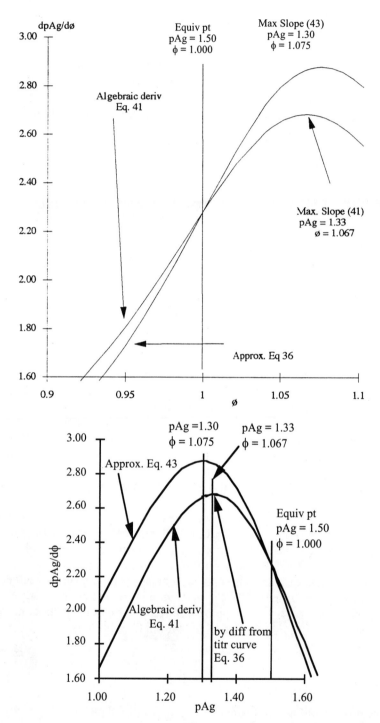

FIGURE 6.12. Slope of the titration curve of Fig. 6.11. The maximum slope of the full equation (41) occurs at pAg = 1.33, ϕ = 1.067, and the maximum slope of the approximate equation (43) occurs at pAg = 1.30, ϕ = 1.075.

COMPLICATIONS IN SOLUBILITY CALCULATIONS

For the titration of a monovalent ion with a divalent ion, such as silver with sulfate ion or chloride with mercurous ion, the end point comes before the equivalence point, and $\phi_{ep} - 1$ is given by the negative of the expression in Eq. (37) or (39).

COMPLICATIONS IN SOLUBILITY CALCULATIONS

In the previous discussions of this chapter, the examples used were chosen to be realistic in the sense that the solid phase yielded primarily just two ions in solution. This is a relatively small fraction of all the cases that can be encountered in analytical chemistry, geochemistry, environmental chemistry, and other fields. Factors that complicate solubility calculations can be broadly divided into ionic strength, which has already been discussed, acid–base reactions, and complex formation.

Acid–Base Effects on Solubility

If the anion of the insoluble salt is also the conjugate base of a weak acid, then the solubility will depend on pH even if the total concentrations of anion and cation are held constant. The most obvious case is a hydroxide salt, such as $Mg(OH)_2$, where the pH dependence is explicitly displayed in the solubility product

$$[Mg^{2+}][OH^-]^2 = K_{so}$$

Substituting the ion product of water, $[H^+][OH^-] = K_w$, you get

$$[Mg^{2+}] = K_{so}[H^+]^2/K_w^2$$

$$\log[Mg^{2+}] = \log S = -pK_{so} + 2\,pK_w - 2\,pH$$

That is, the log of the solubility varies linearly with pH, and the slope of that line is 2, as can be seen from Fig. 6.13, which was calculated with $pK_{so}^\circ = 10.74$. If activity coefficients are included [note $pH = -\log[H^+]\,\gamma_+$ and $\gamma_{2+} = (\gamma_+)^4$; γ_+ was calculated from the Davies equation (15) using $I = 3S$ (HCl added to reach low pH: $[Cl^-] = 2[Mg^{2+}]$).

$$\log S = -pK_{so}^\circ + 2\,pK_w^\circ - 2\,pH - 6\log\gamma_+$$

The solubility in pure water is found where $[OH^-] = 2S$, at $pH = 10.56$.

Salt of a Weak Acid

Values of K_{so} and K_a for some representative salts are given in Table 6.2 (ions of equal charge), Table 6.3 (ions of unequal charge), and Table 6.4 (salts of polyprotic acids).

For a somewhat more complicated example, consider the slightly soluble salt MA, where A is the anion of a weak acid HA:

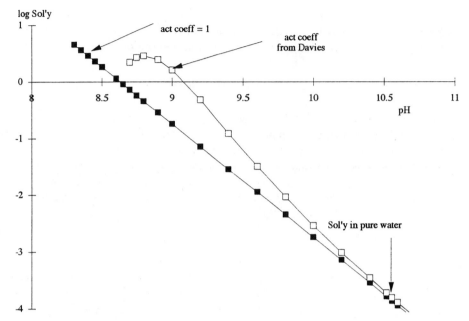

FIGURE 6.13. Solubility of $Mg(OH)_2$ as a function of pH. For pH below 10, including activity coefficients makes a significant difference. For pH < 8.4, the solubility is too large for meaningful activity coefficient estimation, and depends strongly on the anion of the acid used to adjust pH.

TABLE 6.4. Solubility products for salts of weak monoprotic acids,[18] at 25°C, $I = 0$

Salt	Ions of equal charge	
	pK_{so}	pK_a
AgAc	2.71±0.05	4.757±0.002
$CuCrO_4$	5.44	6.51±0.02
Hg_2SO_4	6.13±0.04	1.99±0.01
$PbSO_4$	7.79±0.02	1.99±0.01
$BaCrO_4$	9.97±0.01	6.50±0.02
$BaSO_4$	9.96±0.03	1.99±0.01
$PbCrO_4$	12.6	6.50±0.02

TABLE 6.5. Solubility products of weak monoprotic acids at 25°C, $I = 0$

Ions of unequal charge		
Ag_2SO_4	4.83±0.03	1.99±0.01
BaF_2	5.76	3.17±0.02
MgF_2	8.18	3.17±0.02
CaF_2	10.41	3.17±0.02
Ag_2CrO_4	11.92±0.03	6.50±0.02

COMPLICATIONS IN SOLUBILITY CALCULATIONS

$$[M^+][A^-] = K_{so}$$
$$[H^+][A^-] = K_a[HA]$$
$$C_A = [HA] + [A^-]$$

Decreasing pH by adding a strong acid will increase [H$^+$], increase the ratio [HA]/[A^-], and if C_A is constant, decrease [A^-] to compensate, causing the solubility [M^+] to increase. This sequence of events can also be represented by the reaction

$$H^+ + MA(s) \rightleftharpoons M^+ + HA$$

Recall that for a monoprotic acid of total concentration C_A

$$[A^-] = \frac{C_A K_a}{[H^+] + K_a} \tag{48}$$

In general $C_A = C_A^o + S$, where C_A^o is the total concentration of A introduced in addition to the salt MA—as when NaA or HA is added to produce a common-ion effect. The general expression for solubility of MA as a function of pH is therefore

$$S = [M^+] = \frac{K_{so}}{[A^-]} = \frac{K_{so}([H^+] + K_a)}{(C_A^o + S)K_a} \tag{49}$$

If no external source of A is present, $C_A^o = 0$, $C_A = S$, and

$$S = \sqrt{\frac{K_{so}([H^+] + K_a)}{K_a}} \tag{50}$$

Example 7. Find the solubility of silver acetate AgAc in dilute HNO$_3$ as a function of pH.[20] From Table 6.2, pK_{so} = 2.71 (see Fig. 6.8 earlier in this chapter for evaluation of pK_{so} of AgAc) and pK_a = 4.75. Since $C_A^o = 0$ in Eq. (49), Eq. (50) applies. Set [H$^+$] = C_{HNO_3} for a first approximation without activity coefficients

$$S = 10^{-1.20}\sqrt{\frac{C_{HNO_3}}{10^{-4.75}} + 1}$$

This equation is plotted in Fig. 6.14.

A preliminary calculation with Eq. (50) and the data given makes it apparent that the ionic strength over the whole range is rather large (approximately equal to S) and so a calculation including activity coefficients is appropriate. Repeating the derivation of Eq. (50), but including activity coefficients in the equilibrium equations, gives

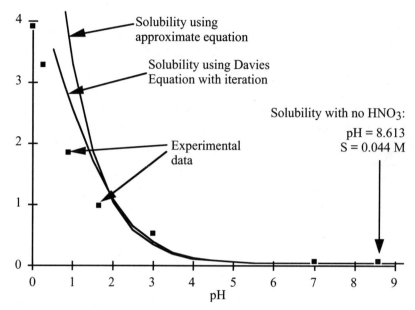

FIGURE 6.14. Solubility of AgAc (M) in added HNO_3 as a function of pH. S is calculated from the approximate Eq. (50) and also from Eq. (51), making use of the Davies equation to estimate activity coefficients. Experimental data are from Hill and Simmons (1909). Solubility without HNO_3 is obtained from Fig. 6.15.

$$S = \sqrt{K_{so}\left(\frac{10^{-pH}}{K_a \gamma_+ \gamma_o} + 1\right)} \qquad (51)$$

The Davies equation gives an estimate of γ_+ and γ_-. In addition, a reasonable estimate of $\log \gamma_o$ is $0.1I$. The result is not much different from Eq. (50), as can be seen in Fig. 6.14. Also plotted on Fig. 6.14 are some experimental data which agree for pH < 2.[21]

Example 8. Calculate the solubility of AgAc in pure water, and the pH of the saturated solution. In addition to Eq. (51), which is based on the solubility product and the protonation equilibrium of acetate, the "pure water" condition requires a charge balance to be satisfied:

$$[Ag^+] + [H^+] = [Ac^-] + [OH^-]$$

Since the solubility[22] S' is equal both to $[Ag^+]$ and to total $[HAc] + [Ac^-]$,

$$S' = \frac{SK_a}{[H^+] + K_a} + \frac{K_w}{[H^+]} - [H^+]$$

COMPLICATIONS IN SOLUBILITY CALCULATIONS

Canceling the term $S'K_a$ and expanding gives:

$$S' = \left(\frac{K_w}{[H^+]} - [H^+]\right)\left(1 + \frac{K_a}{[H^+]}\right)$$

The equilibrium constant expressions are modified to include activity coefficients by substituting $[H^+] = (10^{-pH})/\gamma_+$, $K_w = K_w^o/(\gamma_+\gamma_-)$ and $K_a = K_a^o\gamma_o/(\gamma_+\gamma_-)$

$$S' = \left(\frac{K_w^o 10^{+pH}}{\gamma_-} - \frac{10^{-pH}}{\gamma_+}\right)\left(1 + \frac{K_a^o\gamma_o 10^{+pH}}{\gamma_-}\right)$$

This equation, labeled "S from charge balance," is plotted together with the values of S obtained from Eq. (51) on Fig. 6.15. The intersection of the two curves corresponds to the solubility of AgAc in pure water: $S = 0.063$ M, pH = 8.675.

Salts of polyprotic acids (Table 6.4) can be treated by the same techniques. The concentration of the particular ion that occurs in the salt is obtained by the methods of Chapter 5.

For example, in calculating the solubility of calcite, $CaCO_3$, in a closed vessel ("closed" to avoid the question of CO_2 gas escaping or having to be controlled—these factors are discussed Chapter 9), the concentration solubility product

$$[Ca^{2+}][CO_3^{2-}] = K_{so}$$

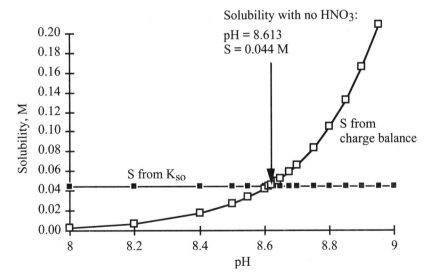

FIGURE 6.15. Solubility of AgAc versus pH showing intersection corresponding to pure water.

TABLE 6.6. Solubility products of polyprotic acids[23] at 25°C, $I = 0$

Salt	pK_{so}	pK_{a1}	pK_{a2}	pK_{a3}
$Ca_3(PO_4)_2(\beta)$	26.0	2.148±0.001	7.199±0.002	12.3±0.02
$CaHPO_4$	6.58±0.03			
$MgCO_3$	7.46	6.363±0.005	10.33±0.01	
$MgCO_3(H_2O)_3$	4.67			
$MgCO_3(H_2O)_5$	4.54			
$CaCO_3$(calcite)	8.48±0.02	6.363±0.005	10.33±0.01	
$CaCO_3$(aragonite)	8.30±0.02			
$MgNH_4PO_4$	12.6	2.148±0.001	7.199±0.002	12.3±0.02
	for NH_4^+	9.244±0.005		

is combined with the expression for $[CO_3^{2-}]$ as a function of pH:

$$[CO_3^{2-}] = \frac{C_T K_{a1} K_{a2}}{[H^+]^2 + K_{a1}[H^+] + K_{a1}K_{a2}} \qquad (52)$$

where $C_T = [CO_2] + [HCO_3^-] + [CO_3^{2-}]$ is the total of all dissolved carbonate species, and K_{a1} and K_{a2} are the concentration acidity constants of CO_2 (see Table 6.6). An example is shown in Fig. 6.16.

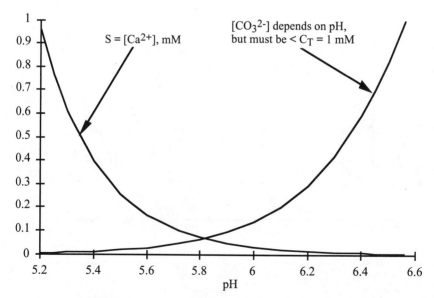

FIGURE 6.16. Dependence of $CaCO_3$ solubility on pH for $C_T = 1$ mM.

Effect of Complex Formation on Solubility

The most difficult equilibrium effects on solubility to predict come from the formation of complexes between cations and anions in solution. These are presented in detail in Chapter 7, but their effect on solubility is easy to generalize. Just as the reaction of acid with the anion of a salt increases that salt's solubility; so the reaction of a complexing agent with the cation increases its solubility.

Significant effects of complex formation on solubility range from the simple interaction of the salt's own cation and anion, such as

$$AgCl(s) + Cl^- \rightleftharpoons AgCl_2^- \quad \text{and}$$
$$AgCl_2^- + Cl^- \rightleftharpoons AgCl_3^{2-}, \quad \text{etc.}$$

to the reaction of ferric iron in swamp water with hundreds of different organic (humic) acids, each with somewhat different complexation strengths and acidity constants. Clearly, an accurate prediction of solubility in the presence of many acids and complexing agents is a daunting task, and requires a wide base of equilibrium data.[24]

PROBLEMS

1. Use the solubility products in Table 6.1 to calculate the solubility of:
   (a) $PbSO_4$ in pure water. Compare with handbook value of 0.00425 g/100 mL at 25°C.
   (b) MgF_2 in pure water. Compare with handbook value of 0.0076 g/100 mL at 18°C, decreasing slightly with increasing temperature.
   (c) $La(IO_3)_3$ in pure water. Compare with handbook value of 1.7 g/100 mL at 25°C.

2. Use the solubility products in Table 6.1 to calculate the solubility of:
   (a) $PbSO_4$ in 0.05 M $Pb(NO_3)_2$;
   (b) $PbSO_4$ in 10^{-4} M $NaNO_3$;
   (c) $PbSO_4$ in 10^{-3} M Na_2SO_4;
   (d) MgF_2 in 0.10 M KF;
   (e) MgF_2 in 10^{-6} M $Mg(NO_3)_2$.

3. Calculate the fraction of chloride in a 1.0×10^{-5} M solution of NaCl that could be precipitated by adding:
   (a) the stoichiometric amount of 0.1 M $AgNO_3$ (to make final $[NO_3^-] = [Cl^-]$ without significant dilution);
   (b) a 10% excess of $AgNO_3$ ($[NO_3^-] = 1.10[Cl^-]$);
   (c) a 100% excess of $AgNO_3$ ($[NO_3^-] = 2.0[Cl^-]$).

4. A 0.451 g sample of AgCl is washed prior to ignition and weighing. The washing solution usually contains NH_4Cl or HNO_3, which evaporate during ignition. Calculate the fraction of the sample that will be retained if it is washed with each of the following solutions:
   (a) 200 mL pure water;
   (b) 150 mL 0.10 M NH_4Cl, followed by 50 mL pure water;
   (c) 50 Ml 0.001 M $AgNO_3$ followed by 3 washings of 50 mL HNO_3 each.

5. Calculate the weight of precipitate (if any) produced when the following solutions are mixed:
   (a) 25 mL of 0.050 M $Sr(NO_3)_2$ and 10 mL of 0.15 M Na_2SO_4;
   (b) 1.0 mL of 0.10 M $AgNO_3$ and 100 mL of 0.005 M HCl;
   (c) 0.25 g KF and 250 mL of 0.050 M $MgCl_2$.

6. Thallium is to be precipitated as TlCl. What volume of 1.00 M HCl should be added to 10.0 mL of a solution containing 383 mg thallium in order that precipitation should be 99.5% complete?

7. Calculate the solubility of $Mg(OH)_2$ in a hydrochloric acid solution whose initial pH is 3.50. What is the pH of the saturated solution?

8. Calculate the volume of 0.10 M acetic acid required to dissolve 5.83 mg of $Mg(OH)_2$ and the pH of the saturated solution.

9. Find the solubility of $Mg(OH)_2$ in a buffer consisting of 0.01 M NH_4Cl and 0.01 M ammonia. Neglect the possible formation of $Mg-NH_3$ complexes.

10. Find the solubility of $Mg(OH)_2$ in 0.001 M NaOH. Note that the complex $MgOH^+$ forms with an equilibrium constant $[MgOH^+] = 10^{2.58}[Mg^{2+}][OH^-]$, and hence is not completely negligible. Estimate the error in solubility resulting from neglecting the formation of $MgOH^+$.

11. Consider a general ionic salt M_zX_y. Let its solubility in pure water be S_o. Show that the solubility S in a solution containing Na_zX of analytical concentration C is always less than or equal to S_o for an positive real value of C, providing that the only ions in the solution are M^{y+}, X^{z-}, Na^+, H^+, and OH^-.

12. A solution is 0.010 M in $BaCl_2$ and 0.010 in $SrCl_2$. If 100 mL of this solution is mixed with 200 mL of 0.010 M Na_2SO_4, what fraction of each ion is precipitated?

13. A solution of 0.010 M in NaCl and 1.0×10^{-3} M in NaBr. Concentrated $AgNO_3$ is added. Does AgCl or AgBr precipitate first? What is the maximum degree of purity the solution of the second ion could attain?

14. 1.15 g of $PbSO_4$ is shaken with 100 mL of 0.10 M KIO_3, and allowed to come to equilibrium. What fraction of the $PbSO_4$ has been converted to solid $Pb(IO_3)_2$?

15. 0.001 M chloride is titrated with 0.1 M silver nitrate. How precisely must pAg be measured in order for the titration error to be less than 0.1%?

16. If pAg can be measured to ±0.01 units, what is the lowest concentration of chloride that can be titrated (with 0.1 M silver ion) and still have a titration error < 1%?

17. Derive an equation for the titration error that results when you take the end point at the point of maximum slope for the titration of chloride with mercurous ion. $[Hg_2^{2+}]$ is measured using a mercury electrode, as a function of volume of chloride added. Evaluate the error if 0.1 M Cl$^-$ is titrated with 0.1 M Hg_2^{2+}. From Table 6.1, $[Hg_2^{2+}][Cl^-]^2 = K_{so} = 10^{-17.88}$.

18. A mixture of chloride, bromide, and iodide can be titrated with silver nitrate to three successive end points, using a silver electrode as a potentiometric detector. Derive an equation for pAg as a function of titrant volume added. Evaluate pAg at the three equivalence points if 100 mL of a mixture of 0.01 M chloride, 0.001 M bromide, and 0.0001 M iodide is titrated with 0.1 M $AgNO_3$. From the full titration curve, do you seen any difficulties in determining any of the end points? See Table 6.1 for K_{so} values.

19. Derive an equation for pAg as a function of V for the Mohr titration of Cl$^-$ (C_o, V_o) with Ag$^+$ (C, V) in the presence of chromate (C_{In}, V_o). Can you find a plausible value for C_{In} so that the first appearance of the red precipitate Ag_2CrO_4 coincides with the equivalence point of the reaction between Ag$^+$ and Cl$^-$? (What concentration of Ag$^+$ corresponds to "first appearance"? A small excess will certainly be required; assume that this is 0.02 C_{In}.)

20. In the Volhard method for determining chloride, a measured amount of standard silver nitrate solution is added to the unknown, and the excess silver nitrate is titrated with potassium thiocyanate (after the solid silver chloride has been removed, so that it will not react during the back titration). The end point of the back titration is obtained by using ferric ion as an indicator, since a slight excess of thiocyanate produces a brilliant red ferric complex.

    In a typical titration, 20.00 mL of 0.100 M NaCl is treated with 30.00 mL of 0.100 M $AgNO_3$. The solution is boiled to coagulate the AgCl; the precipitate is filtered off and washed with 30 mL dilute nitric acid; and the washings are added to the filtrate. Calculate the error introduced by washing the precipitate, assuming solid AgCl is in equilibrium with 30 mL of 0.01 M HNO_3 and part of the excess of 0.1 M $AgNO_3$ retained by the precipitate (say 0.2 mL).

    Then 1.0 mL of 0.50 M ferric nitrate is added to the filtrate, and it is titrated with 0.0200 M KSCN unitil the red color first appears [SCN$^-$] = 10^{-5}. Estimate the error due to using the FeSCN end point for the back titration. Note that the equivalence point is [Ag$^+$] = [SCN$^-$] = $\sqrt{K_{so}}$ = $10^{-6.0}$, since pK_{so} = 12.0 for AgSCN (Table 6.1). If the smallest volume difference that can be estimated is 0.01 mL, how do the errors due to volume measurements compare with these other potential errors?

21. Calculate the solubility of silver acetate in 0.01 M nitric acid, and the pH of the saturated solution. Weak complexes can form: $[\text{AgAc(aq)}] = 10^{+0.73}[\text{Ag}^+][\text{Ac}^-]$, $[\text{AgAc}_2^-] = 10^{+0.64}[\text{Ag}^+][\text{Ac}^-]^2$ (at $I = 0$, 25°C from Smith and Martell, Vol. 4). How much difference does including these make?

22. Derive an equation analogous to Eq. (49), applying to a salt MX_2, where X is the anion of a monoprotic acid.

23. Calculate the solubility of CaF_2 in pure water and the pH of the saturated solution. Compare the value obtained using $pK_a = 3.17$ for HF with the value obtained assuming F^- is a strong anion like Cl^- (i.e., $pK_a \ll 0$).

24. 10 mL of 0.01 M $MgCl_2$ and 100 mL of 0.001 M HF are mixed. Calculate the number of moles of MgF_2 precipitated and the concentration of Mg^{2+} remaining in the solution. If the solution is diluted from 110 mL to 1 liter, what fraction of the precipitate dissolves?

25. Lead can be determined gravimetrically as the formate $Pb(CHO_2)_2$, which is then ignited to give the oxide. How many mL of 1.0 M formic acid should be added to 50 mL of solution containing 215 mg Pb, in order that precipitation should be 99.9% complete?

26. Can the solubilities of lead formate ($pK_{so} = 6.70$, $pK_a = 3.75$) and copper iodate ($pK_{so} = 7.13$, $pK_a = 0.79$) be made equal by adjusting pH? If so, at what pH value are the solubilities equal? How does this pH value depend on the ionic strength?

27. Derive an equation for the solubility of silver acetate in a strong acid such as HNO_3 of concentration C. Use pK_a as a function of ionic strength (Fig. 4.1), the formation constants of the Ag^+–Ac^- complexes, if needed (Problem 21), and the following data (from Hill and Simmons, 1909) to obtain a value for pK_{so}. Compare with Fig. 6.14.

HNO_3 (mole/L)	AgAc(sol'y)(g/L)
0.000	11.130
0.500	85.310
1.000	161.900
2.000	307.400
4.020	549.300
5.030	656.000
6.440	792.200

28. Derive an equation for the solubility of silver acetate in acetic acid of concentration C. Use the following data (Linke and Seidell, p. 27) together with the ionic strength dependence of pK_a (Fig. 4.1), and the formation constants of the Ag^+–Ac^- complexes, if needed (Problem 21) to evaluate pK_{so}. Compare with Figs. 6.7 and 6.8,

HAc (g/L)	AgAc(sol'y)(g/L)
0.000	11.130
1.000	10.730
2.000	10.320
2.980	9.980
4.190	9.520
4.980	9.190
5.990	8.720
6.800	8.290
8.010	7.730
8.970	7.310
9.960	6.780
11.020	6.150
12.320	5.330
12.970	4.960
13.970	4.290
14.960	3.430
15.930	2.480
17.280	1.090

29. Excess solid $CaHPO_4$ is equilibrated with one liter of water. Calculate the number of moles of $CaHPO_4$ that dissolve, and the pH of the saturated solution. How is this answer affected by including the ion pairs

$$[CaHPO_4(aq)] = 10^{+2.7} [Ca^{2+}][HPO_4^{2-}]$$
$$[CaH_2PO_4^+] = 10^{+1.08} [Ca^{2+}][H_2PO_4^-]$$

in the mass and charge balances?

30. A solution containing 0.002 M Ca^{2+} is mixed with an equal volume of solution containing 0.002 M phosphate. As pH is varied, determine whether $CaHPO_4$ or $Ca_3(PO_4)_2$ is precipitated, calculate the solubility of the precipitated salt, and find the pH where both solids are in equilibrium with the solution.

31. Calculate the solubility of $CaCO_3$ (calcite) in 0.01 M $NaHCO_3$, in a closed system (to avoid the possible loss of CO_2 to the gas phase).

32. Calculate the solubility of $CaCO_3$ in pure water in a closed system.

33. Calculate the solubility of $CaCO_3$ in various concentrations C of $NaHCO_3$ in a closed system. Compare the simplest model (only CO_3^{2-} and HCO_3^-) with a more complicated model including activity coefficients and the ion-pair equilibrium

$$[CaHCO_3^+] = 10^{+1.0} [Ca^{2+}][HCO_3^-]$$

34. Magnesium can be determined gravimetrically[25] by precipitation as $MgNH_4PO_4(H_2O)_6$ and ignition of the precipitate to give $Mg_2P_2O_7$, which is

then weighed. In the procedure, the sample is approximately 0.2 g MgO, which is dissolved in HCl and neutralized with NH_3 to the methyl red indicator end point (pH = 5) and diluted to 150 mL. To this, 2 g of $(NH_4)_2HPO_4$ are added. Excess NH_3 is then added to make the NH_3 concentration 0.5 M. What fraction of Mg (from the original 0.2 g MgO) fails to be precipitated by this method?

NOTES

1. W. Nernst, "Über Gegenseitige Beeinflussung der Löslichkeit von Salzen," *Z. physik Chem.* **4**:372, 1899. In this paper Nernst presented evidence based on the solubility of silver acetate in silver nitrate and solutions, that the same equilibrium laws apply to the solubility of salts as apply to homogeneous systems. The solubility product law was stated explicitly here for the first time. Surprisingly, the soublity product of silver acetate is not listed in the standard references by Smith and Martell and Sillén and Martell. It is derived later in this chapter, pp. 213–215.
2. Selected from R. M. Smith and A. E. Martell, *Critical Stability Constants* (1976–1989) and L. G. Sillén and A. E. Martell, *Stability Constants* (1964, 1971).
3. Hg_2^{2+} is a single divalent ion.
4. Similar equations govern the case where barium is in excess: $S = [SO_4^{2-}]$, and $S + P =$ total sulfate.
5. M. L. Salutsky "Precipitates: Their Formation, Properties, and Purity," Chapter 18 of *Treatise on Analytical Chemistry*, edited by I. M. Kolthoff, P. J. Elving, and E. B. Sandell, New York, Interscience Encyclopedia, 1959; A. G. Walton, *The Formation and Properties of Precipitates*, New York, Interscience Publishers, 1967.
6. As you probably already know, modern analysis almost never relies on gravimetric techniques, but employs spectrophotometry, atomic absorption spectroscopy, x-ray fluorescence, plasma spectroscopy, and other techniques to determine elemental composition.
7. The higher points (marked 3 M and 4 M $HClO_4$) can be fit if the 0.2 factor of the Davies equation is adjusted to 0.35. Compare with Fig. 7.12.
8. You can also express $[Fe^{3+}] = 10^{-3pH} K_{so}^\circ/(K_w^\circ)^3 \gamma_{3+}$. If pH is held constant, $[Fe^{3+}]$ varies as $1/\gamma_{3+}$, or $(\gamma_+)^{-9} = 25$ for $I = 1$ or $\gamma_+ = 0.7$. See pp. 267–270.
9. W. F. Linke and A. Seidell, *Solubilities of Inorganic and Metal–Organic Compounds*, 4th Ed., Van Nostrand, NY, 1958. Vol. 1, pp. 26–36.
10. In 1962 I made a crude extrapolation of these data to get $pK_{so}^\circ = 2.40$, which was listed in Table 6-2 of *Ionic Equilibrium* (1964), p. 195.
11. Ref. 1: W. Nernst, *Z. physik. Chem.* **4**:371, 1889; Ref. 2: J. Jaques, *Trans. Faraday Soc.* **5**:235, 1910; Ref. 3: F. H. MacDougall, *J. Am. Chem. Soc.* **46**:730–37, 1942.
12. Macdougall and Allen (1942).
13. pK_{so} (DH), 90% confidence limits on the mean. For pK_{so} (Davies) the average is 2.59 ± 0.05 in $AgNO_3$; 2.44 ± 0.11 in NaAc.
14. Unsymmetrical salts such as Ag_2CrO_4, Ag_2SO_4, and Hg_2Cl_2 are dealt with in the next section.
15. But for those forming an unsymmetrical salt such as M_2X the point of maximum slope and the equivalence point are different (see below).

16. Compare with the discussion in the preceding section following Eq. (22) and also the discussion of titration error in Chapter 4.
17. Eq. (42) can also be obtained from Eq. (40) by setting $C \gg ([M] + 2K_{so}/[M]^2)$.
18. Selected from Smith and Martell and Sillén and Martell. pK_{so} for AgAc is derived in Fig. 6.8, p. 215. For sulfates, K_a is the ionization constant of HSO_4^-; for chromates, K_a is the ionization constant of $HCrO_4^-$. In concentrated solutions, the dimerization of $HCrO_4^-$ must be taken into account: $[Cr_2O_7^{2-}] = 10^{+1.52} [HCrO_4^-]^2$
19. See Fig. 6.8.
20. HNO_3 is used instead of HCl because HCl reacts with Ag^+ to precipitate AgCl. Weak complexes can form: $[AgAc(aq)] = 10^{+0.73}[Ag^+][Ac^-]$, $[AgAc_2^-] = 10^{+0.64}[Ag^+][Ac^-]^2$ (at $I = 0.25°C$ from Smith and Martell, Vol. 4); these are ignored for the moment but will be discussed later.
21. A. E. Hill and J. D. Simmons, *J. Am. Chem. Soc.* **31**:821–39, 1909; data on solubility of AgAc in HNO_3 at 25°C, quoted by Linke and Seidell, p. 27. pH was estimated as $-\log[\gamma_{\pm HNO_3} (C_{HNO_3} - S)]$, where C_{HNO_3} is the concentration of the acid, S is the molar solubility of AgAc, and $\gamma_{\pm HNO_3}$ was taken from Robinson and Stokes, p. 491.
22. This equation applies also to unsaturated solutions of AgAc of concentration S'. The condition of saturation is supplied by K_{so} as given by Eq. (49). See *Ionic Equilibrium*, 1964, pp. 198–201.
23. Selected from Smith and Martell and Sillén and Martell. Many of these salts have several crystal forms, sometimes with substantially different solubilities. Several examples are given in Table 6.4. Almost all cations and anions of polyprotic acids form ion pairs in solution, and in some cases the ion pairs contribute significantly to the solubility; these complications are discussed in later chapters.
24. The very important practical problem of obtaining a good precipitate for filtration, coarse enough to be retained, yet free of foreign ions, is outside the scope of this book. This topic is well covered in standard reference works on analytical chemistry. Similarly, the fractionation of truly soluble, colloidal, and crystalline materials in natural and polluted waters is very important, but outside the scope of this book.
25. A. W. Taylor, A. W. Frazier, and E. L. Gurney "Solubility products of magnesium ammonium and magnesium potassium phosphate," *Trans. Faraday Soc.* **59**:1580–85, 1963.

7

COMPLEX FORMATION

Stepwise Formation of Inorganic Complexes in Solution
  Mononuclear Equilibria
  Distribution Diagrams
    Species Fraction versus Ligand Concentration
    Logarithmic Plot
    Cumulative Distribution Diagram
    Mass Balance on Ligand In Cd^{2+}–Cl^- System
Ionic Strength Dependence of Formation Constants
Effect of Complex Formation on Solubility
  Silver Chloride in Excess Chloride
Hydrolysis of Metal Ions
  Mercury–Hydroxide Complexes
  Aluminum Hydroxide Complexes
Iron–Hydroxide Complexes
Case Study: Acid Mine Drainage—Pyrite Oxidation
  A Chemical Model for Pyrite Oxidation Products
  Field Data
Problems

STEPWISE FORMATION OF INORGANIC COMPLEXES IN SOLUTION

Any species in solution formed by the combination of two or more simpler species—which can also exist independently in the solution—is called a *complex*. A complex may be positive, negative, or uncharged. Organic complexes are described in Chapter 8.

Acid–base systems are special cases in which the hydrogen ion is one of the simple species, and the mathematics of complex formation equilibria, mass balances, and charge balances includes all we have already presented regarding acid–base systems. One of the significant formal differences between acid–base systems and complex formation systems is the way in which reactions are written. Acid–base equilibria are conventionally written as dissociations

$$HA \rightleftharpoons H^+ + A^-$$

$$[H^+][A^-] = K_a [HA]$$

but complex formation equilibria are conventionally written as associations:

$$Cd^{2+} + Cl^- \rightleftharpoons CdCl^+$$

$$[CdCl^+] = K_1 [Cd^{2+}][Cl^-]$$

Although these two conventions may initially cause confusion, careful reference to the actual equilibrium form (as Smith and Martell do in *Critical Stability Constants*) should prevent mistakes from this source.

Metal ion complexes were first recognized as units in crystalline salts, such as $Cu(NH_3)_4^{2+}$ in the salt $CuSO_4(NH_3)_4$, but they also exist in solution. The extent to which metals form complexes in solution varies widely. The alkali and alkaline earth metals form only weak complexes, best described as ion pairs since the cation and anion show only a little greater affinity than can be accounted for by electrostatic interaction. At the other extreme are very stable covalent complexes, such as $Co(NH_3)_6^{3+}$: The ammonia in this molecule does not react even with concentrated sulfuric acid.[1]

The most straightforward type to consider is the mononuclear complex, consisting of a central metal ion to which is bound a number of neutral or anionic groups called *ligands*: ML_n. The number of ligands attached to the central ion is called its *coordination number* (n), and this number, together with the geometry of the complex, is determined by the electronic structure of the field created by the metal ion and its ligands.[2]

In aqueous solution, most metal ions are never really "uncomplexed," since they are associated with a number of water molecules. Some are firmly bonded, such as $Al(H_2O)_6^{3+}$ and $Cu(H_2O)_4^{2+}$, and can be identified in crystals by x-ray diffraction.

If equilibrium is very slow to be established [minutes to hours in the case of $Co(NH_3)_5(H_2O)^{3+}$), the composition of the complex can be determined by direct chemical analysis. X-ray diffraction can tell the nature of complexes in crystals, and these can be inferred to be present in solution. Raman, infrared, and ultraviolet spectroscopic analysis of solutions can reveal intermediate species. Because complexes carry a different charge from the parent ions, their formation affects the conductance of electrolytes and such data can be analyzed to yield equilibrium constants. Complex formation also affects colligative properties such as freezing point, boiling point, and vapor pressure, and although these effects are of historic importance, they are not usually used now for quantitative evaluation of equilibria.

The most widely used method is potentiometric titration, using the glass electrode to determine pH (and hence [OH$^-$] or the speciation of acidic or basic ligands). Electrodes selective to other ions such as Cd^{2+}, Pb^{2+}, F^-, Cl^-, etc., have been used in titrations over a range of metal ion concentrations to establish equilibrium constants for complexes containing these species. Closely related are polarographic, amperometric, solublity, liquid partition, and ion-exchange techniques. Calorimetric measurements can help establish the enthalpy, and hence the temperature dependence, of complex equilibria. A large body of equilibrium data has been obtained by these methods

R. M. Smith and A. E. Martell, *Critical Stability Constants*,[3] is the best place to get values for equilibrium constants at 25°. But for the methodology, as well as other temperatures and solvents, one must consult the original references. For work published before 1970, the two volumes of *Stability Constants* by Sillén and Martell[4] are helpful in giving not only the reference and conditions but a code for method and the actual values obtained by each study.[5]

Mononuclear Equilibria

Like polyprotic acids, most complexes involve a number of intermediate species, but unlike many polyprotic acids, many metal ion complexes have overlapping stepwise equilibria, and this makes computation more difficult. However, for a mononuclear system of complexes with the form ML_n, once the ligand concentration is established, the rest of the concentrations can be easily calculated. Furthermore, as will be shown, the relative amount of each complex is dependent only on the ligand and not on the metal ion concentration—provided the complexes are all mononuclear.

Because there is always excess water present in aqueous solution, it is conventional to omit the water molecules in formulas for complexes. For example, hydrated copper ion is written Cu^{2+}, not $Cu(H_2O)_4^{2+}$ or $Cu(H_2O)_6^{2+}$, and the first hydrolysis product of aluminum ion is written $AlOH^{2+}$, not $Al(H_2O)_5OH^{2+}$.

Table 7.1 lists the formation constants for a few mononuclear complex ion systems with chloride as ligand. The constants are overall formation constants:

$$[CdCl^+] = \beta_1 [Cd^{2+}][Cl^-]$$

$$[CdCl_2] = \beta_2 [Cd^{2+}][Cl^-]^2$$

TABLE 7.1. Log of formation constants for chloride complexes with Hg^{2+}, Ag^+, and Cd^{2+} at 25°C[7]

Metal	Medium	β_1	β_2	β_3	β_4
Hg^{2+}	$I = 0.5$	6.74±0.1	13.22±0.2	14.2±0.1	15.2±0.1
Hg++	$I = 3.0$	7.15±0.08	13.99±0.01	15.1±0.4	16.1±0.1
Ag^+	$I \to 0$	3.31±0.00	5.25±0.01		
Ag^+	$I = 5.0$	3.70	5.62	6.4	6.1
Cd^{2+}	$I \to 0$	1.98±0.03	2.6±0.1	2.4±0.1	1.7
Cd^{2+}	$I = 4.0$	1.66±0.1	2.4±0.1	2.8±0.3	2.2±0.3

STEPWISE FORMATION OF INORGANIC COMPLEXES IN SOLUTION

$$[CdCl_3^-] = \beta_3 \, [Cd^{2+}][Cl^-]^3$$

$$[CdCl_4^{2-}] = \beta_4 \, [Cd^{2+}][Cl^-]^4$$

Data for these four equilibria above have been measured[6] using a cadmium amalgam electrode or polarography to measure Cd^{2+} activity in potentiometric titrations with various amounts of cadmium and chloride. Accurate measurement of the full range of complexes requires a medium of high ionic strength so that changes in the metal and ligand concentrations do not affect their activity coefficients. The ionic medium for the Cd-Cl studies was typically $NaClO_4$ or $LiClO_4$ at ionic strengths from 0.5 to 4.0. Data were also extrapolated to $I = 0$ from measurements at higher ionic strength, and these values depend somewhat on the type of extrapolation used.

Distribution Diagrams

Several ways of presenting the equilibrium data for mononuclear complexes are in common use:

- Plots of species concentration $[ML_n]$ versus ligand concentration $[L]$
- Plots of the fraction $\alpha_n = [ML_n]/[M]_{total}$
- Log or semilog plots of the above
- Plots of the cumulative fractions α_1, $\alpha_1 + \alpha_2$, etc. versus $\log[L]$ so that a vertical line at a given value of $[L]$ is divided into segments equal to α_1, α_2, α_3, etc. (see example below).

The simplicity and generality of these plots depends on a special property of the mononuclear equilibria: The fraction α_n depends only on ligand concentration $[L]$ and not on the total metal concentration, as will be shown below.

Species Fraction versus Ligand Concentration. Consider first the plot of species fraction vs log of ligand concentration for the cadmium chloride complexes. The fractions are defined as

$$\alpha_0 = [Cd^{2+}]/[Cd]_T \qquad (1)$$

$$\alpha_1 = [CdCl^+]/[Cd]_T \qquad (2)$$

$$\alpha_2 = [CdCl_2]/[Cd]_T \qquad (3)$$

$$\alpha_3 = [CdCl_3^-]/[Cd]_T \qquad (4)$$

$$\alpha_4 = [CdCl_4^{2-}]/[Cd]_T \qquad (5)$$

The total cadmium concentration is the sum of all the cadmium-containing species:

$$[Cd]_T = [Cd^{2+}] + [CdCl^+] + [CdCl_2] + [CdCl_3^-] + [CdCl_4^{2-}] \qquad (6)$$

Each of the species can be expressed, via the equilibrium constants, in terms of $[Cd^{2+}]$ and $[Cl^-]$, as described above:

$$[CdCl^+] = \beta_1 [Cd^{2+}][Cl^-] \tag{7}$$

$$[CdCl_2] = \beta_2 [Cd^{2+}][Cl^-]^2 \tag{8}$$

$$[CdCl_3^-] = \beta_3 [Cd^{2+}][Cl^-]^3 \tag{9}$$

$$[CdCl_4^{2-}] = \beta_4 [Cd^{2+}][Cl^-]^4 \tag{10}$$

Substituting Eq. 6 in Eqs. (1)–(5) and substituting Eqs. (7)–(10) in that result gives the following results. Note that $[Cd^{2+}]$, which appears in both numerator and denominator, cancels.[8]

$$\alpha_0 = \frac{1}{1 + \beta_1[Cl^-] + \beta_2[Cl^-]^2 + \beta_3[Cl^-]^3 + \beta_4[Cl^-]^4} \tag{11}$$

$$\alpha_1 = \frac{\beta_1[Cl^-]}{1 + \beta_1[Cl^-] + \beta_2[Cl^-]^2 + \beta_3[Cl^-]^3 + \beta_4[Cl^-]^4} \tag{12}$$

$$\alpha_2 = \frac{\beta_2[Cl^-]^2}{1 + \beta_1[Cl^-] + \beta_2[Cl^-]^2 + \beta_3[Cl^-]^3 + \beta_4[Cl^-]^4} \tag{13}$$

$$\alpha_3 = \frac{\beta_3[Cl^-]^3}{1 + \beta_1[Cl^-] + \beta_2[Cl^-]^2 + \beta_3[Cl^-]^3 + \beta_4[Cl^-]^4} \tag{14}$$

$$\alpha_4 = \frac{\beta_4[Cl^-]^4}{1 + \beta_1[Cl^-] + \beta_2[Cl^-]^2 + \beta_3[Cl^-]^3 + \beta_4[Cl^-]^4} \tag{15}$$

Some sample values of β were given in Table 7.1. For these calculations, the data obtained at ionic strength 3.0 M and 25°C were used: $\log \beta_1 = 1.5$, $\log \beta_2 = 2.2$, $\log \beta_3 = 2.3$, $\log \beta_4 = 1.6$. The results are shown in Fig. 7.1.

Logarithmic Plot. A similar diagram (Fig. 7.2) is obtained by plotting $\log \alpha_i$ vs $-\log[Cl^-]$. This plot gives more detail in the region where α_i is small compared to 1. If the vertical scale is shifted so that the horizontal axis intercepts it at $\log[Cd]_t$ instead of zero, then this plot gives the actual concentrations $\log[CdCl_i]$ instead of α_i vs $-\log[Cl^-]$ (Fig. 7.3).

Cumulative Distribution Diagram. Still another way of displaying these equilibria is in a cumulative distribution diagram. In Fig. 7.4, the curve farthest to the right is simply α_0, the same curve as the one labeled $[Cd^{2+}]$ in Fig. 7.1. The curve next to the left is $\alpha_0 + \alpha_1$, and the difference between the two is α_1. This region

STEPWISE FORMATION OF INORGANIC COMPLEXES IN SOLUTION 243

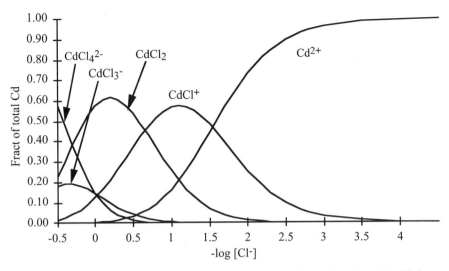

FIGURE 7.1. Fraction of total cadmium at $I = 3.0$ and 25°C as a function of $-\log[Cl^-]$.

between the two curves is labeled [CdCl$^+$], indicating that if a vertical line is drawn anywhere, the portion that lies in this region will equal α_1, the fraction of Cd present as CdCl$^+$. Similarly, the third curve from the right is $\alpha_0 + \alpha_1 + \alpha_2$, and a vertical line in the region between the second and third curves is α_2; the region is labeled [CdCl$_2$]. The fourth curve is $\alpha_0 + \alpha_1 + \alpha_2 + \alpha_3$, and the region between the third

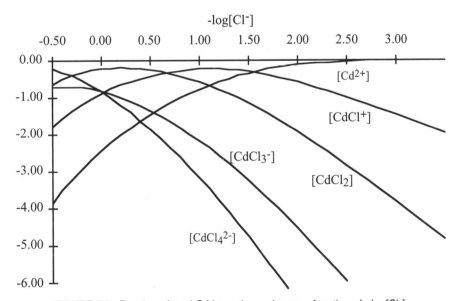

FIGURE 7.2. Fraction of total Cd in each species as a function of $-\log[Cl^-]$.

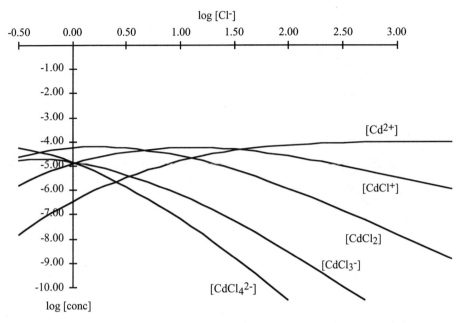

FIGURE 7.3. Distribution of actual species concentrations for $[Cd]_t = 10^{-4}$ M.

and fourth curves is labeled $[CdCl_3^-]$. The remaining area of the graph is labeled $[CdCl_4^{2-}]$.

For example, look at the vertical line on Fig. 7.4 drawn at $-\log[Cl^-] = 0.50$. It is divided into five segments by the curves, two large ones and three small ones:

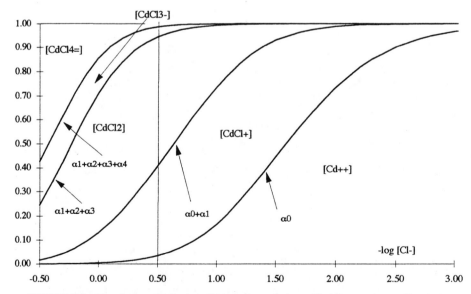

FIGURE 7.4. Cumulative distribution diagram for cadmium chloride complexes. See text.

STEPWISE FORMATION OF INORGANIC COMPLEXES IN SOLUTION

Species$_i$	α_i
Cd^{2+}	0.03
$CdCl^+$	0.37
$CdCl_2$	0.55
$CdCl_3^-$	0.04
$CdCl_4^{2-}$	0.01
Total	1.00

Thus the vertical width of a region at any value of ligand concentration shows immediately the relative amount of metal ion in that complex.

Mass Balance on Ligand in Cd^{2+}–Cl^- System. In all the above calculations, we have taken free chloride concentration $[Cl^-]$ as the independent variable. In many situations, it is the total chloride $[Cl]_t$ that is adjusted or known, and therefore a mass balance on chloride needs to be added to the mass balance on Cd and the various equilibria.

Example 1. Calculate the concentrations of all species in a solution containing 0.01 mole per liter of the salt $CdCl_2$ (this indicates that the starting material contains two moles of chloride per mole of cadmium and is distinct from the uncharged complex $CdCl_2$ in solution). To simplify the problem assume that sufficient HNO_3 has been added to prevent significant formation of Cd hydroxide complexes. The mass balances are

$$[Cd]_T = 0.01 = [Cd^{2+}] + [CdCl^+] + [CdCl_2] + [CdCl_3^-] + [CdCl_4^{2-}]$$

$$[Cl]_T = 2\,[Cd]_T = 0.02 = [Cl^-] + [CdCl^+] + 2\,[CdCl_2] + 3\,[CdCl_3^-] + 4\,[CdCl_4^{2-}]$$

The complexes are given in terms of $[Cd^{2+}]$ and $[Cl^-]$ by the equilibria, Eqs. (19)–(22). The full set of equations is fourth order in $[Cl^-]$ but can be simplified by using only the largest terms.

As a first approximation, assume that complex formation is slight, so that

$$[Cd^{2+}] = 0.01 \quad \text{and} \quad [Cl^-] = 0.02$$

With the trial values values, the equilibria (Table 7.1, constants for I = 0) give

$$[CdCl^+] = 10^{+1.98}\,[Cd^{2+}][Cl^-] = 0.0191$$

$$[CdCl_2] = 10^{+2.6}\,[Cd^{2+}][Cl^-]^2 = 0.0016$$

$$[CdCl_3^-] = 10^{+2.4}[Cd^{2+}][Cl^-]^3 = 2.00 \times 10^{-5}$$

$$[CdCl_4^{2-}] = 10^{+1.7}\,[Cd^{2+}][Cl^-]^4 = 8.02 \times 10^{-8}$$

From this rough calculation it seems clear that [CdCl$^+$] is comparable in size to [Cd^{2+}] and cannot be neglected, but that neither [CdCl$_3^-$] nor [CdCl$_4^{2-}$] will have much influence on the mass balance. [CdCl$_2$] is intermediate, accounting for about 16% of the mass balance on Cd^{2+}.

The next approximation is to include [CdCl$^+$] and [CdCl$_2$]:

$$[Cd]_T = 0.01 = [Cd^{2+}] + [CdCl^+] + [CdCl_2] + \cdots$$

$$[Cl]_T = 2\,[Cd]_T = 0.02 = [Cl^-] + [CdCl^+] + 2\,[CdCl_2] + \cdots$$

$$[Cd^{2+}]\{1 + 10^{+1.98}[Cl^-] + 10^{+2.6}[Cl^-]^2\} = 0.01$$

$$[Cl^-] + [Cd^{2+}]\{10^{+1.98}[Cl^-] + (2)(10^{+2.6})[Cl^-]^2\} = 0.02$$

The problem has been reduced to two equations linear in [Cd^{2+}] and quadratic in [Cl$^-$].

One standard approach to such a situation (see Chapter 1) is to solve one of the equations for [Cd^{2+}] in terms of [Cl$^-$] and substitute in the other:

$$[Cd^{2+}] = \frac{0.01}{1 + 10^{+1.98}[Cl^-] + 10^{+2.6}[Cl^-]^2}$$

$$[Cl^-] + \frac{0.01(10^{+1.98}[Cl^-] + 2 \times 10^{+2.6}[Cl^-]^2)}{1 + 10^{+1.98}[Cl^-] + 10^{+2.6}[Cl^-]^2} = 0.02$$

Simplifying this equation results in

$$10^{+2.6}[Cl^-]^3 + 10^{+1.98}[Cl^-]^2 + (1 - (0.01)(10^{+1.98}))[Cl^-] = 0.02$$

This is a cubic in [Cl$^-$], which can be solved by a number of methods, described in Chapter 1. To obtain a better approximation for the root of $f(x) = 0$, choose a first approximation x_0 and obtain a better approximation from Newton's method (p. 27):

$$x_1 = x_0 - \frac{f(x_0)}{f'(x_0)}$$

Setting $x =$ [Cl$^-$] and evaluating the coefficients gives

$$f(x) = 398\,x^3 + 95.5\,x^2 + 0.045\,x - 0.02 = 0$$

$$f'(x) = 1194\,x^2 + 191\,x + 0.045$$

It makes little difference what the first approximation is, so long as it is in a reasonable range. Since [Cl$^-$]$_T$ = 0.02, then x must be less than this, but must also be positive: $0 < x \leq 0.02$. Choose $x_0 = 0.02$, and get

$$x_1 = 0.02 - \frac{f(0.02)}{f'(0.02)} = 0.02 - \frac{0.0223}{4.34} = 0.0149$$

A second iteration gives $x = 0.0139$, and further iterations converge on $x = [Cl^-] = 0.01385$. From the third equation on p. 246 above,

$$[Cd^{2+}] = \frac{0.01}{1 + 10^{+1.98}(0.01385) + 10^{+2.6}(0.01385)^2} = 0.00417$$

The same result can be obtained with the secant method or the "goal seek" function of a spreadsheet program, as described in Chapter 1.

Example 2. Another approach is to interpolate the two original quadratic equations until they are balanced.

$$f_1 = [Cd^{2+}](1 + 10^{+1.98}[Cl^-] + 10^{+2.6}[Cl^-]^2) = 0.01$$

$$f_2 = [Cl^-] + [Cd^{2+}]\{10^{+1.98}[Cl^-] + (2)(10^{+2.6})[Cl^-]^2\} = 0.02$$

This is easily done with a spreadsheet program. Beginning with the upper limits for each of the variables $[Cd^{2+}] = 0.01$ and $[Cl^-] = 0.02$, the two functions are calculated and the values of the variables adjusted until $f_1 = 0.01$ and $f_2 = 0.02$, as shown in the following table.

$[Cd^{2+}]$	$[Cl^-]$	f_1 (=0.01?)	f_2 (=0.02?)	
0.01	0.02	0.0307	0.0423	
0.01	0.016	0.0263	0.0333	
0.01	0.012	0.0220	0.0246	
0.01	0.01	0.0199	**0.0203**	*
0.008	0.014	0.0193	0.0259	
0.008	0.012	0.0176	**0.0221**	*
0.006	0.014	0.0145	0.0230	
0.006	0.012	0.0132	**0.0196**	*
0.004	0.014	**0.0097**	**0.0200**	*
0.00415	**0.0139**	**0.0100**	**0.0200**	*

The results are the same as were obtained using Newton's method or the "goal seek" function on the combined equations.

A final check of the answers is to calculate the concentrations of all species and to verify the mass balances directly.

$[Cl^-] = 0.01385$ from either Newton's method or interpolation

$[Cd^{2+}] = [Cd]_T/(1+\beta_1[Cl^-] + \beta_2[Cl^-]^2 + \beta_3[Cl^-]^3 + \beta_4[Cl^-]^4) = 0.004167$

$[CdCl^+] = [Cd^{2+}]\beta_1[Cl^-] = 0.00551$

$[CdCl_2] = [Cd^{2+}]\beta_2[Cl^-]^2 = 0.000318$

$[CdCl_3^-] = [Cd^{2+}]\beta_3[Cl^-]^3 = 2.78 \times 10^{-6}$

$[CdCl_4^{2-}] = [Cd^{2+}]\beta_4[Cl^-]^4 = 7.69 \times 10^{-9}$

$[Cd]_{T\ check} = [Cd^{2+}] + [CdCl^+] + [CdCl_2] + [CdCl_3^-] + [CdCl_4^{2-}] = 0.0100$

$[Cl]_{T\ check} = [Cd^{2+}] + [CdCl^+] + 2[CdCl_2] + 3[CdCl_3^-] + 4[CdCl_4^{2-}] = 0.0200066$

The difference between $[Cl]_{T\ check}$ and the actual $[Cl]_T$ is a result of neglecting $[CdCl_3^-]$ and $[CdCl_4^{2-}]$ in the approximate mass balance. The approximations result in an error of only 0.03%.

Example 3. An interesting system to compare with $CdCl_2$ is $HgCl_2$, where the complexes are much stronger and not evenly spaced in ligand concentration. The algebraic forms are the same, but the constants are different (see Table 7.1). A distribution diagram shows this clearly (Compare Fig. 7.5 with Fig. 7.4).

Unlike the Cd–Cl complexes (Fig. 7.4), which are relatively evenly spaced, and significant only if $[Cl^-] > 10^{-3}$, the Hg–Cl complexes (Fig. 7.5) are significant for $[Cl^-] > 10^{-9}$. The complex $HgCl_2$ itself is dominant from $[Cl^-] = 10^{-2}$ to $10^{-6.5}$, whereas $HgCl^+$ occupies only a small region between $HgCl_2$ and Hg^{2+}.

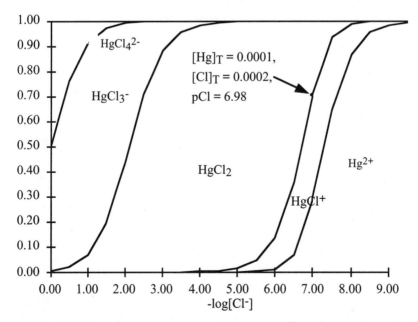

FIGURE 7.5. Distribution diagram for mercuric chloride species. Formation constants at $I = 3.0$, 25°C are taken from Table 7.1.

The calculation for a fixed total concentration 1×10^{-4} M of the salt $HgCl_2$ proceeds in the same way as the calculation for a fixed concentration of $CdCl_2$ did above:

$$f_1 = [Hg]_T = [Hg^{2+}](1 + 10^{+1.98}[Cl^-] + 10^{+2.6}[Cl^-]^2) = 0.0001$$

$$f_2 = [Cl]_T = [Cl^-] + [Hg^{2+}]\{10^{+1.98}[Cl^-] + (2)(10^{+2.6})[Cl^-]^2\} = 0.0002$$

These equations can be solved either by substuting $[Hg^{2+}]$ from the first equation into the second and using Newton's method, or adjusting $[Hg^{2+}]$ and $[Cl^-]$ simultaneously until f_1 and f_2 achieve the correct values (on the right-hand side). By either method, you will obtain $[Cl^-] = 1.05 \times 10^{-7} = 10^{-6.98}$, which occurs at the right-hand (smaller $[Cl^-]$) end of the $HgCl_2$ field and is marked on Fig. 7.5 with an arrow. The Hg species have concentrations $[Hg^{2+}] = 2.82 \times 10^{-5}$, $[HgCl^+] = 4.17 \times 10^{-5}$, $[HgCl_2] = 3.02 \times 10^{-5}$. The these values add up to $[Hg]_T = 1 \times 10^{-4}$; the higher complexes $[HgCl_3^-]$ and $[HgCl_4^{2-}]$ have concentrations less than 10^{-9}.

IONIC STRENGTH DEPENDENCE OF FORMATION CONSTANTS

For some systems, a considerable amount of equilibrium data is available, for all the formation constants, at several ionic strengths.[9] The cadmium–chloride system is one of these, and will be used as an example in this section.

Log β_1 for the formation of $CdCl^+$ (at 25°) is plotted as a function of ionic strength in Fig. 7.6. The experimental data were selected by Smith and Martell. The dependence on ionic strength is obtained from the equilibrium constant equation:

$$\beta_1^\circ = \frac{[CdCl^+]\gamma_+}{[Cd^{2+}]\gamma_{++}[Cl^-]\gamma_-} = \beta_1 \frac{\gamma_+}{\gamma_{++}\gamma_-} \tag{16}$$

$$\log \beta_1 = \log \beta_1^\circ + \log \gamma_{++} + \log \gamma_- - \log \gamma_+ \tag{17}$$

From the Davies equation (p. 49, where z is the charge on the ion and I is ionic strength),

$$\log(\gamma_z) = -0.5z^2 \left(\frac{\sqrt{I}}{1+\sqrt{I}} - 0.2I \right) \tag{18}$$

$$\log \gamma_- = \log \gamma_+, \quad \log \gamma_{++} = 4 \log \gamma_+$$

$$\log \beta_1 = \log \beta_1^\circ + 4 \log \gamma_+ \tag{19}$$

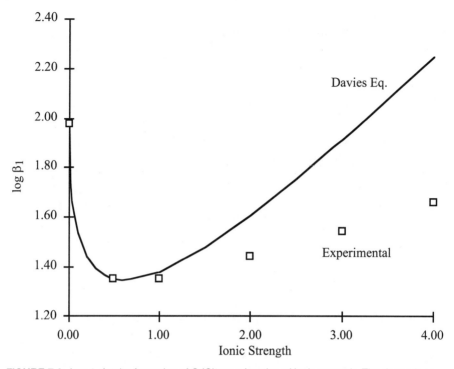

FIGURE 7.6. Log β_1 for the formation of $CdCl^+$ as a function of ionic strength. The dependence predicted by the Davies equation (continuous line) falls higher than the experimental values.

This function (continuous line in Fig. 7.6) falls higher than the experimental values, implying that the combination of activity coefficients as predicted by the Davies equation (18) is high compared to the experimental measurements. In Fig. 7.10 below, two other functions for log γ_+ are demonstrated, one of which gives a much better interpolation function.

However, as you will see in Figs 7.7–7.9, the Davies equation produces an excellent inerpolation function for β_2, β_3, and β_4,

For the second complex in the series, the equilibrium constant expression is:

$$\beta_2^o = \frac{[CdCl_2]\gamma_o}{[Cd^{2+}]\gamma_{++}[Cl^-]^2\gamma_-^2} = \beta_2 \frac{\gamma_o}{\gamma_{++}\gamma_-^2} \qquad (20)$$

Using the Davies Equation to evaluate $\log(\gamma_{++}) = 4 \log(\gamma_+)$ and $\log(\gamma_-^2) = 2 \log(\gamma_+)$, and setting $\log(\gamma_o) = +0.1I$

$$\log \beta_2 = \log \beta_2^o + 6 \log \gamma_+ - 0.1\ I \qquad (21)$$

IONIC STRENGTH DEPENDENCE OF FORMATION CONSTANTS

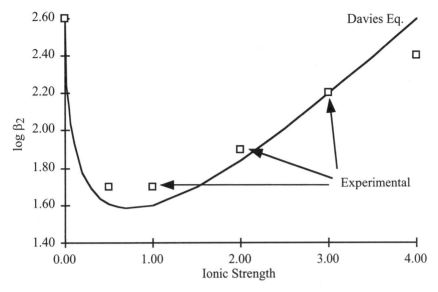

FIGURE 7.7. Log β_2 for the formation of $CdCl_2$ as a function of ionic strength. Note the good agreement between the experimental values and the Davies equation.

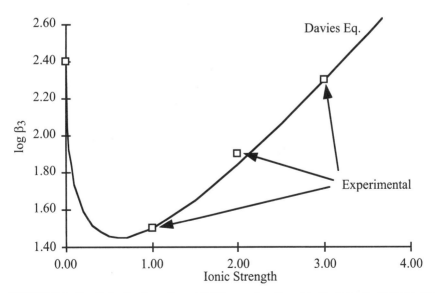

FIGURE 7.8. Log β_3 for the formation of $CdCl_3^-$ as a function of ionic strength. Note the good agreement between the experimental values and the Davies equation.

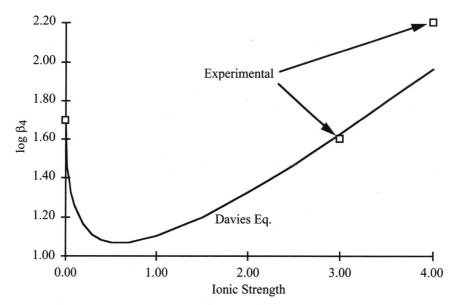

FIGURE 7.9. Log β_4 for the formation of $CdCl_4^{2-}$ as a function of ionic strength. Note the good agreement between the experimental values and the Davies equation.

This function is plotted together with the experimental data for β_2 in Fig. 7.7. The fit is as good as one could expect, and much better than Fig. 7.6.

Figure 7.8 compares the experimental data for β_3 with the Davies equation:

$$\beta_3^\circ = \frac{[CdCl_3^-]\gamma_-}{[Cd^{2+}]\gamma_{++}[Cl^-]^3\gamma_-^3} = \beta_3 \frac{\gamma_-}{\gamma_{++}\gamma_-^3} \tag{22}$$

$$\log \beta_3 = \log \beta_3^\circ + 6 \log \gamma_+ \tag{23}$$

where the γ_z are calculated from Eq. (18). Again, the fit is as good as could be expected.

Finally, in Fig. 7.9, the comparison for β_4 is made with:

$$\beta_4^\circ = \frac{[CdCl_4^{2-}]\gamma_=}{[Cd^{2+}]\gamma_{++}[Cl^-]^4\gamma_-^4} = \beta_3 \frac{\gamma_=}{\gamma_{++}\gamma_-^4} \tag{24}$$

$$\log \beta_4 = \log \beta_4^\circ + 4 \log \gamma_+ \tag{25}$$

where γ_z is calculated from Eq. (18). The fit is not bad, considering the lack of data and the high ionic strength.

Thus Fig. 7.6–7.9 have shown that the Davies equation fits the experimental data for the formation of Cd–Cl complexes with 2, 3, or 4 chlorides, but not for one chloride. Is it possible to fit the data for β_1 with a one-parameter equation, analogous to the Davies equation? One possibility is to combine the Debye–Hückel theory with a salting-out effect, according to Guggenheim's theory[10]:

$$\log(\gamma_z) = -0.5z^2 \left(\frac{\sqrt{I}}{1+\sqrt{I}} - bI \right) \quad (26)$$

The Davies equation is a special case with, $b = 0.2$; if this is reduced to 0.14, the fit is much better, as can be seen in Fig. 7.10. While this parameter adjustment is certainly reasonable, and gives a more accurate representation of the experimiental data, it defeats the most important property of the Davies equation—no adjustable parameters.

The Debye–Hückel equation with adjustable ion size parameter a

$$\log(\gamma_z) = -0.5z^2 \left(\frac{\sqrt{I}}{1 + 0.33a\sqrt{I}} \right) \quad (27)$$

which is frequently used in equilibrium calculations on complex mixtures, does not fit nearly so well. The lowest curve on Fig. 7.10 represents Eq. (28),

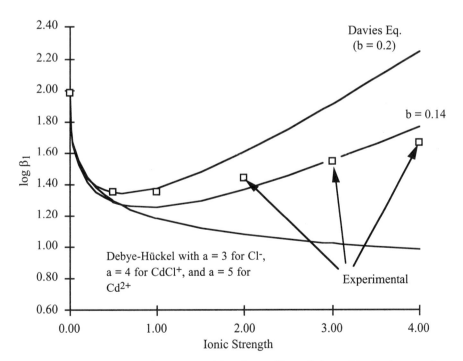

FIGURE 7.10. Comparison of experimental data for β_1 with predictions of the Davies equation, a modified Davies equation, and the Debye–Hückel equation with Kielland's ion-size parameters.

$$\log \beta_1 = \log \beta_1^\circ + \log \gamma_{Cl^-} - \log \gamma_{CdCl^+} + \log \gamma_{Cd^{2+}} \quad (28)$$

These activity coefficients were obtained from Eq. (27) with $a = 3$ for Cl^-, $a = 4$ for $CdCl^+$, and $a = 5$ for Cd^{2+}—the values given in Kielland's table[11] (See Chapter 2, p. 47):

$$\log \beta_1 = \log \beta_1^\circ - \frac{0.5\sqrt{I}}{1+(0.33)(3)\sqrt{I}} + \frac{0.5\sqrt{I}}{1+(0.33)(4)\sqrt{I}} - \frac{(0.5)(4)\sqrt{I}}{1+(0.33)(5)\sqrt{I}} \quad (29)$$

One parameter that is not sensitive to the form of the ionic activity coefficient expression is the stepwise equilibrium constant $K_3 = \beta_3/\beta_2$

$$K_3^\circ = \frac{\beta_3^\circ}{\beta_2^\circ} = \frac{[CdCl_3^-]\gamma_{CdCl_3^-}}{[CdCl_2]\gamma_o[Cl^-]\gamma_{Cl^-}} = 10^{-0.2 \pm 0.1} \quad (30)$$

where values of β_3° and β_2° are taken from Table 7.1. If the activity coefficients of $CdCl_3^-$ and Cl^- (both small singly charged negative ions) are close to each other in magnitude, they cancel and leave only the activity coefficient of uncharged $CdCl_2$, which can be approximated by a function linear in ionic strength. As showon in Fig. 7.11, the experimental data can therefore be fitted with a straight line that has intercept $\log K_3^\circ = -0.20$ and slope $b' = +0.1$ (the same as the salting-out coefficient used in the Davies equation). Note that a better fit would be obtained with a constant term between -0.1 and 0.0 (broken line in Fig. 7.11).

In general, rough calculations (± 0.5 log units) can be made with data from any ionic strength near that of the solution under consideration. Sometimes the error may be reduced by using the Davies, Guggenheim, or Debye–Hückel equation to inter-

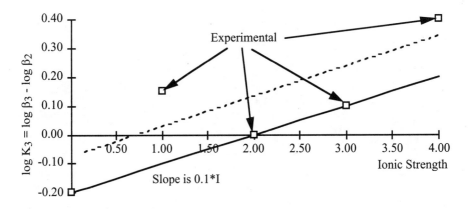

FIGURE 7.11. Log $K_3 = -0.20 + 0.1I$. A better fit would be obtained with a constant term between -0.1 and 0.0 (broken line).

polate. For careful calculations, the full set of available data should be plotted as a function of ionic strength, and the best interpolation made to the ionic strength of the solution under consideration.

Example 4. The final polish to be given to Examples 1 and 2 is to adjust the stability constants for ionic strength. Those calculations were done with constants extrapolated to $I = 0$, but the actual ionic strength of 0.01 M $CdCl_2$ is significantly higher than that. This is a region and a system for which the Davies equation is an accurate predictor of the ionic strength dependence, as was demonstrated in Figs. 7.6–7.10. The ionic strength is estimated using the concentrations calculated in the first part of this example:

$$I = \tfrac{1}{2}(4[Cd^{2+}] + [CdCl^+] + [Cl^-] + \cdots) = 0.0180$$

Note that $[CdCl_2]$ is omitted because it is uncharged; the two higher complexes are omitted because their concentrations are very small.

With this ionic strength, the activity coefficients can be estimated. Since the Davies equation depends only on the ionic charge, all the various activity coefficients can be lumped into one factor that is a multiple of $\log \gamma_+$, as in Eqs. (16)–(25).

$$\log(\gamma_+) = 0.5\left(\frac{\sqrt{I}}{1 + \sqrt{I}} - 0.2I\right) = -0.0574$$

$$\log \beta_1 = 1.98 + 4 \log \gamma_+ = 1.75 \tag{19}$$

$$\log \beta_2 = 2.6 + 6 \log \gamma_+ - 0.1I = 2.25 \tag{21}$$

$$\log \beta_3 = 2.4 + 6 \log \gamma_+ = 2.06 \tag{23}$$

$$\log \beta_4 = 1.7 + 4 \log \gamma_+ = 1.47 \tag{25}$$

Note that although the ionic strength is not large, it has a significant effect on the β_i.

The final stage is to repeat the calculations with the new values of β_i, which yield the following results:

$$[Cl^-] = 0.01507$$

$$[Cd^{2+}] = [Cd]T/(1 + \beta_1[Cl^-] + \beta_2[Cl^-]^2 + \beta_3[Cl^-]^3 + \beta_4[Cl^-]^4) = 0.005291$$

$$[CdCl^+] = [Cd^{2+}]\beta_1[Cl^-] = 0.00449$$

$$[CdCl_2] = [Cd^{2+}]\beta_2[Cl^-]^2 = 0.000216$$

$$[CdCl_3^-] = [Cd^{2+}]\beta_3[Cl^-]^3 = 2.06 \times 10^{-6}$$

$$[CdCl_4^{2-}] = [Cd^{2+}]\beta_4[Cl^-]^4 = 8.08 \times 10^{-9}$$

$[Cd]_{T\ check} = [Cd^{2+}] + [CdCl^+] + [CdCl_2] + [CdCl_3^-] + [CdCl_4^{2-}] = 0.0100$

$[Cl]_{T\ check} = [Cd^{2+}] + [CdCl^+] + 2[CdCl_2] + 3[CdCl_3^-] + 4[CdCl_4^{2-}] = 0.0200052$

[Cl⁻] is about 9% higher, [Cd^{2+}] is about 27% higher, and the complexes are all lower in concentration than in the first approximation.

EFFECT OF COMPLEX FORMATION ON SOLUBILITY

In Chapter 6, the solubility of simple salts was calculated, the effect of common ions presented. A detailed explanation of precipitation titrations and their errors was followed by the effects of ionic strength on solubility. At the end of the chapter, the effect of pH on those salts that had ions that were weak bases or acids was demonstrated using the salt silver acetate as an example.

In this chapter, the effects of complex formation will be presented. These are quite analogous to the effects of pH on salts of weak acids or bases, in that the formation of a complex ion removes from the solubility equilibrium one of the salt ions, resulting in greater solubility. In general, if a salt MX is dissolved in a solution containing the ligand Y, the equilibria are

$$[M^+][X^-] = K_{so} \tag{31}$$

$$[MY] = K_1 [M^+][Y^-] \tag{32}$$

The mass balance on M gives its solubility

$$[M]_T = [M^+] + [MY] + \cdots = [M^+](1 + K_1[Y^-] + \cdots) \tag{33}$$

$$[M]_T = \frac{K_{so}}{[X]}(1 + K_1[Y] + \cdots) \tag{34}$$

where "$\cdots$" represents the possible presence of other complexes. From Eq. (33), it is easy to see that any increase in the concentration of the ligand Y produces an increase in the solubility $[M]_T$.

If X is the same as Y, that is, when the complexes are formed between the metal ion and the anion of the slightly soluble salt, an increase in $[Y]$ or $[X]$ at first causes a decrease in solubility $[M]_T$ because of the common-ion effect, and then an increase because of the formation of M–Y complexes.

Silver Chloride in Excess Chloride

An important example of this is the solubility of AgCl in excess chloride.

Example 5. Calculate the solubility of AgCl in NaCl solutions of concentration 1 to 10⁻⁶ M. In Chapter 6 (Fig. 6.6, p. 212) you saw that almost all the experimental data for pK_{so} as a function of ionic strength fell above the Davies equation line and even farther above the Debye–Hückel equation line. For interpolation in this example, it is convenient to adjust the linear term in the Davies–Guggenheim equation to 0.35 I instead of 0.2 I:

$$\log(\gamma_\pm) = -0.5\left(\frac{\sqrt{I}}{1+\sqrt{I}} - 0.35I\right)$$

This gives a better fit to the experimental data (Fig. 7.12).

Considering the extraordinary number of studies on the solubility of AgCl,[12] data on the formation of $AgCl_3^{2-}$ and $AgCl_4^{3-}$ are rather sparse. The best data, as compiled by Smith and Martell (1986), are shown in Fig. 7.13. Only a few points are noted as "experimental data." Most of the points are interpolation curves based on the Davies equation, which predicts

$$\log \beta_1^\circ = 3.31 \; @ \; 25°C \quad \log \beta_1 = \log \beta_1^\circ + 2 \log \gamma_+$$
$$\log \beta_2^\circ = 5.25 \quad \log \beta_2 = \log \beta_2^\circ + 2 \log \gamma_+$$
$$\log \beta_3 = \log \beta_3^\circ = 6.4$$
$$\log \beta_4^\circ = 6.7 \quad \log \beta_4 = \log \beta_4^\circ - 4 \log \gamma_+$$

Since β_3 is essentially independent of ionic strength, extrapolation to $I = 0$ is simple, but cannot accomodate both log $\beta_3 = 6.4$ at $I = 5$ and 5.04 at $I = 0.2$. One solution is to ignore the lower value; another is to add a term: log β_3 = log β_3° + 0.27I to get a line of positive slope, as shown in Fig. 7.13. For β_4, the equation above permits

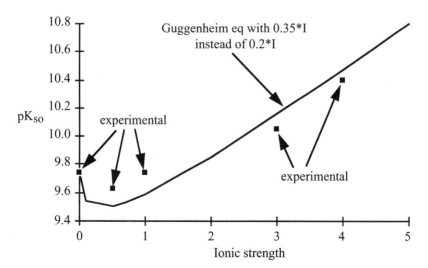

FIGURE 7.12. Solubility product of AgCl as a function of ionic strength.

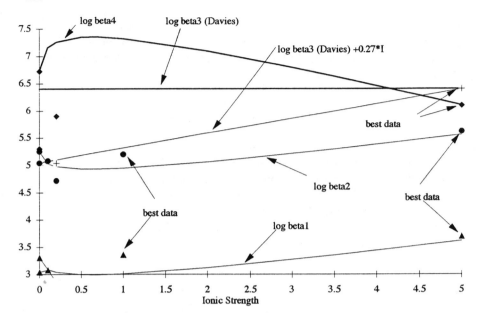

FIGURE 7.13. Formation constants of Ag–Cl complexes as a function of ionic strength. The curves for β_3 and β_4 are adjusted to fit the data at $I = 5$. Other data (I. Leden, *Svensk Kem Tidskr* **64**:249, 1952) at $I = 0.2$ are $\beta_3 = 5.04$ and $\beta_4 = 5.90$, quite a bit lower than those on this graph.

extrapolation from 6.10 at $I = 5$ to 6.72 at $I = 0$. Other data for these constants are either not at constant ionic strength or less reliable.

The solubility $S = [Ag]_T$ is given directly by the mass balance on Ag and the equilibria for the solubility of AgCl and the complexes:

$$[Ag]_T = [Ag^+] + [AgCl] + [AgCl_2^-] + [AgCl_3^{2-}] + [AgCl_4^{3-}]$$

$$[Ag]_t = \frac{K_{so}}{[Cl^-]}(\beta_1[Cl^-] + \beta_2[Cl^-]^2 + \beta_3[Cl^-]^3 + \beta_4[Cl^-]^4)$$

In some literature, the combination $K_{sn} = K_{so}\beta_n$ etc. is used to simplify the equation slightly:

$$[Ag]_t = K_{so}[Cl^-]^{-1} + K_{s1} + K_{s2}[Cl^-] + K_{s3}[Cl^-]^2 + K_{s4}[Cl^-]^3 = \sum_{n=0}^{4} K_{sn}[Cl^-]^{n-1}$$

This equation is plotted in Fig. 7.14. For comparison, the first term only is also plottted. Note that although the common-ion effect predicts a solubility of 10^{-10} in 1 M chloride, the influence of the Ag–Cl complexes produces a solubility of $10^{-2.2}$.

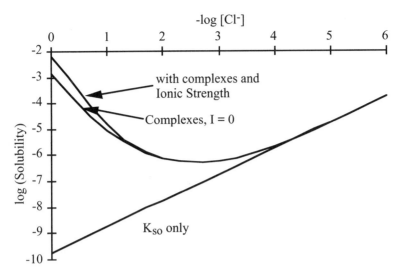

FIGURE 7.14. Solubility of AgCl showing the effect of complexes at $[Cl^-] > 10^{-3.5}$. Curves were calculated with constants for $I = [Cl^-]$, using ionic strength dependence of the various constants as given above. (If I is assumed to be zero for all constants, the solubility at $[Cl^-] > 0.1$ is calculated to be lower, and at $[Cl^-] = 1$, $S = 10^{-2.8}$, not $10^{-2.2}$.)

HYDROLYSIS OF METAL IONS

The formation of hydroxide complexes of metal ions follows the same pattern as has been demonstrated for chloride complexes, and the primary difference is that pH is the master variable. Although hydroxide complexes of alkali and alkaline earth metals are weak and monoprotic, some transition metals tend to yield structurally more complicated complexes and to have insoluble oxide or hydroxide phases. In Table 7.2 are listed some constants for mercury (primarily mononuclear), aluminum (several polynuclear complexes), and iron, which has a polynuclear complex, $Fe_2(OH)_2^{4+}$, that becomes dominant at high Fe^{3+} concentrations in acid solutions.

Mercury–Hydroxide Complexes

Mercuric ion reacts strongly with water in dilute solutions to form hydroxide complexes $HgOH^+$ and $Hg(OH)_2$; in concentrated solutions, the precipitate HgO is formed.[16] Figure 7.15 shows the distribution of Hg^{2+} in solutions more dilute than about 10^{-4} M as a function of pH.[17]

The distribution diagram of Fig. 7.15 is very similar to those presented earlier for chloride complexes:

$$\alpha_0 = \frac{1}{1 + \beta_1[OH^-] + \beta_2[OH^-]^2} \tag{35}$$

TABLE 7.2. Equilibrium constants for hydroxide complexes of Hg^{2+}, Fe^{3+} and Al^{3+}

	log K for Hg–OH complexes at 25°C, at various ionic strengths[13]					
Equil.	0	0.1	0.5	1.0	2.0	3.0
$\log \beta_1$	10.60		10.04	10.14		10.67±0.07
$\log \beta_2$	21.8±0.1	21.2	21.2	21.3	21.6	22.13±0.02
$\log \beta_3$	20.9					20.9
$\log \dfrac{[M_2L]}{[M]^2[L]}$	10.7					11.5
$\log \dfrac{[M_3L_3]}{[M]^3[L]^3}$	35.6					36.1
$\log K_{so}(\text{HgO, red})$	−25.44					−25.6

Note: M_2L/M^2L is the equilibrium constant for the reaction $2\,Hg^{2+} + OH^- \rightleftharpoons Hg_2OH^{3+}$, etc.

	log K for Al^{3+}–OH complexes at 25°C, at various ionic strengths.[14]			
Equil	0	0.1	0.5	1.0	2.0
$\log \beta_1$	9.01±0.04	8.47±0.02	8.21±0.05	8.31	
$\log \beta_2$	17.8	16.8±0.01			
$\log \beta_3$	25.5	24.7			
$\log \beta_4$	33.4±0.4		31.5±0.1		3.0

HYDROLYSIS OF METAL IONS

$\log \dfrac{[M_2L_2]}{[M]^2[L]^2}$	20.3			20.0	
$\log \dfrac{[M_3L_4]}{[M]^3[L]^4}$	42.1	42.0	41.4	42.5	
$\log \dfrac{[M_{13}L_{32}]}{[M]^{13}[L]^{32}}$	349.2		330.2	336.5	
$\log K_{so}$ (Al(OH)$_3$, α)	−33.5				

log K for Fe^{3+}–OH complexes at 25°C, at various ionic strengths[15]

Equil	0	0.1	0.5	1.0	2.0	3.0
$\log \beta_1$	11.81±0.03	11.27±.01	11.10±0.03	11.10±0.10	11.22	11.21±0.08
$\log \beta_2$	23.3		21.7	22.0±0.1		22.1
$\log \beta_3$	<28.8					
$\log \beta_4$	34.4					
$\log \dfrac{[M_2L_2]}{[M]^2[L]^2}$	25.1±0.05	24.7	24.8	25.0±0.1	25.2	25.4±0.1
$\log \dfrac{[M_3L_4]}{[M]^3[L]^4}$	49.7					51.0
$\log K_{so}$ for M(OH)$_3$(s)	−38.8±0.02					−38.6
$\log K_{so}$ for MOOH(α)	−41.5					−41.1
$\log K_{so}$ for M$_2$O$_3$(α)	−42.7					

Note: Three solid phases are listed that have the same solubility product form $K_{so} = [Fe^{3+}][OH^-]^3$ but K_{so} has different values. The amorphous Fe(OH)$_3$ listed first has the highest solubility; α-Fe$_2$O$_3$ has the lowest. Additional value: $\log \beta_1$ at $I = 0.7$ is 11.08±0.04.

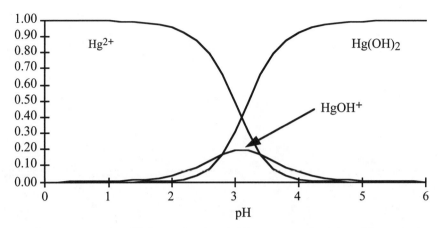

FIGURE 7.15. Distribution of Hg^{2+} as a function of pH in unsaturated solutions. The y axis gives the fractions $\alpha_0 = [Hg^{2+}]/[Hg]_T$, $\alpha_1 = [HgOH^+]/[Hg]_T$, $\alpha_2 = [Hg(OH)_2]/[Hg]_T$.

$$\alpha_1 = \frac{\beta_1[OH^-]}{1 + \beta_1[OH^-] + \beta_2[OH^-]^2} \tag{36}$$

$$\alpha_2 = \frac{\beta_2[OH^-]^2}{1 + \beta_1[OH^-] + \beta_2[OH^-]^2} \tag{37}$$

These equations apply to unsaturated solutions. If a precipitate is present, the additional relationship $[Hg^{2+}][OH^-]^2 = K_{so}$ must be satisfied. How can you tell whether a precipitate is present or not? Calculate the various species as if the solution were unsaturated, and compare the value obtained for $[Hg^{2+}]$ with the value that would apply in a saturated solution: $K_{so}/[OH^-]^2$.

Example 6. Is a 1 mM Hg(II) solution at pH = 1 saturated or unsaturated with respect to HgO? Refer to Fig. 7.16. A solution containing 1 mM Hg total, at pH = 1, has

$$[Hg^{2+}] = \alpha_0[Hg]_t = 0.99 \text{ millimolar}$$

At this pH, the ion product is $[Hg^{2+}][OH^-]^2 = (10^{-3})(10^{-13})^2 = 10^{-29}$, very much smaller than the solubility product $K_{so} = 10^{-25.44}$. Therefore, the solution is unsaturated.

Example 7. Find the distribution of species at pH = 3.0, and test for saturation.

$$[Hg^{2+}] = \alpha_0[Hg]_T$$

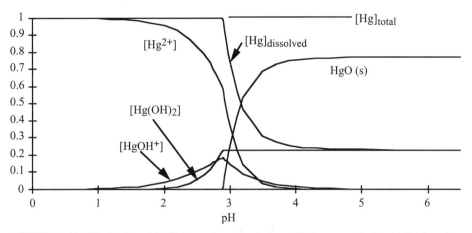

FIGURE 7.16. Distribution of Hg^{2+} (conc. in millimolar) as pH changes. Begin with $[Hg]_T = 1$ mM in strongly acidic solutions. Note the formation of $HgOH^+$ and then the precipitation of $HgO(s)$ at about pH = 2.9. $[Hg(OH)_2]$ reaches a constant value of 0.76 mM at higher pH, while the other species approach zero.

$$[HgOH^+] = \alpha_1[Hg]_T$$
$$[Hg(OH)_2] = \alpha_2[Hg]_T$$

Begin with the supersaturated solution: a provisional value of $[Hg]_T = 1$ mM

$$[Hg^{2+}]_{supersat} = \alpha_0[Hg]_T = 0.49 \text{ mM}$$

But at this pH, the solubility product gives

$$[Hg^{2+}]_{sat} = K_{so}/[OH^-]^2 = 0.36 \text{ mM}$$

which is smaller. Therefore, the solution is saturated, and the second value is the correct one for $[Hg^{2+}]_{sat}$. This value gives

$$[Hg]_{T(sol)} = [Hg^{2+}]_{sat}/\alpha_0 = 0.36/0.49 = 0.737 \text{ mM}$$

Of the original 1.000 mM, 0.737 mM remains in solution, and the rest $(1.000 - 0.737 = 0.229)$ is present as precipitate

$$[Hg^{2+}]_{sat} = 0.363 \text{ mM}$$
$$[HgOH^+] = \alpha_1[Hg]_{T(sol)} = 0.145 \text{ mM}$$
$$[Hg(OH)_2] = \alpha_2[Hg]_{T(sol)} = 0.229 \text{ mM}$$

and 0.263 millimoles of precipitate per liter.

Example 8. Find the distribution of species and amount of precipitate formed at pH = 6.0. Here the solubility product gives

$$[Hg^{2+}] = K_{so}/[OH^-]^2 = 10^{-25.44}/(10^{-8})^2 = 3.6 \times 10^{-10} \text{ M} = 3.6 \times 10^{-7} \text{ mM}$$

From this,

$$[HgOH^+] = K_1[Hg^{2+}][OH^-] = 10^{10.6}(3.6 \times 10^{-7})(10^{-8}) = 1.4 \times 10^{-4} \text{ mM}$$

$$[Hg(OH)_2] = \beta_2 K_{so} = 0.771 \text{ mM}$$

independent of pH. Thus the amount of precipitate per liter is $1.000 - 0.771 - 0.00014 - 3.6 \times 10^{-7} = 0.229$ millimoles, essentially independent of pH.

Example 9. In Table 7.2, two other equilibria are quoted, which have not been used in this discussion. "$\log(M_2L/M^2.L) = 10.7$" is interpreted as

$$\frac{[Hg_2OH^{3+}]}{[Hg^{2+}]^2[OH^-]} = 10^{10.7}$$

and "$\log(M_3L_3/M^3.L^3) = 35.6$" is interpreted as

$$\frac{[Hg_3(OH)_3^{3+}]}{[Hg^{2+}]^3[OH^-]^3} = 10^{35.6}$$

How do they influence the results of the calculations above?

These equilibria can be combined with the preceding equilibria to obtain a logarithmic concentration diagram, shown in Fig. 7.17. As in Fig 7.16, there is a discontinuity in slope when the precipitate forms, marked "sat'n" on Fig. 7.17. As before, Hg^{2+} is the dominant species at low pH, and $Hg(OH)_2$ is the dominant species at high pH. The polynuclear complexes reach their greatest influence around pH = 1.2, where the HgO precipitate is beginning to form. Even there, they constitute only 0.1% of the total mercury species. [18]

Figure 7.18, which covers the pH range from 6 to 14, shows how $Hg(OH)_3^-$ becomes an important species at high pH.

Aluminum Hydroxide Complexes

As can be seen from Table 7.2, aluminum forms a series of mononuclear and polynuclear complexes quite analogous to iron. Using the same mass balance equations as for $Fe(OH)_3$ (see below), with slightly different equilibrium constants, the saturation point can be determined from the intersection of the curves for $[Al^{3+}]$ calculated for unsaturated and saturated solutions. For unsaturated solutions,

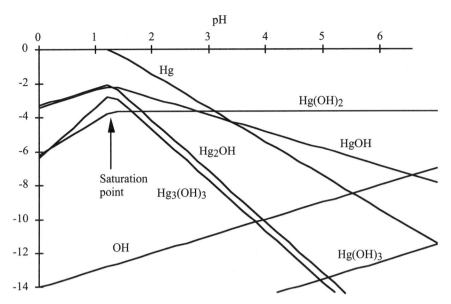

FIGURE 7.17. Approximate logarithmic concentrations of Hg species in 1.0 M total Hg solution as a function of pH. The polynuclear complexes Hg_2OH^{3+} and $Hg_3(OH)_3^{3+}$ achieve their maximum influence around the saturation point, pH = 1.2.

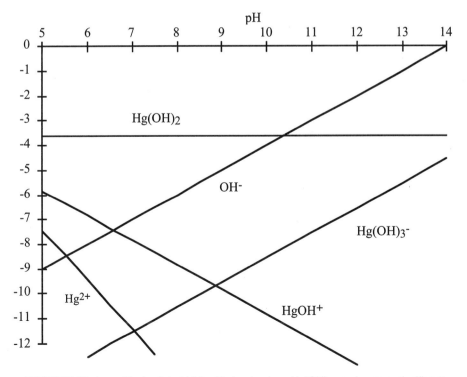

FIGURE 7.18. Logarithmic plot at high pH showing how $Hg(OH)_3^-$ can become significant.

$$[Al^{3+}]_{unsat} = \frac{Al_T}{Den} \qquad (38)$$

$$Den = 1 + \beta_1[OH^-] + \beta_2[OH^-]^2 + \beta_3[OH^-]^3 + \beta_4[OH^-]^4 + 2\beta_{22}[Al^{3+}][OH^-]^2$$
$$+ 3\beta_{34}[Al^{3+}]^2[OH^-]^4 + 13\beta_{13,32}[Al^{3+}]^{12}[OH^-]^{32} \qquad (39)$$

Since $[Al^{3+}]$ is an important factor in the denominator, Eq. (39) must be solved by iteration. One way to do this is to set $[Al^{3+}] = Al_T$ for the first iteration, and then to insert the result of this calculation in the denominator of the next iteration, until the results converge. Another approach is to use Newton's method to solve for $[Al^{3+}]$ in the full polynomial

$$[Al^{3+}]f[OH^-] + 2\beta_{22}[Al^{3+}]^2[OH^-]^2 + 3\beta_{34}[Al^{3+}]^3[OH^-]^4$$
$$+ 13\beta_{13,3}[Al^{3+}]^{13}[OH^-]^{32} = Al_T$$

where $f[OH^-]$ represents the first five terms of Eq. (39). A numerical example will be found in the discussion of the Fe–OH system below.
For saturated solutions:

$$[Al^{3+}]_{sat} = K_{so}/[OH^-]^3 \qquad (40)$$

For $Al_T = 0.5$ M, the intersection of saturated and unsaturated curves occurs near pH = 2.7 (see Fig. 7.19).

An additional polynuclear complex, $Al_{13}(OH)_{32}^{7+}$ is well documented,[19] a precursor of amorphous $Al(OH)_3$, but it forms only slowly at room temperature.

$$[Al_{13}(OH)_{32}^{7+}] = \beta_{13,32}[Al^{3+}]^{13}[OH^-]^{32} \qquad (41)$$

At $I = 1.0$, $\log(\beta_{13,32}) = 336.5$. The concentration of the complex in saturated and unsaturated solutions is always small compared to Al_T, but reaches a maximum near the saturation point (Fig. 7.20).

In many applications, the mononuclear complexes alone are sufficient to describe the system, even in pH and total Al ranges where the polynuclear species might be expected.[20]

Figure 7.21 shows the solubility of $Al(OH)_3$ as a function of pH over the whole range. Note that the general shape of the curves is similar to those for $Fe(OH)_3$ in Fig. 7.22, but the saturation pH is higher, the solubility minimum is flatter, and the minimum solubility is not so low. The polynuclear complexes have their maximum influence around the pH of saturation.

HYDROLYSIS OF METAL IONS

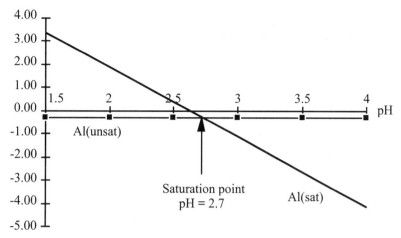

FIGURE 7.19. Plot of log[Al^{3+}] calculated assuming unsaturated solution (constant at −0.3) and assuming saturation with Al(OH)$_3$. The two lines meet at pH = 2.7.

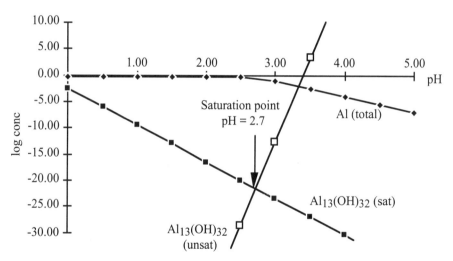

FIGURE 7.20. Wider concentration range showing log[Al$_{13}$(OH)$_{32}^{7+}$] calculated from the values for an unsaturated solution, and for a solution saturated with Al(OH)$_3$. The total dissolved Al in both saturated and unsaturated solutions is shown for comparison. The saturation point is pH = 2.73, as in Fig. 7.19, and the concentration of Al$_{13}$(OH)$_{32}^{7+}$ is about $10^{-21.55}$.

Iron–Hydroxide Complexes

Ferric iron is very common, and forms the oxidized couple of one of the most important oxidation–reduction systems. It is soluble in acid and slightly soluble in highly basic solutions, but is largely precipitated as amorphous Fe(OH)$_3$, α–FeOOH (goethite), or Fe$_2$O$_3$ (hematite) at intermediate pH. The solubility product K_{so} = [Fe^{3+}][OH$^-$]^3 for these three minerals is of the same form, but from Table 7.2, you

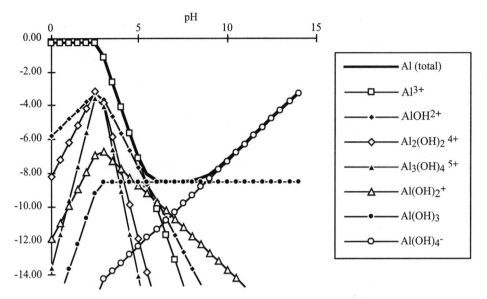

FIGURE 7.21. Aluminum hydroxide species as a function of pH. Total Al was taken to be 0.5 M and constants from Table 7.2 for $I = 1.0$ (where available) were used in the calculations.

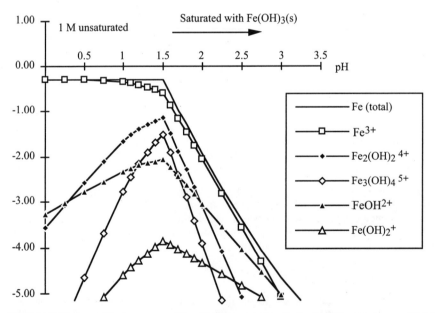

FIGURE 7.22. Logarithmic concentration diagram showing the hydroxide complexes of 0.5 M Fe^{3+} as a function of pH in strong acid media. Equilibrium constants for $I = 3.0$ at 25°C are taken from Table 7.2. To the left the solution is unsaturated (Eqs. 44–47); to the right the solubility product of amorphous $Fe(OH)_3$ ($pK_{so} = 38.8$) is assumed to hold (Eq. 48).

can see Fe_2O_3 (pK_{so} = 42.7) is the least soluble, α–FeOOH (pK_{so} = 41.5) is next most soluble, and freshly precipitated $Fe(OH)_3$ (pK_{so} = 38.8) is the most soluble. In the following examples, the solid phase will be assumed to be $Fe(OH)_3$.

Ferric iron forms a series of four mononuclear and two polynuclear soluble complexes. The polynuclear complex $Fe_2OH_2^{4+}$ can be the dominant ferric hydroxide species at high concentrations in acid solutions. To calculate the mononuclear complex concentrations as a function of pH, you need only follow the many previous examples in this chapter. But when the polynuclear complexes are included, the calculations involve iteration.

In unsaturated or supersaturated solutions, the mass balance on Fe can be expressed as follows:

$$Fe_T = [Fe^{3+}] + [FeOH^{2+}] + [Fe(OH)_2^+] + [Fe(OH)_3] + [Fe(OH)_4^-]$$
$$+ 2\,[Fe_2(OH)_2^{4+}] + 3\,[Fe_3(OH)_4^{5+}] \tag{42}$$

Substituting the equilibrium expressions gives:

$$Fe_T = [Fe^{3+}](1 + \beta_1[OH^-] + \beta_2[OH^-]^2 + \beta_3[OH^-]^3 + \beta_4[OH^-]^4\}$$
$$+ \{\beta_{22}[Fe^{3+}]^2[OH^-]^2 + \beta_{34}\,[Fe^{3+}]^3[OH^-]^4) \tag{43}$$

Note that the last two terms, representing the polynuclear complexes, are nonlinear (quadratic and cubic) in $[Fe^{3+}]$, and this means that the denominator of the normal expression for a mononuclear complex is *not* independent of the metal ion concentration. For example:

$$[Fe^{3+}]_{unsat} = \frac{Fe_T}{Den} \tag{44}$$

where the denominator is

$$Den = 1 + \beta_1[OH^-] + \beta_2[OH^-]^2 + \beta_3[OH^-]^3 + \beta_4[OH^-]^4 + 2\,\beta_{22}[Fe^{3+}][OH^-]^2$$
$$+ 3\,\beta_{34}[Fe^{3+}]^2[OH^-]^4 \tag{45}$$

As with the Al–OH system, the initial value of $[Fe^{3+}]$ in Eq. (45) must be guessed. As will be seen, anything less than Fe_T is a good choice for unsaturated solutions. Then a new value of $[Fe^{3+}]$ is obtained from Eq. (44), and the process repeated until it converges.

Once a satisfactory value of $[Fe^{3+}]$ is obtained, the other concentrations can easily be obtained from the equilibria, for example:

$$[FeOH^{2+}] = \beta_1[Fe^{3+}][OH^-] = \beta_1[OH^-]Fe_t/Den \tag{46}$$

$$[Fe_2(OH)_2^{4+}] = \beta_{22}[Fe^{3+}]^2[OH^-]^2, \quad \text{etc.} \tag{47}$$

A plot of the various concentrations as a function of pH is given in Fig. 22 for $Fe_t = 0.5$ M using constants for ionic strength 3.0 from Table 7.2.[21]

At higher pH, the solution becomes saturated with ferric hydroxide and the solubility product is obeyed at equilibrium. This actually makes the calculations easier, since $[Fe^{3+}]$ becomes a simple function of pH instead of the nonlinear function of Eqs. (44) and (45).

$$[Fe^{3+}]_{sat} = K_{so}/[OH^-]^3 = (K_{so}/K_w^3)[H^+]^3 \tag{48}$$

From this, the other concentrations can be calculated directly. The point at which the equilibrium system becomes saturated can be found from the intersection of $[Fe^{3+}]_{unsat}$ (the function of pH given by Eqs. 44 and 45) and $[Fe^{3+}]_{sat}$ (the function of pH given by Eq. 48). This intersection is shown in Fig. 7.23 to occur at pH = 1.5 for 0.5 M Fe^{3+} at $I = 3.0$.

At considerably higher pH, the concentration of soluble species decreases, to a minimum of about 10^{-10} M at pH = 8.5; then increases with the formation of anionic $Fe(OH)_4^-$. This is shown in Fig. 7.24. Measuring the solubility near this minimum (and hence the constant β_3) in real solutions is less accurate than in other regions because of complications from colloidal ferric hydroxide.

Case Study: Acid Mine Drainage—Pyrite Oxidation

Much of mining waste (tailings) consists of sulfide minerals, from the pyrite discarded in the Pennsylvania coal mines to the copper ores of Colorado and Utah. These sulfide minerals react with water and oxygen to produce an acid solution that can cause serious water contamination in the mine tailing area and downstream.

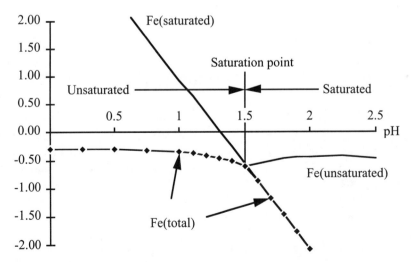

FIGURE 7.23. Intersection of $[Fe^{3+}]$ functions for saturated and unsaturated solutions at pH = 1.5.

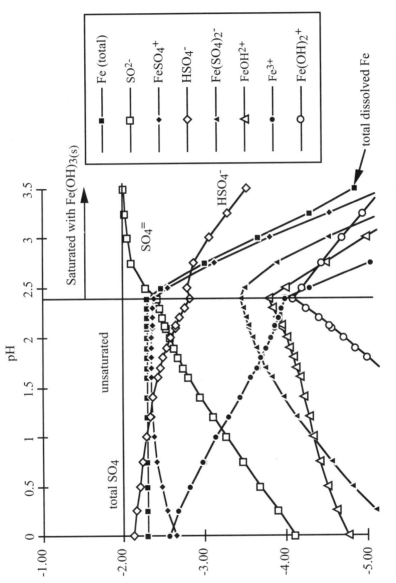

FIGURE 7.24. Distribution of species in a solution containing 0.005 M total Fe^{3+} and 0.01 M total SO_4^{2-}. The solution is saturated with $Fe(OH)_3$ for pH > 2.1. Calculations used constants from Table 7.3 for $I = 0$.

In the presence of oxygen and water, pyrite (FeS_2) and related minerals such as marcasite (also FeS_2) and arsenopyrite ($FeAsS_2$) oxidize to produce ferric ion, ferric hydroxide, and sulfuric acid:

$$FeS_2 \text{ (s)} + {}^{15}/_4 O_2 + \tfrac{1}{2} H_2O \rightarrow Fe^{3+} + 2SO_4^{2-} + H^+$$

$$Fe^{3+} + 2SO_4^{2-} + H^+ + 3H_2O \rightleftharpoons Fe(OH)_3(s) + 2SO_4^{2-} + 4H^+$$

The slow step of oxidation involves microbial action, which produces Fe^{2+} and SO_4^{2-} from FeS_2, followed by microbial or chemical oxidation of Fe^{2+} to Fe^{3+}, as well as direct action of the released Fe^{3+} on FeS_2. Each of these processes produces acidity.[22]

A Chemical Model for Pyrite Oxidation Products. As $Fe(OH)_3$ precipitates, more acid is released, which tends to reduce pH and buffer the system near the precipitation point. Similarly, the equilibrium between SO_4^{2-} and HSO_4^- tends to buffer the pH near $pK_a = 2$. In the previous section you saw some of the Fe^{3+}–OH^- equilibria; this model also contains the Fe^{3+}–SO_4^{2-} and the SO_4^{2-}–HSO_4^- equilibria.

Table 7.3 contains some data for low ionic strength, partially repeating Table 7.2, but without the less soluble phases of ferric oxide, and with the addition of formation constants for the ferric–sulfate system.

In unsaturated or supersaturated solutions, the mass balance on Fe can be expressed as follows:

$$[Fe^{3+}]_T = [Fe^{3+}] + [FeOH^{2+}] + [Fe(OH)_2^+] + [Fe(OH)_3] + [Fe(OH)_4^-]$$
$$+ 2[Fe_2(OH)_2^{4+}] + 3[Fe_3(OH)_4^{5+}] + [FeSO_4^+] + [Fe(SO_4)_2^-] \tag{49}$$

Substituting the equilibrium expressions gives:

TABLE 7.3. Log K for Fe^{3+}–OH and SO_4^{2-} complexes at 25°C, at low ionic strengths[23]

Equil at $I =$	0	0.1	0.5
Fe^{3+}, OH^-: log β_1	11.81±0.03	11.27±.01	11.10±0.03
log β_2	23.3		21.7
log β_3	< 28.8		
log β_4	34.4		
log $\dfrac{[M_2L_2]}{[M]^2[L]^2}$	25.1±0.05	24.7	24.8
log $\dfrac{[M_3L_4]}{[M]^3[L]^4}$	49.7		
log K_{so} ($M(OH)_3(s)$)	−38.8±0.02		
Fe^{3+}, SO_4^{2-}: log β_1	4.04±0.1		2.24±0.1
log β_2	5.38		
Fe^{3+}, HSO_4^-: log β_1	0.6 to 2.5[24]		
H^+, SO_4^{2-}: pK_a	1.99±0.01	1.55±0.05	1.32±0.06

HYDROLYSIS OF METAL IONS

$$[Fe^{3+}]_t = [Fe^{3+}](1 + \beta_1[OH^-] + \beta_2[OH^-]^2 + \beta_3[OH^-]^3 + \beta_4[OH^-]^4 + \beta_1^S[SO_4^{2-}]$$
$$+ \beta_2^S[SO_4^{2-}]^2) + (\beta_{22}[Fe^{3+}]^2[OH^-]^2 + \beta_{34}[Fe^{3+}]^3[OH^-]^4) \quad (50)$$

As before, the last two terms, representing the polynuclear complexes, are nonlinear (quadratic and cubic) in $[Fe^{3+}]$, and this means that the denominator of the normal expression for a mononuclear complex is *not* independent of the metal ion concentration. For example:

$$[Fe^{3+}] = \frac{Fe_T}{Den} \quad (51)$$

where the denominator is

$$Den = 1 + \beta_1[OH^-] + \beta_2[OH^-]^2 + \beta_3[OH^-]^3 + \beta_4[OH^-]^4 + \beta_1^S[SO_4^{2-}] + \beta_2^S[SO_4^{2-}]^2$$
$$+ 2\beta_{22}[Fe^{3+}][OH^-]^2 + 3\beta_{34}[Fe^{3+}]^2[OH^-]^4 \quad (52)$$

One approach to solving this set of equations begins with Eqs. (51) and (52), and evaluates $[Fe^{3+}]$ and $[SO_4^{2-}]$ by iteration.

The mass balance on sulfate is simpler, but closely linked to the Fe balance:

$$[SO_4]_T = [SO_4^{2-}] + [Fe(SO_4)^+] + [Fe(SO_4)_2^-] + [HSO_4^-] \quad (53)$$

$$[SO_4^{2-}] = \frac{[SO_4^{2-}]_T}{1 + \beta_1^S[Fe^{3+}] + \beta_2^S[Fe^{3+}][SO_4^{2-}] + [H^+]/K_a} \quad (54)$$

The initial values of $[Fe^{3+}]$ and $[SO_4^{2-}]$ must be guessed. Constraints are $[Fe^{3+}]_o \leq [Fe^{3+}]_T$ and $[SO_4^{2-}]_o \leq [SO_4^{2-}]_T$ in unsaturated solutions and $[Fe^{3+}]_o \leq K_{so}/[OH^-]^3$ in saturated solutions.

For example, set the initial value of $[Fe^{3+}] = [Fe^{3+}]_T$, then obtain $[SO_4^{2-}]$ from Eq (54). Then a new value of $[Fe^{3+}]$ is obtained from Eqs. (51) and (52), and the process repeated until it converges. Once satisfactory values of $[Fe^{3+}]$ and $[SO_4^{2-}]$ are obtained, the other concentrations can easily be obtained from the equilibria, for example:

$$[FeOH^{2+}] = \beta_1[Fe^{3+}][OH^-] \quad (55)$$

$$[FeSO_4^+] = \beta_1^S[Fe^{3+}][SO_4^{2-}], \quad etc. \quad (56)$$

A plot of the various concentrations as a function of pH is given in Fig. 7.25 for $[Fe^{3+}]_T = 0.005$ M and $[SO_4]_T = 0.01$ M using constants for $I = 0$ from Table 7.3. Five species comprise more than 10% of the total in solutions with pH < 2:

$$[SO_4^{2-}], [FeSO_4^+], [HSO_4^-], [Fe^{3+}], [Fe(SO_4)_2^-]$$

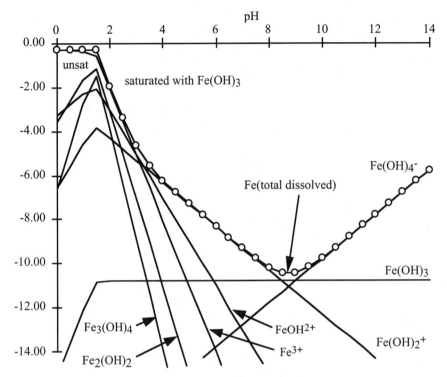

FIGURE 7.25. Logarithmic diagram for 0.5 M Fe^{3+} as a function of pH. The calculations are the same as Fig. 7.22, except for emphasizing the high pH region. Note that the minimum solubility of $Fe(OH)_3$ occurs at about pH = 8.5, and is of the order of 10^{-10} M.

Smaller, but still significant, are $[FeOH^{2+}]$ and $[Fe(OH)_2^+]$. Smaller than 0.1% of $[Fe]_T$ are $[Fe_2(OH)_2^{4+}]$ and $[Fe_3(OH)_4^{5+}]$—but seven species reach at least 0.3% of $[Fe^{3+}]_T$ or $[SO_4]_T$ at the saturation point.

A refinement would be to calculate concentrations of all species and find the ionic strength, which could then be used to calculate activity coefficients in an iterative procedure. The ionic strength of the solutions considered in Examples 10–12 is about 0.02 M, which gives $\log \gamma_+ = -0.05$ and $\log \gamma_{++} = -0.20$. The ions most affected by ionic strength are $[Fe_2(OH)_2^{4+}]$ with $\log \gamma_+ = -0.80$ and $[Fe_3(OH)_4^{5+}]$ with $\log \gamma_+ = -1.25$, but these ions are present in quite low concentrations (see Examples 10 and 12). If ionic strength and activity coefficients are to be included, the solutions at pH < 2.4 should contain excess Cl⁻ and solutions at pH > 2.4 should contain excess Na⁺ to account for solutions added to adjust pH.

Example 10. Calculate the concentrations of the various species in 0.005 M total Fe^{3+} and 0.01 M SO_4^{2-} at pH = 2.0. Begin with $[Fe^{3+}] = [Fe]_T = 0.005 = 10^{-2.30}$ and substitute this in Eq. (54) to get $[SO_4^{2-}] = 10^{-5.11}$, which is clearly too small. However, this value in Eq.(51) gives a revised $\log[Fe^{3+}] = -4.43$ and $\log[SO_4^{2-}] = -3.36$. Successive iterations give $\log[Fe^{3+}] = -3.12, -3.59, -3.70, -3.76, -3.79$,

HYDROLYSIS OF METAL IONS

−3.80, −3.81, −3.81 and log[SO_4^{2-}] = −3.02, −2.81, −2.69, −2.63, −2.60, −2.59, −2.58, −2.58.

With these convergent answers you can readily calculate the concentrations of the other species (compare Fig. 7.25):

$$\log[Fe^{3+}] = -3.82$$

$$\log[SO_4^{2-}] = -2.57$$

$$\log[OH^-] = -12.00, \quad \text{since pH} = 2.0$$

$$\log[FeOH^{2+}] = -4.01$$

$$\log[Fe(OH)_2^+] = -4.72$$

$$\log[Fe(OH)_3] = -11.02$$

$$\log[Fe(OH)_4^-] = -17.42$$

$$\log[Fe_2(OH)_2^{4+}] = -6.53$$

$$\log[Fe_3(OH)_4^{5+}] = -9.75$$

$$\log[HSO_4^-] = -2.58$$

$$\log[FeSO_4^+] = -2.35$$

$$\log[Fe(SO_4)_2^-] = -3.58$$

Checking the mass balances, you will find

$$\log[SO_4^{2-}]_T = 1.99$$

$$\log[Fe^{3+}]_T = 2.30$$

Note that at this pH, the predominant iron-containing species is $FeSO_4^+$; and Fe^{3+} as well as $Fe(SO_4)_2^-$ are both less than 10% of the total iron.

Example 11. Is the solution of Example 10 saturated or unsaturated? The ion product is $[Fe^{3+}][OH^-]^3 = 10^{-3.82-3(12.00)} = 10^{-39.2}$, which is less than the solubility product $K_{so} = 10^{-38.8}$. The solution is unsaturated.

At pH higher than about 2, the solution becomes saturated with ferric hydroxide and the solubility product is obeyed at equilibrium. This actually makes the calculations easier, since $[Fe^{3+}]$ becomes a simple function of pH instead of the nonlinear mass balance function of Eqs. (51) and (52).

$$[Fe^{3+}]_{sat} = K_{so}/[OH^-]^3 = (K_{so}/K_w^3)[H^+]^3 \tag{57}$$

From this value of $[Fe^{3+}]_{sat}$, $[SO_4^{2-}]$ and the other concentrations can be calculated directly.

Example 12. Find the concentrations of the major species at pH = 3. Note $[OH^-] = 10^{-11.00}$. Provisionally, assume that the solution is saturated, so that $[Fe^{3+}] = 10^{-38.8+3(11.00)} = 10^{-5.80}$. Substituting this value in Eq. (54) gives $[SO_4^{2-}] = 10^{-2.00}$; a few iterations gives $10^{-2.05}$.

The various species concentrations are (compare with Example 10 and Fig. 7.25):

$$\log[Fe^{3+}] = -5.80$$
$$\log[SO_4^{2-}] = -2.05$$
$$\log[OH^-] = -11.00, \quad \text{since pH} = 3.0$$
$$\log[FeOH^{2+}] = -4.99$$
$$\log[Fe(OH)_2^+] = -4.70$$
$$\log[Fe(OH)_3] = -10.00$$
$$\log[Fe(OH)_4^-] = -15.40$$
$$\log[Fe_2(OH)_2^{4+}] = -8.50$$
$$\log[Fe_3(OH)_4^{5+}] = -11.74$$
$$\log[HSO_4^-] = -3.81$$
$$\log[FeSO_4^+] = -2.35$$
$$\log[Fe(SO_4)_2^-] = -4.52$$

Checking the mass balances, you will find

$$\log[SO_4^{2-}]_T = -2.00$$
$$\log[Fe^{3+}]_{T,\,diss} = -3.66, \quad \text{total of dissolved iron species}$$
$$P = [Fe]_T - [Fe^{3+}]_{T,\,diss} = 10^{-2.30} - 10^{-3.66} = 10^{-2.32}$$

Note that only 4.4% of the total iron ($10^{-3.66\,+\,2.30} = 10^{-1.36} = 0.044$) remains in solution. The exact values of these results will change somewhat if all the constants are corrected to the appropriate ionic strength, 0.02 M. (Recall $\log \gamma_+ = -0.06$, $\log \gamma_{++} = -0.24$, etc.)

The exact point at which the equilibrium system becomes saturated can be found from the intersection of $[Fe^{3+}]_{unsat}$ (the function of pH given by Eqs. 51 and 52) and $[Fe^{3+}]_{sat}$ (the function of pH given by Eq. 57). In this system, it occurs at pH = 2.40. Such an intersection was shown in Fig. 7.23 to occur at pH = 1.5 when $[Fe^{3+}]_T = 0.5$ M, $[SO_4^{2-}] = 0$ and I = 3.0. The difference is partly because of dilution but mainly because of complexation between Fe^{3+} and SO_4^{2-}.

A further constraint could be provided by the stoichiometry of the pyrite oxidation reaction in the absence of other acids or bases:

$$[Fe^{3+}]_T = [H^+]_T \quad \text{and} \quad [SO_4^{2-}]_T = 2\,[H^+]_T$$

where $[H^+]_T$ is given by the proton condition with baseline Fe^{3+}, SO_4^{2-}, and H_2O:

$$[H^+]_T = [H^+] + [HSO_4^-] - [OH^-] - [FeOH^{2+}] - 2[Fe(OH)_2^+] - 4[Fe(OH)_4^-]$$
$$-2[Fe_2(OH)_2^{4+}] - 4[Fe_3(OH)_4^{5+}]$$

In an alternative computational approach, these constraints can be applied to the general Eqs. (49)–(57) in the format of Table 7.4. In this calculation method, the mass balances giving $[Fe^{3+}]_T$, $[SO_4^{2-}]_T$, and $[H^+]_T$ are explictly displayed as sums, and the free variables pH, $[Fe^{3+}]$, and $[SO_4^{2-}]$ are adjusted[25] to balance the stoichiometry constraints $[SO_4^{2-}]_T - [SO_4^{2-}]_o = 0$, $[Fe^{3+}]_T - [H^+]_T = 0$ and $[SO_4^{2-}]_T - 2[H^+]_T = 0$. As noted before, the total sulfate is a measure of the degree to which the reaction has proceeded, even though much of it is complexed with iron in the unsaturated solutions.

A few results of such calculations are given in Table 7.5. (The first sample calculation has the same total Fe and SO_4 as Fig. 7.25 and the examples, but differs slightly because of the ionic strength corrections.)

Note that the free Fe^{3+} is only a small fraction of the total, whereas free sulfate is about half of the total. All three of these examples are very close to saturation, which implies that the oxidation products tend to hold a composition close to that of saturated $Fe(OH)_3$.

This system will be explored further in the context of automated computation methods in Chapter 12.

Field Data. There is a definite correlation between pH and $[SO_4^{2-}]$ in data from various mine drainages and pit lakes, but it does not correspond to a simple stoichiometry, such as was assumed in the previous discussion, ranging from $[SO_4^{2-}] = \frac{1}{2}[H^+]$ to $[SO_4^{2-}] = 2[H^+]$.

In Figs. 7.26 and 7.27, data from abandoned pit-mine lakes in Montana and Nevada,[26] and data from diverse mine drainages in Colorado,[27] show a correlation that can be summarized by

$$[SO_4^{2-}] = 32[H^+] - 00134, \quad \text{(Fig. 7.26), or by}$$

$$\log[SO_4^{2-}] = -0.33\,\text{pH}, \quad \text{(Fig. 7.27)}$$

Even though the rate of $[H^+]$ production from sulfide mineral oxidation may be linearly related to the rate of $[SO_4^{2-}]$ production, the surrounding rocks react with much of the $[H^+]$, leaving the $[SO_4^{2-}]$ unreacted in solution (hence the factor 32 instead of 2 in the correlation of Fig. 7.26).

TABLE 7.4. Calculation of pH, [Fe^{3+}], and [SO$_4^{2-}$] in a solution with [SO$_4^{2-}$]$_T$ = 0.005

	A	B	C	D	E	F	G
2	**Pyrite oxidation products:**						
3	FeS2 + (15/4) O2 + (1/2) H2O = Fe3+ + 2 SO4= + H+						
4							
5	[SO4]set =	0.005	Set	Init =	1		
6	Dissolved Species list with Equilibrium constants at 25 °C			I =	0.0067	est =IF(Ini>0,I,0.01), I = D74	
7	H+ master var ([H+] =10^-pH/y)			log y+ = Iy=	-0.0384	=-0.5*(SQRT(I)/(1+SQRT(I))+0.2*I)	
8	pH =	2.6398	adjusted below				Davies
9	Fe3+	0.000189	adjusted below	Smith & Martell			I = 0.1
10	SO4=	0.002569	adjusted below	@ I = 0	Davies Eq	Correct to I	
11	OH-	=Kw/H		log Kw =	-13.997	-13.92	=D10-2*Iy	-13.74
12	[FeOH]2+	=b1*Fe*OH		log b1 =	11.81	11.58	=D11+Iy*(9+1-4)	11.03
13	[Fe(OH)2]+	=b2*Fe*OH^2		log b2 =	22.30	21.92	=D12+Iy*(9+2-1)	21.00
14	[Fe(OH)4]-	=b4*Fe*OH^4		log b4 =	34.40	33.94	=D13+Iy*(9+4-1)	32.84
15	[Fe2(OH)2]4+	=b22*Fe^2*OH^2		log b22 =	25.10	24.95	=D14+Iy*(2*9+2-16)	24.58
16	[Fe3(OH)4]5+	=b43*Fe^3*OH^4		log b43 =	49.70	49.47	=D15+Iy*(3*9+4-25)	48.92
17	HSO4-	=K1*H*SO4		log K1 =	1.99	1.84	=D16+Iy*(1+4-1)	1.47
18	FeSO4+	=c1*Fe*SO4		log c1 =	4.04	3.58	=D17+Iy*(9+4-1)	2.48
19	[Fe(SO4)2]-	=c2*Fe*SO4^2		log c2 =	5.38	4.77	=D18+Iy*(9+2*4-1)	3.30
20							
21	Precipitate						
22	Fe(OH)3(s)	Kso = Fe*OH^3		log Kso =	-38.80	-38.34	=D21-Iy*(9+3)	
23							
24	pH	**2.6398**	Adjust to make 2*[H]t - [SO4]t = 0			**0.000000**	=2*Ht-SO4t
25	H	2.504003E-03	=10^(-pH-Iy)				
26	OH	4.800E-12	=10^E11/H				
27							

HYDROLYSIS OF METAL IONS

28	Mass Balances				
29	[Fe3+]	**1.890E-04**	Adjusted (<3.94e-04)to [Fe]T-0.5*[SO4]T=0	**0.00000**	=Fet-0.5*SO4t
30	[FeOH]2+	3.444E-04	=10^E12*Fe*OH		
31	[Fe(OH)2]+	3.586E-05	=10^E13*Fe*OH^2		
32	[Fe(OH)4]-	8.713E-16	=10^E14*Fe*OH^4		
33	2*[Fe2(OH)2]4+	1.454E-05	=2*10^E15*Fe^2*OH^2		
34	3*[Fe3(OH)4]5+	3.168E-07	=3*10^E16*Fe^3*OH^4		
35	FeSO4+	1.841E-03	=10^E18*Fe*SO4		
36	[Fe(SO4)2]-	7.262E-05	=10^E19*Fe*SO4^2		
37					
38	[Fe]T	**2.498E-03**	=SUM(B29:B36)		
39					
40	SO4	**0.002569**	adjust to make [SO4]T-[SO4]set= 0	**0.00000**	=SO4t-SO4set
41	HSO4-	4.412E-04	=10^E17*H*SO4		
42	FeSO4+	1.841E-03	=10^E18*Fe*SO4		
43	2*[Fe(SO4)2]-	1.452E-04	=2*10^E19*Fe*SO4^2		
44					
45	[SO4]T	**4.996E-03**	in unsol'n [SO4]T would be 2*[Fe]T =	4.995E-03	0.01%
46			In sat'd sol , [Fe]ppt = [Fe]T-0.5*[SO4]T =	-5.725E-07	undersat
47	Proton balance: Zero level is H2O, Fe3+, SO4=				
48	H	2.504E-03	= H		
49	HSO4-	4.412E-04	=10^E17*H*SO4		
50	-OH	-4.800E-12	= -B26 = -10^E11/H		
51	-[FeOH]2+	-3.444E-04	= -B30 = - 10^E12*Fe*OH		
52	-2*[Fe(OH)2]+	-7.172E-05	=-2*B31 =-2*10^E13*Fe*OH^2		
53	-4*[Fe(OH)4]-	-3.485E-15	=-4*B32 = -4*10^E14*Fe*OH^4		
54	-2*[Fe2(OH)2]4+	-2.909E-05	=-B33 = -2*10^E15*Fe^2*OH^2		
55	-4*[Fe3(OH)4]5+	-1.267E-06	=-(4/3)*B34 =-4*10^E16*Fe^3*OH^4		
56					
57	[H]T	**2.499E-03**		Calc'd value	
58			In sat or unsat sol'n 2*Ht - SO4t should = 0	1.215E-06	0.05%

TABLE 7.5. Sample results obtained by the calculation shown in Table 7.4; concentrations in mM

$[SO_4^{2-}]_T$	$[SO_4^{2-}]_f$	$[Fe^{3+}]_T$	$[Fe^{3+}]_f$	$[H^+]_T$	pH	Sat?
10	4.32	5.00	0.337	5.00	2.420	supersat
5	2.57	2.50	0.189	2.50	2.640	undersat
1	0.78	0.505	0.035	0.498	3.102	undersat

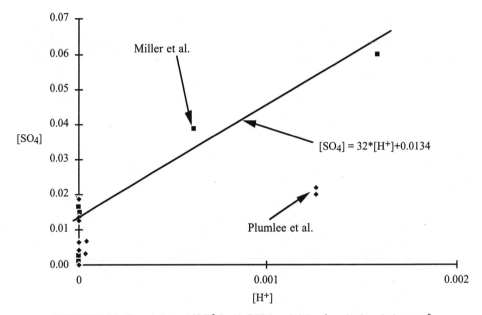

FIGURE 7.26. Correlation of $[SO_4^{2-}]$ with $[H^+]$ for pit lakes[1] and mine drainages.[2]

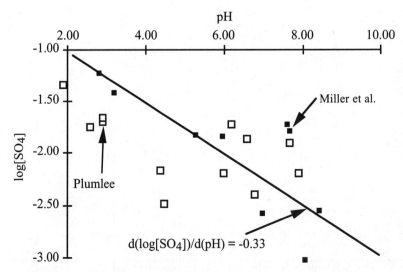

FIGURE 7.27. Correlation of $\log[SO_4^{2-}]$ with pH for the same data as Fig. 7.26.

PROBLEMS

Use equilibrium constants from Table 7.1 and Table 7.2 where needed.[28]

1. Calculate the concentrations of all species and the ligand number $\bar{n}$ in acid cadmium chloride solutions of the following concentrations:
   (a) 0.010 M $CdCl_2$.
   (b) 1.0×10^{-5} M $CdCl_2$.
   (c) 0.001 M $Cd(NO_2)_2$ and 0.002 M HCl. Note that the complex with the largest concentration is not necessarily the stoichiometric $CdCl_2$.
   (d) 0.10 M $CdCl_2$ and 0.10 M HCl.
   (e) 1.0 M $CdCl_2$ and 2.0 M HCl. Estimate the ionic strength and obtain concentration constants at that ionic strength from Figs. 7.1–7.4.

2. Calculate a curve of $-\log[Cd^{2+}]$ versus volume added for a titration where 100 mL of 1.0 M $Cd(NO_3)_2$ is titrated with 4.0 M HCl. Estimate the ionic strength at each point and obtain concentration constants at that ionic strength from Figs. 7.1–7.4.

3. Calculate the concentrations of all species and the ligand number $\bar{n}$ in acid mercuric chloride solutions of the following concentrations (assume NaCl has been added to make $I = 0.5$):
   (a) 0.01 M $HgCl_2$.
   (b) 0.001 M $Hg(NO_2)_2$ aand 0.002 M HCl.
   (c) 0.01 M $Hg(NO_2)_2$ and 1.0×10^{-5} M HCl.
   (d) 0.001 M $Hg(NO_2)_2$ and 0.1 M HCl.
   (e) 3.0 M $HgCl_2$ and 3.0 M HCl. Use $I = 3$ values from Table 7.1.

4. A method for the determination of chloride by titration with mercuric nitrate makes use of the exceptional stability of $HgCl_2$. Sodium nitroprusside $Na_2Fe(CN)_5NO$ is added as an indicator. The end point is determined by the appearance of a white precipitate of $HgFe(CN)_5NO$, for which $pK_{so} = 8.6$. Calculate a titration curve of $-\log[Hg^{2+}]$ versus volume added for the titration of 100 mL of 0.01 M NaCl with 0.01 M $Hg(NO_3)_2$. What is the titration error if nitroprusside is present at 10^{-3} M and the end point is taken when 10% of the nitroprusside is precipitated?

5. Calculate the solubility of AgCl and the concentrations of the dissolved silver species in the following solutions:
   (a) 1.0 M HCl.
   (b) 0.01 M HCl.
   (c) 1.0×10^{-4} M HCl.
   (d) 1.0×10^{-6} M HCl.

6. Calculate the two upper curves shown in Fig. 7.14 for AgCl solubility in chloride solutions, comparing the "$I = 0$" values (obtained by setting all activity co-

efficients to 1.0) with the values obtained by estimating the ionic strength and interpolating the concentration equilibrium constants at that ionic strength.

7. Calcium fluoride is slightly soluble in water, yielding Ca^{2+}, F^-, and in acid solutions the protonated fluoride species HF and HF_2^-. Constants at $I = 0$ are:

$$CaF_2(s) \rightleftharpoons Ca^{2+} + 2F^-, \quad pK_{so} = 10.57$$
$$H^+ + F^- \rightleftharpoons HF, \quad \log K_1 = 3.19$$
$$H^+ + HF \rightleftharpoons HF_2^-, \quad \log K_2 = 0.53$$

Compare the calculated solubility of CaF_2 in 0.1 M HF using the above equilibria with the model where a complex CaF^+ with $\log \beta_1 = 1$ is introduced.

8. Calcium sulfate forms complexes with its anion, with protons in acid solutions, and with hydroxide in basic solutions. Calculate the solubility of $CaSO_4$ as a function of pH from 1 to 13. Note where $Ca(OH)_2(s)$ forms. Constants at $I = 0$ are:

$$CaSO_4(s) \rightleftharpoons Ca^{2+} + SO_4^{2-}, \quad pK_{so} = 4.37$$
$$Ca^{2+} + SO_4^{2-} \rightleftharpoons CaSO_4(aq), \quad \log K_1 = 2.31$$
$$HSO_4^- \rightleftharpoons H^+ + SO_4^{2-}, \quad pK_a = 1.89$$
$$Ca^{2+} + OH^- \rightleftharpoons CaOH^+, \quad \log K_1' = 1.37$$
$$Ca(OH)_2(s) \rightleftharpoons Ca^{2+} + 2\,OH^-, \quad \log K_{so}' = 5.03$$

9. A method of analysis for mercuric ion is based on the formation of undissociated $HgI_2(aq)$ and solid HgI_2. A sample containing Hg^{2+} in nitric acid solution is titrated with KI and the conductiity of the solution is monitored. H^+ and NO_3^- will be present in essentially constant concentration, affected only by dilution. $[K^+]$ will increase as the titrant is added, and $[Hg^{2+}]$ will decrease as iodide complexes with it. When two equivalents of iodide have been added, the conductivity reaches a minimum; excess iodide produces HgI_3^- and HgI_4^{2-}, which increase the conductivity. Let a volume V of 0.1 M KI be added to 100 mL of 0.01 M $Hg(NO_3)_2$ acidified with 0.01 M HNO_3. Calculate the total concentration of positive charge (as a surrogate for conductivity) as a function of V. What is the titration error if the end point is taken at the minimum conductivity? Constants at $I = 0.5$ for Hg^{2+}–I^- complexes are: $\log \beta_1 = 12.87$, $\log \beta_2 = 23.82$, $\log \beta_3 = 27.46$, $\log \beta_4 = 23.85$.

10. In a variant of the titration of Hg^{2+} with KI, the end point is taken when a red precipitate of HgI_2 appears.[29] What titration error results from this end point? $\log K_{s2} = \log K_{so}\beta_2 = -3.78$; hence $pK_{so} = 27.60$. See Problem 9 for other constants.

11. The Liebig titration[30] for determination of cyanide ion by titration with silver ion relies on the formation of the soluble complex $Ag(CN)_2^-$ so long as CN^- is in excess. At the end point, a white precipitate of $AgCN(s)$ appears. Let V_o mL of KCN with concentration C_o be titrated with V mL of $AgNO_3$. The mass balances are:

$$\frac{CV}{V+V_o} = [Ag^+] + P + [Ag(CN)_2^-] + [Ag(CN)_3^{2-}] + [Ag(CN)_4^{3-}]$$

$$\frac{C_oV_o}{V+V_o} = [CN^-] + [HCN] + P + 2[Ag(CN)_2^-] + 3[Ag(CN)_3^{2-}] + 4[Ag(CN)_4^{3-}]$$

where P moles of AgCN precipitate form per liter of solution. The first equivalence point, with the stoichiometry $Ag(CN)_2^-$, occurs when $2CV = C_oV_o$; hence the fraction titrated is defined to be $\phi = 2CV/C_oV_o$.[31] A second equivalence point, corresponding to the formation of AgCN, occurs at $\phi = 2$. Show that in the region around the first equivalence point, where $P = 0$ by definition, and the concentrations $[Ag^+]$, $[Ag(CN)_3^{2-}]$, and $[Ag(CN)_4^{3-}]$ are small compared to $[Ag(CN)_2^-]$, the fraction titrated is given as a function of $[CN^-]$ by:

$$\phi = \frac{1 - \dfrac{[CN^-]}{C_o}\left(1 + \dfrac{[H^+]}{K_a}\right)}{1 - \dfrac{[CN^-]}{2C}\left(1 + \dfrac{[H^+]}{K_a}\right)}$$

Show also that

$$[Ag^+] = \left(\frac{\phi CC_o}{2C + \phi C_o}\right)\left(\frac{1}{\beta_2[CN^-]^2}\right)$$

where $K_a = 10^{-9.32}$ is the acidity constant of HCN, and $\beta_2 = 10^{+20.5}$ is the formation constant for $Ag(CN)_2^-$. Other constants that may be useful include $pK_{so} = 15.92$, $\log \beta_3 = 21.8$, $\log \beta_4 = 20.8$. Plot a curve of pAg versus ϕ at pH = 10.0, and calculate the titration error if the end point is taken where the first precipitate appears.[32]

12. Calculate the pH and concentrations of the mercuric species in an aqueous solution saturated with HgO. (See Table 7.2 for constants.)

13. A solution initially containing 0.1 M HNO_3 is saturated with HgO. What is the pH of the final solution?

14. A solution containing 1 M $HgCl_2$ is titrated with base. At what pH does HgO precipitate?

15. Calcium hydroxide is moderately soluble in water, giving Ca^{2+}, $CaOH^+$, and OH^- (see Problem 8 for constants). Calculate the pH of saturated $Ca(OH)_2$.

16. Estimate the solubility of $Fe(OH)_3$ in 0.01 M $HClO_4$. Constants are found in Table 7.2.

17. Estimate the solubility of $Fe(OH)_3$ in 0.01 M H_2SO_4 and the pH of the resulting solution. In addition to the constants in Table 7.2, note that at $I = 0$, the pK_a for $HSO_4^- = 1.89$, and the formation constants of Fe^{3+}–SO_4^{2-} complexes are: $\log \beta_1 = 4.04$, $\log \beta_2 = 5.30$.

18. The chemical basis of the photographic fixing bath is the solubility of undeveloped AgBr in sodium thiosulfate ($Na_2S_2O_3$). Estimate the solubility of AgBr in $Na_2S_2O_3$ of concentration C. The formation constants for Ag^+–$S_2O_3^{2-}$ complexes are $\log \beta_1 = 8.8$, $\log \beta_2 = 13.46$, $\log \beta_3 = 14.16$; the solubility product of AgBr is $pK_{so} = 12.34$; and the formation constants for Ag^+–Br^- complexes are $\log \beta_1 = 4.38$, $\log \beta_2 = 7.34$, $\log \beta_3 = 8.00$, $\log \beta_4 = 8.73$. Note that as AgBr dissolves, $[Br^-]$ increases.

19. If a photographic fixing bath contains 750 mL of 0.5 M $Na_2S_2O_3$, and a typical film after development contains 150 mg of AgBr, how many films can be fixed in the bath? Why is the recommended number much less than this?

20. When ammonia is added to a silver nitrate solution, at first Ag_2O precipitates because of the increase in pH and then dissolves because of the formation of Ag^+–NH_3 complexes. Draw a diagram of $\log C_{Ag}$ versus $\log C_{NH_3}$, showing the region within which Ag_2O precipitates.[33] The solubility product of Ag_2O is defined as $K_{so} = [Ag^+][OH^-] = 10^{-7.71}$; the formation constants for Ag^+–OH^- complexes are $\log \beta_1 = 2.0$, $\log \beta_2 = 6.0$; and the formation constants for Ag^+–NH_3 complexes are $\log \beta_1 = 3.32$, $\log \beta_2 = 7.23$.

21. Using the equilibrium constants for $I = 0$ from Table 7.2, calculate the concentrations of all aluminum-containing species in a solution containing $[Al^{3+}]_T = 0.01$ at pH = 2. Test to see that the solution is unsaturated. Compare with Fig. 7.21.

22. Interpolate the equilibrium constants for Al–OH complexes in Table 7.2, using the Davies equation, with suitable adjustment of the constant term, to obtain each as a function of ionic strength. Calculate the concentrations of all species (including the $[Cl^-]$ necessary to adjust the pH) in a solution containing $[Al^{3+}]_T = 0.10$ at pH = 2. From these results calculate a preliminary value for the ionic strength and revise the previous answers by iteration. Compare with Fig. 7.21.

23. Repeat the calculations described in Problem 22 at a series of pH values from 0 to 5. To reach the higher pH values, excess NaOH will have to be added; this will add a term in $[Na^+]$ to the ionic strength. At what pH is neither acid nor

base necessary (i.e., $[Na^+] = [Cl^-] = 0$)? At each pH, test for saturation with $Al(OH)_3$.

24. Aluminum (M) forms strong complexes with fluoride (Smith and Martell, Vols. 4 and 6):

Expression for K	log K ($I = 0$)	log K ($I = 0.5$)
$[MF]/[M][F^-]$	7.0	6.11 ± 0.03
$[MF_2]/[M][F^-]^2$	12.6	11.2 ± 0.1
$[MF_3]/[M][F^-]^3$	16.7	15.0 ± 0.3
$[MF_4]/[M][F^-]^4$	19.1	18.0 ± 0.8
$[HF]/[H^+][F^-]$	3.17 ± 0.02	2.93 ± 0.03

Interpolate these constants using the Davies equation, combine them with the constants in Table 7.2, and calculate the solubility of $Al(OH)_3$ in a solution containing 0.2 M NaF at pH = 4.0. Compare with the solubililty in 0.2 M NaCl. Iterate the ionic strength calculation if necessary.

25. Find the pH at which $Al(OH)_3$ is just barely saturated in a medium containing 0.5 M NaF.

26. Using the equilibrium constants for $I = 0$ from Table 7.2, calculate the concentrations of all iron-containing species in a solution containing $[Fe^{3+}] = 0.01$ at pH values from 0 to 3. At each pH calculate the ionic strength (including acid or base necessary to reach that pH). Interpolate the equilibrium constants in Table 7.2 (using the Davies equation with adjustment of the constant term), and iterate to obtain revised concentrations. Compare with Fig. 7.22.

27. In Table 7.2, solubility products for $Fe(OH)_3(s)$, $FeOOOH(\alpha)$, and $Fe_2O_3(\alpha)$ are given at I = 3.0. Figure 7.22 was calculated with $pK_{so}= 38.8$. Show how Fig. 7.22 would change if a different iron hydroxide solid phase were assumed.

28. Calculate the minimum solubility of $Fe(OH)_3(s)$ and the pH at which it occurs. Use log $\beta_3 = 28.8$ (as in Table 7.1 and Fig. 7.24). The database for the program PHREEQC lists log $K = -12.56$ for the reaction

$$Fe^{3+} + 3\ H_2O = Fe(OH)_3(aq) + 3\ H^+$$

How would the use of this constant change the minimum solubility and its pH?

29. Compare the speciation at pH = 2 of 0.005 M Fe^{3+} in non-complexing medium with the speciation in the presence of 0.01 M SO_4^{2-} (see Fig. 7.25). Check for saturation in both cases.

30. Following the scheme of Tables 7.4 and 7.5, with $[SO_4^{2-}]_T = 2\ [Fe^{3+}]_T$, verify the results for $[SO_4^{2-}]_T = 10$ mM (supersat) and 5 mM (undersat). Interpolate to find the precise saturation point.

31. Mix the acid mine water containing $[SO_4^{2-}]_T = 1$ mM and $[Fe^{3+}] = 0.5$ mM with a natural stream water saturated with solid $CaCO_3$ and with $CO_2(g)$ at 0.01 atm. See Chapter 9 for equations and equilibrium constants.

32. There is considerable disagreement in the literature about the association constant between Fe^{3+} and HSO_4^-. As noted in Table 7.3, Smith and Martell (Vol. 6) list $\log \beta_1 = 0.6$ at $I = 3.0$, but the PHREEQC database (see Table 12.9) lists $\log \beta_1 = 2.47$ at $I = 0.34$ In Fig. 7.25, Table 12.2, and Fig. 12.1, the species $FeHSO_4^{2+}$ is omitted. Estimate what change there would be in $[Fe^{3+}]$ and in $[SO_4^{2-}]$ in the acid mine water of Problem 30, if $[FeHSO_4^+]$ were included with $\beta_1 = 0.6$ and with $\beta_1 = 2.5$.

NOTES

1. Alfred Werner, *Selected Papers*, edited by G. B. Kauffman. Dover, New York; 1968, 190 pp; A. B. Lamb and A. T. Larson, *J. Am. Chem. Soc.* **42**:2024, 1920, measured $\beta_6 = [Co(NH_3)_6^{3+}]/[Co^{3+}][NH_3]^6 = 10^{+33.6}$; hence equal concentrations of $[Co(NH_3)_6^{3+}]$ and $[Co^{3+}]$ are in equilibrium with $[NH_3] = 10^{-5.6}$. Slow kinetics also contribute to the complex's lack of reactivity.

2. C. K. Jorgensen, Modern aspects of ligand field theory. North Holland, Amsterdam, 1971, 538 pp; L. E. Orgel, *Introduction to Transition Metal Chemistry: Ligand-Field Theory*. Wiley, New York, 1960, 180 pp.

3. Vols. 1–6, Plenum Press, New York (1974, 1975, 1976, 1977, 1982, 1989). R. M. Smith and A. E. Martell, *NIST Critically Selected Stability Constants of Metal Complexes Database* Version 3.7. National Inst. Standards and Technology, Gaithersburg, MD, 1997.

4. Chemical Society Special Publications 17 (1964) and 25 (1971).

5. See also A. E. Martell and R. J. Motekaitis, *Determination and Use of Stability Constants*, VCH Publishers, 1992; J. N. Butler, *Ionic Equilibrium*, first edition, Addison-Wesley, 1964, pp. 34–59; F. J. C. Rossotti and H. Rossotti, *Determination of Stability Constants*, McGraw Hill, 1961.

6. C. E. Vanderzee and H. J. Dawson, *J. Am. Chem. Soc.* **75**:5659–1953. See other references in Smith and Martell, also Sillén and Martell.

7. Selected from Smith and Martell, Vols. 4, 5, and 6.

8. The general theorem is easily proved for a series of complexes ML_k, where k ranges from 0 (metal ion alone) to n, the maximum coordination number of the complex series. The mass balance on M is

$$[M]_T = [M] + [ML] + [ML_2] \cdots = \Sigma[ML_k] = \Sigma\beta_k[M][L]^k = [M]\Sigma\beta_k[L]^k$$

$$\alpha_i = \frac{[ML_i]}{[M]_t} = \frac{\beta_i[M][L]^i}{[M]\sum_k \beta_k[L]^k} = \frac{\beta_i[L]^i}{\sum_k \beta_k[L]^k}$$

Because $[M]$ is linear in both numerator and denominator, it cancels, and the resultant expresssion depends only on $[L]$ and the equilibrium constants.

9. See D. R. Turner, M. Whitfield, and A. G. Dickson, "The equilibrium speciation of dissolved components in freshwater and seawater at 25°C and 1 atm pressure." *Geochim. Cosmochim. Acta* **45**:855–81, 1981, for other examples and approaches.

10. E. A. Guggenheim, *Thermodynamics*, 4th ed. North-Holland, Amsterdam, 1959; *ibid.*, 5th Ed., 1967.

11. J. Kielland, *J. Am. Chem. Soc.* **59**:1675, 1937. See also J. N. Butler, *Ionic Equilibrium*, 1964, pp. 434–35. Recall that the use of different size parameters for different ions is thermodynamically inconsistent (see discussion of J Kielfand's table in Chapter 2, p. 46).

12. Over 100 references between 1900 and 1963 (Sillén and Martell, Spec. Pub. 17, 1964); 40 between 1964 and 1968 (Sillén & Martell, Spec. Pub. 25, 1971; still more after that).

13. From Smith and Martell, *Critical Stability Constants*, Vol. 5, Plenum Press, 1982. These differ significantly from the earlier values reported in Sillen and Martell and in Smith and Martell, Vol. 4. There are no new values for Hg–OH listed in Vol. 6 (1989).

14. Smith and Martell, Vols. 4 and 6. Plenum Press, 1976, 1989. Solubility data are listed only in Vol. 4. The most recent complex formation data are listed in Vol. 6. No Al^{3+} data are listed in Vol. 5.

15. Smith and Martell, Vols. 4, 5, 6. Plenum Press, 1976, 1982, 1989. Solubility data are listed only in Vol. 4. The most recent and reliable complex formation data are listed in Vol. 6.

16. In the aqueous environment, mercury ocurs not only as these ions, but also as the chloride complexes, sulfide complexes, solid HgS, organics such as CH_3Hg [C. T. Driscoll et al., *Environ. Sci. Techol.* **28**:136A–143A, 1994; T. C. Hutchinson and K. M. Meema, eds. *Lead, Mercury, Cadmium and Arsenic in the Environment* (SCOPE 31), Wiley, NY, 1987], and elemental mercury (M. Amyot et al., *Environ. Sci. Technol.* **28**:2366–71, 1994).

17. In equilibrium with solid HgO, $[Hg(OH)_2] = K_{so} \beta_2 = 10^{-3.64}$ at $I = 0$. At total Hg concentrations below this, there will be no precipitate.

18. The saturation point and the relative amount of polynuclear complexes changes with total Hg concentration. For example, if total Hg = 0.001 (as in Fig. 7.16), the saturation point is at pH = 2.9; the maximum polynuclear species occur at that pH but are only 0.01% of the total.

19. C. F. Baes and R. E. Mesmer, *The Hydrolysis of Cations*, Wiley, New York, 1976.

20. J. N. Butler, appendix to "Aluminum and gallium arrest formation of cerebrospinal fluid by the mechanism of OH^- depletion," by B. P. Vogh, D. R. Godman, and T. H. Maren, *J. Pharmacology and Exp. Therapeutics* **233**:715–21, 1985.

21. These constants were available and relatively accurate. The ionic strength of 0.5 M $FeCl_3$, fully dissociated in strong acid, would be $0.5[(0.5)(9) + (1.5)(1)] = 3.0$ M, but this would be reduced by the formation of hydroxide complexes. At high pH, complete reaction of 0.5 M $FeCl_3$ with NaOH would produce 0.5 M solid $Fe(OH)_3$ (which would not contribute to the ionic strength) plus NaCl with net ionic strength 1.5.

22. P. C. Singer and W. Stumm, *Science* **167**:3921, 1970; F. Walsh, *Water Pollution Microbiology*, Vol. 1, R. Mitchell, Ed., 1971; W. Stumm and J. J. Morgan, *Aquatic Chemistry*, 1st Edition, Wiley, 1970; R. V. Nicholson, "Iron-sulfide oxidation mechanisms: Laboratory studies," in *Short Course Handbook on Environmental Chemistry of Sulfide Mine Wastes*, J. L. Jambor and D. W. Blowes, Eds., Mineralogical Association of Canada, 1994, pp. 163–83.

23. Smith and Martell, Vols. 4, 5, 6. Plenum Press, 1976, 1982, 1989. Solubility data are listed only in Vol. 4. The most recent and reliable complex formation data are listed in Vol. 6 (see also Table 7.2).
24. Smith and Martell (Vol. 6) list log β_1 = 0.6 at I = 3.0, but the PHREEQC database (see Table 12.9) lists log β_1 = 2.47 at I = 0. In Fig. 7.25 and Table 12.2 and Fig. 12.1, the species $FeHSO_4^{2+}$ is omitted.
25. It is often easiest to simply adjust the variables manually until the balance occurs. A somewhat more sophisticated approach would be to use the Newton–Raphson (multivariable) method on the three constraints in the form $f(x,y,z) = 0$.
26. G. C. Miller, W. B. Lyons, and A. Davis, "Understanding the Water Quality of Pit Lakes," *Environ. Sci. Technol.,* 1996.
27. G. S. Plumlee et al., "Empirical studies of diverse mine drainages in Colorado: Implications for the prediction of mine–drainage chemistry, Billings Montana Symposium on Planning, Rehabilitation and Treatment of Disturbed Lands, 1993. U.S. Geological Survey MS 973, Federal Center, Denver, CO 80225.
28. Constants given in the problems were selected from L. G. Sillén and A. E. Martell, *Stability Constants*, Chemical Society Special Publications 17 (1964) and 25 (1971) unless otherwise noted.
29. J. Marozeau, *Pharm. Chim.* **18**:302, 1832, cited by H. A. Laitinen, *Chemical Analysis*, McGraw-Hill, New York, 1960, p. 224.
30. J. Liebig, *Ann. Chem. Liebigs,* **77**:102, 1851, cited by H. A. Laitinen, *op. cit.,* pp. 225–27
31. Note that for this reaction the equivalent weight of silver is twice (not half) its atomic weight.
32. See J. N. Butler, *Ionic Equilibrium*, Addison-Wesley, Reading MA, 1964, pp. 280–83, and Laitinen, *op. cit.,* pp. 225–27, for details.
33. See J. N. Butler, *Ionic Equilibrium,* 1964, pp. 303-8, especially Fig. 8-15.
34. From K. W. Sykes, *Chem. Soc. London Spec. Pub* 1, pp. 64–70, 1954 or 1957, as quoted by S. V. Mattigod and G. Sposito, *Soil Sci. Soc. Am. J.* **41**:1092–97, 1977. Other values are 1.78 at I = 0.15 (Sykes, quoted by Sillén and Martell, 1964) and 0.78 at I = 1.2 (M. W. Lister and D. E. Rivington, *Can. J. Chem.*, **33**:1591, 1603, 1955, quoted by Sillén and Martell, 1964.

8

ORGANIC COMPLEXES

Formation of Amino Acid Complexes
Chelates
    EDTA Complexes
    Conditional Stability Constants
    Metal Buffers
Complexometric Titrations
    Metal Indicators
Solvent Extraction
Problems

FORMATION OF AMINO ACID COMPLEXES

The chief difference between inorganic and organic complexes is the competition between complexed metal ions and protons for the ligand. Typically, the strongest complexes are formed when the organic ligand is fully deprotonated. Another difference is that when a number of reactive groups exist on the same organic molecule, it can form "chelates" where several ligand groups attached to the same molecule also satisfy the coordination sites on a metal ion. To illustrate the first of these concepts, consider the interaction of Cu^{2+} with glycine. "HG" represents H_2N-CH_2-COOH[1]:

$$G^- + H^+ \rightleftharpoons HG, \quad \log K_{1(H)} = 9.78 \quad (1)$$

$$HG + H^+ \rightleftharpoons H_2G^+, \quad \log K_{12(H)} = 2.35 \quad (2)$$

$$G^- + Cu^{2+} \rightleftharpoons CuG^+, \quad \log K_{1(Cu)} = 8.62 \quad (3)$$

$$CuG^+ + G^- \rightleftharpoons CuG_2, \quad \log K_{2(Cu)} = 6.97 \quad (4)$$

$$Cu^{2+} + OH^- \rightleftharpoons CuOH^+, \quad \log K_{1(OH)} = 6.66 \quad (5)$$

$$H_2O \rightleftharpoons H^+ + OH^-, \quad \log K_w = -14.00 \quad (6)$$

Concentrations of the nine species H^+, OH^-, G^-, HG, H_2G^+, Cu^{2+}, $CuOH^+$, CuG^+, and CuG_2 are related by the six equilibria above plus two mass balances (G_T = total G; Cu_T = total Cu) and a charge balance:

$$G_T = [G^-] + [HG] + [H_2G^+] + [CuG^+] + 2[CuG_2] \quad (7)$$

$$Cu_T = [Cu^{2+}] + [CuG^+] + [CuG_2] + [CuOH^+] \quad (8)$$

$$[H^+] + [H_2G^+] + [CuG^+] + 2[Cu^{2+}] + [CuOH^+] = [G^-] + [OH^-] \quad (9)$$

This complicated set of equations is most easily investigated under conditions where $[H^+]$ is fixed by a buffer or adjusted by titration with strong acid or base, and where Cu_T is small compared to either G_T or the buffer concentration. Under these conditions, Eq. (9) may be set aside in favor of the specified pH, and Eq. (7) becomes

$$G_T = [G^-] + [HG] + [H_2G^+] + \cdots \quad (10)$$

$$G_T = [HG]\left(\frac{1}{K_{1(H)}[H^+]} + 1 + K_{12(H)}[H^+]\right) + \cdots \quad (11)$$

A logarithmic diagram illustrating this is shown in Fig. 8.1. Note that the ligand G^- is the principal species only for pH above 10.

If copper is introduced to the system, say at $Cu_T = 1 \times 10^{-4}$ M, the fraction complexed by G^- depends on pH as well as on G_T. In particular, the mass balance on Cu can be factored in the same way as we have done for polyprotic acids:

$$[Cu^{2+}] = \frac{Cu_T}{(1 + K_{1(Cu)}[G^-] + K_{1(Cu)}K_{2(Cu)}[G^-]^2 + K_{1(OH)}K_w/[H^+])} \quad (12)$$

where

$$[G^-] = \frac{G_T}{(1 + K_{1(H)}[H^+] + K_{12(H)}[H^+]^2)} \quad (13)$$

Thus both $[Cu^{2+}]$ and $[G^-]$ are functions of pH, and the other species can be evaluated using the equations corresponding to the equilibria Eq. (1)–(9).

The logarithmic concentrations of the various species are plottted in Fig 8.2. Note that Cu^{2+} is the dominant copper species for pH < 3, CuG^+ for pH between 3 and 5, and CuG_2 for pH >5.

CHELATES

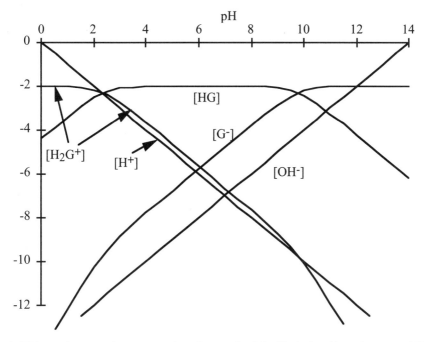

FIGURE 8.1. Logarithmic concentration diagram for 0.01 M glycine. Note that neutral HG dominates over the pH range from 3 to 9, and that the ligand G^- becomes important only for pH > 10.

At pH > 12.1, $Cu(OH)_2(s)$ precipitates. This pH limit is found by combining Eq. (12) with the solubility product:

$$[Cu^{2+}][OH^-]^2 = K_{so} = 10^{-19.32} \quad (14)$$

$$1 + K_{1(Cu)}[G^-] + K_{1(Cu)}K_{2(Cu)}[G^-]^2 + \frac{K_{1(OH)}K_w}{[H^+]} - \frac{Cu_t K_w^2}{K_{so}[H^+]^2} = 0 \quad (15)$$

Given Cu_T and G_T, $[G^-]$ is obtained as a function of pH from Eq. (13), and substituted in Eq. (15) to obtain a polynomial in $[H^+]$. This can be solved by adjusting pH until Eq. (15) is satisfied.[2]

CHELATES

Amino acids and polyamines contain several groups that are good electron donors and can coordinate to metal ions forming complexes. A complex formed by coordi-

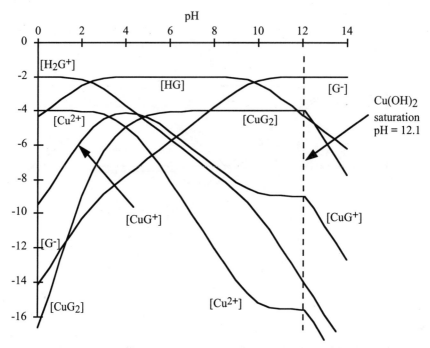

FIGURE 8.2. Logarithmic diagram for a solution containing 10^{-2} M glycine and 10^{-4} M Cu^{2+}. CuG^+ exceeds Cu^{2+} for pH > 3.5, and CuG_2 dominates above pH = 5. Above pH = 12.1, $Cu(OH)_2(s)$ precipitates.

nation of several groups on the same ligand to the same metal atom is known as a *chelate*.[3] Chelates are normally considerably more stable than the coresponding complex formed from separate ligands.

For example, hexammine nickel(II) is considerably less stable than tris(ethylenediamine) nickel(II) (Fig. 8.3 and Table 8.1).

For example, if $[NH_3] = 0.01$,

$$[Ni(NH_3)_6^{2+}]/[Ni^{2+}] = 10^{8.74}[NH_3]^6 = 10^{-3.26}$$

In contrast, if ethylenediamine [en] = 0.01,

$$[Ni(en)_3^{2+}]/[Ni^{2+}] = 10^{18.61}[en]^3 = 10^{+12.61}$$

nearly a factor of 10^{16} larger.

Part of this difference in stability is simply an entropy effect: When one coordination point of a bidentate ligand is attached to the metal ion, the other cooordination point is close by and hence more likely to be attached. In addition, there can be an energetic increase in stability resulting from the formation of five-membered chelate rings.

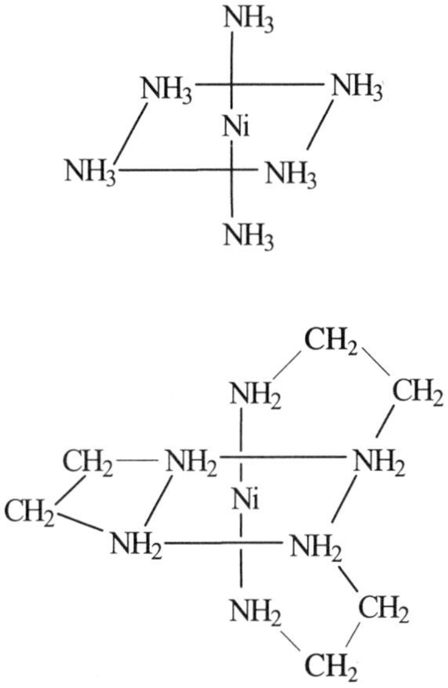

CHELATES

[Structural diagram of Ni(NH₃)₆ complex]

[Structural diagram of Ni(ethylenediamine)₃ complex]

FIGURE 8.3. Structures[21] for complex formation compared for ammonia and ethylenediamine.

EDTA Complexes

The most important chelating agents for practical applications are those that completely satisfy the coordination requirements of a given metal ion, forming a 1:1 complex. Perhaps the most widely investigated and used compound of this type is ethylene diamine tetra-acetic acid, EDTA.

EDTA is a tetraprotic acid. As the structure of Fig. 8.4 shows, the completely deprotonated form can coordinate at four oxygens and two nitrogens, satisfying six coordination valences of a metal ion simultaneously, as well as forming stable five-membered chelate rings. Figure 8.5 shows the distribution of its acidic and basic forms as a function of pH, and Table 8.2 gives equilibrium constants for some EDTA complexes.

TABLE 8.1. Equilibrium constants[21] for complex formation for ammonia and ethylenediamine

	Ammonia		Ethylenediamine	
	$\log K_1$	2.80	$\log K_1$	7.66
	$\log K_2$	2.24		
	$\log K_3$	1.73	$\log K_2$	6.40
	$\log K_4$	1.19		
	$\log K_5$	0.75	$\log K_3$	4.55
	$\log K_6$	0.03		
Overall	$\log \beta_6$	8.74	$\log \beta_3$	18.61

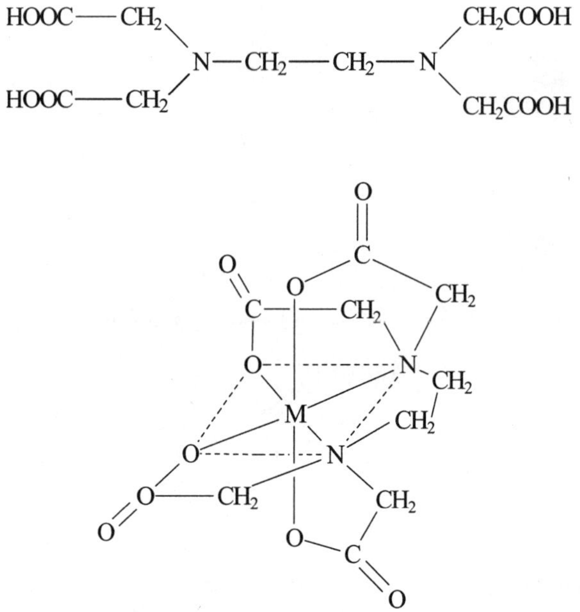

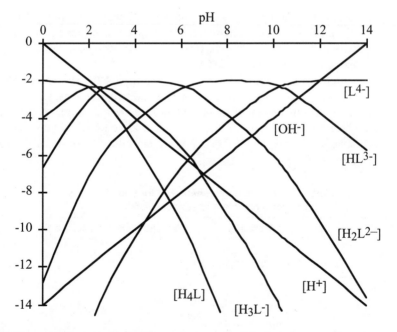

FIGURE 8.4. Structure of EDTA bonded to a six-coordinate metal ion M. The paralellogram around M is meant to indicate the horizontal coordination plane.

FIGURE 8.5. Distribution of acid–base species in 0.01 M EDTA. The fully deprotonated species L^{4-} becomes dominant only above pH = 10.

TABLE 8.2. Equilibrium constants for EDTA (L^{4-})[4]

Complex	log (overall formation constant)
HL^{3-}	10.19
H_2L^{2-}	16.32
H_3L^-	19.01
H_4L	21.01
MgL^{2-}	8.85
$MgHL^-$	13.13
CaL^{2-}	10.65
$CaHL-$	14.07
FeL^-	25.1
$FeHL$	26.4

The principal factor affecting the formation of chelates in practical situations is the pH of the solution. Chelating agents are polyprotic acids, and usually only the fully deprotonated form makes strong complexes with a metal ion. This means that chelate complexes are most stable in strongly basic solutions. Sometimes partially protonated complexes of the form MHL or hydrolyzed basic complexes of the form MOHL are formed, but these are usually less stable than the simple 1:1 chelate ML.

Most metals precipitate as hydroxides in strongly alkaline media, and a buffer medium is often chosen so that it acts as a subsidiary complexing agent to prevent precipitation of the hydroxide. Typical buffers contain ammonia, phosphate, or citrate. Of course, these introduce additional equilibria to be taken into account in equilibrium calculations. In a practical system, therefore, one must consider the metal ion and its hydrolysis products, the acid–base equilibria of the chelating agent, the interaction of the metal ion with the buffer components, the acid–base equilibria of those buffer components, and the interaction of the metal ion with the chelating agent.

Conditional Stability Constants

Fortunately, the concentration of free metal ion in chelate systems is normally very low, and polynuclear complexes are therefore not important. A system of mononuclear complexes can be simplified by factoring out the metal ion concentration and the ligand concentration (see note 8, p. 286) to give a simple form called the *conditional stability constant.*

This is how it is done: Divide all the species in solution into three groups: $[M]'$ contains the metal ion but not the ligand, $[L]'$ contains the ligand but not the metal ion, and $[ML]'$ which contains both metal and ligand. For this example, the buffer is taken to be NH_4^+–NH_3, and charges are omitted for simplicity.

$$[M]' = [M] + [M\mathrm{NH}_3] + [M(\mathrm{NH}_3)_2] + \cdots + [M\mathrm{OH}] + [M(\mathrm{OH})_2] + \cdots$$

$$[M]' = [M](\beta_1^{\mathrm{NH}_3}[\mathrm{NH}_3] + \beta_2^{\mathrm{NH}_3}[\mathrm{NH}_3]^2 + \cdots + \beta_1^{\mathrm{OH}}[\mathrm{OH}] + \beta_2^{\mathrm{OH}}[\mathrm{OH}]^2 + \cdots) \quad (16)$$

$$[M]' = [M]A_M$$

where[5] A_M depends on pH and NH_3 concentration but not on $[M]$. Similarly,

$$[L]' = [L] + [\mathrm{H}L] + [\mathrm{H}_2L] + \cdots$$

$$[L]' = [L](1 + \beta_1[\mathrm{H}] + \beta_2[\mathrm{H}]^2 + \cdots) \quad (17)$$

$$[L]' = [L]A_L$$

$$[ML]' = [ML] + [MHL] + [MOHL] + \cdots$$

$$[ML]' = [ML](1 + K_{MHL}[\mathrm{H}] + K_{MOHL}[\mathrm{OH}]) \quad (18)$$

$$[ML]' = [ML]A_{ML}$$

In general A_M is a function of NH_3 concentration and pH; A_L, and A_{ML} are functions only of pH, which in turn is a function of the buffer composition. So long as the complexes are mononuclear, these functions are independent of the total concentration of M or L.

Figure 8.6 shows the buffer index for 0.1 M NH_3–NH_4^+ solutions, with a maximum at pH = pK_a = 9.25. Figure 8.7 shows the function A_M over the pH range from 8 to 11.4.[6] Including NH_3 increases A_M about 15%, and raises slightly the pH at which $\mathrm{Mg(OH)}_2$ precipitates (Fig. 8.8).

The function A_L is shown in Fig. 8.9.[7] This quantity varies over a much larger range than A_M.

Let the true equilibrium constant for formation of the complex ML from the free ions M and L be K:

$$[ML] = K[M][L] \quad (19)$$

Then the conditional constant K' is defined in terms of the total concentrations (Eqs. 16–18) to be

$$K' = \frac{[ML]'}{[M]'[L]'} = \frac{[ML]A_{ML}}{[M]A_M[L]A_L} = \frac{A_{ML}}{A_M A_L}K \quad (20)$$

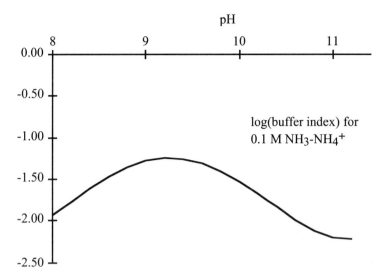

FIGURE 8.6. Log of buffer index (dC_b/dpH) for NH_3–NH_4^+ buffer in the pH range 8 to 11.4. The maximum buffer index occurs where pH = pK_a = 9.25. See Chapter 4.

The pH dependence of K' (Fig. 8.9) is governed primarly by A_L. While the ammonia buffer helps to keep the pH constant during the titration, it has little effect on the stability of the complex or the precipitation of $Mg(OH)_2$, since Mg^{2+} complexes with NH_3 are weak compared to those with EDTA.

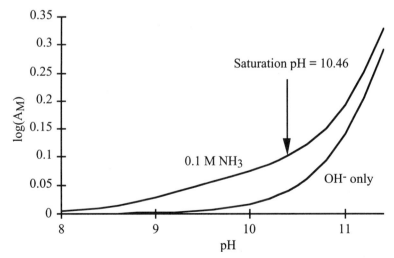

FIGURE 8.7. The function A_M for Mg^{2+} over the pH range from 8 to 11. 0.1 M NH_3 buffer is compared with the solution with no buffer except OH^-. Saturation points correspond to $[Mg]_t$ = 0.01 and are taken from Fig. 8.8. At lower $[Mg]_T$, the saturation pH is higher.

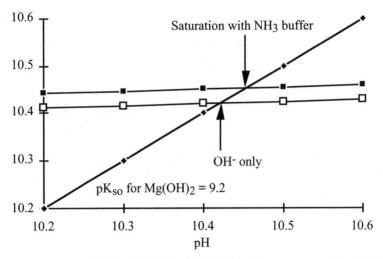

FIGURE 8.8. Saturation for Mg(OH)$_2$ in 0.01 M Mg^{2+}–0.1 M NH$_3$ occurs at pH = 10.46. Without NH$_3$, the saturation point is pH = 10.42. pK_{so} = 9.2 used for these calculations is for the "active" solid; Brucite has pK_{so} = 11.15 and correspondingly lower pH at saturation: 9.43 to 9.45.

From Fig. 8.8, Mg(OH)$_2$ would be expected to precipitate above pH about 10.4. Fortunately, this is close to the pH where most of the EDTA is in the fully deprotonated form, which produces the maximum value for log K' in Fig. 8.10. Above pH = 10.5, if the solution were to remain unsaturated, a decrease in K' would result from the formation of MgOH$^+$.

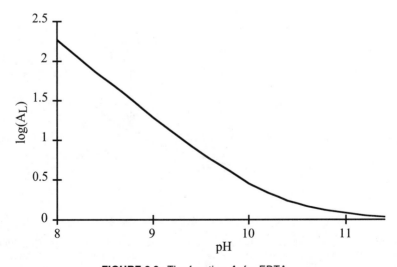

FIGURE 8.9. The function A_L for EDTA.

CHELATES

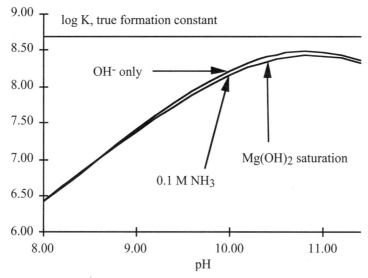

FIGURE 8.10. Conditional formation constant for MgEDTA in solutions containing 0.1 M NH_3 buffer, and solutions without NH_3, compared with the "true" formation constant for the free ions.

Metal Buffers

A solution containing a metal ion and a complexing agent (in excess) can fix the free concentration of that metal ion at a stable value.

Example 1. Consider a solution containing 0.010 M Mg^{2+} and 0.020 M EDTA, buffered to pH = 10.0 with 0.1 M NH_3. Find the concentration of free Mg^{2+} at equilibrium. Using the conditional constant already derived (Eqs. 16–20 and Figs 8.7, 8.9, and 8.10) at pH = 10,

$$\log K' = \log(K_{MgL}) + \log(A_{ML}) - \log(A_{M(NH_3)}) - \log(A_L)$$
$$= 8.69 + 3.7 \times 10^{-7} - 0.076 - 0.450 = 8.16$$

Note that the terms $\log(A_{ML})$ and $\log(A_{M(NH_3)})$ have very little influence on the final value. The conditional equilibrium is

$$[ML]' = 10^{+8.16} [M]'[L]'$$

The mass balances are

$$[L]' + [ML]' = 0.020$$
$$[M]' + [ML]' = 0.010$$

Since an excess of EDTA has been added, $[M]'$ is small compared to $[ML]'$ and the mass balances yield

$$[ML]' = 0.010$$
$$[L]' = 0.020 - [ML]' = 0.010$$
$$[M]' = 10^{-8.16} [ML]'/[L]' = 10^{-8.16}$$

To relate this value to the free Mg^{2+} concentration, use Eq. (16) (The numerical value for $A_{M(NH_3)}$ at pH = 10 will be found in Fig. 8.7 or in the equation above for log K'):

$$[M]' = [Mg^{2+}] A_{M(NH_3)}$$
$$[Mg^{2+}] = 10^{-8.16} 10^{-0.076} = 10^{-8.24}$$

The rest of the Mg^{2+} is bound up as complexes, mainly with EDTA.

Thus the presence of excess EDTA has served to fix the concentration of free magnesium ions at this particular value. If more magnesium is added, it will react with the EDTA, and the concentration of free ion will change only slightly. For example:

$$[ML]' = 0.011$$
$$[L]' = 0.009$$
$$[M]' = 10^{-8.16} [ML]'/[L]' = 10^{-8.16} \, 10^{+0.087} = 10^{-8.07}$$
$$[Mg^{2+}] = 10^{-8.07} \, 10^{-0.076} = 10^{-8.15}$$

$[Mg^{2+}]$ has changed from $10^{-8.24}$ (5.75×10^{-9}) to $10^{-8.15}$ (7.07×10^{-9}), an increase of 1.32×10^{-9}, very small compared to the change in total magnesium concentration, which increased by 10^{-3}.

This action is analogous to the action of an acid–base buffer, and hence solutions containing an excess of chelating agent are known as *metal buffers*.

COMPLEXOMETRIC TITRATIONS

Because of the high stability of chelate complexes, titration of metal ions with chelating agents can be a very precise method of analysis. The calculation of theoretical titration curves, using the conditional stability constant formalism, is similar to the acid–base titrations discussed in Chapter 4 or the precipitation titrations discussed in Chapter 6.

EDTA is usually obtained as its disodium salt Na_2H_2L, and the reaction of this substance with a metal ion liberates protons:

$$M^{2+} + H_2L^{2-} \rightleftharpoons ML^{2-} + 2H^+$$

COMPLEXOMETRIC TITRATIONS

If a titration is carried out in an unbuffered solution, the acidity increases as EDTA is added, causing the equilibrium to be shifted to the left, and resulting in a poor end point. A buffer is therefore essential for accurate complexometric titration.

The general equations for the titration curve are easily expressed in terms of the conditional equilibrium constant:

$$[ML]' = K'[M]'[L]' \qquad (21)$$

where the primed concentrations include whatever other complexes may result from the buffer, or other components of the solution that do not change during the titration. If V_o mL of a solution containing metal ion M of total concentration C_o are titrated with V mL of complexing agent of concentration C, the mass balances are

$$[M]' + [ML]' = \frac{C_o V_o}{V + V_o} \qquad (22)$$

$$[L]' + [ML]' = \frac{CV}{V + V_o} \qquad (23)$$

Eliminating $[ML]'$ between these equations and solving for V gives

$$[L]' - [M]' = \frac{CV - C_o V_o}{V + V_o} \qquad (24)$$

$$V = V_o \left(\frac{C_o - [M]' + [L]'}{C + [M]' - [L]'} \right) \qquad (25)$$

Substituting Eq. (21) in Eq. (22) gives $[L]'$ as a function of $[M]'$

$$[L]' = \frac{CV}{V + V_o} \left(\frac{1}{1 + K'[M]'} \right) \qquad (26)$$

Equations (25) and (26) define a titration curve of V versus $[M]'$ with K', C, C_o, and V_o as given parameters (see Fig. 8.11). This is best solved by iteration.[8]

An alternative is to express Eq. (24) in terms of the fraction titrated,

$$\phi - 1 = \frac{V + V_o}{C_o V_o} ([L]' - [M]') \qquad (27)$$

where $\phi = CV/C_o V_o$. In this case also, expressing ϕ as a function of $[M]'$ over the whole range is not simple; but if you consider only the region near the equivalence point, where

$$\phi \approx 1 \quad \text{and} \quad V \approx C_o V_o / C \qquad (28)$$

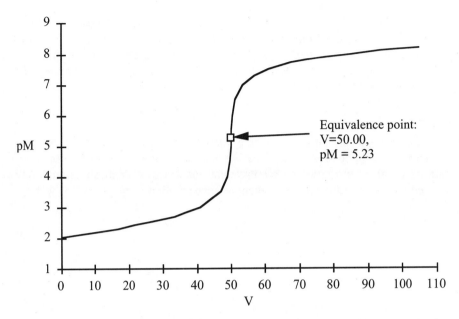

FIGURE 8.11. Titration curve, showing $pM(-\log[Mg^{2+}]')$ as a function of volume in the titration of 50 mL 0.010 M Mg^{2+} with 0.010 M EDTA in 0.1 M ammonia buffer at pH = 10.0.

and note that $[ML]'$ is large compared to either $[M]'$ or $[L]'$, the problem is more tractable. From either Eq. (22) or (23)

$$[ML]' = \frac{CC_o}{C + C_o} + \cdots \qquad (29)$$

Combining Eqs. (21), (27), and (29) gives an equation for the titation curve near the equivalence point:

$$\text{titr error} = \phi_{ep} - 1 = \frac{1}{K'[M]'_{ep}} - \frac{C_o + C}{CC_o}[M]'_{ep} \qquad (30)$$

where $[M]'_{ep}$ is the total concentration of all complexes containing M except those with L, at the chosen end point of the titration. The equivalence point is easily obtained from Eq. (30) with $\phi_{ep} = 1$:

$$[M]' = \sqrt{\frac{CC_o}{K'(C + C_o)}} \qquad (31)$$

Differentiating Eq. (30) and applying Eq. (31), you can obtain an expression for the sharpness index at the equivalence point:

$$\eta = \frac{d\text{pM}}{d\phi} = 0.217\sqrt{\frac{K'CC_o}{C + C_o}} \tag{32}$$

Note that when $C = C_o = 0.01$, for the sharpness index to be greater than 10^3, K' must be greater than $10^{+9.7}$. If K' is smaller than $10^{+5.7}$, η will be less than 10, and a practical titration is not possible.

Example 2. Calculate the titration curve when 50 mL of 0.010 M Mg^{2+} in 0.10 M ammonia buffer of pH = 10 is titrated with 0.010 M EDTA. From Fig. 8.10, the conditional equilibrium constant is $K' = 10^{+8.16}$. Equations (25) and (26) give the curve of Fig. 8.11. Note that Eq. (31) gives an equivalence point at $[M] = 10^{-5.23}$, which agrees at $V = 50.00$ with the calculations displayed in Fig. 8.11. The sharpness index is $\eta = 185$, an intermediate practical value.

The titration curve of Fig. 8.11 implies that there is a method, analogous to pH measurement, that can determine Mg^{2+} throughout the course of the titration. This is in fact true. For certain metal ions, such as Mg^{2+}, Ca^{2+}, Cu^{2+}, and Pb^{2+}, ion-selective electrodes based on liquid ion exchangers are available. For others, such as Pb^{2+}, Cu^{2+}, and Cd^{2+}, solid membrane ion-selective electrodes are based on a mixture of the metal sulfide with Ag_2S.[9]

Metal Indicators

An older way to detect the end point in a complexometric titration is to use a highly colored chelate of the appropriate stability, such as Eriochrome Black T. This compound is a triprotic acid H_3In (Fig. 8.12) that complexes with many divalent ions including Ca^{2+}, Mg^{2+}, Cd^{2+}, Co^{2+}, Cu^{2+}, Pb^{2+}, and Zn^{2+}. This dye forms a red chelate MIn^- with the metal ions, which contrasts nicely with the blue color of the free dye HIn^{2-}.

If the indicator concentration is small compared to that of EDTA, its equilibrium will be controlled by the equilibrium of the metal ion with EDTA. The smallest titration error will result if there are equal amounts of free indicator and complex at the equivalence point of the titration. When substituted in the indicator equilibrium expression,

$$[MIn^-]' = K'_{In}[M]'[In^{3-}]'$$

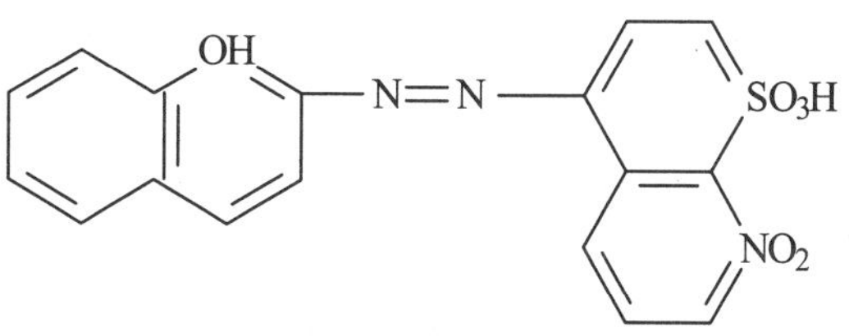

FIGURE 8.12. Structure of Eriochrome Black T (H_3L): 3-hydroxy-4-(1-hydroxy-2-naphthylazo)-8-nitronaphthalene-1-sulfonic acid.

If the end point is taken to be where there are equal concentrations of red and blue forms,

$$[\text{In}^{3-}]' = [M\text{In}^-]'$$

this gives

$$K'_{\text{In}}\,[M]' = 1$$

At the end point of the EDTA titration, $[M]' = [M]'_{\text{ep}}$, and the optimum value of K'_{In} is

$$K'_{\text{In}} = 1/[M]'_{\text{ep}}$$

Example 3. If Eriochrome Black T is used as an indicator in the titration of 0.01 M Mg^{2+} with 0.01 M EDTA (Fig. 8.11), what is the optimum pH for the titration?

The equilibria relevant to the Eriochrome Black T–Mg^{2+} system are[10]:

$$H^+ + \text{In}^{3-} \rightleftharpoons H\text{In}^{2-}, \quad \log \beta_1 = 11.55$$

$$2H^+ + \text{In}^{3-} \rightleftharpoons H_2\text{In}^-, \quad \log \beta_2 = 17.85$$

$$Mg^{2+} + \text{In}^{3-} \rightleftharpoons Mg\text{In}^-, \quad \log K = 7.0$$

By analogy with A_L of Eq 17, a pH-dependent function A_{In} can be defined:

$$[\text{In}]' = [\text{In}](1 + \beta_1[H] + \beta_2[H]^2 + \cdots)$$

$$[\text{In}]' = [\text{In}]A_{\text{In}}$$

The function A_{Mg} is given by Eq. (16) and Fig. 8.7, and the function $A_{Mg\text{In}}$ can be taken to be unity in the absence of any information about protonated or deprotonated complexes. K'_{In} is given by Eq. (20) to be

$$K'_{\text{In}} = K_{\text{In}}\,\frac{A_{Mg\text{In}}}{A_{Mg}A_{\text{In}}}$$

and is plotted as a function of pH in Fig. 8.13. Also on Fig. 8.13 is the pH dependence of the equivalence point for the Mg–EDTA titration (Eq. 31). The point where the two curves cross is the pH where the red and blue forms of the indicator are equal in concentration and the equivalence point of the titration is also attained, and could be considered an "optimum" pH. This optimum will vary somewhat if the ratio of red and blue forms at the visual or spectrophotometric end point is not exactly equal to one, but the titration error for any actual condition can be obtained from

COMPLEXOMETRIC TITRATIONS

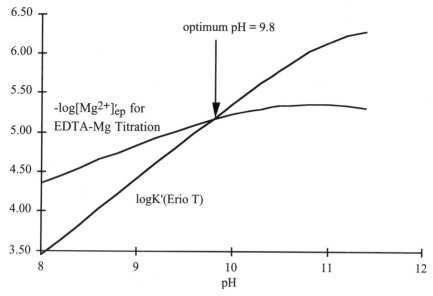

FIGURE 8.13. Conditional formation constant for Eriochrome Black T compared with the predicted end point (Eq. 31) for the titration of 0.01 M Mg^{2+} with 0.01 M EDTA (Fig. 8.11).

Eq. (30). For example, at pH = 10.2, $\log K'_{In} = 5.54$ for the indicator. If the red and blue forms are equal in concentration, $[In]' = [MgIn]'$ and $[Mg]' = 1/K'_{In} = 10^{-5.54}$. For the Mg–EDTA complex, $\log K' = 8.27$ (Eqs. 16–20 and Fig. 8.10) Substituting these values in Eq. (30) gives

$$\text{titr error} = \phi_{ep} - 1 = \frac{1}{K'[M]'_{ep}} - \frac{C_o + C}{CC_o}[M]'_{ep}$$

$$= \frac{1}{10^{+8.27-5.54}} - \frac{0.02}{0.0001} 10^{-5.54} = 1.3 \times 10^{-3}$$

Note that at the equivalence point (Eq. 31):

$$[Mg]' = \sqrt{\frac{0.0001}{(0.02)(10^{8.27})}} = 10^{-5.29}$$

The indicator end point is at $[Mg]' = 10^{-5.54}$, slightly smaller than the equivalence point, and hence a little farther along the titration curve of Fig. 8.11, yielding a positive titration error.

In the standard procedure, the sample concentration is adjusted to be between 0.005 and 0.01 M and is initially buffered to pH = 10. During the titration, protons are released, and the pH drops. When the equivalence point is reached, the pH is between 9.0 and 9.5, nearly the optimum (Fig. 8.13).

If the pH falls too low, K' for the Mg–EDTA complex decreases and the titration curve (Fig. 8.11) flattens out, making the end point less sharp. If the pH is initially too high, $Mg(OH)_2$ may precipitate. A calculation such as shown in Fig. 8.8 says that precipitation of amorphous $Mg(OH)_2$ occurs above about pH = 10.45, but as the precipitate ages and crystallizes, its solubility decreases, and it will redissolve only slowly as EDTA is added. The result is a negative titration error of uncertain magnitude. Therefore, time is of the essence if the initial pH is higher than 10.

SOLVENT EXTRACTION

One of the important techniques for quantitative analysis, as well as for determining the stability constants of organic complexes, is extraction—the equilibrium between an aqueous solution containing a metal ion and an organic ligand, and an organic solvent that preferentially extracts the uncharged complex from the aqueous solution.[11]

The general method is to equilibrate an aqueous solution with an organic solvent, and then to measure the distribution ratio (or "extraction coefficient") of total metal concentration in each phase:

$$D = \frac{[M]_{(tot,org)}}{[M]_{(tot,aq)}} \tag{33}$$

The more fundamental equilibrium is that of an uncharged metal–organic species MX_n between the organic and aqueous phases[12]:

$$[MX_n]_{aq} \rightleftharpoons [MX_n]_{org}$$

$$P = \frac{[MX_n]_{org}}{[MX_n]_{aq}} \tag{34}$$

the equilibrium constant is called the *partition coefficient P*. In addition, the uncharged acid HX or other uncharged species or ion pairs can also be partitioned between the aqueous and organic phases.

A simple example is one in which the only species in the organic phase is a single neutral complex such as MX_n. Then

$$[M]_{tot,org} = [MX_n]_{org} + \cdots$$

In the aqueous phase,

$$[M]_{tot,aq} = [M]_{aq} + [MX]_{aq} + \cdots + [MX_n]_{aq} + [MOH] + \cdots$$

$$[X]_{tot,aq} = [X]_{aq} + [HX]_{aq} + \cdots + [MX]_{aq} + \cdots + n[MX_n]_{aq}$$

$$D = \frac{P[MX_n]_{aq}}{[M]_{aq} + [MX]_{aq} + \cdots + [MX_n]_{aq} + [MOH] + \cdots} \quad (35)$$

$$D = \frac{P\beta_n[X]_{aq}^n}{1 + \beta_1[X]_{aq} + \cdots + \beta_n[X]_{aq}^n + \beta_{OH}[OH^-] + \cdots} \quad (36)$$

The distribution ratio is primarily a function of the ligand concentration $[X]$, which in turn usually depends on pH; and if MX_n is the predominant species in both the aqueous and organic phases, $D = P$. A closely related function, the fraction of metal extracted into the organic phase, is

$$x_{extr} = \frac{[M]_{tot,org}}{[M]_{tot,org} + [M]_{tot,aq}} = \frac{D}{1+D} \quad (37)$$

The following two examples are based on equilibrium data for copper extraction by the complexing agent dithizone (1,5-diphenyl thiocarbazone). This reagent is fairly insoluble (10^{-6} M) in water[13] but forms an intense green solution in chloroform or carbon tetrachloride. When the dithizone-containing organic phase is equilibrated with an aqueous solution, an orange or red metal–dithizone complex is formed that contrasts vividly with the original solution, and hence has been the basis for a number of colorimetric methods for trace metals.[14]

Example 4. 100 mL of 0.1 M HCl containing 2×10^{-7} M aqueous Cu^{2+} is extracted with an equal volume of 0.008 M dithizone (H_2L) in CCl_4. Find the fraction extracted as a function of pH. Some equilibria are[15]

$$CuL(aq) \rightleftharpoons CuL(org), \quad \log K_1 = 4.30$$
$$Cu(HL)_2(aq) \rightleftharpoons Cu(HL)_2(org), \quad \log K_2 = 4.85$$
$$H_2L(aq) \rightleftharpoons H_2L(org), \quad \log K_3 = 4.03$$
$$H^+(aq) + HL^-(aq) \rightleftharpoons H_2L(aq), \quad \log K_4 = 4.55$$
$$Cu^{2+}(aq) + 2\,HL^-(aq) \rightleftharpoons Cu(HL)_2(aq), \quad \log K_5 = 22.3$$
$$Cu^{2+}(aq) + HL(aq) \rightleftharpoons CuL(aq) + H^+(aq), \quad \log K_6 = 12.61$$

The acidity constant of dithizone in water is $pK_a = \log K_4 = 4.55$, so that HL^- and H_2L are the aqueous acid–base species under normal conditions. A total of 9 species are involved in the complete model: Cu^{2+}(aq), $Cu(HL)_2$(aq), CuL(aq), H^+(aq), HL^-(aq), H_2L(aq), H_2L(org), $Cu(HL)_2$(org), CuL(org). $[L^{2-}]$ is very small since $K_{a2} > 15$, and hydroxo complexes of Cu^{2+} are omitted. There are no data on the formation of $CuHL^+$ or $CuHL_2^-$; presumably these are small compared to the others.

The mass balances in the aqueous phase are

$$[Cu^{2+}]_{tot,aq} = [Cu^{2+}]_{aq} + [Cu(HL)_2]_{aq} + [CuL]_{aq} + \cdots$$

$$[L]_{t,\,aq} = [H_2L]_{aq} + [HL^-]_{aq} + \cdots$$

The mass balances in the organic phase consist only of the uncharged species:

$$[Cu^{2+}]_{tot,org} = [Cu(HL)_2]_{org} + [CuL]_{org}$$

$$[L]_{t,\,org} = [H_2L]_{org} + 2[Cu(HL)_2]_{org} + [CuL]_{org}$$

These total concentrations in the organic phase are not independent of the input data, but are determined by the aqueous concentrations and the equilibria. If equal volumes of solution are mixed, the total concentrations in both phases when equilibrium is reached are

$$[Cu]_t = [Cu^{2+}]_{tot,aq} + [Cu^{2+}]_{tot,org} = 2 \times 10^{-7}/2 = 10^{-7.0}$$

$$[L]_t = [L]_{t,\,aq} + [L]_{t,\,org} = (0.008)/2 = 10^{-2.4}$$

The final relation is a charge balance. In a solution composed of C_{HCL} and C_{H_2L}, this is

$$[H^+]_{aq} = [HL^-]_{aq} + [Cl^-] + [OH^-]$$

The total concentrations $[Cu^{2+}]_{tot,aq}$, $[L]_{t,\,aq}$, and C_{HCL} are input data. The total number of unknowns is therefore 13: the nine species listed above, plus $[Cu^{2+}]_{tot,aq}$, $[Cu^{2+}]_{tot,org}$, $[L]_{t,\,aq}$, and $[L]_{t,\,org}$. The six equilibria plus the 7 mass/charge balances provide exactly enough independent equations to fully determine the unknowns.

At this point, some simplifications are in order. As a first approximation, consider that the solution is strongly acid (pH = 1); hence $[Cl^-] \gg [HL^-] \gg [OH^-]$, $[H^+]_{aq} = C_{HCL} = 0.1$, and $[H_2L] \gg [HL^-]$.

Since all three of the partition coefficients (K_1, K_2, and K_3) are about the same order of magnitude, the concentrations of [CuL], [Cu(HL)$_2$] and [H$_2$L] in the organic phase will be in about the same proportions as in the aqueous phase, and none can be obviously neglected. As a first simplification, guess that $[Cu(HL)_2] \gg [CuL] \gg [Cu^{2+}]$ in both the aqueous and organic phases. This leaves simplified mass and charge balances:

$$[Cu(HL)_2]_{aq} + [Cu(HL)_2]_{org} + \cdots = 10^{-7.0}$$

$$[H_2L]_{aq} + [H_2L]_{org} + \cdots + [Cu(HL)_2]_{aq} + 2[Cu(HL)_2]_{org} = 10^{-2.4}$$

Substituting the first mass balance in the second gives a slightly simpler result:

$$[H_2L]_{aq} + [H_2L]_{org} + \cdots + [Cu(HL)_2]_{org} = 10^{-2.4} - 10^{-7.0}$$

SOLVENT EXTRACTION

The next stage is to substitute the equilibria in these balance equations:

$$[CuL]_{org} = K_1[CuL]_{aq}, \quad \log K_1 = 4.30$$
$$[Cu(HL)_2]_{org} = K_2[Cu(HL)_2]_{aq}, \quad \log K_2 = 4.85$$
$$[H_2L]_{org} = K_3[H_2L]_{aq}, \quad \log K_3 = 4.03$$
$$[H_2L]_{aq} = K_4[H^+]_{aq}[HL^-]_{aq}, \quad \log K_4 = 4.55$$
$$[Cu(HL)_2]_{aq} = K_5[Cu^{2+}]_{aq}[HL^-]_{aq}^2, \quad \log K_5 = 22.3$$
$$[CuL]_{aq}[H^+]_{aq} = K_6[Cu^{2+}]_{aq}[HL^-]_{aq}), \quad \log K_6 = 12.61$$

Reactions 2, 3, and 4, together with the charge balance and the mass balances on Cu and L, give two equations in two unknowns:

$$[Cu(HL)_2]_{aq} + \cdots = 10^{-7}/(1 + K_2) = 10^{-11.85}$$
$$[H_2L]_{aq}(1 + K_3) + \cdots + K_2[Cu(HL)_2]_{aq} = 10^{-2.4} - 10^{-7.0}$$

which reduce to one equation in $[H_2L]$:

$$[H_2L]_{aq} = (10^{-2.4} - 10^{-7.0} - (10^{+4.85})(10^{-4.03}) = 10^{-6.43}$$

From this result the other concentrations can be obtained:

$$[H_2L]_{org} = K_3[H_2L]_{aq} = (10^{+4.03})(10^{-6.43}) = 10^{-2.40} \; (\approx [L]_{tot})$$
$$[HL^-]_{aq} = [H_2L]_{aq}/(K_4[H^+]_{aq}) = (10^{-6.43})(10^{-4.55})(10^{+1}) = 10^{-9.98}$$
$$[Cu(HL)_2]_{aq} = 10^{-7}/(1 + 10^{+4.85}) = 10^{-11.85}$$
$$[Cu(HL)_2]_{org} = K_2[Cu(HL)_2]_{aq} = (10^{+4.85})(10^{-11.85}) = 10^{-7.00} \; (\approx [Cu]_{tot})$$
$$[Cu^{2+}]_{aq} = [Cu(HL)_2]_{aq}/(K_5[HL^-]_{aq}^2) = (10^{-11.85})(10^{-22.3})(10^{+(2)(9.98)}) = 10^{-14.19}$$
$$[CuL]_{aq} = K_6[Cu^{2+}]_{aq}[HL^-]_{aq}/[H^+]_{aq}) = (10^{+12.61})(10^{-14.19})(10^{-9.98})(10^{+1.0}) = 10^{-10.56}$$
$$[CuL]_{org} = K_1[CuL]_{aq} = (10^{+4.30})(10^{-10.56}) = 10^{-6.26}$$

Now check the mass balances:

$$[L]_t = [H_2L]_{aq}+[H_2L]_{org}+[HL^-]_{aq}+2[Cu(HL)_2]_{aq}+2[Cu(HL)_2]_{org}+[CuL]_{aq}+[CuL]_{org}$$
$$= 10^{-6.43}+10^{-2.40}+10^{-9.98}+10^{-11.55}+10^{-6.97}+10^{-10.56}+10^{-6.26} = 10^{-2.40} \quad \text{OK}$$

$$[Cu]_t = [Cu^{2+}]_{aq} + [Cu(HL)_2]_{aq} + [Cu(HL)_2]_{org} + [CuL]_{aq} + [CuL]_{org}$$
$$= 10^{-14.19}+10^{-11.85}+10^{-7.00}+10^{-10.56}+10^{-6.26} = 10^{-6.19} \quad \text{should be } 10^{-7.00}$$

This last mass balance was not quite correct because the species $[CuL]_{org}$ was omitted from the approximate mass balances used to obtain these answers.

Finally, we are ready to calculate the fraction of copper extracted into the organic phase. Use the results of the approximate calculation to determine $[Cu]_t$:

$$x_{extr} = \frac{[Cu(HL)_2]_{org} + [CuL]_{org}}{[Cu]_t} = \frac{10^{-7.00} + 10^{-6.26}}{10^{-6.19}} = 0.994$$

The next phase in this calculation is to include all the species in the model. In addition, the next example shows how the distribution of species is affected by changes in pH.

Example 5. 100 mL of 0.1 M NaCl with various concentrations of HCl containing 10^{-7} M aqueous Cu^{2+} is extracted with 10 mL of 0.01% (0.004 M) dithizone (1,5-diphenyl thiocarbazone, H_2L) in CCl_4 to form a complex, make a more complete model, including the other equilibria and the mass balance on both phases, and calculate the fraction extracted into the organic phase as a function of pH ($-\log[H^+]$ for simplicity).

Follow the derivations in the previous example, with pH, $[Cu]_t$, and $[L]_t$ as independent variables, but including all species in the model:

$$[Cu]_t = [Cu^{2+}]_{tot,aq} + [Cu^{2+}]_{tot,org}$$
$$= [Cu^{2+}]_{aq} + [Cu(HL)_2]_{aq} + [Cu(HL)_2]_{org} + [CuL]_{aq} + [CuL]_{org}$$
$$= [Cu^{2+}]_{aq}\{1 + (1 + K_2)K_5[HL^-]^2 + (1 + K_1)K_6[HL^-]/[H^+]\}$$

$$[L_t] = [L]_{tot,aq} + [L]_{t,org}$$
$$= [H_2L]_{aq} + [H_2L]_{org} + [HL] + 2[Cu(HL)_2]_{aq} + 2[Cu(HL)_2]_{org} + [CuL]_{aq} + [CuL]_{org}$$

$$[L]_t = [HL]\{1 + K_4[H^+](1 + K_3)\} \left\{ \frac{[Cu]_t\left(2K_5[HL^-]^2(1 + K_2) + K_6\frac{[HL^-]}{[H^+]}(1 + K_1)\right)}{1 + K_5[HL^-]^2(1 + K_2) + K_6\frac{[HL^-]}{[H^+]}(1 + K_1)} \right\}$$

$[H^+]$, $[Cu]_t$, and the equilibrium constants are part of the input data. This equation is solved for $[HL^-]$, using "goal seek" or other approximation methods to find a value of $[HL^-]$ such that the right-hand side equals the known value of $[L]_t$. Convergence is rapid.

Then $[Cu^{2+}]$ is found from its mass balance:

$$[Cu^{2+}] = \frac{[Cu]_t}{1 + K_5[HL^-]^2(1 + K_2) + K_6\frac{[HL^-]}{[H^+]}(1 + K_1)}$$

The remainder of the species can be found directly from the appropriate equilibria (above). Results are summarized in Table 8.3 and displayed in Fig. 8.14.

SOLVENT EXTRACTION

TABLE 8.3. Spreadsheet calculation of species distribution in dithizone extraction example.

logK1 =	4.3	Ltmp-Ltot	=HL*(1+(1+10^logK3)*10^(logK4-pH))+					
logK2 =	4.85		Cutot*(2*10^logK5*HL^2*(1+10^logK2)+(1+10^logK1)*10^(logK6+pH)*HL)/					
logK3 =	4.03		(1+(1+10^logK2)*10^logK5*HL^2*(1+10^logK1)*10^(logK6+pH)*HL)-Ltot					
logK4 =	4.55							
logK5 =	22.3	Cu2+	=LOG(Cutot/(1+(1+10^logK2)*10^logK5*HL^2+(1+10^logK1)*10^(logK6+pH)*HL))					
logK6 =	12.61							
Cutot =	9.0909E-08	=1e-7*100/110	logCutot	-7.041				
Ltot =	0.00036364	=0.004*10/110	LogLtot	-3.439				
		Log of conc>>			(aq)		(org)	
pH	HL(goal see)	Ltmp-Ltot	logHL	H2L(aq)	H2L (org)	Cu2+	Cu(HL)2	CuL (aq)	CuL (org)	D	%X
0	9.561E-13	0	-12.019	-7.469	-3.439	-11.939	-13.678	-8.828	-11.349	-7.049	16080	0.99994
1	9.563E-12	0	-11.019	-7.469	-3.439	-13.939	-13.678	-8.828	-11.349	-7.049	20138	0.99995
2	9.564E-11	0	-10.019	-7.469	-3.439	-15.939	-13.678	-8.828	-11.349	-7.049	20189	0.99995
3	9.564E-10	5.421E-20	-9.019	-7.469	-3.439	-17.939	-13.678	-8.828	-11.349	-7.049	20190	0.99995
4	9.563E-09	0	-8.019	-7.469	-3.439	-19.939	-13.678	-8.828	-11.349	-7.049	20190	0.99995
5	9.561E-08	0	-7.019	-7.469	-3.439	-21.939	-13.678	-8.828	-11.349	-7.049	20190	0.99995
6	9.539E-07	0	-6.021	-7.471	-3.441	-23.938	-13.679	-8.829	-11.349	-7.049	20189	0.99995
7	9.319E-06	0	-5.031	-7.481	-3.451	-25.928	-13.689	-8.839	-11.348	-7.048	20184	0.99995
8	7.572E-05	0	-4.121	-7.571	-3.541	-27.836	-13.778	-8.928	-11.347	-7.047	20140	0.99995
9	2.635E-04	0	-3.579	-8.029	-3.999	-29.374	-14.233	-9.383	-11.343	-7.043	20018	0.99995
10	3.503E-04	0	-3.456	-8.906	-4.876	-30.496	-15.107	-10.257	-11.342	-7.042	19961	0.99995
11	3.623E-04	0	-3.441	-9.891	-5.861	-31.510	-16.092	-11.242	-11.341	-7.041	19954	0.99995
			=LOG(B36)	=logK3+H2Laq	=logK4-pH+logHL	=logK5+Cu+2*logHL	=logK2+CuHL2aq	=logK1+CuLaq			
				=logK4-pH+logHL				=logK6+Cu+logHL+pH			

311

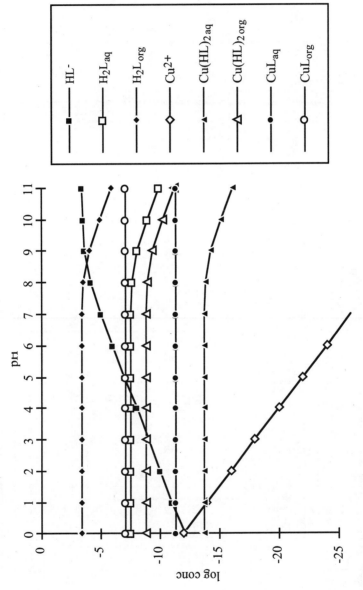

FIGURE 8.14. Species concentration in the dithizone extraction example as a function of pH.

The most obvious result is that most of the species concentrations are essentially independent of pH in the acid and neutral region. Only [HL^-] increases linearly with pH and [Cu^{2+}] decreases linearly with pH. The other species are buffered by the relatively large amount of complex which is held in the organic phase, and only when [HL^-] exceeds [H_2L] at about pH > 8.5 do the organic phase concentrations decline with increasing pH. Over most of the pH region, the distribution coefficient $D = 2 \times 10^4$ and the fraction X of Cu^{2+} in the organic phase is 0.9995.

Four types of extraction systems are surveyed by Dyrssen et al., and only the first was discussed here:

1. Chelating acid HA in the organic phase, acid H^+ in the aqueous phase;
2. Chelating acid HA in organic phase, acid H^+ and anion L^- in aqueous phase;
3. Chelating acid HA and adduct-forming ligand L in organic phase, acid H^+ in aqueous phase;
4. Salt of long-chain base or cation BH^+X^- in organic phase, corresponding acid HX in aqueous phase.

There are many other variants on this theme. The inclusion of equilibrium solid phases of sparingly soluble organic reagents and metal–organic compounds can also help stabilize the concentration of the species of interest over a variation in solution composition.

PROBLEMS

1. Calculate the conditional formation constant for Mg–EDTA as a function of pH in unbuffered medium and in 0.1M NH_3 buffer (Fig. 8.10). Use equilibrium constants listed in Table 8.2 and in the footnote to Fig. 8.7.

2. The conditional stability constant for the Mg complex of EDTA at pH = 9.0 in 0.10 M ammonia buffer is $10^{+7.38}$ (see Fig. 8.10). If equal volumes of 0.01 M EDTA and buffered 0.01 M Mg^{2+} are mixed, calculate the fraction of Mg present as MgEDTA, and the fraction present as uncomplexed Mg^{2+}.

3. Calculate the concentrations at pH = 9 of free Ca^{2+} and free Mg^{2+} in a solution containing 0.01 M each of total Ca and Mg, and 0.02 M total EDTA. Use equilibrium constants for H^+, Ca^{2+}, and Mg^{2+} with EDTA from Table 8.2.

4. Find the solubility of $BaSO_4$ in 0.01 M EDTA buffered to pH = 10.0. The conditional stability constant for the barium complex with EDTA at pH = 10.0 is $10^{+7.30}$, and pK_{so} for $BaSO_4$ is 10.0.

5. The scarlet complex of nickel with dimethylglyoxime (NiA_2) is the basis of a classic gravimetric method for nickel. This precipitate dissolves in acids because A^- is protonated. Some relevant equilibria are:

$$HA(s) \rightleftharpoons HA \text{ (aq)}, \quad \log K = -2.26$$
$$HA \text{ (aq)} \rightleftharpoons H^+ + A^-, \quad \log K = -10.67$$
$$NiA_2(s) \rightleftharpoons NiA_2 \text{ (aq)}, \quad \log K = -6.00$$
$$NiA_2 \text{ (aq)} \rightleftharpoons Ni^{2+} + 2\, A^-, \quad \log K = -17.83$$

Calculate the solubility of the complex in an unbuffered solution as a function of pH. What is the lowest pH at which 99.9% of the nickel in a 0.10 M solution will be precipitated?

6. In the standard gravimetric procedure for nickel, the dimethylglyoxime complex is precipitated from an ammonia solution. Define a conditional solubility product as

$$K'_s = [Ni^{2+}]'[A^-]'^2$$
$$[Ni^{2+}]' = [Ni^{2+}] + [NiNH_3^{2+}] + \cdots + [NiOH^+] + \cdots$$
$$[A^-]' = [A^-] + [HA]$$

For the Ni^{2+}–NH_3 complexes,[16] $\log \beta_1 = 3.0$, $\log \beta_2 = 5.18$, $\log \beta_3 = 6.82$, $\log \beta_4 = 7.98$. Other data are given in Problem 5. Calculate K'_s as a function of pH in 0.1 M NH_3 buffer. Is there an optimum pH at which K'_s is a minimum? If so, find that value.

7. Calculate the titration curve of Fig. 8.11 using data in the text (and K' from Problem 1). How does the curve differ if pH = 9.0 instead of 10.0?

8. If $\log[Mg^{2+}]_{ep}$ can be determined with a precision of ±0.1 units, what is the corresponding error in ϕ for the titration curve of Fig. 8.11 (pH = 10)? How does the titration error differ if pH = 9.0 instead of 10.0?

9. Draw a diagram showing which species of Eriochrome Black T predominate at various values of pH and $\log[Mg^{2+}]$. The species HIn^- and $MgIn^-$ are red, HIn^{2-} is blue and In^{3-} is orange. Equilibrium constants are given in the text. On the diagram draw a boundary showing where $Mg(OH)_2$ precipitates, and where $[Mg^{2+}]_{ep}$ occurs for $C = C_o = 10^4$, 10^{-3}, and 10^{-2}. For what range of pH and C is this titration effective?

10. Ferric ion can be titrated with EDTA, using thiocyanate as an indicator. The end point, when the deep red color of the ferric thiocyanate complexes disappears, is at about $[FeSCN^{2+}] = 10^{-5}$. Calculate the conditional formation constant for the Fe^{3+}–EDTA complex in the pH range from 1 to 3, and determine the optimum pH for titration of 0.01 M Fe^{3+} with 0.01 M EDTA, using 10^{-3} M KSCN as indicator. Equilibrium constants for Fe^{3+}–EDTA are given in Table 8.2; those for Fe^{3+}–OH complexes are given in Table 7.2. For Fe^{3+}–SCN in $I = 0.1$ $LiClO_4$,[17] $\log \beta_1 = 2.17$, $\log \beta_2 = 3.60$. Note that Fe^{3+} is sufficiently small at the end point that the polynuclear Fe–OH complexes can be neglected.

11. Nitrilotriacetic acid $N(CH_2COOH)_3$ or NTA is a common ingredient in laundry detergents. It complexes calcium and magnesium and hence softens the water. It can also mobilize lead from plumbing by the same mechanism. Here are some equilibrium constants[18] for $I = 0.1$, 20°C. (NTA is H_3L):

Equilibrium	log K
$H^+ + L^{3-} \rightleftharpoons HL^{2-}$	9.71 ± 0.02
$H^+ + HL^{2-} \rightleftharpoons H_2L^-$	2.47 ± 0.02
$H^+ + H_2L^- \rightleftharpoons H_3L$	1.75 ± 0.05
$Ca^{2+} + L^{3-} \rightleftharpoons CaL^-$	6.46 ± 0.01
$Pb^{2+} + L^{3-} \rightleftharpoons PbL^-$	11.47 ± 0.04
$Ca^{2+} + CO_3^{2-} \rightleftharpoons CaCO_3(s)$	7.53[19]
$Pb^{2+} + CO_3^{2-} \rightleftharpoons PbCO_3(s)$	12.00[20]

Assume the wash water contains 100 ppm calcium (1 mM) and has pH = 9, and that the waste pipes are encrusted with a mixture of $CaCO_3$ and $PbCO_3$.

(a) What is the concentration of dissolved lead in the presence of 1 mM NTA?

(b) Compare with the situation where NTA is absent.

(c) Compare with the situation where concentrated detergent (0.1 M NTA) is poured down the drain.

12. Using the equilibrium constants for $Al(OH)_3$ and $Fe(OH)_3$ from Chapter 7, calculate the solubility of these solid phases in a solution containing 0.1 M EDTA at pH = 3, 6, and 9.

13. Look up the stability constants for several EDTA analogues in Smith and Martell. Compare the fraction of Mg^{2+} and Fe^{3+} complexed by these agents at pH = 10.

NOTES

1. Glycine exists in neutral solution principally as $^+H_3N-CH_2-COO^-$—see Chapter 5. These constants (at 25°C, $I = 0$) were selected from Sillén and Martell, *Stability Constants,* 1964, p. 377. Cu^{2+} also forms a solid hydroxide $Cu(OH)_2$ with log $K_{so}= -19.32$ and a series of polynuclear hydroxide complexes. See Smith and Martell, Vol. 6, 1989, for recent values.

2. Use Newton's method, the secant method, or the automatic "goal seek" function. All work best with nearly linear functions or with a first approximation close to the final answer. See Chapter 1.

3. The term *chelate* is derived from a Greek word meaning "crab's claw." Chelating agents are termed *bidentate* ("two–toothed") if they can occupy two coordination valences of a metal, *tridentate* ("three–toothed") if they can occupy three coordination valences, etc.

4. At 25°C, $I = 0.1$. From Smith and Martell, Vol. 6, 1989.

5. Gerold Schwarzenbach, who introduced this formalism (*Complexometric Titrations,*

London, Methuen, 1957), used the symbol α_H for A_L and β_A for A_M. Anders Ringbøm, another early researcher in the field, ("Complexation Reactions," Chap. 14 in *Treatise on Analytical Chemistry*, I. M. Kolthoff, P. J. Elving, and E. B. Sandell, Eds., Interscience Encyclopedia, New York, 1959) used $\alpha_{L(H)}$ for A_L and $\alpha_{M(NH_3)} + \alpha_{M(OH)}$ for A_M. I took the liberty of changing the notation in order to avoid confusion with α_n for the fraction present as a given complex and β_n for the overall formation constant of that complex. This formalism is also known as "side reaction coefficients" (A. Dickson, private communication).

6. Equilibrium constants used to evaluate A_M were: for Mg–NH$_3$ (Bjerrum, 1941): log $\beta_1 = 0.23$, log $\beta_2 = 0.08$, log $\beta_3 = -0.34$, log $\beta_4 = -1.04$, log $\beta_5 = -1.99$, log $\beta_6 = -3.29$; Mg–OH: log $K_1 = 2.58$ (Stock and Davies, *Trans Faraday Soc.* **44**:856, 1948. Note Smith and Martell, Vol. 5, gives log $K_1 = 2.8$ at $I = 0$.

7. Equilibrium constants used to evaluate A_L are listed in Table 8.2.

8. To plot the curve of Fig. 8.11, Eqs. (25) and (26) were solved simultaneously by iteration, starting with an assumed value of $[M]'$ and $V = V_o$ in Eq. (26), which gave a value for $[L]'$. These values of $[M]'$ and $[L]'$ were then substituted in Eq. (25) to give a revised value of V. This procedure was followed for each point on the curve, and converged in less than 10 iterations. Attempting to substitute $[L]'$ algebraically in Eq. (25) results in a messy quadratic in V.

9. See, for example: Richard A. Durst, Ed., *Ion–Selective Electrodes*, National Bureau of Standards special publication 314, Washington, DC: U.S. Govt. Printing Office, 1969; Henry Freiser, *Ionic-Selective Electrodes in Analytical Chemistry*, Plenum Press, New York, 1978–1980. Arthur K. Covington, Ed., *Ion-Selective Electrode Methodology*, CRC Press, Boca Raton, FL, 1979; Karl Cammann, *Working with Ion-Selective Electrodes: Chemical Laboratory Practice*, translated from the German by Albert H. Schroeder, Springer-Verlag, Berlin and New York, 1979; Daniel Ammann, *Ion-Selective Microelectrodes: Principles, Design, and Application*, Springer-Verlag, Berlin and New York, 1986. For the use of ion-selective electrodes in thermodynamic studies, see J. N. Butler and R. N. Roy, Chapter 4, in *Activity Coefficients in Electrolyte Solutions*, Ed. K. S. Pitzer, CRC Press, 1991, pp. 155–208.

10. Sillén and Martell, *Stability Constants*, The Chemical Society, London, Special Publications No. 17 (1964) and No. 25 (1971). Table 1094.

11. This section is primarily based on H. A. Laitinen, *Chemical Analysis*, McGraw Hill, NY, 1960, pp 258–75; D. Dyrssen, D. Jagner, and F. Wengelen, *Computer Calculation of Ionic Equilibria and Titration Procedures*, Almqvist & Wiksell, Stockholm (also Wiley, N.Y.), 1968, Chapter 6; F. J. C. Rossotti and H. Rossotti, *The Determination of Stability Constants*, McGraw-Hill, NY, 1961, Chapter 11.

12. Since P describes an equilibrium between two phases, it is an equilibrium constant, a ratio of activities that is rigorously constant, but since activity coeffficients of uncharged species are either small or rendered constant by a constant-ionic-strength medium, concentrations are normally used for these equilibria.

13. The solubility of HL and CuL_2 are given by Smith and Martell, Vol. 3, p. 319, to be

$$\log[\text{H}L]/[\text{H}L(s)] = -6.61; \quad \log[\text{Cu}L_2]/[\text{Cu}L_2(s)] = -8.40$$

14. E. B. Sandell, *Colorimetric Determinations of Traces of Metals*, 3rd ed., Interscience, NY, 1959. Dithizone is given far more detailed coverage (pp. 139–74) than the 15 other organic reagents discussed.

15. Equilibrium constants for this example are taken from R. W. Geiger and E. B. Sandell, *Anal. Chim. Acta* **8**:197, 1953, quoted by Sillén and Martell, *Stability Constants*, Spec. Pub. 17, 1964, and by Laitinen, *Chemical Analysis,* p. 264. In the original paper, and in *Stability Contants*, reactions 3 and 6 were not given explicitly; instead the two composite reactions

$$Cu^{2+} (aq) + 2\ H_2L(org) = Cu(HL)_2(org) + 2\ H^+ (aq) \qquad \log K_x = 10$$

$$Cu^{2+} (aq) + Cu(HL)_2(org) = 2\ CuL(org) + 2\ H^+ (aq) \qquad \log K_y = 6.66$$

were given. These can be combined with reactions #1, 2#, #4, and #5 to give the simpler reactions #3 and #6: #3 = (#x − #5 − #2 + 2*#4)/2. #6 = (#y + #2 + #5)/2 − #1.

16. Sillen and Martell, 1971: 1M NH_4NO_3 data from Fridman et al., *Zhur. neorg. Khim.* **11**: 1641, 1966.

17. V. N. Vasil'eva and V. P. Vasil'ev, *Isvest. V.U.Z. Khim.* **9**:185, 1966, quoted by Sillén and Martell, Spec. Publ. No. 25, 1971.

18. L. G. Sillén and A. E. Martell, *Stability Constants*, Chem. Soc. Spec. Pub. No. 25, 1971. Table 448.

19. pK_{so}° is 8.43 at 20°C. See Table 9.1. At $I = 0.1$, the Davies equation gives $\log \gamma_{++} = -0.45$ and hence $pK_{so} = 7.53$ at $I = 0.1$. See Fig. 9.6.

20. At $I = 0.1$, 25°C. Smith and Martell, Vol. 6.

21. J. Bjerrum, *Metal Ammine Formation in Aqueous Solution*, P. Hasse & Son, Copenhagen, 1941, reprinted 1957.

9

OXIDATION–REDUCTION EQUILIBRIA

Oxidation–Reduction Reactions
  Oxidation–Reduction versus Complex Formation
Electrochemical Cells
  IUPAC Sign Conventions
    Concentration Dependence of Potential
    Cell Discharge: From Open Circuit to Short Circuit
Redox (Half-Cell) Potentials
  The Redox Potential $p\varepsilon$ and Its Analogy to pH
  Activity versus Potential at Constant pH
  The Electron Balance
  Formal Potential
Effect of pH on Potential
  Formal Potential Including OH Complexes
  pH–$p\varepsilon$ Diagrams
  Manganese Redox Reactions
Influence of Complexing Agents on Fe(II)/Fe(III)
  Fe^{3+}/Fe^{2+} with EDTA
  Fe^{3+}/Fe^{2+} with Phenanthroline
Redox Titrations
  Equivalence-Point Potential for Symmetric Titrations
  Unsymmetrical Titrations: Ferrous–Dichromate
Can There Be a Unique Redox Potential in an Aqueous Solution?
Problems

OXIDATION–REDUCTION REACTIONS

Oxidation and reduction are fundamental concepts in chemistry, but are difficult to express quantitatively without resorting to arbitrary definitions. Oxidation is often defined as a loss of electrons, as illustrated by the conversion of a metal to its ion:

$$Ag(s) \rightleftharpoons Ag^+ + e^-$$

and because free electrons do not normally accumulate,[1] oxidation must be accompanied by an equivalent reduction (such as $O_2 + 4H^+ + 4e^- \rightleftharpoons 2H_2O$).

However, in most systems the element that is oxidized or reduced may not be so clearly defined. In the electrolytic oxidation of cyanide ion, for example:

$$CN^- + H_2O \rightleftharpoons CNO^- + 2H^+ + 2e^-$$

How can you decide whether it is carbon, nitrogen, hydrogen, or oxygen that has lost the two electrons?

The usual way out of this dilemma is to arbitrarily assign the electrons shared by two atoms to the more *electronegative*[2] of the two, or to divide the electrons equally between two atoms of the same electronegativity. The formal charge remaining on the atom when this division is made is known as its *oxidation number*. For example, the oxidation number of Ag(s) is 0, and of Ag^+ is +1, without ambiguity. Since oxygen is more electronegative than hydrogen, the oxidation number of hydrogen in water is +1, and that of oxygen is –2. Note that since the H_2O molecule is neutral, the net oxidation number is $2(+1) - 2 = 0$. The net oxidation number remains the same regardless of how the electrons are divided up.

The electronegativity convention is convenient but arbitrary. A simplified version is to assign H = +1, O = –2, N = –3 and adjust the others so that the sum of oxidation numbers of elements in a species must equal the formal charge on that species. The electrons in a bond may be divided in any way desired, but this division must be done consistently within any set of oxidation–reduction reactions. Oxidation can then be defined as an increase in oxidation number and reduction as a decrease in oxidation number.

For example, if the three electrons shared by C and N in cyanide CN^-, are assigned to nitrogen, this gives N an oxidation number of –3 (as in NH_3). Since the net charge (overall oxidation number) must be –1, C has an oxidation number[3] of +2. The oxidation numbers in CNO^- can be similarly assigned: O = –2, N = –3, C = +4, giving a net charge of –1. Under these assumptions, N, H, and O do not change in oxidation number, but C changes from +2 to +4.

To take a contrasting example, according to the electronegativity convention, the dissociation of water to its ions

$$H_2O \rightleftharpoons H^+ + OH^-$$

does not involve an oxidation or a reduction. The oxidation number of oxygen remains −2, and the oxidation number of hydrogen remains +1. However, breaking the covalent bond between hydrogen and oxygen actually increases the electron density around the oxygen and decreases the electron density around the hydrogen ion. If oxidation is defined rigorously as a loss of electrons, the hydrogen has been partially oxidized and the oxygen has been partially reduced.

Oxidation–Reduction Versus Complex Formation

On the basis of arbitrarily defined oxidation numbers, it is possible to divide all chemical reactions into two classes: those that do not involve a change in oxidation number (acid–base and complex formation) and those that do. Again, the electron density around various atoms certainly changes in acid–base or complex-formation reactions, but the arbitrary division of electrons in calculating the oxidation number of elements hides those changes.

A more practical distinction is based on speed of reaction. Acid–base reactions typically take place in milliseconds, and many complex formation reactions take place within the time of mixing solutions. Slower reactions include the formation of polynuclear species and the equilibration of solutions with solid phases. In contrast, very few oxidation–reduction reactions take place within the few minutes required to mix solutions, and many occurring in nature require microbial mediation to catalyze them. Indeed, the earth's environment is far from equilibrium with respect to oxygen, nitrogen, and carbon compounds—most living organisms, including humans, depend on this energy imbalance for their existence.[4]

An example from the laboratory is potassium permanganate solution. At equilibrium aqueous $KMnO_4$ produces MnO_2 and O_2. The equilibrium constant for the reaction

$$4MnO_4^- + 4 H^+ \rightleftharpoons 4 MnO_2(s) + 3 O_2(g) + 2 H_2O$$

can be calculated from redox potentials (see later in this chapter) to be

$$[H^+][MnO_4^-] = 10^{-23.8} P_{O_2}^{3/4}$$

If P_{O_2} is about 0.2 atm and $[H^+]$ is about 10^{-6} molar, and the solution is in equilibrium with solid MnO_2, this equilibrium expression yields $[MnO_4^-] = 10^{-17.3}$. Yet permanganate solutions of 0.1 molar concentration are sufficiently stable to be used as a standard oxidation titrant for several days before they begin to decompose.[5]

ELECTROCHEMICAL CELLS

In Chapter 1, a brief mention was made of how equilibrium constants are related to free energy, enthalpy, and entropy changes. Because oxidation–reduction reactions were frequently studied using electrochemical cells, their equilibrium constants

ELECTROCHEMICAL CELLS 321

may be found tabulated as electrochemical potentials,[6] as well as free energy changes[7] and equilibrium constants[8] for half-reactions involving electrons (normally written as reductions; see below).

These various quantities are interrelated by the thermodynamic equation[9]

$$\Delta G = -nFE \tag{1}$$

where E is the potential ($E°$ is the standard potential) of an electrochemical cell in which a particular reaction takes place;

n is the number of electrons transferred in that reaction;

F is the Faraday constant (the charge on Avogadro's number of electrons, 96,497 coulombs, 96.497 kJ/volt, or 23.06 kcal/volt); and

ΔG is the free energy ($\Delta G°$ is the standard free energy) of that reaction.

Example 1. The Daniell cell consists of a zinc electrode in contact with zinc sulfate solution and a copper electrode in contact with copper sulfate solution, and may be represented as

$$-Zn/Zn^{2+}//Cu^{2+}/Cu^{+}$$

where "/" represents a phase boundary. "//" represents a liquid junction between the copper and zinc sulfate solutions.[10] In the physical cell, the copper electrode is always positive and the zinc electrode is always negative. The electromotive force (potential of the cell extrapolated to zero current) is 1.10 volts at 25°C. When the cell discharges spontaneously, the greater the current drawn, the smaller the potential—it reaches zero when the cell is short-circuited (Fig. 9.1).

The reactions that take place in the discharging cell produce electrons at the zinc electrode (hence it is negatively charged) and consume electrons at the copper electrode (hence it is positively charged):

$$Zn(s) \rightleftharpoons Zn^{2+} + 2\ e^{-}$$
$$Cu^{2+} + 2\ e^{-} \rightleftharpoons Cu(s)$$

The overall reaction is the sum of these:

$$Zn(s) + Cu^{2+} \rightleftharpoons Zn^{2+} + Cu(s)$$

IUPAC Sign Conventions

While there is no question which electrode is positive in the laboratory, the representation of a cell on paper follows an arbitrary set of conventions developed by the International Union of Pure and Applied Chemistry (IUPAC)[11]:

1. The electrode potential is given the sign of the right-hand electrode in the diagram. For example, the diagram

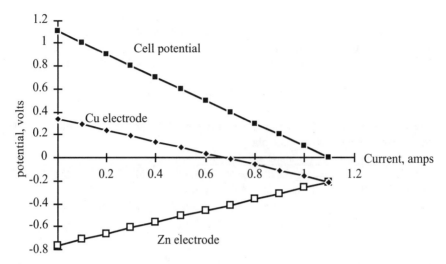

FIGURE 9.1. Current–potential curves for the Daniell cell. Internal resistance is assumed to be 1 ohm, equally divided between the two electrodes. Overvoltage is neglected. Single electrode contributions are measured with respect to the standard hydrogen electrode (see p. 325).

$$-Zn/Zn^{2+}//Cu^{2+}/Cu+$$

is given the potential +1.10 V, whereas

$$+Cu/Cu^{2+}//Zn^{2+}/Zn-$$

is given the potential −1.10 V. The actual cell has potential 1.10 V without any sign.

2. The reaction corresponding to the diagram is found by imagining electrons flowing from left to right in the external circuit. The first diagram above corresponds to the spontaneous discharge reaction

$$Zn(s) + Cu^{2+} \rightleftharpoons Zn^{2+} + Cu(s)$$

The second diagram corresponds to the reverse reaction, which would take place if an external current source were applied to the cell:

$$Zn^{2+} + Cu(s) \rightleftharpoons Zn(s) + Cu^{2+}$$

3. Thus a cell diagram, potential, and cell reaction can all be related by the same thermodynamic equations. For the Daniell cell, Eq. 1 yields

$$E = +1.10 \text{ V}, \quad n = 2$$

ELECTROCHEMICAL CELLS

$$\Delta G = -nFE = -2 \ (96.50 \ \text{kJ/V})(1.10 \ \text{V}) = -212 \ \text{kJ/mol}$$

or $-2 \ (23.06 \ \text{kcal/V})(1.10 \ \text{V}) = -50.7$ kcal/mol (spontaneous). For the reverse reaction, $E = -1.10$ and $\Delta G = +50.7$ kcal/mol (nonspontaneous).

Concentration Dependence of Potential. Electrode potentials depend on the concentration of dissolved species according to the Nernst equation.[12] This fundamental relation is derived from the chemical potential or partial molal free energy. For the reaction

$$aA + bB \rightleftharpoons cC + dD$$

the free energy change ΔG (at constant temperature and pressure) is

$$\Delta G = \Delta G° + RT \ln \frac{\{C\}^c \{D\}^d}{\{A\}^a \{B\}^b} \tag{2}$$

where R is the gas constant (8.314 J/mol K or 1.987 cal/mol K, T is the temperature in K, $\{A\}$ represents the activity of species A, etc. (under any specified conditions, not only at equilibrium). $\Delta G°$ is the standard free energy change, which can be obtained from tabulated free energies of formation:

$$\Delta G° = c \ \Delta G°_{f,c} + d \ \Delta G°_{f,d} - a \ \Delta G°_{f,a} - b \ \Delta G°_{f,b} \tag{3}$$

Since the condition for equilibrium is $\Delta G = 0$,

$$\Delta G° = -RT \ln \left\{ \frac{\{C\}^c \{D\}^d}{\{A\}^a \{B\}^b} \right\}_{\text{equil}} \tag{4}$$

$$\Delta G° = -RT \ln K \tag{5}$$

where K is the equilibrium constant for the reaction.

In terms of potentials (use Eq. 1),

$$E = E° - \frac{RT}{nF} \ln \left(\frac{\{C\}^c \{D\}^d}{\{A\}^a \{B\}^b} \right) = E° - \frac{2.303 RT}{nF} \log \left(\frac{\{C\}^c \{D\}^d}{\{A\}^a \{B\}^b} \right) \tag{6}$$

At 25°C, $2.303 RT/F = 0.05915$ V.

Example 2. For the Daniell cell, with the reaction

$$\text{Zn(s)} + \text{Cu}^{2+} \rightleftharpoons \text{Zn}^{2+} + \text{Cu(s)}$$

the activities of the pure metals are unity, and the activity of the dissolved species are their concentrations multiplied by activity coefficients:

$$E = E° - \frac{0.059}{2} \log \frac{[Zn^{2+}]\gamma_{Zn^{2+}}}{[Cu^{2+}]\gamma_{Cu^{2+}}}$$

E is the standard potential (1.10 V) when all reactants and products are at unit activity.[13]

Cell Discharge: From Open Circuit to Short Circuit. There are two equilibrium conditions that are commonly encountered with electrochemical cells. The equilibrium potential at open circuit is called the *electromotive force* or simply the *potential* and is given by Eq. 6. When the cell is short-circuited, its internal reaction goes to equilibrium, and this condition is found by setting $E = 0$, under which condition the concentrations are governed by

$$E° = \frac{RT}{nF} \ln \frac{\{C\}^c\{D\}^d}{\{A\}^a\{B\}^b} = \frac{2.303RT}{nF} \log \frac{\{C\}^c\{D\}^d}{\{A\}^a\{B\}^b} = \frac{0.059}{n} \log K \quad (7)$$

For the Daniell cell, this short-circuit equilibrium condition is

$$1.10 = \frac{0.059}{2} \log \frac{[Zn^{2+}]\gamma_{Zn^{2+}}}{[Cu^{2+}]\gamma_{Cu^{2+}}}$$

or $[Zn^{2+}]/[Cu^{2+}] = 10^{+37.3}$.

Example 3. Suppose a cell is set up with equal concentrations of the two ions $[Zn^{2+}]° = [Cu^{2+}]°$; as it discharges, $[Zn^{2+}]$ increases and $[Cu^{2+}]$ decreases; the second term in the equation above becomes more negative, and hence E becomes more negative. This is the same direction of change as was predicted from the cell resistance at higher current (Fig. 9.1), but this change persists even when the current is interrupted. If the cell is 50% discharged ($x = 0.5$),

$$E = E° - \frac{0.059}{2} \log \frac{[Zn^{2+}]°(1+x)\gamma_{Zn^{2+}}}{[Cu^{2+}]°(1-x)\gamma_{Cu^{2+}}}$$

$$= 1.10 - \frac{0.059}{2} \log \frac{1+x}{1-x} = 1.07 \text{ V}$$

This process is illustrated in Fig. 9.2. As x increases toward 1.00, the cell potential sharply decreases. When the cell is short-circuited ($E = 0$),

$$\log \frac{1+x}{1-x} = \frac{(2)(1.10)}{0.059} = 37.3$$

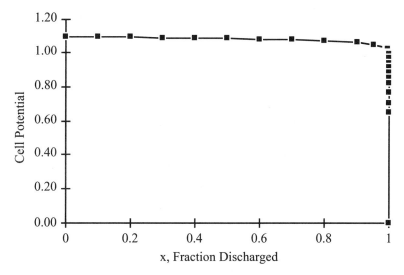

FIGURE 9.2. Potential of the Daniell cell as it is discharged. 99% discharge decreases the potential from 1.10 to 0.96 volts. At $E = 0$, only $10^{-37.3}$ of the original Cu^{2+} remains.

which gives $1 - x = 10^{-37.3}$. Only one part in $10^{37.3}$ of the original copper ion remains when equilibrium is reached. This conclusion can also be reached through the Nernst equation as above.

REDOX (HALF-CELL) POTENTIALS

Tabulation of electrochemical potentials is simplified by the introduction of half-cell or redox potentials. Oxidation–reduction reactions are divided into two parts by setting the standard electrode potential for hydrogen reduction equal to zero.[14] The IUPAC convention is to tabulate these values as reduction potentials; then the numbers for $E°$ have the same sign as the polarity of the electrode in a cell where hydrogen is the other electrode.

Example 4. The standard potentials of the zinc and copper electrodes are the potentials of the cells where the right-hand electrode is a hydrogen electrode:

$$Zn/Zn^{2+}/H^+/H_2/Pt, \quad E = -0.76 \text{ V}$$

$$Zn + 2H^+ \rightleftharpoons H_2 + Zn^{2+}$$

which is abbreviated as the half-cell reaction

$$Zn \rightleftharpoons Zn^{2+} + 2e^-, \quad E = -0.76 \text{ V}$$

Likewise,

$$Cu/Cu^{2+}/H^+/H_2/Pt, \quad E = +0.34 \text{ V}$$

$$Cu + 2H^+ \rightleftharpoons H_2 + Cu^{2+}$$

$$Cu \rightleftharpoons Cu^{2+} + 2e^-, \quad E = +0.34 \text{ V}$$

If the zinc electrode is placed in a solution containing zinc ion, the reduced species is zinc metal, with activity = 1. The oxidized species is Zn^{2+}, with activity = $[Zn^{2+}]\gamma_{++}$ and $n = 2$:

$$E = E^\circ_{Zn} + 0.0295 \log([Zn^{2+}]\gamma_{++}), \quad E^\circ_{Zn} = -0.76$$

The higher the activity of the oxidized species, the more positive the electrode potential. The higher the activity of the reduced species, the more negative the electrode potential.[15]

The dependence of half-cell potentials on current is shown schematically in Fig. 9.1. Note that when the cell is short-circuited, the Cu and Zn potentials have the same value.

The general format of a redox potential is

$$E = E^\circ + \frac{2.303 RT}{nF} \log \frac{\{ox\}}{\{red\}} \tag{8}$$

where {ox} represents the product of the activities of the oxidized species, and {red} represents the product of activities of reduced species.

A short list of common redox potentials is given in Table 9.1.

These potentials are used with Eq. (6) or (8). The standard conditions are 1 atm for gases, 1 mol/L for dissolved species (including H^+), and unity for H_2O and pure solids.

Example 5. Here is a more complicated example than the Daniell cell: Calculate the equilibrium constant for the reaction given at the beginning of this chapter:

$$4 \text{ MnO}_4^- + 4 \text{ H}^+ \rightleftharpoons 4 \text{ MnO}_2(s) + 3 \text{ O}_2(g) + 2 \text{ H}_2O$$

In Table 9.1, you can find three half-reactions that can be combined to make the desired equilibrium:

$$\text{MnO}_4^- + 8 \text{ H}^+ + 5 e^- \rightleftharpoons \text{Mn}^{2+} + 4 \text{ H}_2O$$

$$\text{MnO}_2(s) + 4\text{H}^+ + 2 e^- \rightleftharpoons \text{Mn}^{2+} + 2 \text{ H}_2O$$

$$\text{O}_2(g) + 4 \text{ H}^+ + 4 e^- \rightleftharpoons 2 \text{ H}_2O$$

REDOX (HALF-CELL) POTENTIALS   327

TABLE 9.1. Some common redox potentials at 25°C[16]

Reduction reaction	$E°$ (volts)	$p\varepsilon°$
$Na^+ + e^- \rightleftharpoons Na(s)$	−2.714	−45.9
$Zn^{2+} + 2e^- \rightleftharpoons Zn(s)$	−0.763	−12.9
$Fe^{2+} + 2e^- \rightleftharpoons Fe(s)$	−0.44	−7.4
$V^{3+} + 2e^- \rightleftharpoons V^{2+}$	−0.255	−4.3
$2H^+ + 2e^- \rightleftharpoons H_2(g)$	0.00	0.0
$Cu^{2+} + e^- \rightleftharpoons Cu^+$	+0.16	+2.7
$AgCl(s) + e^- \rightleftharpoons Ag(s) + Cl^-$	+0.22	+3.7
$Cu^{2+} + 2e^- \rightleftharpoons Cu(s)$	+0.340	+5.75
$Cu^+ + e^- \rightleftharpoons Cu(s)$	+0.520	+8.8
$Fe^{3+} + e^- \rightleftharpoons Fe^{2+}$	+0.771	+13.0
$Hg_2^{2+} + 2e^- \rightleftharpoons 2Hg$	+0.796	+13.46
$Ag^+ + e^- \rightleftharpoons Ag(s)$	+0.799	+13.5
$Hg^{2+} + 2e^- \rightleftharpoons Hg$	+0.911	+15.4
$O_2(g) + 4H^+ + 4e^- \rightleftharpoons 2H_2O$	+1.229	+20.8
$MnO_2(s) + 4H^+ + 2e^- \rightleftharpoons Mn^{2+} + 2H_2O$	+1.23	+20.8
$Cl_2 + 2e^- \rightleftharpoons 2Cl^-$	+1.358	+23.0
$MnO_4^- + 8H^+ + 5e^- \rightleftharpoons Mn^{2+} + 4H_2O$	+1.51	+25.5
$Co^{3+} + e^- \rightleftharpoons Co^{2+}$	+1.92	+32.5

If the second reaction is reversed and added to the first, Mn^{2+} cancels out, leaving a three-electron reduction of MnO_4^- to MnO_2. The three electrons are supplied by the oxidation of water to O_2. For each reaction, the potential is multiplied by the number of electrons to produce $nE°$, a quantity proportional to the free energy[17]

$$MnO_4^- + 8H^+ + 5e^- \rightleftharpoons Mn^{2+} + 4H_2O, \quad (5)(+1.51) = 7.55$$

$$Mn^{2+} + 2H_2O \rightleftharpoons MnO_2(s) + 4H^+ + 2e^-, \quad (2)(-1.23) = -2.46$$

Combining these gives the desired reduction of MnO_4^- to MnO_2:

$$MnO_4^- + 4H^+ + 3e^- \rightleftharpoons MnO_2(s) + 2H_2O, \quad nE° = +7.55 - 2.46 = 5.09$$

Next combine this with the oxygen reaction

$$\tfrac{3}{4}[2H_2O \rightleftharpoons O_2(g) + 4H^+ + 4e^-], \quad (3)(-1.229) = -3.69$$

$$MnO_4^- + H^+ \rightleftharpoons MnO_2(s) + \tfrac{1}{2}H_2O + \tfrac{3}{4}O_2(g)$$

$$nE° = 5.09 - 3.69 = +1.40$$

Here $n = 3$, and hence $E° = +0.47$. From Eq. (7) ($\log K = nE°/0.059 = 23.8$), the equilibrium concentrations (activity of $MnO_2 = 1.0$) are related by

$$P_{O_2}^{3/4} = 10^{23.8}[MnO_4^-][H^+]$$

If $P_{O_2} = 0.2$, $[H^+] = 10^{-7}$,

$$[MnO_4^-] = 10^{-23.8} P_{O_2}^{3/4}[H^+]^{-1} = 10^{-17.3}$$

The Redox Potential pε and Its Analogy to pH

You may already have noted that the factor $2.303RT/F$ continues to occur in the equations (such as Eq. 6) relating concentrations to potential. Sillén[18] (and others before him) suggested that the potential scale be changed from volts to units of $2.303RT/F$.

$$p\varepsilon = EF/(2.3RT) \quad \text{and} \quad p\varepsilon^\circ = E^\circ F/(2.3RT) \tag{9}$$

Then a change of a factor of ten in a concentration would change the new scale by one logarithmic unit. Values of $p\varepsilon^\circ$ are listed together with E° in Table 9.1.

Example 6. The oxygen potential is given by (see Table 9.1)

$$E = 1.23 \text{ V} + \frac{2.3RT}{4F} \log\{P_{O_2}[H^+]^4 \gamma_+^4\}$$

$$p\varepsilon = 20.75 + \tfrac{1}{4}\log\{P_{O_2}[H^+]^4\gamma_+^4\} = 20.75 - pH + \tfrac{1}{4}\log\{P_{O_2}\}$$

Thus when pH increases by one unit, pε decreases by one unit; when P_{O_2} increases by a factor of 10, pε increases by 0.25 units.

The function $p\varepsilon$ plays the role of a master variable in redox equilibria, just as pH plays the role of the master acidity variable in acid–base equilibria. Here are some analogies ({} stands for activity of the species in brackets)

Acid–base reaction	Redox reaction
$HA \rightleftharpoons H^+ + A^-$	$Fe^{3+} + e^- \rightleftharpoons Fe^{2+}$
$H_2O + H^+ = H_3O^+$ ($K = 1$)	$\tfrac{1}{2}H_2 \text{(g)} = H^+ + e^-$ ($K = 1$)
$\{H^+\}\{A^-\} = K_a\{HA\}$	$\{Fe^{2+}\} = K\{Fe^{3+}\}\{e^-\}$
$pH = -\log\{H^+\}$, $pK_a = -\log K_a$	$p\varepsilon = -\log\{e^-\}$, $p\varepsilon^\circ = -\log K$
$pH = pK_a + \log[\{A^-\}/\{HA\}]$	$p\varepsilon = p\varepsilon^\circ + \log[\{Fe^{3+}\}/\{Fe^{2+}\}]$

Further analogies can be made.[19]

Activity Versus Potential at Constant pH

Because most redox reactions involve H^+, it is convenient to begin by holding pH constant and observing the various concentrations as potential changes.

Example 7. Show how the concentrations of iron species in acid solution vary with pε. For simplicity, consider only the species Fe^{3+}, Fe^{2+}, and Fe(s), with total

REDOX (HALF-CELL) POTENTIALS 329

concentration in solution limited to $C = 0.1$ M. The mass balance and Nernst equations are: in solution,

$$[Fe^{3+}] + [Fe^{2+}] = C$$

$E = 0.771 + 0.059 \log\{[Fe^{3+}]/[Fe^{2+}]\}, \quad p\varepsilon = 13.03 + \log\{[Fe^{3+}]/[Fe^{2+}]\}$

$E = -0.44 + 0.059 \log[Fe^{2+}], \quad p\varepsilon = -7.44 + \log[Fe^{2+}]$ in equil. with Fe(s)

If Fe(s) is present,

$$\log[Fe^{2+}] = p\varepsilon + 7.44$$

$$\log[Fe^{3+}] = p\varepsilon - 13.03 + \log[Fe^{2+}] = 2\, p\varepsilon - 5.59$$

If only the solution is present, $[Fe^{3+}] = C - [Fe^{2+}]$, or $[Fe^{2+}] = C - [Fe^{3+}]$ and

$$p\varepsilon = 13.03 + \log\{[Fe^{3+}]/[Fe^{2+}]\} = 13.03 + \log(C - [Fe^{2+}]) - \log[Fe^{2+}]$$

$$p\varepsilon = 13.03 + \log\{[Fe^{3+}]/[Fe^{2+}]\} = 13.03 + \log[Fe^{3+}] - \log(C - [Fe^{3+}])$$

None of these equations depends on pH, a simplification resulting from assuming an acid solution. (In Example 10, the sulfate complexes will be included, and in Example 11, the hydroxide complexes.)

At low $p\varepsilon$, Fe(s) predominates with activity 1.0, but at $p\varepsilon > -9.4$, the 0.1 mole of Fe(s) added per liter dissolves, only the solution phase is present, and $[Fe^{2+}]$ predominates for $p\varepsilon < 13$. At $p\varepsilon > 13$, $[Fe^{3+}]$ predominates. These questions are plotted in Fig. 9.3. Including activity coefficients would make small changes.[20]

The Electron Balance

Analogous to the proton condition, which provides the last simultaneous equation to define an acid–base equilibrium system, one can construct an electron balance for redox systems, in which the electrons lost by one set of species are balanced against the electrons gained by another set of species.

Example 8. Find the equilibrium concentrations when equal volumes of 0.010 M $Hg(NO_3)_2$ and $Fe(NO_3)_2$ are mixed in a solution sufficiently acid to prevent hydrolysis. The standard potentials[21] are

$$Fe^{3+} + e^- \rightleftharpoons Fe^{2+}, \quad E° = +0.771$$

$$2\, Hg^{2+} + 2e^- \rightleftharpoons Hg_2^{2+}, \quad E° = +1.026$$

The equilibrium constant for the reaction is

$$Hg^{2+} + Fe^{2+} \rightleftharpoons \tfrac{1}{2}Hg_2^{2+} + Fe^{3+}$$

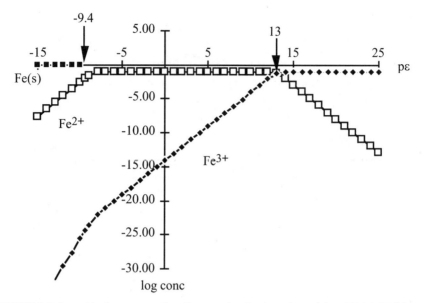

FIGURE 9.3. Logarithmic concentration diagram showing how the activity of Fe(s), [Fe^{2+}], and [Fe^{3+}] vary with potential. Total iron in all oxidation states is 0.1 mole/L. pε = 14.7 corresponds to the answer to Example 8.

$$\log K = (16.9)(1)(1.026 - 0.771) = 4.31$$

$$[Fe^{3+}][Hg_2^{2+}]^{1/2} = 10^{+4.31}[Hg^{2+}][Fe^{2+}]$$

The mass balances are

$$[Fe^{2+}] + [Fe^{3+}] = 0.005$$

$$[Hg^{2+}] + 2[Hg_2^{2+}] = 0.005$$

A charge balance provides no useful information since it introduces the additional variables [H$^+$] and [OH$^-$], and so we are left with three equations in four unknowns. The final relation is supplied by another stoichiometric condition—the electron balance: The formation of each Hg$_2^{2+}$ ion involves the loss of two electrons and the formation of each Fe^{3+} ion involves the gain of one electron:

$$2\,[Hg_2^{2+}] = [Fe^{3+}]$$

Note that this balance could also be deduced from the stoichometry of the balanced reaction. The solution to this problem is now a straightforward substitution of the mass balances in the equilibrium expression, and solution of the final nonlinear equation by iteration or curve-crawling[22]:

REDOX (HALF-CELL) POTENTIALS

$$[Fe^{3+}][Fe^{3+}]^{\frac{1}{2}} (0.5)^{\frac{1}{2}} = 10^{+4.31} [Fe^{2+}]^2$$

$$[Fe^{3+}]^{1.5} = 10^{+4.46}\{0.005 - [Fe^{3+}]\}^2$$

$[Fe^{3+}] = 4.89 \times 10^{-3}$, $[Fe^{2+}] = 1.09 \times 10^{-4}$, $[Hg^{2+}] = 1.09 \times 10^{-4}$, $[Hg_2^{2+}] = 2.45 \times 10^{-4}$. These values satisfy the equilibrium expression as well as the mass and electron balances (they will, of course, be somewhat different if activity coefficients are included in the calculations). The redox potential can be obtained from either couple; for example

$$p\varepsilon = p\varepsilon° + \log \frac{[Fe^{3+}]}{[Fe^{2+}]} = 13.03 + \log \frac{4.89 \times 10^{-3}}{1.09 \times 10^{-4}} = 14.7$$

This point is shown on Fig. 9.3, slightly above $p\varepsilon°$.

Example 9. Obtain an electron balance for the five manganese species in equilibrium with MnO_2 in acid solution.

$$MnO_2 + 4H^+ + 2\,e^- \rightleftharpoons Mn^{2+} + 2H_2O$$

$$MnO_2 + 4H^+ + e^- \rightleftharpoons Mn^{3+} + 2H_2O$$

$$MnO_2 + H_2O \rightleftharpoons MnO_3^- + 2\,H^+ + e^-$$

$$MnO_2 + 2\,H_2O \rightleftharpoons MnO_4^{2-} + 4\,H^+ + 2\,e^-$$

$$MnO_2 + 2\,H_2O \rightleftharpoons MnO_4^- + 4\,H^+ + 3\,e^-$$

For very Mn^{2+} ion formed, the system gains two electrons. For every Mn^{3+} ion formed, the system gains one electron. For every MnO_3^- ion formed, one electron is lost. For every MnO_4^{2-} ion formed, two electrons are lost, and for every MnO_4^- ion formed, three electrons are lost. Balancing the total number of electrons gained against the total number lost gives

$$2\,[Mn^{2+}] + [Mn^{3+}] = [MnO_3^-] + 2\,[MnO_4^{2-}] + 3\,[MnO_4^-]$$

An added oxidizing agent would introduce another reaction

$$ox + e^- \rightleftharpoons red$$

For each additional reduced species (over red°, that which was originally there) one electron is added to the system; hence a term appears on the left side of the electron balance:

$$red - red° + 2\,[Mn^{2+}] + [Mn^{3+}] = [MnO_3^-] + 2\,[MnO_4^{2-}] + 3\,[MnO_4^-]$$

Similarly, an added reducing agent would produce the electron balance

$$2\,[Mn^{2+}] + [Mn^{3+}] = [MnO_3^-] + 2\,[MnO_4^{2-}] + 3\,[MnO_4^-] + ox - ox°$$

Formal Potential

The standard redox potential $E°$ or $p\varepsilon°$ is determined by the free activities of the reactants and products of the redox reaction, extrapolated to $I = 0$. However, it is rare that the only species in solution are the ones listed in the redox tables. In most practical situations, it is not the free ionic activity of a species that is known, but the total concentration including complexes and ion pairs, and it is convenient to express the redox potential in terms of these total concentrations. This approach is analogous to the use of total concentrations to express the acid–base equilibria in seawater (Chapter 10). The relation between the standard potential $p\varepsilon_f°$ and the formal potential $p\varepsilon_t°$ involves activity coefficients and complex formation equilibria with the various ligands present.

The redox potential is determined by the free ion activities:

$$p\varepsilon = p\varepsilon_f° + \log\left(\frac{[Fe^{3+}]_f \gamma_{3+}}{[Fe^{2+}]_f \gamma_{2+}}\right) \tag{10}$$

$p\varepsilon_f° = 13.03$ from Table 9.1. This equation may also be written as a formal potential in terms of total concentrations, with the activity coefficients absorbed into $p\varepsilon_t°$ (compare footnote 22):

$$p\varepsilon = p\varepsilon_t° + \log\left(\frac{[Fe^{3+}]_t}{[Fe^{2+}]_t}\right) \tag{11}$$

$p\varepsilon$ is the same in both cases; hence

$$p\varepsilon_t° = p\varepsilon_f° + \log\left(\frac{[Fe^{3+}]_f \gamma_{3+}}{[Fe^{3+}]_t}\right) - \log\left(\frac{[Fe^{2+}]_f \gamma_{2+}}{[Fe^{2+}]_t}\right) \tag{12}$$

To relate the free concentrations to the total concentrations, employ the mass balances

$$[Fe^{3+}]_t = [Fe^{3+}]_f(1 + \beta_1[L] + \beta_2[L]^2 + \cdots) \tag{13}$$

$$[Fe^{2+}]_t = [Fe^{2+}]_f(1 + \beta_1'[L] + \beta_2'[L]^2 + \cdots) \tag{14}$$

$$p\varepsilon_t° = p\varepsilon_f° + \log\left(\frac{1 + \beta_1'[L] + \beta_2'[L]^2 + \cdots}{1 + \beta_1[L] + \beta_2[L]^2 + \cdots}\right) + \log\left(\frac{\gamma_{3+}}{\gamma_{2+}}\right) \tag{15}$$

where $[L]$ represents the concentration of a ligand such as OH^-, SO_4^{2-}, or an organic complexing agent (see pp. 342–348).

REDOX (HALF-CELL) POTENTIALS

Example 10. Find the formal potential $p\varepsilon_t^\circ$ for Fe^{3+}/Fe^{2+} in 0.01 M sodium sulfate. The stability constants of Fe with sulfate at $I = 0$ are[23] (hydroxide complexes are treated later; see also Table 7.3)

$$\text{for } Fe^{3+}: \quad \log \beta_1 = 4.04$$
$$\text{for } Fe^{3+}: \quad \log \beta_2 = 5.38$$
$$\text{for } Fe^{2+}: \quad \log \beta_1' = 1.0$$

The mass balances are:

$$[Fe^{3+}]_t = [Fe^{3+}]_f (1 + \beta_1[SO_4^{2-}] + \beta_2[SO_4^{2-}]^2)$$
$$[Fe^{2+}]_t = [Fe^{2+}]_f (1 + \beta_1'[SO_4^{2-}])$$

The redox potential as tabulated ($p\varepsilon^\circ = 13.03$) is for free ions:

$$p\varepsilon = p\varepsilon_f^\circ + \log \frac{[Fe^{3+}]_f \gamma_{3+}}{[Fe^{2+}]_f \gamma_{2+}}$$

but the same $p\varepsilon$ applies, with a different $p\varepsilon_t^\circ$, if the right-hand side is expressed in terms of total concentrations:

$$p\varepsilon = p\varepsilon_t^\circ + \log \frac{[Fe^{3+}]_t}{[Fe^{2+}]_t}$$

$$= p\varepsilon_t^\circ + \log \frac{[Fe^{3+}]_f \gamma_{3+} (1 + \beta_1[SO_4^{2-}] + \beta_2[SO_4^{2-}]^2)}{[Fe^{2+}]_f \gamma_{2+} (1 + \beta_1'[SO_4^{2-}])}$$

Setting these two expressions equal and cancelling the free iron concentrations gives

$$p\varepsilon_f^\circ = p\varepsilon_t^\circ + \log \frac{(1 + \beta_1[SO_4^{2-}] + \beta_2[SO_4^{2-}]^2)\gamma_{3+}}{(1 + \beta_1'[SO_4^{2-}])\gamma_{2+}}$$

Inserting numerical values (the Davies equation at $I = 0.03$) gives $\gamma_+ = 10^{-0.03}$ from

$$p\varepsilon_t^\circ = 13.03 - \log\left(\frac{1 + 10^{4.04}(0.01) + 10^{5.38}(0.01)^2}{1 + 10^{1.0}(0.01)}\right) - \log \frac{(10^{-0.03})^9}{(10^{-0.03})^4}$$

$$= 13.03 - 2.09 + 0.15 = 11.09$$

Thus the formal potential on 0.01 M Na_2SO_4 is more than 2.0 units lower than $p\varepsilon^\circ$ for the free ions. Most of this difference results from sulfate complexes rather than ionic strength.

EFFECT OF pH ON POTENTIAL

For many redox reactions, both protons and electrons are involved, and hence the concentration of the various species is a function of both potential and pH. In addition, ligands present in the solution, from OH⁻ to large organic molecules, are generally affected by pH.

Formal Potential Including OH Complexes

Example 11. Estimate the concentrations of Fe^{3+}, $FeOH^{2+}$, $Fe(OH)_2$, Fe^{2+}, $FeOH^+$ in equilibrium with $Fe(OH)_3(s)$ at pH = 4.0 as a function of potential. The redox potential (Table 9.1) is

$$Fe^{3+} + e^- \rightleftharpoons Fe^{2+}, \quad p\varepsilon^\circ = +13.03$$

For simplicity, activity coefficients will be taken equal to 1.0. From Table 7.2, the hydrolysis equilibria of Fe^{3+} at $I = 0$ are

$\log \beta_1 = +11.81$, $[FeOH^{2+}] = \beta_1[Fe^{3+}][OH^-] = 10^{-7.0}$

$\log \beta_2 = +23.3$, $[Fe(OH)_2^+] = \beta_2[Fe^{3+}][OH^-]^2 = 10^{-5.5}$

$\log \beta_3 = +28.8$, $[Fe(OH)_3] = \beta_3[Fe^{3+}][OH^-]^3 < 10^{-10}$

$\log \beta_4 = +34.4$, $[Fe(OH)_4^-] = \beta_4[Fe^{3+}][OH^-]^4 = 10^{-14.4}$

$\log K_{so} = -38.8$, $[Fe^{3+}][OH^-]^3 = K_{so}$

The corresponding equilibria for Fe^{2+} are[24]

$\log \beta_1' = +4.7$, $[FeOH^+] = \beta_1'[Fe^{2+}][OH^-] = 10^{-1.03-p\varepsilon}$

$\log \beta_2' = +7.4$, $[FeOH_2] = \beta_2'[Fe^{2+}][OH^-]^2 = 10^{-8.33-p\varepsilon}$

$\log \beta_3' = +9.3$, $[FeOH_3^-] = \beta_3'[Fe^{2+}][OH^-]^3 = 10^{-16.43-p\varepsilon}$

$\log \beta_4' = +8.9$, $[FeOH_4^{2-}] = \beta_4'[Fe^{2+}][OH^-]^4 = 10^{-26.83-p\varepsilon}$

$\log K_{so}' = -14.4$, $[Fe^{2+}][OH^-]^2 = K_{so}'$

Fixing pH = 4.0 fixes $[OH^-] = 10^{-10}$, and if this solution is in equilibrium with $Fe(OH)_3$,

$$[Fe^{3+}] = K_{so}/[OH^-]^3 = 10^{-38.8+3(10)} = 10^{-8.8}$$

$[Fe^{2+}]$ is then obtained from

$$E = E^\circ + (RT/F) \ln[Fe^{3+}]/[Fe^{2+}], \text{ or}$$

EFFECT OF pH ON POTENTIAL

$$p\varepsilon = p\varepsilon° + \log[Fe^{3+}]/[Fe^{2+}]$$

$$\log[Fe^{2+}] = \log[Fe^{3+}] + p\varepsilon° - p\varepsilon = -8.8 + 13.03 - p\varepsilon = 4.27 - p\varepsilon$$

Concentrations of the various hydroxide complexes can then be obtained from the Fe and OH concentrations as a function of pH and $p\varepsilon$.

These quantities are plotted in Fig. 9.4. Note that because the Fe(III) species are controlled by the solubility of $Fe(OH)_3$, their concentration does not change with potential at constant pH. The Fe(II) species (particularly Fe^{2+}) are largest at the most negative $p\varepsilon$, but at high $p\varepsilon$ the Fe(III) species predominate, the largest of which (at this pH) is $Fe(OH)_2^+$. Polynuclear species of Fe(III) also occur, and their equilibrium constants are given in Table 7.2. They would show up as additional horizontal lines, but would not be predominant because of the low concentration ($< 10^{-5}$) of dissolved Fe(III).

pH–pε Diagrams

Closely related to the diagrams just developed, diagrams indicating the predominant species on a plot of potential versus pH have been found useful as summaries of thermodynamic data in geochemistry, corrosion, and environmental science.[25] Such a map, showing the overview of logarithmic concentration diagrams, can be constructed by imagining a three-dimensional diagram with $p\varepsilon$ and pH on the hori-

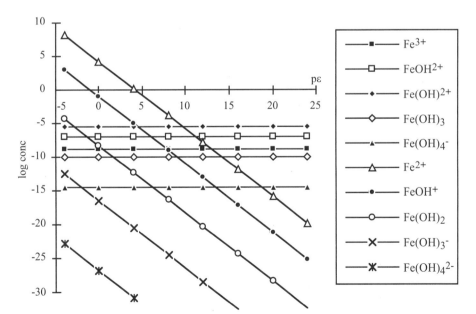

FIGURE 9.4. Concentrations of Fe(II) and Fe(III) species at pH = 4 as a function of $p\varepsilon$. The horizontal axis (activity = 1) corresponds to $Fe(OH)_3(s)$.

zontal plane and concentration on the vertical axis. A view of this from above would show polygonal areas where certain species are dominant.

Example 12. Find $p\varepsilon_t^\circ$ for Fe^{3+}/Fe^{2+} as a function of pH in a noncomplexing medium. This involves only the formation of the hydroxo complexes and $Fe(OH)_3(s)$. In unsaturated solution,

$$\frac{[Fe^{3+}]_t}{[Fe^{3+}]_f} = 1 + \beta_1[OH^-] + \beta_2[OH^-]^2 + \beta_3[OH^-]^3 + \beta_4[OH^-]^4$$

$$+ \beta_{22}[Fe^{3+}]_f[OH^-]^2 + \beta_{34}[Fe^{3+}]_f[OH^-]^2$$

If saturated solutions, $[Fe^{3+}]_f$, and hence $[Fe^{3+}]_t$, are not independent but are determined by the solubility product and $[OH^-]$:

$$[Fe^{3+}]_f = \frac{K_{so}}{[OH^-]^3}$$

and $[Fe^{3+}]_t$ is given by the mass balance above.

The constants can be found in Example 11, or in Table 7.2. At $I = 0$,

$\log \beta_1 = 11.8$, $\quad \log \beta_4 = 34.4$

$\log \beta_2 = 23.3$, $\quad \log \beta_{22} = 25.1$

$\log \beta_3 < 28.8$, $\quad \log \beta_{34} = 49.7$

$\log K_{so} = -38.8$ for freshly precipitated $Fe(OH)_3(s)$

The corresponding mass balance and equilibrium constants for Fe^{2+} are:

$$[Fe^{2+}]_t = [Fe^{2+}]_f(1 + \beta_1'[OH^-] + \beta_2'[OH^-]^2 + \beta_3'[OH^-]^3 + \beta_4'[OH^-]^4 + \cdots)$$

$\log \beta_1' = 4.5$, $\quad \log \beta_3' = 10.0$

$\log \beta_2' < 7.4$, $\quad \log \beta_4' = 9.6$

$\log K_{so}' = -15.1$ for $Fe(OH)_2(s)$

From these data, $p\varepsilon_t^\circ$ can be calculated as a function of pH, as shown in Fig. 9.5. the curve is in three sections. The low-pH branch corresponds to an unsaturated solution in equilibrium with Fe(II) and Fe(III) species. The presence of polynuclear species causes a small dependence on Fe(III) concentration:

$$p\varepsilon = p\varepsilon_t^\circ + \log \frac{[Fe^{3+}]_f(1 + \beta_1[OH^-] + \beta_2[OH^-]^2 + \beta_{22}[Fe^{3+}]_f[OH^-]^2 + \cdots)}{[Fe^{2+}]_f(1 + \beta_1'[OH^-] + \cdots)}$$

$$= p\varepsilon_f^\circ + \log \frac{[Fe^{3+}]_f}{[Fe^{2+}]_f}$$

EFFECT OF pH ON POTENTIAL

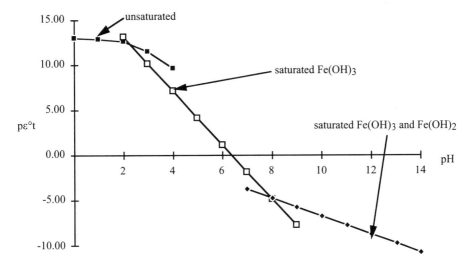

FIGURE 9.5. $p\varepsilon$ for Fe^{3+}/Fe^{2+} as a function of pH calculated using $I = 0$ constants.

$$p\varepsilon_t^\circ = p\varepsilon_f^\circ + \log \frac{(1 + \beta_1[OH^-] + \beta_2[OH^-]^2 + \beta_{22}[Fe^{3+}]_f[OH^-]^2 + \cdots)}{(1 + \beta_1'[OH^-] + \cdots)}$$

In the second section, at pH above about 2.2, the solution is saturated with $Fe(OH)_3$, and $[Fe^{3+}]_f = K_{so}/[OH^-]^3$. $[Fe^{2+}]_t$ is given by the same expression as before

$$[Fe^{2+}]_t = [Fe^{2+}]_f(1 + \beta_1[OH^-] + \cdots)$$

Hence $p\varepsilon$ depends on $[Fe^{2+}]$ and pH, but $p\varepsilon_t^\circ$ depends on $[Fe^{3+}]$ and pH:

$$p\varepsilon = p\varepsilon_t^\circ + \log \frac{K_{so}/[OH^-]^3}{[Fe^{2+}]_f(1 + \beta_1'[OH^-] + \cdots)} = p\varepsilon_f^\circ + \log \frac{[Fe^{3+}]_f}{[Fe^{2+}]_f}$$

$$p\varepsilon_t^\circ = p\varepsilon_f^\circ + \log \frac{(1 + \beta_1'[OH^-] + \cdots)[OH^-]^3[Fe^{3+}]_f}{K_{so}}$$

Finally, the third branch of the curve in Fig. 9.5 is where both $Fe(OH)_3(s)$ and $Fe(OH)_2(s)$ are present. The free ion activities are given by

$$[Fe^{3+}\}_f = K_{so}/[OH^-]^3$$

$$[Fe^{2+}]_f = K_{so}'/[OH]^2$$

$$p\varepsilon = p\varepsilon_f^\circ + \log \frac{[Fe^{3+}]_f}{[Fe^{2+}]_f} = p\varepsilon_f^\circ + \log\left(\frac{K_{so}}{K_{so}'}\right) - \log[OH^-]$$

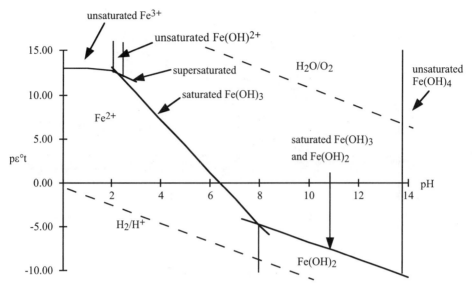

FIGURE 9.6. pH versus pε for the Fe(II)/Fe(III) system.

Note that pε is dependent on pH but not on either of the total iron concentrations, but that $p\varepsilon_t$ defined as above would include terms for both [Fe^{2+}] and [Fe^{3+}].

From Fig. 9.5 it is only a short step to a pε–pH map, as shown in Fig. 9.6 for the Fe(II)/Fe(III) system. The boundaries correspond to an activity of 1; positively charged solution species at lower concentrations than 1 M would be represented by a boundary shifted to more negative values of pε. Areas are labeled with the formula of the species that is present in the highest concentration. In the left portion of the map (pε « 0 at pH = 0; pε « –14 at pH = 14) water is reduced; the rate increases as pε becomes more negative but depends on the hydrogen overvoltage. In the upper right portion (pε » +21 at pH = 0; pε » +7 at pH = 14) water is oxidized. The rate increases as pε becomes more positive but depends on the oxygen overvoltage.

Manganese Redox Reactions

Manganese is an important redox system. In geochemistry, the mineral pyrolusite [MnO_2, Mn(IV)] indicates an oxidizing geological environment; rhodochrosite [$MnCO_3$, Mn(II)] indicates a reducing environment. In analytical chemistry, permanganate [MnO_4^-, Mn(VII)] is a standard oxidizing reagent. MnO_2 is the oxidizing component in the Leclanché dry-cell battery.

Example 13. Show the dependence of the various manganese species on pε if all are in equilibrium with $MnO_2(s)$. Compare with the stoichiometric Example 9

EFFECT OF pH ON POTENTIAL

above, where some reactions are listed and an electron balance is developed. The reduction potentials for four of those reactions are[26,27]:

$$MnO_2 + 4H^+ + 2e^- \rightleftharpoons Mn^{2+} + 2H_2O, \quad E° = +1.23, \, p\varepsilon° = 20.8$$

$$MnO_2 + 4H^+ + e^- \rightleftharpoons Mn^{3+} + 2H_2O, \quad E° = +0.96, \, p\varepsilon° = 16.2$$

$$MnO_4^{2-} + 4H^+ + 2e^- \rightleftharpoons MnO_2 + 2H_2O, \quad E° = +2.27, \, p\varepsilon° = 38.4$$

$$MnO_4^- + 4H^+ + 3e^- \rightleftharpoons MnO_2 + 2H_2O, \quad E° = +1.70, \, p\varepsilon° = 28.7$$

In more basic solutions, the hydroxide complexes of Mn(II) and Mn(III) must be included:

for Mn^{2+}–OH^-: $\log \beta_1 = 3.4$, $\log \beta_4 = 7.7$, $\log K_{so} = -12.8$ at $I = 0$[28]

for Mn^{3+}–OH^-: $\log \beta_1' = 14.6$, $\log \beta_2' = 28.5$ at $I = 3.0$[29]

Additional reactions might include elemental Mn(s) and Mn(OH)$_3$(s), but neither of these is stable in the presence of MnO$_2$.

The concentrations [Mn^{2+}], [Mn^{3+}], [MnO$_4^{2-}$], and [MnO$_4^-$] are obtained from the Nernst equations with the activity of MnO$_2$ and H$_2$O equal to unity and with pH = $-\log[H^+]$. For the first reaction:

$$p\varepsilon = p\varepsilon° + \frac{1}{n}\log\left(\frac{ox}{red}\right) = p\varepsilon° + \tfrac{1}{2}\log\frac{[H^+]^4}{[Mn^{2+}]} = p\varepsilon° - 2pH - \tfrac{1}{2}\log[Mn^{2+}]$$

$$\log[Mn^{2+}] = 2(p\varepsilon° - p\varepsilon) - 4\,pH$$

For the other three, similar manipulations yield

$$\log[Mn^{3+}] = p\varepsilon° - p\varepsilon - 4\,pH$$

$$\log[MnO_4^{2-}] = 2(p\varepsilon - p\varepsilon°) + 4\,pH$$

$$\log[MnO_4^-] = 3(p\varepsilon - p\varepsilon°) + 4\,pH$$

The hydroxide complexes are obtained as in Chapter 7. Note $\log[OH^-] = -pK_w + pH$ ($pK_w = 14.0$)

$$\log[MnOH^+] = \log \beta_1 + \log[Mn^{2+}] - pK_w + pH$$

$$\log[Mn(OH)_4^{2-}] = \log \beta_4 + \log[Mn^{2+}] - 4pK_w + 4\,pH$$

$$\log[Mn(OH)^{2+}] = \log \beta_1' + \log[Mn^{2+}] - pK_w + pH$$

$$\log[Mn(OH)_2^+] = \log \beta_2' + \log[Mn^{2+}] - 2pK_w + 2\,pH$$

These various lines are shown in Figs. 9.7–9.9 for pH = 0, 7, and 14. MnO$_2$, with activity = 1, is a horizontal line at 0 on the log scale. As expected, species of the

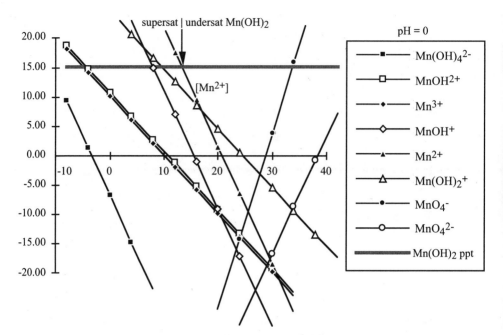

FIGURE 9.7. Logarithmic diagram of manganese species activity in equilibrium with $MnO_2(s)$ at pH = 0. $a_{MnO_2} = 1$ is the horizontal axis. The shaded bar represents the value of $[Mn^{2+}]$ in equilibrium with $Mn(OH)_2(s)$.

lower oxidation states Mn(II) and Mn(III) fall at lower pε values, and their concentrations decrease with increasing pε; species of the higher oxidation states $[MnO_4^{2-}]$ and $[MnO_4^-]$ fall at higher pε, and their concentrations increase with increasing pε. A simple interpretation of the graph of pH = 0 (Fig. 9.7) is that at pε < 18, MnO_2 is converted to Mn^{2+}, from pε = 18 to 25, MnO_2 is converted to $Mn(OH)^{2+}$, from pε = 25 to 28, MnO_2 is the most stable species, and for pε > 28, MnO_2 is converted to MnO_4^-.

The solid phase $Mn(OH)_2$ precipitates when $\log[Mn^{2+}] \geq \log K_{so} - 2\log[OH^-]$. At pH = 0, this occurs at $\log[Mn^{2+}] = +15$, an unrealistically high value; but at pH = 7 (Fig. 9.8), the precipitation limit is at $\log[Mn^{2+}] = +0.8$, and at pH = 14 (Fig. 9.9) it is at $\log[Mn^{2+}] = -13$. Thus the predominant species at pH = 7, in order of increasing pε, are Mn^{2+}, $Mn(OH)_2(s)$, $MnO_2(s)$, MnO_4^-. At pH = 14, the order is $Mn(OH)_4^{2-}$, MnO_2, MnO_4^-.

Figure 9.10 shows a map of the largest concentrations at any value of pH and pε, analogous to Fig. 9.6 for the Fe system.

Mn^{2+} cannot be reduced to Mn(s) under most conditions; hydrogen will be evolved instead. But in practice Mn metal can be prepared by electrolytic deposition from a concentrated manganese sulfate–ammonium sulfate solution at pH = 5.

In acid solutions of low oxidizing power, Mn^{2+} predominates; in basic solutions, solid $Mn(OH)_2$ predominates. Aqueous $Mn(OH)_2$ complex never becomes predomi-

EFFECT OF pH ON POTENTIAL

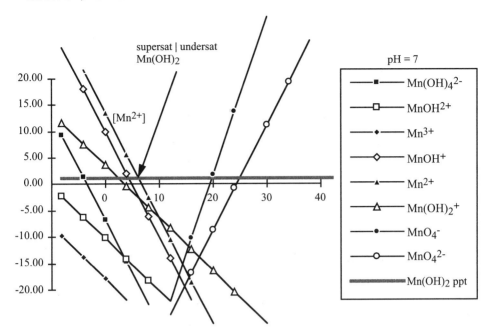

FIGURE 9.8. Logarithmic diagram of manganese species activity in equilibrium with $MnO_2(s)$ at pH = 7. $a_{MnO_2} = 1$ is the horizontal axis. The shaded bar represents the value of $[Mn^{2+}]$ in equilibrium with $Mn(OH)_2(s)$.

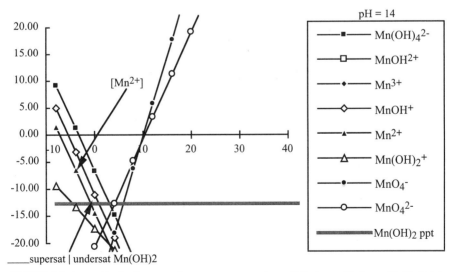

FIGURE 9.9. Logarithmic diagram of manganese species activity in equilibrium with $MnO_2(s)$ at pH = 14. $a_{MnO_2} = 1$ is the horizontal axis. The shaded bar represents the value of $[Mn^{2+}]$ in equilibrium with $Mn(OH)_2(s)$.

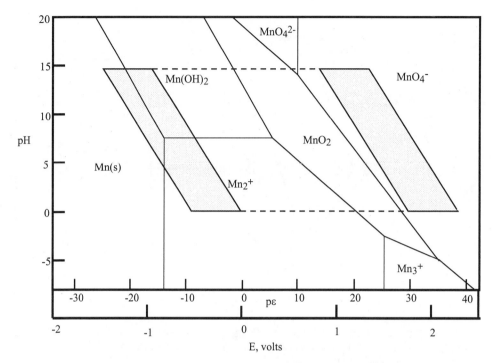

FIGURE 9.10. Predominance-area (pH vs. pε) diagram for the manganese system. Shaded areas represent hydrogen and oxygen overvoltages of 0 to 0.6 volts.

nant; it is restricted to concentrations of less than 10^{-6} molar by precipitation of $Mn(OH)_2(s)$. $MnOH^+$ likewise does not become predominant, but is important near the border between Mn^{2+} and $Mn(OH)_2(s)$. $Mn(OH)_3^-$ becomes important only in very basic solutions.

In solutions of moderate oxidizing power, solid MnO_2 predominates at all pH values. MnO_4^{2-} can be obtained in strong alkalies, and Mn^{3+} is important in highly oxidizing acid solutions. In strongly oxidizing media, MnO_4^- is the predominant species.

Diagrams such as this can give an overall picture of oxidation–reduction systems, but quantitative calculations cannot be done easily because the mass balances and the electron balance are usually not expressible as a simple intersection of two lines.

Influence of Complexing Agents on Fe(II)/Fe(III)

The formal potential of a redox couple, just like any other equilibrium constant, is affected by complex formation between the active species and other species in the solution. We have already defined the formal potential and noted that for Fe^{3+}/Fe^{2+}, $E_t°$ or $p\varepsilon_t°$ decreases with increasing pH, because hydroxide complexes with Fe^{3+} are stronger than those with Fe^{2+}. This section explores some examples of other complexing agents.

As derived above (Eqs. 13–15, p. 332), the free ion concentrations are related to the total concentrations by mass balances:

$$[Fe^{3+}]_t = [Fe^{3+}]_f(1 + \beta_1[L] + \beta_2[L]^2 + \cdots)$$

$$[Fe^{2+}]_t = [Fe^{2+}]_f(1 + \beta_1'[L] + \beta_2'[L]^2 + \cdots)$$

In general, the total concentration $[M]_t$ is always greater than (or equal to) $[M]_f$.

Here L is some ligand, β_n is the formation constant of $Fe(III)L_n$, and β_n' is the formation constant of $Fe(II)L_n$. If $Fe(OH)_3(s)$ precipitates, $[Fe^{3+}] = K_{so}/[OH^-]^3$, and if $Fe(OH)_2$ precipitates, $[Fe^{2+}] = K_{so}'/[OH^-]^2$ (see Figs. 9.5 and 9.6). As long as the complexes are mononuclear, the relation between $p\varepsilon_f°$ and $p\varepsilon_t°$ is independent of the metal ion concentration. For most ligands, the complexes with Fe^{3+} are stronger than those with Fe^{2+}, and hence $p\varepsilon°_t$ in the presence of ligand tends to be lower (more negative) than in a noncomplexing medium at the same pH.[30]

$$p\varepsilon_t° = p\varepsilon_f° + \log\left(\frac{1 + \beta_1'[L] + \beta_2'[L]^2 + \cdots}{1 + \beta_1[L] + \beta_2[L]^2 + \cdots}\right) + \log\frac{\gamma_{3+}}{\gamma_{2+}}$$

Fe^{3+}/Fe^{2+} with EDTA

Example 14. Find $p\varepsilon°_t$ for Fe^{3+}/Fe^{2+} as a function of pH in the presence of 0.01 M EDTA. The relevant equilibrium constants can be found in Table 8.1.

Complex	log(overall formation constant)[31]
HL^{3-}	10.19
H_2L^{2-}	16.32
H_3L^-	19.01
H_4L	21.01
$Fe(III)L^-$	25.1
$Fe(III)HL$	26.4
$Fe(III)OHL^{2-}$	3.73
$Fe(II)L^{2-}$	14.30
$Fe(II)HL^-$	2.8
$Fe(II)OHL^{2-}$	5.23

The mass balances are (charges omitted from simplicity):

$$[L]_t = [L] + [HL] + [H_2L] + [H_3L] + [H_4L] + [Fe(III)L] + [Fe(III)HL]$$
$$+ [Fe(III)OHL] + [Fe(II)L] + [Fe(II)HL] + [Fe(II)OHL] + \cdots$$

$$[Fe^{3+}]_t = [Fe^{3+}] + [FeOH^{2+}] + [Fe(OH)_2^+] + \cdots + [Fe(III)L] + [Fe(III)HL]$$
$$+ [Fe(III)OHL]$$

$$[Fe^{2+}]_t = [Fe^{2+}] + [FeOH^+] + [Fe(OH)_2] + \cdots + [Fe(II)L] + [Fe(II)HL]$$
$$+ [Fe(II)OHL]$$

So long as the complexes are mononuclear, the ratio of each species to the free ion is dependent only on pH and ligand concentration. For example:

$$\frac{[Fe(III)HL]}{[Fe^{3+}]_f} = \beta[H^+][L^{4-}]$$

In the spreadsheet calculation (Table 9.2), the first group of data represents the various complexes containing L; the values are the ratio of the species concentrations to $[L^{4-}]$; for example,

$$\frac{[Fe(III)HL]}{[L^{4-}]} = \beta[H^+][Fe^{3+}]_f, \quad \text{etc.}$$

The second group of data represents the various complexes containing Fe^{3+}, and the third the complexes containing Fe^{2+}. These calculations are iterative, since the first group depends on knowing $[Fe^{3+}]_f$ and $[Fe^{2+}]_f$; calculation of these (in the second and third groups) depends on $[L^{4-}]$ as calculated in the first group.[32]

In Table 9.2 the saturation indices (SI) for $Fe(OH)_3(s)$ and $Fe(OH)_2(s)$ are calculated as a function of pH. SI for $Fe(OH)_3(s)$ is negative (undersaturated) for pH < 9.5 and positive for pH > 9.5. SI for $Fe(OH)_2(s)$ is negative over the entire pH range. $p\varepsilon_t^\circ$, which is equal to $p\varepsilon$ when $[Fe^{3+}]_t = [Fe^{2+}]_t$, can be obtained for the unsaturated solutions from the calculated values of $[Fe^{3+}]_f$ and $[Fe^{2+}]_f$ in Table 9.2:

$$p\varepsilon_t^\circ = p\varepsilon = p\varepsilon_f^\circ + \log[Fe^{3+}]_f - \log[Fe^{2+}]_f$$

Because of the mononuclear nature of the complexes, this $p\varepsilon$ is independent of total dissolved Fe. In the saturated solution, $[Fe^{3+}]_f = K_{so}/[OH^-]^3$ and

$$p\varepsilon = p\varepsilon_f^\circ + \log(K_{so}) - 3\log[OH^-] - \log[Fe^{2+}]_f$$

This value of $p\varepsilon$ depends on $[Fe^{2+}]_f$.[33] These functions are plotted in Fig. 9.11. Note that $p\varepsilon$ decreases with pH at low pH, but then remains essentially constant to pH = 9.5, where precipitation occurs. At that point, the presence of EDTA does not affect Fe^{3+}. Since essentially 100% of Fe^{2+} is present as the EDTA complex at high pH (See Table 9.2 and Fig. 9.11), its contribution to $p\varepsilon$ does not change with pH.

Fe^{3+}/Fe^{2+} with Phenanthroline. The reagent 1,10–phenanthroline is noted for its intensely red complex with Fe^{2+} "ferroin" and has been long used for the colorimetric determination of iron as well as a redox indicator.

EFFECT OF pH ON POTENTIAL

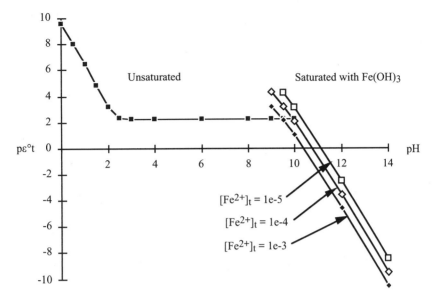

FIGURE 9.11. $p\varepsilon_f^\circ$ for Fe^{3+}/Fe^{2+} in 0.01 M EDTA as a function of pH. In the unsaturated region, $p\varepsilon$ is independent of total dissolved iron.

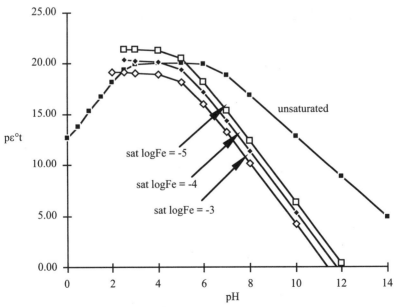

FIGURE 9.12. Redox potential for Fe^{3+}/Fe^{2+} in the presence of 1,10-phenanthroline. In the unsaturated region, $[Fe^{3+}]_t$ and $[Fe^{2+}]_t$ have been taken to be equal. In the saturated solutions, $[Fe^{3+}]_f = K_{so}/[OH^-]^3$.

TABLE 9.2. Calculation sheet for EDTA (Example 14)

pe for Fe3+/Fe2+-EDTA						Iter=	1.00				
	log [Fe2]t=log[Fe3]t =		-3.00								
	log[L]t = -2.00										
	pH		0.00	2.00	4.00	6.00	0.00	8.00	10.00	12.00	
	log OH	=-14+pH	-14.00	-12.00	-10.00	-8.00	0.00	-6.00	-4.00	-2.00	
species	log beta	log sp/[L]	log sp/[L]								
L	0.00		0	0.00	0.00	0.00	0.00	0.00	0.00	0.00	
HL	10.19	=$B$8-pH	10.19	8.19	6.19	4.19	0.19	2.19	0.19	-1.81	
H2L	16.32	=$B$9-2*pH	16.32	12.32	8.32	4.32	0.32	0.32	-3.68	-7.68	
H3L	19.01	=$B$10-3*pH	19.01	13.01	7.01	1.01	-4.99	-4.99	-10.99	-16.99	
H4L	21.01	=$B$11-4*pH	21.01	13.01	5.01	-2.99	-10.99	-10.99	-18.99	-26.99	
Fe3L	25.10	=$B$12+logFe3	18.74	12.32	7.44	3.66	-0.50	1.30	-0.50	-0.90	
Fe3HL	26.40	=$B$13-pH+logFe3	20.04	11.62	4.74	-1.04	-9.20	-5.40	-9.20	-11.60	
Fe3OHL	3.73	=$B$14+pH-14+logFe3	-16.63	-21.05	-23.93	-25.71	-25.87	-26.07	-25.87	-24.27	
Fe2L	14.30	=$B$15+logFe2	11.30	11.27	7.44	3.66	-0.50	1.30	-0.50	-0.90	
Fe2HL	2.80	=$B$16-pH+logFe2	-0.20	-2.23	-8.06	-13.84	-22.00	-18.20	-22.00	-24.40	
Fe2OHL	5.23	=$B$17+pH-14+logFe2	-11.77	-9.80	-11.63	-13.41	-13.57	-13.77	-13.57	-11.97	
log[L]t/[L]		=LOG(10^C60+10^C61+...	21.06	13.40	8.44	4.66	0.50	2.30	0.50	0.10	
log[L]		=IF(Iter=1,logLt-C72,logLt)	-23.06	-15.40	-10.44	-6.66	-2.50	-4.30	-2.50	-2.10	
	Log beta	logsp/[Fe3]	logsp/[Fe3]								
Fe3	0.00	0	0.00	0.00	0.00	0.00	0.00	0.00	0.00	0.00	
Fe3OH	11.81	=$B$24-14+pH	-2.19	-0.19	1.81	3.81	7.81	5.81	7.81	9.81	
Fe3(OH)2	23.30	=$B$25-2*14+2*pH	-4.70	-0.70	3.30	7.30	15.30	11.30	15.30	19.30	
Fe3L	25.10	=$B$26+logL	2.04	9.70	14.66	18.44	22.60	20.80	22.60	23.00	
Fe3HL	26.40	=$B$27-pH+logL	3.34	9.00	11.96	13.74	13.90	14.10	13.90	12.30	
Fe3OHL	3.73	=$B$28+pH-14+logL	-33.33	-23.67	-16.71	-10.93	-2.77	-6.57	-2.77	-0.37	

EFFECT OF pH ON POTENTIAL

log[Fe3]t/[Fe3]		=LOG(10^C76+10^C77+...	3.36	9.78	14.66	18.44	20.80	22.60	23.00	
log[Fe3]		=IF(lter=1,logFet-C83,logFet)	-6.36	-12.78	-17.66	-21.44	-23.80	-25.60	-26.00	
SI (pKso=38.8)		=logFe3+3*pH-3*14+38.8	-9.56	-9.98	-8.86	-6.64	-3.00	1.20	6.80	
log[Fe3]sat		=-38.8-3*pH+3*14						-26.80	-32.80	
	Log beta	logsp/[Fe2]	logsp/[Fe2]							
Fe2	0.00	0	0.00	0.00	0.00	0.00	0.00	0.00	0.00	
Fe2OH	4.70	=$B$37+pH-14	-9.30	-7.30	-5.30	-3.30	-1.30	0.70	2.70	
Fe2(OH)2	7.40	=$B$38+2*pH-2*14	-20.60	-16.60	-12.60	-8.60	-4.60	-0.60	3.40	
Fe2L	14.30	=$B$39+logL	-8.76	-1.10	3.86	7.64	10.00	11.80	12.20	
Fe2HL	2.80	=$B$40-pH+logL	-20.26	-14.60	-11.64	-9.86	-9.50	-9.70	-11.30	
Fe2OHL	5.23	=$B$41+pH-14+logL	-31.83	-22.17	-15.21	-9.43	-5.07	-1.27	1.13	
log[Fe2]t/[Fe2]		=LOG(10^C89+10^C90+10^C	0.00	0.03	3.86	7.64	10.00	11.80	12.20	
log[Fe2]		=IF(lter=1,logFet-C96,logFet)	-3.00	-3.03	-6.86	-10.64	-13.00	-14.80	-15.20	
SI(pK'so=14.4)		=logFe2+2*pH-2*14+14.4	-16.60	-12.63	-12.46	-12.24	-10.60	-8.40	-4.80	
With [Fe3+]t = [Fe2+]t, pe = pe°t = pe°f + log{[Fe3]/[Fe2]}										
pH		=pH	0.00	2.00	4.00	6.00	8.00	10.00	12.00	
pe°t		=13+logFe3-logFe2	9.64	3.26	2.20	2.20	2.20	2.20		
pe°t (sat)		=13+logFe3sat-logFe2						1.00	-4.60	

Unlike EDTA and many other chelating agents, complexes of phenanthroline, methyl-substituted phenanthrolines, and 2,2′-bipyridyl are stronger with Fe^{2+} than with Fe^{3+}. Research on the thermodynamic and kinetic properties of these complexes began in the late 1940s[34] and has continued to the present. Some equilibrium constant values are given in Table 9.3.

The calculation procedure is identical to that for EDTA, and the results are displayed in Fig. 9.12. Note that $p\varepsilon°_t$ for the unsaturated solution increases with pH, then goes through a flat maximum, which is intercepted by the curves for solutions saturated with $Fe(OH)_3$; the lower the total Fe content, the higher the pH at which saturation occurs. At very high pH, the dinuclear ferric phenanthroline complexes $Fe_2L_4(OH)_2^{4+}$ and $Fe_2L_4(OH)^{3-}$ have been reported,[36] but no values for the dimerization of FeL_2 to Fe_2L_4 are available.

TABLE 9.3. Formation constants for 1,10-phenanthroline in 0.1 M KCl[35]

Species	log of formation constant
HL	4.93
H_2L	1.8
$Fe^{2+}L$	5.85
$Fe^{2+}L_2$	11.15
$Fe^{2+}L_3$	21.1
$Fe^{3+}L$	6.5
$Fe^{3+}L_2$	11.4
$Fe^{3+}L_3$	14.1

Redox Titrations

Titrations based on oxidation–reduction reactions have long been used as accurate methods of analysis. Permanganate is one of the classic oxidants because it acts as its own indicator. The titration is carried out until the violet color of MnO_4^- is permanent. Another option is to use an organic indicator such as methylene blue or Ferroin, whose oxidized and reduced forms are of different color, and for which equilibrium is rapidly established. However, potentiometric methods of end-point detection are generally applicable to redox titrations, and are usually more precise than indicators.

EFFECT OF pH ON POTENTIAL

In the chapter on acid–base titrations, the estimation of titration error was discussed in detail; this requires a quantitative prediction of the shape of the titration curve. Once the titration curve is known, then the difference in potential between the end point (the experimental value) and the equivalence point (the theoretical value) can be converted to a difference in volume, and hence to a percentage error.

The general principles of redox titration curves are very similar to those of acid–base titrations, the primary difference being that the basic equation is the electron balance rather than the charge balance.

Example 15. The titration of Fe^{2+} with Ce^{4+} in acid solutions is a well-established method:

$$Fe^{2+} + Ce^{4+} \rightleftharpoons Fe^{3+} + Ce^{3+}$$

If V mL of C molar Ce^{4+} is added to V_o mL of C_o molar Fe^{2+}, the mass balances are

$$[Ce^{3+}] + [Ce^{4+}] = [Ce^{3+}](1 + 10^{p\varepsilon - 24.3}) = CV/(V + V_o) \quad (a)$$

$$[Fe^{2+}] + [Fe^{3+}] = [Fe^{3+}](1 + 10^{-p\varepsilon + 11.5}) = C_o V_o/(V + V_o) \quad (b)$$

At $I = 0$, $E°$ for Ce^{4+}/Ce^{3+} is +1.72 V ($p\varepsilon° = 29.1$),[37] but in 1 M sulfuric acid, the formal potential is +1.44 V ($p\varepsilon = 24.3$).[38]

$$Ce^{4+} + e^- \rightleftharpoons Ce^{3+}$$

$$p\varepsilon = 24.3 + \log\left(\frac{[Ce^{4+}]}{[Ce^{3+}]}\right) \quad \text{or} \quad [Ce^{4+}] = [Ce^{3+}]10^{p\varepsilon - 24.3} \quad (c)$$

The standard potential of Fe^{3+}/Fe^{2+} at $I = 0$ is +0.771 V[39] ($p\varepsilon = 13.0$), but the formal potential of Fe^{3+}/Fe^{2+} in 0.5 M H_2SO_4 (approx. $I = 1$) is +0.67 V[40] ($p\varepsilon = 11.4$).

$$Fe^{3+} + e^- \rightleftharpoons Fe^{2+}$$

$$p\varepsilon = 11.5 + \log\left(\frac{[Fe^{3+}]}{[Fe^{2+}]}\right) \quad \text{or} \quad [Fe^{3+}] = [Fe^{2+}]10^{p\varepsilon - 11.5} \quad (d)$$

The final equation is the electron balance; for every mole of Fe oxidized, a mole of Ce is reduced:

$$[Fe^{3+}] = [Ce^{3+}] \quad (e)$$

The problem is defined by five equations (a)–(e) for five unknowns: $[Ce^{3+}]$, $[Ce^{4+}]$, $[Fe^{2+}]$, $[Fe^{3+}]$, and V. Most of the work has already been done above, and substitution in the electron balance gives

$$\left(\frac{C_oV_o}{V+V_o}\right)\left(\frac{1}{1+10^{-p\varepsilon+11.5}}\right) = \left(\frac{CV}{V+V_o}\right)\left(\frac{1}{1+10^{p\varepsilon-24.3}}\right)$$

$$\left(\frac{CV}{C_oV_o}\right) = \left(\frac{1+10^{p\varepsilon-24.3}}{1+10^{-p\varepsilon+11.5}}\right)$$

This equation is plotted in Fig. 9.13 for $C = C_o = 0.1$ and $V_o = 50$, by choosing values of pε and calculating V. It is similar to the curve for a strong acid–strong base titration.

Equivalence–Point Potential for Symmetric Titrations. The general titration curve for a one-electron oxidation–reduction process without involvement of H$^+$ is derived by a process similar to the above equations, only with general symbols: pε°(ox) for the formal potential of the oxidizing agent and pε° (red) for the formal potential of the reducing agent.

$$\left(\frac{CV}{C_oV_o}\right) = \left(\frac{1+10^{p\varepsilon-p\varepsilon°(ox)}}{1+10^{-p\varepsilon+p\varepsilon°(red)}}\right)$$

At the equivalence point, $CV = C_oV_o$ and

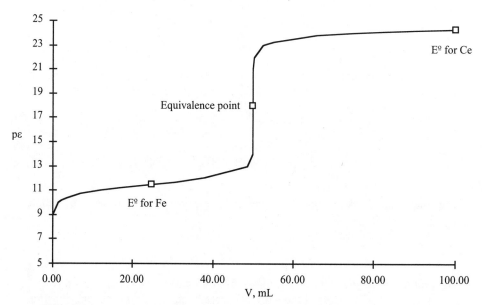

FIGURE 9.13. pε as a function of Ce^{4+} volume for the titration of 50 mL 0.1 M Fe^{2+} with 0.1 M Ce^{4+}. Note the analogy with a strong acid–strong base titration.

$$p\varepsilon - p\varepsilon°(\text{ox}) = -p\varepsilon + p\varepsilon°(\text{red})$$

$$p\varepsilon_{\text{equiv}} = \tfrac{1}{2}[p\varepsilon°(\text{ox}) + p\varepsilon°(\text{red})]$$

Alternatively,

$$E_{\text{equiv}} = \tfrac{1}{2}[E°(\text{ox}) + E°(\text{red})]$$

For the Fe/Ce titration in Example 15, $p\varepsilon_{\text{equiv}} = 17.9$ or $E_{\text{equiv}} = 1.06$ volts. If one or both of the half-reactions involves H^+, there is an additional term involving the pH. If the oxidizing agent contains some of its reduced form, or if the reducing agent contains some of its oxidized form, the electron balance has additional terms representing these impurities, and the equivalence point potential is shifted slightly. Alternatively, the impurity concentrations can be expressed in terms of positive or negative increments of titrant volume, a shift along the titration curve.

Unsymmetrical Titrations: Ferrous–Dichromate.[41] Another classic redox titration is the oxidation of Fe^{2+} with dichromate $Cr_2O_7^{2-}$. This is an unsymmetrical titration, and also involves H^+; the Fe couple transfers one electron, but the reduction of dichromate to Cr^{3+} involves six electrons.[42]

Example 16. If V_o mL of C_o M (0.1 M) Fe^{2+} is titrated with V mL of C M(0.01667) $Cr_2O_7^{2-}$, the mass balances (counting Fe atoms and Cr atoms) in 1 M $HClO_4$ (pH = 1) are similar to Example 15:

$$[Cr^{3+}] + 2[Cr_2O_7^{2-}] = CV/(V + V_o) \quad \text{(a)}$$

$$[Fe^{2+}] + [Fe^{3+}] = C_oV_o/(V + V_o) \quad \text{(b)}$$

The reduction of dichromate proceeds primarily to chromic ion:

$$Cr_2O_7^{2-} + 14\ H^+ + 6\ e^- \rightleftharpoons 2\ Cr^{3+} + 7\ H_2O$$

$$p\varepsilon = p\varepsilon°_{Cr} + \tfrac{1}{6}\log\left(\frac{[Cr_2O_7^{2-}][H^+]^{14}}{[Cr^{3+}]^2}\right) + \tfrac{1}{6}\log\left(\frac{[\gamma_{2-}][\gamma_+]^{14}}{[\gamma_{3+}]^2}\right)$$

If the activity coefficients vary with the square of the charge as in the Davies equation, the last term can be approximated by $(1/6)\log(\gamma_+^{18}/\gamma_+^{18}) = 0$; therefore, $p\varepsilon°$ is approximately independent of ionic strength. At $I = 0$ or 1, $E°_{Cr} = 1.36$ V[43] or $p\varepsilon° = 23.0$. Rearranging the equation gives

$$[Cr^{3+}] = [Cr_2O_7^{2-}]^{1/2}\ [H^+]^7 10^{-3(p\varepsilon - p\varepsilon°_{Cr})} \quad \text{(c)}$$

$$Fe^{3+} + e^- \rightleftharpoons Fe^{2+}$$

$$p\varepsilon = p\varepsilon°_{Fe}\ (I = 0) + \log\left(\frac{[Fe^{3+}]}{[Fe^{2+}]}\right) + \log\left(\frac{[\gamma_{3+}]}{[\gamma_{2+}]}\right)$$

The standard potential of Fe^{3+}/Fe^{2+} at $I = 0$ is $+0.771$ V; $p\varepsilon° = 13.0$. At $I = 1$, $p\varepsilon° = 12.25$, including activity coefficients in $p\varepsilon°$.[44]

$$p\varepsilon = p\varepsilon°_{Fe} \ (I = 0) + \log\left(\frac{[Fe^{3+}]}{[Fe^{2+}]}\right)$$

$$[Fe^{3+}] = [Fe^{2+}] \ 10^{p\varepsilon - p\varepsilon°_{Fe}} \qquad (d)$$

The electron balance for the reaction is obtained by noting that each Cr reduced uses three electrons, and each Fe oxidized produces one. This condition applies throughout the titration.

$$[Fe^{3+}] = 3[Cr^{3+}] \qquad (e)$$

Thus there are five equations (a)–(e) in five unknowns: $[Fe^{2+}]$, $[Fe^{3+}]$, $[Cr^{3+}]$, $[Cr_2O_7^{2-}]$, and V, but because of the nonlinear terms in the chromium potential (Eq. c), the evaluation of this titration curve is much more difficult than for the symmetrical titration of Example 15. The simplest approach is to divide the curve into a region before and a region after the equivalence point.

In the region before the equivalence point, essentially all $Cr_2O_7^{2-}$ is reduced to Cr^{3+}. From (a) and (e),

$$[Cr^{3+}] + \cdots = \frac{CV}{V + V_o} = \tfrac{1}{3}[Fe^{3+}]$$

From the Fe mass balance (b),

$$[Fe^{2+}] = \frac{C_o V_o}{V + V_o} - \frac{3(CV)}{V + V_o}$$

These two equations for $[Fe^{3+}]$ and $[Fe^{2+}]$, when substituted in (d), give $p\varepsilon$ as a function of V. This is plotted as the lower branch of the curve in Fig. 9.14

$$p\varepsilon = p\varepsilon° + \log(3CV) - \log(C_o V_o - 3CV)$$

The equivalence point occurs when

$$CV = \tfrac{1}{3} C_o V_o,$$

Substituting this in (a) and combining the result with (b) and (c) gives the alternative equivalence point condition

$$[Fe^{2+}] = 6[Cr_2O_7^{2-}] \qquad (f)$$

To eliminate as many variables as possible, take

EFFECT OF pH ON POTENTIAL

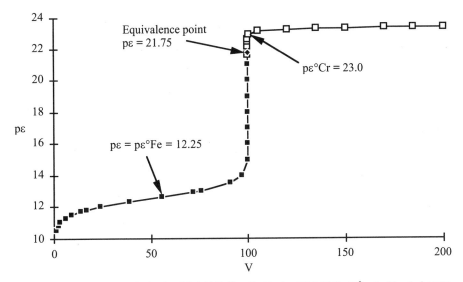

FIGURE 9.14. Titration of 50 mL of 0.1 M Fe^{2+} with V mL of 0.1 M $Cr_2O_7^{2-}$ at pH = 1. Lower branch is based on Fe potential; upper branch is based on Cr potential.

$$7p\varepsilon_{ep} = 6p\varepsilon_{Cr,ep} + p\varepsilon_{Fe,ep} = 6p\varepsilon^\circ_{Cr} + p\varepsilon^\circ_{Fe} + \log\left(\frac{[Fe^{3+}][Cr_2O_7^{2-}][H^+]^{14}}{[Fe^{2+}][Cr^{3+}]^2}\right)_{ep} + \log\left(\frac{\gamma_{3+}\gamma_2\gamma_+^{14}}{\gamma_2\gamma_{3+}^2}\right)$$

Use (e) and (f) to get

$$7p\varepsilon_{ep} = 6p\varepsilon^\circ_{Cr} + p\varepsilon^\circ_{Fe} + \log\left(\frac{3[Cr^{3+}][Cr_2O_7^{2-}][H^+]^{14}}{6[Cr_2O_7^{2-}][Cr^{3+}]^2}\right) + \log\left(\frac{\gamma_{3+}\gamma_2\gamma_+^{14}}{\gamma_2\gamma_{3+}^2}\right)$$

$$p\varepsilon_{ep} = 21.5 - 2pH - \tfrac{1}{7}\log\{2[Cr^{3+}]\gamma_{3+}\}_{ep}$$

$[Cr^{3+}]$ is approximately $C_oV_o/(V + V_o) = 0.05$, and the last term is approximately 0.25. This point, $p\varepsilon_{ep} = 21.75$; noted on Fig. 9.14.

After the equivalence point, $Cr_2O_7^{2-}$ is in excess and $[Fe^{2+}]$ is negligible. $[Cr^{3+}]$ is the amount produced by the first part of the titration, which is equal to $\tfrac{1}{3}[Fe^{3+}]$ by the electron balance.

$$2[Cr_2O_7] = \frac{CV - \tfrac{1}{3}C_oV_o}{V + V_o}$$

$$[Cr^{3+}] = \tfrac{1}{3}\frac{C_oV_o}{V + V_o}$$

$$p\varepsilon = p\varepsilon^\circ_{Cr} + \tfrac{1}{6}\log\left(\frac{[Cr_2O_7^{2-}][H^+]^{14}}{[Cr^{3+}]^2}\right)$$

This function is plotted as the upper branch of the curve in Fig. 9.14. Note that the equivalence point does not occur at the midpoint of the curve (as was the case with the symmetric titration), and that the curvature is much sharper in the region after the equivalence point. Note also that the equivalence point pε depends on pH and on the ionic strength of the medium as well as on the reagent concentrations and dilution factors.

In general the equivalence point of an unsymmetrical titration occurs approximately where

$$p\varepsilon^\circ_{ep} = \frac{m p\varepsilon^\circ_{ox} + n p\varepsilon^\circ_{red}}{m + n}$$

$$E^\circ_{ep} = \frac{m E^\circ_{ox} + n E^\circ_{red}}{m + n}$$

but, as can be seen from Example 16, if there are hydrogen ions involved in the reactions, there will be additional terms in pH, and if one or more species is polynuclear, there will be a dependence on the total concentration.

Another point to be noted is that in an unsymmetrical titration, the inflection point is somewhat closer to the center of the titration curve than is the equivalence point. This phenomenon was seen in the discussion of precipitation titrations, and should be borne in mind when automatic titrators are used to produce a derivative output.[45]

Can There Be a Unique Redox Potential in an Aqueous Solution?[46]

The amazing success of pH as a master variable for acid–base reactions has tempted geochemists and environmental scientists to propose that there might be a redox variable—typically called E_H and measured by a platinum electrode immersed in the solution—which applies to all redox reactions in an aqueous system. As pointed out several times, however, the rate of redox reactions is generally much slower than that of acid–base reactions, and so a failure to reach equilibrium with respect to all redox couples is the rule rather than the exception.

Some relatively fast redox reactions are the basis for analytical methods and electrochemical cells: H^+/H_2, Ag^+/Ag or $AgCl/Ag$, Zn^{2+}/Zn, Cu^{2+}/Cu, Fe^{3+}/Fe^{2+} have all been discussed in this book. We have already noted that the $Cr_2O_7^{2-}/Cr^{3+}$ couple does not have a very reproducible potential, but that the titration with Fe^{2+} works because the Fe^{3+}/Fe^{2+} couple is relatively fast.

But some of the most important environmental redox reactions are very slow: O_2/H_2O, SO_4^{2-}/H_2S, NO_3^-/NH_4^+, CO_2/CH_4, as well as most organic reactions. These do not take place at a significant rate unless they are catalyzed by enzymes or microbial activity.

At the surface of a platinum electrode in an aqueous solution, many reactions take place at different rates, catalyzed by the platinum. For example, in a solution containing 0.02 M Fe^{2+} and 0.02 M Fe^{3+}, the principal reactions are:

$$Fe^{3+} + e^- \rightarrow Fe^{2+}, \quad \text{reduction (cathodic current)}$$

$$Fe^{2+} \rightarrow Fe^{3+} + e^-, \quad \text{oxidation (anodic current)}$$

and when these are in balance, the potential of the electrode is determined by the Fe^{2+}/Fe^{3+} redox couple, $E = E° = +0.77$ volts or $p\varepsilon = +13.0$. Typical data for the kinetics are expressed in terms of an exchange current i_o, potentials, and concentrations:

$$i_c = -i_o[Fe^{3+}] \exp[(E° - E)F/RT] \tag{16}$$

$$i_a = i_o[Fe^{2+}] \exp[(E - E°)F/RT] \tag{17}$$

The net current is $i = i_c + i_a$, and if $[Fe^{3+}] = [Fe^{2+}]$, $i = 0$ when $E = E°$. This is shown in Fig. 9.15.

If $[Fe^{2+}]$ and $[Fe^{3+}]$ are not equal, the current–potential curves have different shapes, and the point where i equals zero shifts: to higher potential if $[Fe^{3+}] > [Fe^{2+}]$, and to lower potential if $[Fe^{2+}] > [Fe^{3+}]$. An example of this is shown in Fig. 9.16. Note that the larger concentration of Fe^{2+} produces a larger anodic (oxidizing) current than in Fig. 9.15. The cathodic (reducing) current is unchanged; hence the potential at which $i = 0$ shifts from $E = 0.77$ to a lower value, $E = 0.75$.

As the concentration of the Fe^{2+} and Fe^{3+} ions is decreased, the currents from this redox couple decrease proportionally, and other reactions become important.

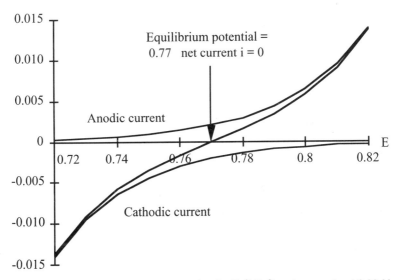

FIGURE 9.15. Cathodic and anodic currents for the Fe^{2+}/Fe^{3+} redox couple at 0.02 M. i_o is assumed to be 0.1 A/cm². The net current is zero at the equilibrium potential.

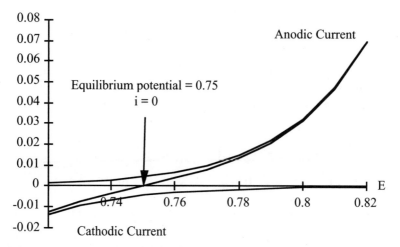

FIGURE 9.16. Cathodic and anodic currents for the Fe^{2+}/Fe^{3+} redox couple with $[Fe^{2+}] = 0.1$ M and $[Fe^{3+}] = 0.02$ M. i_o is assumed to be 0.1 A/cm². The net current is zero at the equilibrium potential $E = 0.75$ V.

Here are some possible reactions at a platinum electrode in a solution in contact with the air:

$$O_2 + 2H^+ + 2e^- \rightarrow H_2O_2, \quad \text{reduction}$$

$$CH_2O + H_2O \rightarrow CO_2 + 4H^+ + 4e^-, \quad \text{oxidation}$$

Here "CH_2O" represents a carbohydratelike organic impurity. At each potential, each reaction has a rate determined by the concentration of reactants and by the degree of catalysis provided by the platinum surface. Because of the low reactant concentrations, the current each reaction can sustain is limited by diffusion, and hence will be constant over a range of potentials. For example, the anodic current of Fe^{2+} oxidation does not increase indefinitely as Figs. 9.15 and 9.16 would imply, but reaches a plateau of potentials slightly higher than the equilibrium potential. Likewise, "CH_2O" oxidizes at a diffusion-limited current (0.0001 in Fig. 9.17) at potentials more positive than about −0.5 V. The sum of cathodic and anodic currents is negative at negative potentials and positive at positive potentials. Between these two extremes is a "mixed potential" where $i = 0$. In general, this is not the equilibrium potential of any of the reactions. A schematic illustration of these factors is shown in Fig. 9.17. The mixed potential is highly sensitive to the relative amounts of the trace reactants and their rates of reaction at the electrode.

Additional reactions could result from trace metals or other trace organics:

$$H_2O_2 + 2H^+ + 2e^- \rightarrow 2H_2O, \quad \text{reduction}$$

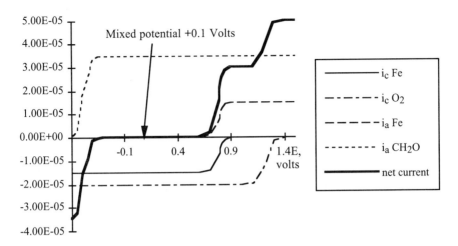

FIGURE 9.17. Schematic diagram of contributions to the potential of a platinum electrode in dilute solution containing Fe^{2+}, Fe^{3+}, O_2, and an oxidizable organic material. Calculations were made from Eqs. (8) and (9) with an additional term to represent diffusion limitation.[50]

$$2H_2O \rightarrow H_2O_2 + 2H^+ + 2e^-, \quad \text{oxidation}$$
$$2H^+ + 2e^- \rightarrow H_2, \quad \text{reduction}$$
$$CO_2 + 4H^+ + 4e^- \rightarrow CH_2O + H_2O, \quad \text{reduction}$$
$$As(V) + 2\ e^- \rightarrow As(III), \quad \text{reduction}$$
$$Mn^{2+} + 2H_2O \rightarrow MnO_2(s) + 4H^+ + 2e^-, \quad \text{oxidation}$$

Lindberg and Runnels[47] surveyed many analyses for E_H and chemical composition of groundwater field samples. From 150,000 samples in the U.S. Geological Survey WATSTORE database, they selected 507 acceptably complete and accurate groundwater analyses. 104 more analyses came from the literature. The value of E_H obtained from the field measurements was compared with the value computed from analytical data for various redox couples. Instead of the expected linear correlation, there was essentially no correlation between the two, as shown in Fig. 9.18.

A prudent approach to measuring redox conditions in an environmental sample, therefore, would involve analysis for all relevant oxidized and reduced forms as well as the measurement of a platinum electrode potential. Examples of reliable measures of redox condition are oxygen partial pressure (in aerobic systems), hydrogen sulfide or methane partial pressure (in anaerobic systems), Fe(III) and Fe(II) or MnO_2 and Mn^{2+} in systems where the concentrations of these components are relatively high. In all cases where platinum electrode data are used to estimate redox condition, the reversibility of the reactions taking place at the electrode needs to be confirmed.

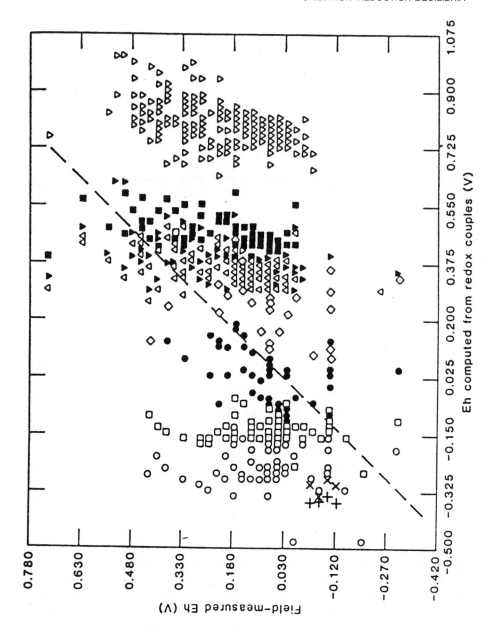

PROBLEMS

Refer to Table 9.1 for redox potential values.

1. Combine the two mercury half-reactions in Table 9.1:

	$E°$	$p\varepsilon°$
$Hg_2^{2+} + 2\,e^- \rightleftharpoons 2Hg$	+0.796	+13.46
$Hg^{2+} + 2e^- \rightleftharpoons Hg$	+0.911	+15.4

   to obtain the redox potential for

   $$2Hg^{2+} + 2e^- \rightleftharpoons Hg_2^{2+}$$

2. Calculate the equilibrium concentrations in solution when 100 mL of 0.010 M $Hg_2(NO_3)_2$ is mixed with 100 mL of 0.010 M $Fe(NO_3)_3$.

3. Calculate the equilibrium concentrations of $[Fe^{3+}]$, $[Fe^{2+}]$, $[Hg_2^{2+}]$, and $[Hg^{2+}]$ in solution when 100 mL of 0.010 M $Fe(NO_3)_3$ is placed in contact with liquid Hg.

4. Draw a logarithmic concentration diagram like Fig. 9.3 showing lines for Cu(s), Cu^+, and Cu^{2+}. Take the total concentration of dissolved Cu to be 0.01 M.

5. Draw a logarithmic diagram showing the ratio of various species activities to Cu(s).

6. The various oxidation states of vanadium are related by the potentials

	$E°$	$p\varepsilon°$
$V^{2+} + 2e^- \rightleftharpoons V(s)$	−1.13	−19.11
$V^{3+} + e^- \rightleftharpoons V^{2+}$	−0.255	−4.31
$VO^{2+} + 2H^+ + e^- \rightleftharpoons V^{3+} + H_2O$	+0.337	+5.70
$VO_2^+ + 2H^+ + e^- \rightleftharpoons VO^{2+} + H_2O$	+1.00	+16.91

FIGURE 9.18. Field values of E_H measured with a platinum electrode versus E as calculated from various redox couples. The dashed line represents the correlation if equilibrium were established. (Reprinted with permission from R. Lindberg and D. W. Runnells.[47] © 1984 American Association for the Advancement of Science.

Symbol	Redox couple
◇	Fe^{3+}/Fe^{2+}
▽	$O_{2\,aq}/H_2O$
○	HS^-/SO_4^{2-}
□	$HS^-/S_{rhombic}$
■	NO_2^-/NO_3^-
▼	NH_4^+/NO_3^-
△	NH_4^+/NO_2^-
+	$CH_{4\,aq}/HCO_3^-$
×	$NH_4^+/N_{2\,aq}$
●	$Fe^{2+}/Fe(OH)_{3(s)}$

Draw a logarithmic diagram showing the activities of the various species in a solution containing 0.01 M total vanadium at pH = 2.

7. Construct pH–potential diagrams like Figs. 9.7–9.10 for the vanadium system.

8. Calculate the formal potential of the reaction Cu(II) + Cl$^-$ + e^- ⇌ CuCl(s) in 1 M HCl.

   Formation constants at $I = 1.0$ are[48]

CuCl(s) ⇌ Cu$^+$ + Cl$^-$	log K_{so} = –6.4
Cu$^+$ + Cl$^-$ ⇌ CuCl(aq)	log β'_1 = +1.4
Cu$^+$ + 2 Cl$^-$ ⇌ CuCl$_2^-$	log β'_2 = +5.8
Cu$^+$ + 3 Cl$^-$ ⇌ CuCl$_3^-$	log β'_2 = +5.3
Cu^{2+} + Cl$^-$ ⇌ CuCl$^+$	log β_1 = –0.2
Cu^{2+} + 2Cl$^-$ ⇌ CuCl$_2$	log β_2 = –0.5

9. Calculate the curve of pε versus V for the titration of 50 mL of 0.10 M Fe^{2+} with 0.10 M MnO$_4^-$ (reduced to Mn^{2+}) in 1.0 M H$_2$SO$_4$. The formal potentials are given in the literature to be $E°_{Fe}$ = +0.68 V (pε = 11.5), $E°_{Mn}$ = +1.51 V (pε = 25.5). The end point is the first sign of the purple permanganate color, at approximately [MnO$_4^-$] = 10^{-5} M.

10. Calculate the curve of pε versus V for the titration of 50 mL of 0.1 M Cu^{2+} with Cr^{2+} in 6 M HCl. Formal potentials are

Cu(II) + e^- ⇌ Cu(I)	$E°$ = +0.45	pε° = 7.6
Cu(I) + e^- ⇌ Cu(s)	$E°$ = 0.0	pε° = 0.0
Cr^{3+} + e^- ⇌ Cr^{2+}	$E°$ = –0.38	pε° = 6.4

    Here Cu(II) represents the sum of all the copper complexes with oxidation state +2, and Cu(I) represents the sum of all the copper complexes with oxidation state +1. Does Cu(s) precipitate?

11. Using the potentials given in Problem 6, calculate the stepwise titration curve of V^{2+} to its higher oxidation states with a strong oxidizing agent, such as MnO$_4^-$.[49]

NOTES

1. In some nonaqueous solvents such as NH$_3$, solvated electrons can reach moderate concentrations, producing the classic blue solution.
2. See L. Pauling, *The Nature of the Chemical Bond*, 3d ed., Cornell Univ. Press, 1960, pp. 79–107.
3. +2 (for C) – 3 (for N) = –1 (net charge).
4. For an extensive discussion on how oxidation and reduction occur in natural waters, see W. Stumm and J. J. Morgan, *Aquatic Chemistry*, 3d ed., Wiley, NY, 1996, pp. 464–98.
5. The decomposition of MnO$_4^-$ is catalyzed by the presence of MnO$_2$, so that if there is

initially present a trace of organic matter, which reacts with MnO_4^- to give MnO_2, the rate of decomposition of permanganate will be much greater. For this reason it is desirable to use very pure water and very clean glassware in preparing standard permanganate solutions for volumetric analysis.

6. See, for example, A. Bard, R. Parsons, and J. Jordan, *Standard Potentials in Aqueous Solution*, M. Dekker, NY, 1985, 834 pp; M. S. Antelman and F. J. Harris, *The Encyclopedia of Chemical Electrode Potentials*, Plenum, NY, 1982, 288 pp. A classic work is W. M. Latimer, *The Oxidation States of the Elements and Their Potentials in Aqueous Solutions*, 2nd ed. (*Oxidation Potentials*, for short), Prentice-Hall, Englewood Cliffs, NJ, 1952, 392 pp. Although the potentials listed by Latimer have the opposite sign from the now-standard reduction potentials, this compilation was the standard reference work on the topic for decades.

7. D. D. Wagman et al., *The NBS Tables of Chemical Thermodynamic Properties*. American Chemical Society and American Institute of Physics for the National Bureau of Standards, Washington, DC, 1982, 392 pp; H. E. Barner and R. V. Scheuerman, *Handbook of Thermochemical Data for Compounds and Aqueous Species*, Wiley, New York, 1978, 156 pp.

8. L. G. Sillén and A. E. Martell, *Stability Constants*, Special Publications No. 17 (1964) and 25 (1971), The Chemical Society, London. Table 1 in each volume gives oxidation–reduction reactions in terms of electrode reduction potential as well as the equilibrium constant for reaction with electrons.

9. See standard works such as G. N. Lewis and M. Randall, *Thermodynamics*, 2nd ed., revised by K. S. Pitzer and L. Brewer, McGraw-Hill, 1961.

10. The cell developed by John F. Daniell in 1836 gave a steady current at 1.08 volts and was used to power railroad signal lights. The design consists of a glass jar with a heavy zinc electrode that dissolved, and a fan of copper below it on which copper crystals were plated. Because $ZnSO_4$ solution is less dense than $CuSO_4$ solution, no separator was necessary—the low-resistance liquid junction was maintained by gravity (hence the name "gravity cell"). Recharging the cell involved simply replacing the zinc electrode and adding copper sulfate crystals to the electrolyte. (*Encyclopedia Britannica*, "Battery," Vol. 3, p. 283 of 1967 ed.)

11. See, for example: G. N. Lewis and M. Randall, *Thermodynamics*, 2nd ed., revised by K. S. Pitzer and L. Brewer, McGraw-Hill, 1961, pp. 354–72; R. G. Bates, *Determination of pH*, 2nd ed., Wiley, NY, 1973, pp. 1–14; more details will be found in R. G. Bates, Part 1, Vol. 1, Chapter 9 of *Treatise on Analytical Chemistry*, I. M. Kolthoff and P. J. Elving, eds., Interscience Publishers, NY, 1959.

12. W. Nernst, *Theoretical Chemistry* (translated from the tenth German edition), Macmillan, New York, 1923. See standard works on thermodynamics (given above); also discussion of pH measurement in Chapter 2.

13. Or whenever the activities of Zn^{2+} and Cu^{2+} are equal. In dilute solutions $\gamma_{Zn^{2+}} \approx \gamma_{Cu^{2+}}$ but in concentrated solutions of equal concentrations, the activities may not be equal. In any case, the major factor affecting E is the ratio of Zn^{2+} to Cu^{2+} in solution.

14. An equivalent assumption is that the standard free energy of formation of the aqueous hydrogen ion is zero at all temperatures.

15. Sometimes you may see the ratio written {red}/{ox}, as Eq. (5) would suggest for a reduction. In such a case, the sign of the log term should be negative (as in Eq. 5) rather than positive. Another confusion results from Latimer's choice to write reactions as

oxidations in his classic book. Oxidation potentials have the opposite sign from the reduction potentials of the IUPAC convention.

16. Selected from A. Bard, R. Parsons, and J. Jordan, *Standard Potentials in Aqueous Solution*, M. Dekker, NY, 1985, 834 pp. $p\varepsilon° = E°F/(2.3RT) = 16.91\ E°$ at 25°C.
17. Simply subtracting potentials will give the right answer only if you are dealing with two half-reactions having the same number of electrons.
18. L. G. Sillén, "Redox diagrams," *J. Chem. Ed.* **29**:600, 1952.
19. W. Stumm and J. J. Morgan, *Aquatic Chemistry*, 3d ed., Wiley, New York, 1996, p. 430.
20. $p\varepsilon = 13.03 + \log\{[Fe^{3+}]\gamma_{3+}/[Fe^{2+}]\gamma_{2+}\}$. If the anion is ClO_4^-, $I = 0.3$ when Fe^{2+} is dominant and $I = 0.6$ when Fe^{3+} is dominant. If the Davies equation is used, with $I = 0.5$ for simplicity, $\log \gamma_{3+} = 9(-0.16)$ and $\log \gamma_{2+} = 4(-0.16)$. Hence $p\varepsilon = 13.03 + \log\{[Fe^{3+}]\gamma_{3+}/[Fe^{2+}]\gamma_{2+}\} = 12.23 + \log\{[Fe^{3+}]/[Fe^{2+}]\}$.
21. Bard, Parsons, and Jordan do not list the reduction potential for Hg^{2+} to Hg_2^{2+}. It must be calculated from two other reduction potentials by combining free energies ($nE°$) to eliminate Hg(1) as follows:

$$Hg^{2+} + 2e^- \rightleftharpoons Hg(1), \quad E_1° = 0.911\ V$$

$$Hg_2^{2+} + 2e^- \rightleftharpoons 2\ Hg(1), \quad E_2° = 0.796\ V$$

$$2\ Hg^{2+} + 2e^- \rightleftharpoons Hg_2^{2+}, \quad E° = \tfrac{1}{2}(4E_1° - 2E_2°) = 1.026\ V$$

22. This equation has two closely spaced roots at 0.00489 and 0.00511. The higher one is > 0.005 and violates the mass balance on Fe; hence it must be discarded. However, Newton's method or a "goal seek" program with too high a starting value could easily produce the incorrect root.
23. For Fe^{3+}–SO_4^{2-} values at $I = 0$ are from Smith and Martell, Vol. 6, 1989. For Fe^{2+}–SO_4^{2-} the β_1 value (at $I = 1$) is from C. F. Wells and M. A. Salam, *J. Chem. Soc.(A)*, 308, 1968, cited by Sillén and Martell, 1971.
24. Smith and Martell, Vol. 5, p. 393.
25. See, for example, M. Pourbaix, *Atlas d'Equilibres Electrochemiques á 25°C*, Gauthier-Villars, Paris, 1963; R. M. Garrels and C. L. Christ, *Solutions, Minerals and Equilibria*, Harper & Row, New York, 1965.
26. From Bard, Parsons, and Jordan, Appendix A.
27. The Mn(V) reduction, $MnO_3^- + 2\ H^+ + e^- \rightleftharpoons MnO_2 + H_2O$, is listed with $E° = 2.5$ in Latimer's *Oxidation Potentials* (1952), but has not survived later and more critical compilations of Sillén and Martell (1964, 1971), and Bard, Parsons, and Jordan (1985).
28. Smith and Martell, Vol. 4, p. 5. Complexes M_2L and M_2L_3 have formation constants 4.1 and 16.4 at $I = 2.0$.
29. Smith and Martell, Vol. 5, p. 394.
30. For a survey of the redox potential of Fe^{3+}/Fe^{2+} with 142 organic ligands as a function of pH, see J. N. Butler, J. Giner, and J. Stark, *Complex Redox Couples for Energy Storage*, Final Report, Project 727-2, Electric Power Research Institute, Palo Alto, CA, 1979, 115 pp. For most, E decreases with increasing pH.
31. Smith and Martell, Vol. 6, 1989, pp. 96–99. Note [FeHL] = β(FeHL)[Fe][H][L], [FeOHL] = β(FeOHL)[Fe][OH][L], etc.

NOTES 363

32. Starting with the unrealistic assumption that the free concentrations equal the total concentrations of each component, convergence is achieved in 10 to 20 iterations.
33. If $[Fe^{3+}]_t$ is defined to include only the species in solution, that value is determined by $[Fe^{3+}]_f$ and pH. If $[Fe^{3+}]_t$ is defined to include precipitated $Fe(OH)_3(s)$ as well, then its value is arbitrary but does not affect pε.
34. T. S. Lee, I. M. Kolthoff, and D. L. Leussing, *J. Am. Chem. Soc.* **70**: 2348–52; 3596–600, 1948.
35. From R. M. Smith and A. E. Martell, *Critical Stability Constants*, Vol. 2, Plenum, NY, 1976, and also from the 1997 National Institutes of Standards and Technology database. These two compilations do not differ significantly. In Sillén and Martell, *Stability Constants*, and several literature reviews, there are a number of inconsistencies that have been resolved since the Smith and Martell volumes were prepared. There appear to be no studies on Fe^{3+}-phenanthroline after S. C. Lahiri and S. Aditya, *Z. Physikal. Chem. N. F.* **41**:173–82, 1964. See especially W. A. E. McBryde, *A Critical Review of Equilibrium Data for Protons and Metal Complexes of 1,10-Phenanthroline, 2,2'-Bipyridyl, and Related Compounds*, Pergamon Press, Oxford, 1978.
36. G. Anderegg, *Helv. Chim. Acta* **45**:1643–57, 1962.
37. Bard, Parsons, and Jordan, 1985, p. 796.
38. G. F. Smith and C. A. Getz, *Ind. Eng. Chem. Anal.* **10**:191, 1938; K. J. Vetter, *Z. Phys. Chem. A* **196**:360, 1951, cited by Sillén and Martell, 1964. Most of the 4.8 unit difference from $I = 0$ can be attributed the formation of ion pairs between Ce^{4+} or Ce^{3+} and sulfate. $K_1(Ce^{3+} - SO_4^{2-}) = 1.51$ at $I = 1.0$ (Smith and Martell, 1974–1989) but $K_1(Ce^{4+} - SO_4^{2-})$ is rather uncertain: 3.3 to 4.8, complicated by ion pairs with 2 or 3 ligands, and also with HSO_4^- involved as a ligand (Sillén and Martell, 1964 and 1971). Roughly,

$$K_1(Ce^{4+} - SO_4^{2-}) - K_1(Ce^{3+} - SO_4^{2-}) = (3.3 \text{ to } 4.8) - 1.5 = (1.8 \text{ to } 3.3)$$

From the Davies equation at $I = 1$, $\log(\gamma_{3+}/\gamma_{4+}) = (-0.15)(3^2 - 4^2) = 1.05$; therefore, the combination of ion pairing plus activity coefficient changes produces difference in pε° of approximately 2.8 to 4.4, compared to the observed difference of 4.8.

39. Bard, Parsons, and Jordan, 1985, p. 793.
40. A. J. Zielen and J. C. Sullivan, *J. Phys. Chem.* **66**:1065, 1962, cited by Sillén and Martell, 1964. As with Ce, the difference in pε° between $I = 0$ and 1 can be estimated from a combination of ion-pairing equilibria: $K_1(Fe^{3+}-SO_4^{2-}) = 2.02$, $K_1(Fe^{2+}-SO_4^{2-}) = 1.0$ and activity coefficients: $\log(\gamma_{2+}/\gamma_{3+}) = (-0.15)(2^2 - 3^2) = 0.75$, to be roughly $\Delta p\varepsilon° = 1.77$, which may be compared with the experimental $\Delta p\varepsilon° = 13.0 - 11.4 = 1.60$.
41. See J. N. Butler, *Ionic Equilibrium*, 1964, pp. 418–20, for an earlier development of these equations.
42. The potential of a platinum electrode in a dichromate solution is poorly reproducible, and does not vary with Cr^{3+} concentration as would be expected. But surprisingly, the dichromate–ferrous reaction follows the curve expected from a reversible couple through the end point, showing time dependence only after the end point is passed. This has been attributed to the relatively high exchange current of the Fe^{3+}/Fe^{2+} couple. See H. A. Laitinen, *Chemical Analysis*, New York, McGraw-Hill, 1960, pp. 352–36; P. K. Winter and H. V. Moyer, *J. Am. Chem. Soc.* **57**:1402, 1935.
43. At $I = 0$ (Bard, Parsons, and Jordan, p. 796).

44. Note that $p\varepsilon^\circ_{Fe}$ is dependent on ionic strength as well as other factors. The Davies equation gives $\log \gamma_+ = -0.15$. The factor $[\gamma_{2+}]/[\gamma_{3+}] = \gamma_+^{(9-4)}$ can be absorbed into $p\varepsilon^\circ$ and hence $p\varepsilon^\circ_{Fe} = 13.0 - 0.75 = 12.25$. This can be compared with 12.37 (1 M HCl), 12.53 (1 M HClO$_4$), and 11.4 (0.5 M H$_2$SO$_4$) (from Sillén and Martell, 1964).

45. For a much more comprehensive and practical discussion of redox titrations, see H. A. Laitinen, *Chemical Analysis*, McGraw Hill, New York, 1960, pp. 326–450. For the application of redox reactions to environmental problems, see W. Stumm and J. J. Morgan, *Aquatic Chemistry*, 3rd ed., Wiley, NY, 1996, Chapters 8, 11, and 15; F. M. M. Morel and J. Hering, *Principles and Applications of Aquatic Chemistry*, Wiley, NY, 1993, Chapter 7.

46. W. Stumm, *Adv. Water Pollut. Res. (Munich)* **1**: 283, 1967; W. Stumm and J. J. Morgan, *Aquatic Chemistry*, 3rd ed., Wiley, NY, 1996, pp. 491–98.

47. R. D. Lindberg and D. D. Runnells, *Science* **225**:925–27, 1984.

48. These values are rather a patchwork quilt. Smith and Martell (Vol. 5, p. 418) give $\log K_{so} = -6.73$ and $I = 0$ and -7.38 at $I = 5$; $\log \beta'_1$ is not listed, but was estimated in *Ionic Equilibrium* (1964, p. 275, $\log K_{s1} = -5.0$); $\log \beta'_2 = 6.06$ at $I = 5$, 25° and 5.5 at $I = 0$, 20°C; $\log \beta'_3 = 5.94$ at $I = 5$. The polynuclear complex Cu$_2$Cl$_4^{2-}$ has formation constant $\log \beta'_{42} = +13.0$ at $I = 5$. Vol. 6 gives $\log \beta_1 = -0.2$ at $I = 1$, $\log \beta_2 = -0.5$ at $I = 0.7$.

49. See J. N. Butler, *Ionic Equilibrium*, 1964, pp. 421–24.

50. For example, $i_c = i_o \exp[-\alpha(E - E^\circ)RT/F]$ for $E \approx E^\circ$; $i_c = i^*_{\lim}\{1 - \exp[(E - E^\circ)RT/F]\}$ for $E \ll E^\circ$. See H. A. Laitinen, *Chemical Analysis*, McGraw-Hill, 1960, pp. 300–4, I. M.Kolthoff and J. J. Lingane, *Polarography*, Interscience, 1952, and other references on polarography or electrochemistry.

10

CARBON DIOXIDE

Importance of Carbon Dioxide
The Basic Equations
  Henry's Law
  Ionization Equilibria
    Ionic Strength and Temperature Dependence of pK_{a1}
    Ionic Strength and Temperature Dependence of pK_{a2}
    Ionic Strength and Temperature Dependence of pK_w
    Solubility Product of Calcium Carbonate
Combining Equilibria with Mass and Charge Balances
  pH Dependence, Constant Partial Pressure
    Atmospheric CO_2 Partial Pressure
    P_{CO_2} from pH and Alkalinity
    $NaHCO_3$ at $P_{CO_2} = 1$ atm
    Other Sources of Alkalinity
Mass Balances in a Closed System
Solutions Saturated with $CaCO_3$ at Constant P_{CO_2}
  High Pressure: $P_{CO_2} = 1$ and 100 atm
  Atmospheric Partial Pressure, $P_{CO_2} = 10^{-3.5}$
  Low Pressure: $P_{CO_2} = 10^{-6}$ atm
Solubility of $CaCO_3$ in Pure Water
  Higher pH
Buffer Capacity
  Types of Buffer Index
  Homogeneous Solution in Closed Systems
  Solution with Gas Phase at Constant P_{CO_2}
  Solution with Solid $CaCO_3$ and Gas Phase at Constant P_{CO_2}
  Buffer Index of Closed System with CO_2 and $CaCO_3$

Correction of Previously Published Equations
Summary of Buffer Capacity Results
Introduction to Case Studies
Six Forms for the Equilibrium Constant
Physiological Fluids
Charge Balance and "Strong Ion Difference"
Models Relating pH to P_{CO_2}
Ion-Pair Models
Precipitation of $CaCO_3$ from Lactated Ringer's Solution
Carbon Dioxide Equilibria in Seawater
Alkalinity Titration in Seawater
Effect of Increased Atmospheric CO_2 on the Oceans
Other Salinities: Ion-Pair Model for Seawater
Variations in the Model
A mixture of Seawater and River Water
Problems

IMPORTANCE OF CARBON DIOXIDE

Carbon dioxide is a diprotic acid, but its applications are far too numerous and important for it to be simply one of the examples in Chapter 5. It occupies a central place in the biosphere and in many of the geological processes that create and erode rocks. Plants, from phytoplankton to trees, absorb CO_2 from the atmosphere and convert it into biomass; respiration by terrestrial and aquatic organisms returns carbon dioxide to the atmosphere. Many marine plants and animals convert CO_2 to calcium carbonate, and when they die the mineral portions of their bodies become reefs, sediment, and limestone. Recently, the impact of increasing CO_2 in the atmosphere on global climate change has become a matter of urgent interest.

Almost all natural waters contain carbonate and carbon dioxide. Water conditioning and wastewater treatment must therefore include processes involving acid–base equilibria of CO_2 and precipitation or dissolution of $CaCO_3$. These topics are considered an essential part of the environmental engineering curriculum, but are not often given a very detailed treatment in engineering textbooks.

Similarly, carbon dioxide is the principal acid–base system in fresh water and seawater. A serious effort has been made by marine chemists to obtain highly accurate data on the equilibria of CO_2 and related systems in seawater, so that small changes in these parameters might be used to help understand global changes in the oceans and atmosphere.

In all animals, O_2 is absorbed and used to oxidize organic matter; the final product is CO_2, which must be discharged from the organism. In human physiology, the acid–base balance of the blood and other fluids is critical to health, and monitoring CO_2 in the blood is a valuable tool for the intensive care of patients.

This chapter sets out the basic equations for carbon dioxide as a gas that dissolves to form a diprotic acid, and for the equilibria of carbon dioxide solutions with al-

THE BASIC EQUATIONS

kali and calcium carbonate. It also gives some detailed examples of how these equations can be applied to problems in physiology and marine chemistry.

THE BASIC EQUATIONS

Henry's Law

Carbon dioxide gas dissolves in water to an extent determined by its partial pressure[1] and by the interaction of dissolved CO_2 with other solutes in the water. The concentration of uncharged (free) CO_2 in solution is normally expressed by Henry's law:

$$[CO_2] = K_H P_{CO_2} \qquad (1)$$

In the presence of other solutes K_H depends on ionic strength, and this dependence is frequently expressed as an activity coefficient for CO_2, γ_o.

$$[CO_2]\,\gamma_o = K_H^\circ P_{CO_2} \qquad (2)$$

At constant temperature, $pK_H = -\log(K_H^\circ) + \log(\gamma_o)$ increases with ionic strength:

$$\log \gamma_o = b\,I \qquad (3)$$

where I is the ionic strength and b varies from 0.11 at 10°C to 0.20 at 330°C in NaCl media from $I = 0$ to 2 M.[4] For example, pK_H° is 1.46 at 25°C, $I = 0$, when P_{CO_2} is in atmospheres.[5] In seawater,[6] $pK_H = 1.54$. At higher temperatures, CO_2 is less soluble, with a minimum of $10^{-2.07}$ at about 175°C. Detailed data are given in Table 10.1 for $I = 0$ and temperatures from 0 to 350°C.[7]

Ionization Equilibria

When CO_2 dissolves in water, it hydrates to yield H_2CO_3. Since this reaction is slow compared to the ionization of H_2CO_3, measurements on the millisecond time scale can distinguish between simple dissolved CO_2 and the hydrated species. This kinetic process is quite important in some physiological systems, and is catalyzed by the enzyme carbonic anhydrase. However, at equilibrium $[H_2CO_3]$ is only about 10^{-3} as large as $[CO_2]$[8] and has no special significance to the acid–base equilibria, since both are uncharged. In this chapter, $[H_2CO_3] + [CO_2]$ will be abbreviated simply as $[CO_2]$[9].

In acid solutions (pH < 5), the principal solution species in equilibrium with CO_2 gas is dissolved CO_2, but at higher pH, it ionizes to produce bicarbonate and carbonate ion, represented by the following concentration equilibria:

$$[H^+][HCO_3^-] = K_{a1}\,[CO_2] \qquad (4)$$

TABLE 10.1. Selected values of equilibrium constants extrapolated to $I = 0$[2,3]

Temp. (°C)	pK_H° (mol/kg atm)	pK_{a1}° (mol/kg)	pK_{a2}° (mol/kg)	pK_w° (mol/kg)	pK_{so}° (mol/kg)^2
0	1.114	6.579	10.625	14.96	8.29
5	1.194	6.518	10.557	14.74	8.31
10	1.270	6.466	10.490	14.53	8.33
15	1.341	6.424	10.430	14.36	8.38
20	1.406	6.389	10.377	14.17	8.43
25	1.464	6.363	10.329	13.997	8.48
30	1.521	6.336	10.290	13.83	8.53
35	1.572	6.317	10.250	13.68	8.58
37	1.591	6.312	10.238	13.62	8.60
40	1.620	6.304	10.220	13.53	8.63
45	1.659	6.295	10.195	13.39	8.69
50	1.705	6.289	10.172	13.26	8.75
100	1.990	6.450	10.160	12.27	9.45
150	2.070	6.770	10.32	11.66	10.75
200	2.050	7.080	10.70	11.28	11.83
250	1.970			11.1	13.20
300	1.840				
350	1.570				

$$[H^+][CO_3^{2-}] = K_{a2}[HCO_3^-] \qquad (5)$$

K_{a1} and K_{a2} depend on ionic strength as well as temperature (see below). Tabulated constants are often extrapolated to $I = 0$, as in Table 10.1.[10] Activity coefficients are introduced to express the ionic strength dependence:

$$[H^+]\gamma_+ [HCO_3^-]\gamma_- = K_{a1}^\circ [CO_2]\gamma_o a_{H_2O} \qquad (6)$$

$$[H^+]\gamma_+ [CO_3^{2-}]\gamma_= = K_{a2}^\circ [HCO_3^-]\gamma_- \qquad (7)$$

Activity coefficients at concentrations below 0.5 M are given approximately by the Davies–Guggenheim equation (see Chapter 2, p. 49).

$$\log(\gamma_z) = -0.5z^2\left(\frac{\sqrt{I}}{1+\sqrt{I}} - bI\right) \qquad (8)$$

where z is the charge on the ion and b is 0.2 for the Davies equation as presented in previous chapters. The activity coefficient of dissolved CO_2 is given by Eq 3 above; a_{H_2O} is approximately equal to the mole fraction of water, 1.00 in dilute solutions. At higher concentrations or in multicomponent electrolytes like seawater, empirical data are necessary for accurate work. As will be seen below, b can be adjusted for a more precise interpolation if data are available at higher ionic strengths.

Ionic Strength and Temperature Dependence of *pK*$_{a1}$. Figure 10.1 compares experimental values of the concentration constant K_{a1} selected by Smith and Martell[2] with the predictions of the Davies equation. Agreement is satisfactory for $I \leq 1.0$, but the experimental value is much higher at $I = 3.0$.

Although solution thermodynamicists prefer to work with molal concentrations of all species, including H$^+$, some oceanographers have adopted a convention in which $-\log[\text{H}^+]\gamma_+$ is replaced by an empirical pH value, such as the National Bureau of Standards (NBS) scale (see pp. 401–404) for other variations on this theme):

$$10^{-\text{pH}} [\text{HCO}_3^-]\gamma_- = K_{a1}^\circ [\text{CO}_2]\gamma_o \tag{9}$$

$$10^{-\text{pH}} [\text{CO}_3^{2-}]\gamma_= = K_{a2}^\circ [\text{HCO}_3^-]\gamma_- \tag{10}$$

Activity coefficients can then be combined with the $I = 0$ equilibrium constants to give:

$$10^{-\text{pH}} [\text{HCO}_3^-] = K_1' [\text{CO}_2] \tag{11}$$

where $\quad K_1' = K_{a1}^\circ \gamma_o/\gamma_- = K_{a1} \gamma_+ \tag{12}$

$$10^{-\text{pH}} [\text{CO}_3^{2-}] = K_2' [\text{HCO}_3^-] \tag{13}$$

where $\quad K_2' = K_{a2}^\circ \gamma_-/\gamma_= = K_{a2} \gamma_+ \tag{14}$

In seawater, pK_1' and pK_2' are significantly lower than in noncomplexing media because the convention is to use total concentrations $[\text{HCO}_3^-]_T$ and $[\text{CO}_3^{2-}]_T$ in the

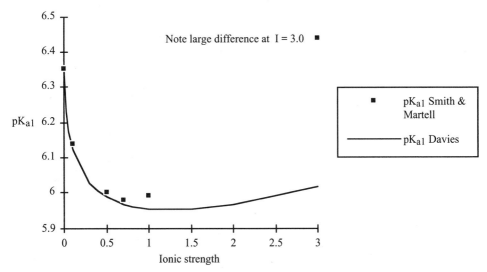

FIGURE 10.1. pK_{a1} as a function of ionic stength. For $I \leq 1$, the experimental data selected by Smith and Martell[2] are well represented by the Davies equation ($b = 0.2$), but the experimental value at $I = 3.0$ is much larger than predicted by the Davies equation (Eq. 8).

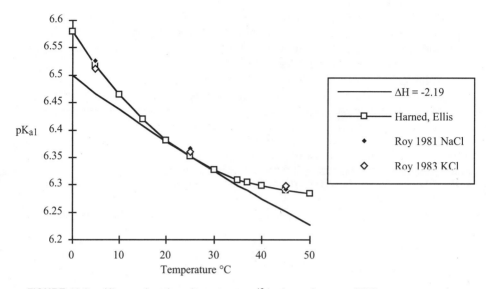

FIGURE 10.2. pK_{a1}° as a function of temperature.[10] In the region near 25°C, a constant value of $dpK_{a1}/dT = 0.0054 \times 10^{-3}$ or $\Delta H = -2.19$ kcal/mole gives a satisfactory approximation to the temperature dependence, but not at temperatures less than 15°C or above 35°C.

equilibria, including ion pairs of $[HCO_3^-]$ and $[CO_3^{2-}]$ with the cations Na^+, Mg^{2+}, and Ca^{2+} in $[HCO_3^-]_T$ and $[CO_3^{2-}]_T$; for example[11]:

$$[HCO_3^-]_T = [HCO_3^-]_f + [NaHCO_3] + [CaHCO_3^+] + [MgHCO_3^+] + \cdots$$

$$[CO_3^{2-}]_T = [CO_3^{2-}]_f + [NaCO_3^-] + [CaCO_3] + [MgCO_3] + \cdots$$

where $[HCO_3^-]_f$ and $[CO_3^{2-}]_f$ represent the free-ion concentrations.

Since $[CO_2]$ is uncharged and forms no significant ion pairs, K_1' increases (pK_1' decreases) as the degree of HCO_3^- ion pairing increases; likewise, since the ion pairs with CO_3^{2-} generally are stronger than those with HCO_3^-, K_2' increases as the degree of ion pairing increases. For example, $pK_1' = 5.97$ and $pK_2' = 9.02$ in standard 35‰ seawater ($I = 0.7$). These are much lower than the $pK_1' = 6.13$ and $pK_2' = 9.86$ in 0.7 M NaCl. These topics are discussed in more detail in Chapter 10.

Figure 10.2 compares experimental values of pK_{a1} as a function of temperature (Table 10.1) with predictions made with a constant value of $d(pK_{a1})/dT = -0.0054$ ($\Delta H = -2.19$ kcal/mole). Constant ΔH gives a satisfactory approximation to the temperature dependence only at temperatures between 15° and 35°C.

Ionic Strength and Temperature Dependence of pK_{a2}. The second acidity constant K_{a2} is plotted as a function of ionic strength in Fig. 10.3. The various experimental results listed by Sillén and Martell[12] and the values selected by Smith and Martell[2] do not agree very well with each other, and most published values are for low I or for seawater. For $I \le 1$ the Davies equation is an adequate interpolation.

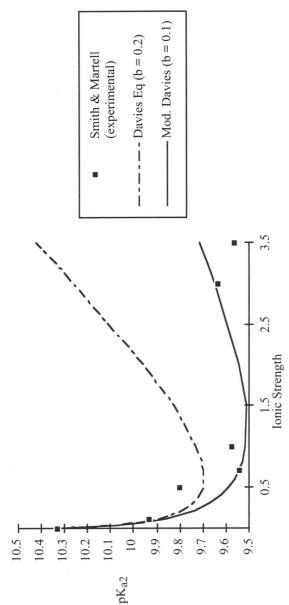

FIGURE 10.3. Second acidity constant K_{a2}. The various experimental results listed by Smith and Martell[2] do not agree very well with each other, but for $I \leq 1$ the Davies equation is an adequate interpolation. At higher ionic strengths, the experimental pK_{a2} is lower than the Davies equation, and a better fit is obtained if b is adjusted from 0.2 to 0.1.

At higher ionic strengths, the experimental pK_{a2} are lower than the Davies equation, but a modified equation with b adjusted to 0.1 fits quite well. The temperature dependence is given in Table 10.1.

Ionic Strength and Temperature Dependence of pK_w. One additional ionization constant is essential for calculations on any acid–base system, the ion product of water. The three versions of the equilibrium are:

$$[H^+][OH^-] = K_w \tag{15}$$

$$[H^+]\gamma_+[OH^-]\gamma_- = K_w^\circ \tag{16}$$

$$10^{-pH}[OH^-]_T = K_w' \tag{17}$$

The temperature dependence[13] of pK_w° is given in Table 10.1, and the ionic strength dependence of K_w is shown in Fig. 10.4. The many studies catalogued by Sillén and Martell (1964 and 1971) and Smith and Martell (1976–1989)[2] cluster above the line representing the Davies equation. A better fit to the data is obtained with $b = 0.25$.

Solubility Product of Calcium Carbonate. The solubility product of $CaCO_3$ is an extremely important parameter for geochemistry and oceanography, and hence has been measured under a wide variety of temperature and ionic strength conditions. For example, calcite is the main component of limestone, which constitutes a large fraction of the earth's near-surface rocks. Calcium carbonate is formed in marine environments by a wide variety of organisms; their detritus is a principal

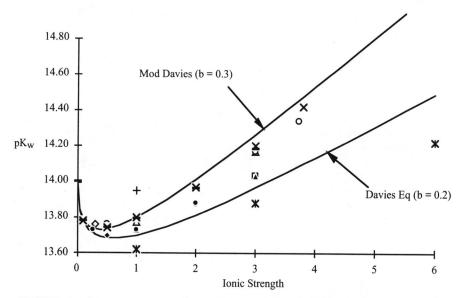

FIGURE 10.4. Ion product of water as a function of ionic strength. Points represent experimental data collected by Sillén and Martell (1964, 1971) and Smith and Martell (1976–1989).

method of transporting carbon dioxide from the atmosphere to the sediments of the deep oceans—and hence of great importance in understanding how CO_2 in the atmosphere is regulated. This, in turn, is a major component of the forces causing global climate change.

Values of pK_{so}° for calcite are listed in Table 10.1 as a function of temperature (Fig. 10.5). The traditional value at 25°C, obtained by interpreting data without consideration of ion pairs, is 8.34. Christ et al.[14] re-evaluated the existing data, using an ion-pairing model, and obtained $pK_{so}^{\circ} = 8.52 \pm 0.04$. The most recent value selected by Smith and Martell (1989) is 8.48 ± 0.02, and the data in Table 10.1 have been adjusted to agree with that.

Figure 10.6 shows the ionic strength dependence of pK_{so} in NaCl media.[15] The Davies equation with the usual coefficient $b = 0.2$ does not fit the data above $I = 0.5$, probably because the ion pairing between Na^+ and CO_3^{2-} is so strong. However, an empirical adjustment to $b = 0.05$ provides an adequate interpolation function over the entire range of ionic strength. Millero et al., who measured the experimental data, used the Pitzer equations[16] to fit that data set over the entire range and estimated $pK_{so}^{\circ} = 8.46 \pm 0.03$.

There are three other mineral forms of calcium carbonate. All are more soluble than calcite.[17] At 25°C, $I = 0$:

$$\text{aragonite, } pK_{so} = 8.30 \pm 0.02$$

$$\text{vaterite, } pK_{so} = 7.91$$

$$\text{hydrocalcite, } pK_{so} = 7.60$$

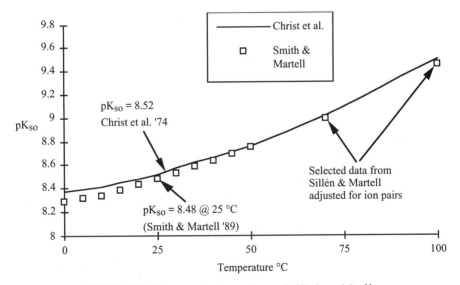

FIGURE 10.5. Temperature dependence of pK_{so}° for calcite.[14]

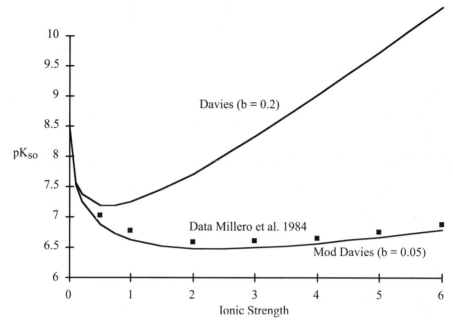

FIGURE 10.6. pK_{so} as a function of ionic strength in NaCl media.[15]

Aragonite is normally the crystal form deposited biogenically; over time it recrystallizes to calcite.

Another important mineral is dolomite, $CaMg(CO_3)_2$:

$$[Ca^{2+}][Mg^{2+}][CO_3^{2-}]^2 = K_{so}, \qquad pK_{so} = 17.0 \pm 0.2^{18}$$

Dolomite is widely distributed, but generally does not occur in recent rocks because of the long time required for its formation from limestone and excess magnesium.

COMBINING EQUILIBRIA WITH MASS AND CHARGE BALANCES

In the first part of this chapter, following the style of the previous discussions on acid–base equilibria (Chapters 3–5) and solubility (Chapter 6), the equilibrium expressions, mass balances, and charge balances will be expressed in terms of concentrations. The equilibrium constants in the discussion below are concentration constants, and "pH" is generally taken to be $-\log[H^+]$. For accurate work at ionic strength above about 10^{-2} M, the concentration constants can be adjusted for ionic strength, as shown in Figs. 10.1, 10.3, 10.4, and 10.6 above, and pH can be calculated as $paH = -\log\{[H^+]\gamma_+\}$, where γ_+ is given by the Davies equation or another theoretical estimate. For the most accurate work, the uncertainty in estimating γ_+ can be avoided by using $[H^+]$ directly as obtained from the concentration equilibria, mass, and charge balances—corresponding to a "pcH" scale (see Chapter 11).

pH Dependence, Constant Partial Pressure

The simplest example is the relationship between carbon dioxide partial pressure and pH at constant temperature. Combining Eqs. (1), (4), and (5) gives

$$[CO_2] = K_H P_{CO_2} \tag{18}$$

$$[HCO_3^-] = \frac{P_{CO_2} K_H K_{a1}}{[H^+]} \tag{19}$$

$$[CO_3^{2-}] = \frac{P_{CO_2} K_H K_{a1} K_{a2}}{[H^+]^2} \tag{20}$$

These can be combined to give total dissolved carbonate $C_T = [CO_2] + [HCO_3^-] + [CO_3^{2-}]$ as a function of P_{CO_2} and pH:

$$C_T = K_H P_{CO_2} \left(1 + \frac{K_{a1}}{[H^+]} + \frac{K_{a1} K_{a2}}{[H^+]^2} \right) \tag{21}$$

Also important is the alkalinity concept, based on the equilibrium between an alkali such as NaOH and carbon dioxide. The charge balance for such a system (see Chapt. 4, p. 116) is

$$[Na^+] + [H^+] = [HCO_3^-] + 2[CO_3^{2-}] + [OH^-]$$

The carbonate alkalinity, A_C, is defined to be equal to the alkali concentration:

$$[Na^+] = A_C = [HCO_3^-] + 2[CO_3^{2-}] + [OH^-] - [H^+]$$

$$A_C = K_H P_{CO_2} \left(\frac{K_{a1}}{[H^+]} + 2 \frac{K_{a1} K_{a2}}{[H^+]^2} \right) + \left(\frac{K_w}{[H^+]} - [H^+] \right) \tag{22}$$

Some marine chemists define carbonate alkalinity to be only $[HCO_3^-] + 2[CO_3^{2-}]$; neglecting the last two terms in Eq. (22) at pH = 8 introduces an error of about 10^{-6} mol/kg, which is close to the minimum measurement error in A_C.

Acidity, sometimes written $[H^+]_T$, is simply the negative of alkalinity—to give the function a positive value at low pH. These various functions of pH are plotted in Fig. 10.7 at constant P_{CO_2}, $10^{-3.5}$ (atmospheric partial pressure[19]).

Atmospheric CO_2 Partial Pressure

Example 1. Find the pH of pure water (alkalinity $A_C = 0$) in equilibrium with atmospheric CO_2, $P_{CO_2} = 10^{-3.5}$ at 25°C. For this system, and from the charge balance (see Eq. 22)

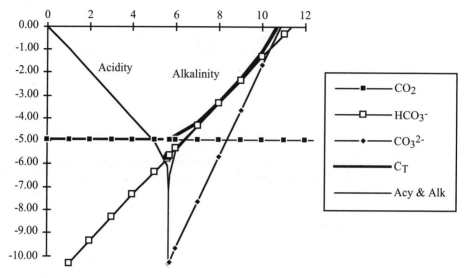

FIGURE 10.7. Logarithmic concentration of carbonate species at 25°C, $I = 0$, $P_{CO_2} = 10^{-3.5}$. See Example 1.

$$[H^+] = [HCO_3^-] + 2[CO_3^{2-}] + [OH^-]$$

$$[H^+] = K_H P_{CO_2}\left(\frac{K_{a1}}{[H^+]} + \frac{K_{a1}K_{a2}}{[H^+]^2}\right) + \frac{K_w}{[H^+]}$$

Refer to Fig. 10.7. The pH at which $A_C = 0$ is somewhat less than 6, where the curves of log(alkalinity) and log(acidity) meet at $-\infty$. Substitute the following values in the above equation: $[H^+] = 10^{-6}$, $K_H = 10^{-1.46}$, $P_{CO_2} = 10^{-3.5}$, $K_{a1} = 10^{-6.36}$, $K_{a2} = 10^{-10.33}$, $K_w = 10^{-14.00}$ to get

$$10^{-6} = 10^{-5.32} + 10^{-9.65} + 10^{-8.00}$$

Note that the first term on the right is much larger than the other two terms, which leads to the approximation

$$[H^+] = \frac{K_{a1}K_H P_{CO_2}}{[H^+]}$$

$$[H^+]^2 = K_{a1}K_H P_{CO_2} = 10^{-11.32}$$

$$pH = 5.66$$

Activity coefficient corrections are not necessary, since $I = \frac{1}{2}([H^+] + [HCO_3^-]) = 10^{-5.66}$ and $\gamma_\pm = 0.998$.

P_{CO_2} from pH and Alkalinity

Example 2. Find the CO_2 partial pressure in equilibrium with a water that has pH = 7.5 and carbonate alkalinity = 0.001 M (i.e., 0.001 M $NaHCO_3$). (The log diagram for this system is similar to Fig. 10.7, but with the lines for CO_2, HCO_3^-, and CO_3^{2-} about 1 unit higher.) Substituting values from Table 10.1 into Eq. (22) (p. 375), and solving for P_{CO_2},

$$P_{CO_2} = \frac{0.001 + 10^{-7.5} - 10^{-6.5}}{10^{-1.46} \dfrac{10^{-6.36}}{10^{-7.5}}(1 + 2 \times 10^{-10.33}/10^{-7.5})} = 0.0021 \text{ atm}$$

Note that many of these terms are small or negligible, and that the approximate form

$$P_{CO_2} = \frac{A_C[H^+]}{K_H K_{a1}} = \frac{(0.001)(10^{-7.5})}{(10^{-1.46})(10^{-6.36})} = 0.0021 \text{ atm}$$

gives the same answer within 0.3%. Since ionic strength is 0.001, the use of activity coefficients also makes little difference.[20]

$NaHCO_3$ at P_{CO_2} = 1 atm

Example 3. Find the pH of 1 M $NaHCO_3$ in equilibrium with 1 atm CO_2. Here you might expect the activity coefficients to make a big difference. Equation (22) still applies, but with $A_C = C_T$. In this combination of mass and charge balances $[HCO_3^-]$ cancels out, and the result reduces to

$$[H^+] + [CO_2] = [CO_3^{2-}] + [OH^-]$$

Figure 10.8 is the same as Fig. 10.7, but with the curves for the carbonate species raised by 3.5 log units. However, $[CO_2]$ and $[CO_3^{2-}]$ still cross at a pH between 8 and 9 (at this pH, both $[H^+]$ and $[OH^-]$ are negligible). The result is

$$K_H P_{CO_2} = \frac{K_{a2} K_{a1} K_H P_{CO_2}}{[H^+]^2}$$

With values of the constants for $I = 0$, and approximate value is obtained

$$[H^+]^2 = K_{a1} K_{a2} = 10^{-16.69}$$
$$\text{pH} = 8.35$$

Note that the algebraic result is independent of P_{CO_2} and the concentration of $NaHCO_3$ so long as $[H^+]$ and $[OH^-]$ are negligible compared to $[CO_2]$ and $[CO_3^{2-}]$.

However, the numerical value of pH is different due to the effect of ionic strength. Since

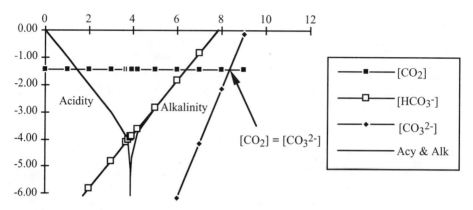

FIGURE 10.8. Logarithmic diagram for $P_{CO_2} = 1.0$. See Example 3. Recall that acidity is the negative of alkalinitiy as defined by Eq. (22).

$$I = \tfrac{1}{2}([\text{Na}^+] + [\text{HCO}_3^-] + \cdots) = 1.00$$

Figure 10.1 gives $pK_{a1} = 5.96$; Fig. 10.3 gives $pK_{a2} = 9.72$. Since $[\text{H}^+]^2 = K_{a1}K_{a2}$,

$$-\log[\text{H}^+] = \tfrac{1}{2}(pK_{a1} + pK_{a2}) = \tfrac{1}{2}(5.96 + 9.72) = 7.84$$

From the Davies equation, $\log \gamma_+ = -0.15$, and

$$\text{pH} = -\log[\text{H}^+] - \log \gamma_+ = 7.99$$

This is significantly lower than the value calculated with $I = 0$ constants.

At constant temperature (hence all the Ks are constant), Eqs. (21) and (22) (p. 375) say that only two of the four functions A_C, C_T, P_{CO_2}, and pH are independent; if two are known, the others can be calculated. Since all four of these parameters can be measured by independent experiments, careful comparison of calculated with observed values provides a check on the accuracy of the measurements and on the assumptions of the equilibrium model. An example of this approach is shown later.

Other Sources of Alkalinity.[21] In most natural waters, carbonate is the main source of alkalinity, but not the only one. Total alkalinity, defined as the amount of acid required to titrate the water from its initial pH to the CO_2 equivalence point (pH ≈ 4.5), includes other acid–base systems. Some of the common species are given in the following equation. Positive terms are basic species present at the starting pH; negative terms are acidic species remaining at the end point:

$$A_T = A_C + [\text{B(OH)}_4^-] + [\text{SiO(OH)}_3^-] + [\text{HS}^-] + [\text{NH}_3] - [\text{HF}] - [\text{HSO}_4^-] + \cdots$$

Additional terms for phosphate or organic materials can be added. This argument leads in general[22] to A_T at any pH value:

COMBINING EQUILIBRIA WITH MASS AND CHARGE BALANCES

$$A_T = A_C + \sum_i [B_i] - \sum_i [HA_i] \qquad (23)$$

where $[B_i]$ are the basic species that are more basic than CO_2 (except for $[HCO_3^-]$, $[CO_3^{2-}]$, $[H^+]$, and $[OH^-]$, which are included in A_C) and $[HA_i]$ are those acidic species that are more acidic than CO_2.[23]

Empirically, the alkalinity titration gives $A_T^{titr} = A_T^\circ - A_T^{ep}$, where A_T° is the value given by Eq. (23) at the start of the titration (e.g., pH = 8), and A_T^{ep} is the value at the equivalence point[24] (typically near $A_C = 0$ or pH $\approx$ 4.5).

$$A_T^{titr} = A_C^\circ - A_C^{ep} + \sum_i [B_i]^\circ - \sum_i [B_i]^{ep} - \sum_i [HA_i]^\circ + \sum_i [HA_i]^{ep}$$

$$A_T^{titr} = A_C^\circ - A_C^{ep} + \sum_i [B_i]^\circ - 0 - 0 + \sum_i [HA_i]^{ep}$$

For seawater, in the initial solution, $[HF]^\circ$ and $[HSO_4^-]^\circ$ are negligible because of the high pH. At the end point $A_C^{ep} \approx 0$, a value that can be calculated from Eq. (22) if pH^{ep} is known. In addition, $[B(OH)_4^-]^{ep}$, $[SiO_3^{2-}]^{ep}$, $[HS^-]^{ep}$, $[NH_3]^{ep}$, etc. are negligible at the end point because of the low pH. The approximate result is

$$A_T^{titr} \approx A_C^\circ + [B(OH)_4^-]^\circ + [SiO(OH)_3^-]^\circ + [HS^-]^\circ + [NH_3]^\circ$$
$$+ [HF]^{ep} + [HSO_4^-]^{ep} + \cdots$$

Example 4. What is the alkalinity obtained by titrating with HCl: a water containing 10^{-3} M carbonate alkalinity, 10^{-4} M total borate ($pK_a = 8.7$) and 10^{-2} M sulfate (pK_a for $HSO_4^- = 1.1$)? The initial pH is 8.3; the end point is 4.5.

Note that at the beginning of the titration

$$[B(OH)_4^-] = C_B K_a/(10^{-pH_0} + K_a) = (10^{-4})(10^{-8.7})/(10^{-8.3} + 10^{-8.7}) = 10^{-4.15}$$

at the end point

$$[HSO_4^-] = C_S \, 10^{-pH_{ep}}/(10^{-pH_{ep}} + K_S) = (10^{-2})(10^{-4.5})/(10^{-4.5} + 10^{-1.1}) = 10^{-5.40}$$
$$A_T = 10^{-3} + 10^{-4.15} + 10^{-5.40} = 1.074 \times 10^{-3}$$

Thus the presence of borate and sulfate increase the measured alkalinity by about 7%.

Mass Balances in a Closed System

In Chapters 4 and 5 we dealt entirely with closed systems, since the acids and bases we were considering were not generally in equilibrium with a gas phase.[25] So far in this chapter we have considered *only* the case where the gas phase was in equilibrium with the solution. However, another important type of constraint on a carbon dioxide system is when the system is closed, and C_T is held constant instead of P_{CO_2}. The mass balance is the same as on p. 375.

$$C_T = [CO_2] + [HCO_3^-] + [CO_3^{2-}] \quad (24)$$

but the equilibria are expressed in terms of one of the concentrations, instead of P_{CO_2}. From Eqs. (4) and (5),

$$C_T = [CO_2]\left(1 + \frac{K_{a1}}{[H^+]} + \frac{K_{a1}K_{a2}}{[H^+]^2}\right) \quad (25)$$

Solving this for $[CO_2]$ and rearranging gives

$$[CO_2] = \frac{C_T[H^+]^2}{([H^+]^2 + K_{a1}[H^+] + K_{a1}K_{a2})} \quad (26)$$

Similarly,

$$[HCO_3^-] = \frac{C_T K_{a1}[H^+]}{([H^+]^2 + K_{a1}[H^+] + K_{a1}K_{a2})} \quad (27)$$

$$[CO_3^{2-}] = \frac{C_T K_{a1} K_{a2}}{([H^+]^2 + K_{a1}[H^+] + K_{a1}K_{a2})} \quad (28)$$

These functions, which are also derived in Chapter 5 (pp. 167–168), are plotted in Fig. 10.9 for $C_T = 10^{-3}$ M.

Comparing Fig. 10.9 with Fig. 10.7, you can see that the major difference is at high pH. At constant P_{CO_2}, $[HCO_3^-]$, $[CO_3^{2-}]$, and hence C_T increase steeply above pH = 6, with more CO_2 being absorbed from the gas phase at higher pH. In the closed system, on the other hand, C_T is constrained to be constant at high pH; $[CO_2]$ and $[HCO_3^-]$ decreases in turn as pH increases, and eventually $[CO_3^{2-}]$ is the dominant species, equal to C_T. (Alkalinity increases above pH = 10 because of the $[OH^-]$ term.)

Solutions Saturated with CaCO$_3$ at constant P_{CO_2}

A substantial proportion of the world's rocks and sediments are composed of calcium carbonate, and therefore the solubility of this mineral is of considerable interest. It also happens to be one of the few minerals that achieves equilibrium quickly with aqueous solutions.

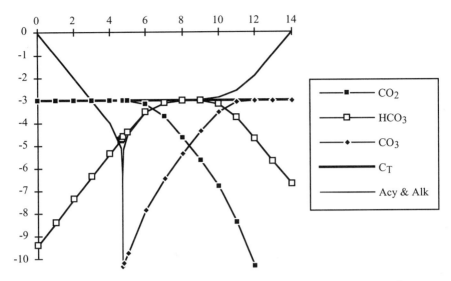

FIGURE 10.9. Carbonate species as a function of pH at 25°C, $I = 0$, $C_T = 10^{-3}$ M.

The general principles of solubility calculations were given in Chapter 6. Calcium carbonate introduces two additional factors: the diprotic acid CO_2 and the moderately strong ion pairs (complexes) $CaHCO_3^+$ and $CaCO_3(aq)$. As you will see, these are most important at high P_{CO_2}. In basic solutions, $CaOH^+$ can also be important. While there are rules of thumb to tell when ion pairs might be important, the rigorous approach is to include them in the model and see what difference it makes to neglect them.

Table 10.2 lists the most important equilibria for the CO_2–$CaCO_3$–H_2O system. The charge balance for aqueous CO_2 + $CaCO_3(s)$ is

$$[H^+] + 2[Ca^{2+}] + [CaHCO_3^+] + [CaOH^+] = [HCO_3^-] + 2[CO_3^{2-}] + [OH^-] \quad (29)$$

Using the equilibria of Table 10.2, each of these terms can be expressed in terms of $[H^+]$ and P_{CO_2}

TABLE 10.2. Equilibrium constants for CO_2–$CaCO_3$[26]

Log of equilibrium constant	25°C, $I = 0$
$[CO_2] = K_H P_{CO_2}$	−1.464
$[HCO_3^-][H^+] = K_{a1}[CO_2]$	−6.363
$[CO_3^{2-}][H^+] = K_{a2}[HCO_3^-]$	−10.329
$[OH^-][H^+] = K_w$	−13.997
$[Ca^{2+}][CO_3^{2-}] = K_{so}$ (calcite)	−8.48
$[CaHCO_3] = K_{CaH}[Ca^{2+}][HCO_3^-]$	+1.26
$[CaCO_3(aq)] = K_{CaC}[Ca^{2+}][CO_3^{2-}]$	+3.15
$[CaOH^+] = K_{CaOH}[Ca^{2+}][OH^-]$	+1.3

$$[H^+] + \frac{K_{so}[H^+]^2}{P_{CO_2}K_HK_1K_2}\left(2 + K_{CaH}\frac{P_{CO_2}K_HK_1}{[H^+]} + K_{CaOH}\frac{K_w}{[H^+]}\right)$$

$$= \frac{P_{CO_2}K_HK_1}{[H^+]}\left(1 + \frac{2K_2}{[H^+]}\right) + \frac{K_w}{[H^+]} \tag{30}$$

If base is added to increase pH, a term in strong base concentration, such as Cb, $[Na^+]$, or [Alk] is added to the left side; if acid is added to decrease pH, a term in strong acid concentration, such as C_a, $[Cl^-]$, or [Acy] is added to the right side.

High Pressure: P_{CO_2} = 1 and 100 atm. To show how the relative amounts of the various species change with pH and P_{CO_2}, logarithmic concentration diagrams are presented in Figs. 10.10–10.13. For Fig. 10.10, P_{CO_2} was set at 1.0 atm. The point of charge balance gives the pH of a saturated aqueous solution of $CaCO_3$ in equilibrium with P_{CO_2} but no other acids or bases. A first approximation to Eq. (29) is given by the three highest lines on Fig. 10.10:

$$2[Ca^{2+}] + [CaHCO_3^+] = [HCO_3^-] + \cdots \tag{31}$$

Substituting the functions used in Eq. (30) gives

$$\frac{2K_{so}[H^+]^2}{P_{CO_2}K_HK_1K_2} + \frac{K_{CaH}K_{so}[H^+]}{K_2} = \frac{P_{CO_2}K_HK_1}{[H^+]} + \cdots \tag{32}$$

Substituting values for the constants at $I = 0$, 25°C from Table 10.2 gives

$$[H^+]^3 = 10^{-17.81} - 10^{-6.87}[H^+]^2$$

Neglecting the second term gives $[H^+] = 10^{-5.94}$; this value when substituted in the second term gives $[H^+]^3 = 10^{-17.81} - 10^{-18.75} = 10^{-17.86}$, or

$$[H^+] = 10^{-5.95}$$

This point is noted by the arrow in Fig. 10.10. Note that the contribution of $[CaHCO_3^+]$ is significant (10%) but small at 1 atm.

At higher pressures it becomes more significant. Figure 10.11 shows a magnified view of the balance point for P_{CO_2} = 100 atm, where $[CaHCO_3^+]$ is about equal to $[Ca^2]$. The pH at the balance point is only approximate, since the ionic strength is about 0.1 and the activity coefficients, particularly of divalent Ca^{2+}, will be significantly lower than 1.0. Set

$$I = \frac{1}{2}(4[Ca^{2+}] + [CaHCO_3^+] + [HCO_3^-])$$
$$= \frac{1}{2}(4 \times 10^{-1.6} + 10^{-1.5} + 10^{-1.15}) = 0.0916$$

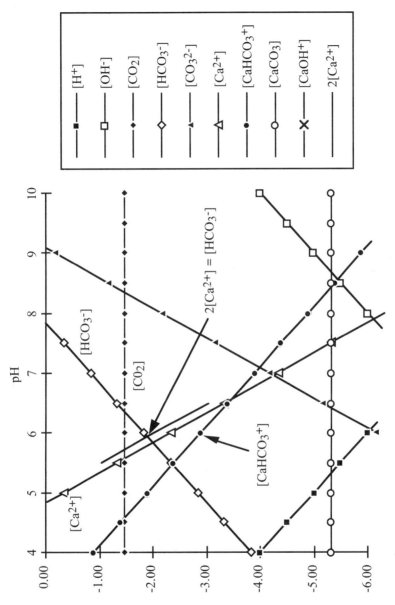

FIGURE 10.10. Log of concentrations of species in equilibrium with $P_{CO_2} = 1$ atm and solid calcite as a function of pH. The charge balance for no added acid or base is $2[Ca^{2+}] = [HCO_3^-] + \cdots$. Note that $[CaHCO_3^+]$ is 10% of $[HCO_3^-]$.

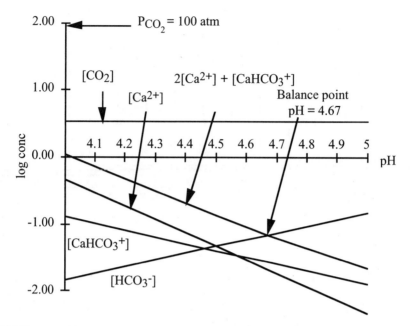

FIGURE 10.11. Magnified view of the balance point at $P_{CO_2} = 100$ atm, where $2[Ca^{2+}] + [CaHCO_3^+] = [HCO_3^-]$. With activity corrections, the balance point is 4.94.

Hence

$$\log \gamma_o = +0.01, \quad \log \gamma_+ = \log \gamma_- = -0.107$$

$$\log \gamma_{++} = \log \gamma_= = -0.428$$

In Eq. (32), substitute

$$K_{so} = K_{so}^o/(\gamma_{++}\gamma_=) = 10^{-8.27}$$

$$K_H = K_H^o/\gamma_o = 10^{-1.47}$$

$$K_1 = K_{a1}^o(\gamma_o/\gamma_+\gamma_-) = 10^{-6.14}$$

$$K_2 = K_{a2}^o/\gamma_= = 10^{-9.90}$$

$$K_{CaH} = K_{CaH}^o \gamma_{++} = 10^{+1.69}$$

$$pH = -\log\{[H^+]\gamma_+\} = -\log[H^+] + 0.107$$

($I = 0$ constants will be found in Table 10.2). The balance point with these corrections is at pH = 4.94.

Atmospheric Partial Pressure, $P_{CO_2} = 10^{-3.5}$. Returning to the average atmospheric partial pressure, $P_{CO_2} = 10^{-3.5}$, $[CaHCO_3^+]$ is much less important, and the simple relationship

COMBINING EQUILIBRIA WITH MASS AND CHARGE BALANCES

$$2[Ca^{2+}] = [HCO_3^-] + \cdots$$

is adequate to find the charge balance point.

$$\frac{2K_{so}[H^+]^2}{P_{CO_2}K_HK_1K_2} = \frac{P_{CO_2}K_HK_1}{[H^+]} + \cdots \tag{33}$$

Substuting $I = 0$ constants in Eq. (33) gives

$$[H^+]^3 = \frac{(P_{CO_2}K_HK_1)^2 K_2}{2K_{so}} = \frac{10^{2(-3.5-1.464-6.363)-10.329}}{10^{0.3-8.48}} = 10^{-24.80}$$

$$[H^+] = 10^{-8.27}$$

This point is shown by the arrow on Fig. 10.12. Note that both $[Ca^{2+}]$ and $[HCO_3^-]$ are approximately 10^{-3}, and hence activity coefficient corrections will be small ($\log \gamma_\pm = 0.015$).

Low Pressure: $P_{CO_2} = 10^{-6}$ atm. With extremely low partial pressures of CO_2, the terms in $[CO_3^{2-}]$ and $[OH^-]$ become important, as is shown in Fig. 10.13 for $P_{CO_2} = 10^{-6}$ atm. Fortunately, the ion-pair terms are small.

$$2[Ca^{2+}] = [HCO_3^-] + 2[CO_3^{2-}] + [OH^-] \tag{34}$$

$$\frac{2K_{so}[H^+]^2}{P_{CO_2}K_HK_1K_2} = \frac{P_{CO_2}K_HK_1}{[H^+]} + \frac{2P_{CO_2}K_HK_1K_2}{[H^+]^2} + \frac{K_w}{[H^+]} + \cdots \tag{35}$$

$$[H^+]^3 = 10^{-29.58} + \frac{10^{-39.84}}{[H^+]} + \cdots \tag{36}$$

Neglecting $[CO_3^{2-}]$, the second term on the right of Eq. (35), gives pH = 9.86; successive approximations using this value for $[H^+]$ in the second term gives

$$[H^+] = 10^{-9.82}$$

This point is shown as an arrow on the magnified chart of Fig. 10.13.

From these examples, it is clear that the simplest approximation is good for $P = 1$ down to almost $P = 10^{-6}$. Equation (35) yields the approximate formula (plotted in Fig. 10.14)

$$[H^+]^3 = \frac{(P_{CO_2}K_HK_1)^2 K_2}{2K_{so}} \tag{37}$$

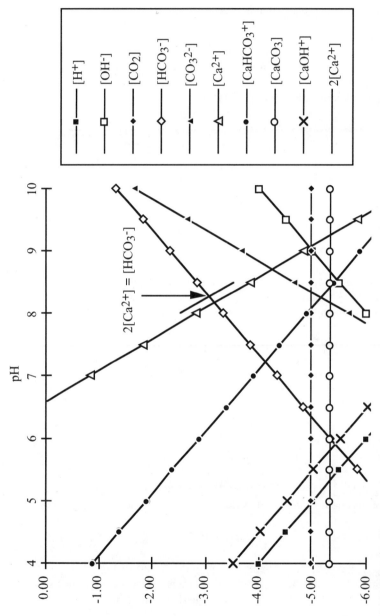

FIGURE 10.12. Log of concentrations of species in equilibrium with $P_{CO_2} = 10^{-3.5}$ atm and solid calcite as a function of pH. The charge balance for no added acid or base is $2[Ca^{2+}] = [HCO_3^-] + \cdots$. Note that $[CO_3^{2-}]$, $[CaCO_3]$ and $[CaHCO_3^+]$ are all small compared to $[HCO_3^-]$.

COMBINING EQUILIBRIA WITH MASS AND CHARGE BALANCES 387

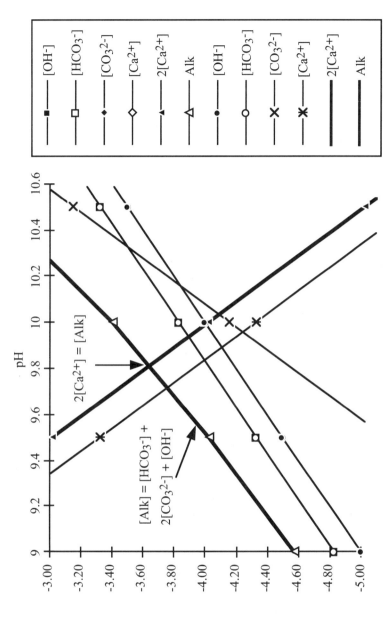

FIGURE 10.13. Log of concentrations of species in equilibrium with $P_{CO_2} = 10^{-6}$ atm and solid calcite as a function of pH. The charge balance for no added acid or base is $2[Ca^{2+}] = [HCO_3^-] + 2[CO_3^{2-}] + [OH^-]$. Note that the scales are magnified compared to Figs. 10.8 and 10.10 so as to make the lines clearer. $[CaCO_3]$, $[CaHCO_3]$, and $[CaOH^+]$ are all small compared to the four major species.

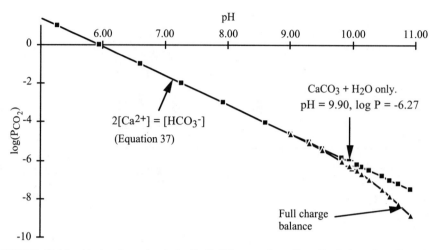

FIGURE 10.14. pH of water saturated with $CaCO_3$ at various P_{CO_2}. Deviations from linearity occur below log $P_{CO_2} = -5$ (and above +1; see Fig. 10.10).

For partial pressures above 1 atm, the term for $[CaHCO_3^+]$ should be included (see Eq. 32); for $P < 10^{-6}$, Eq. (30) is poorly conditioned, with the $[Ca^{2+}]$ and $[CaOH^+]$ terms tending to infinity and the $[HCO_3^-]$ and $[CO_3^{2-}]$ terms tending to zero. This case is discussed in the next section.

Solubility of $CaCO_3$ in Pure Water

The extreme condition, where the system contains only $CaCO_3$ and water, must rely on the mass balances of a closed system, Eqs. (24)–(28). The total carbonate C_T is not known, but depends on the extent to which $CaCO_3$ dissolves:

$$C_T = [Ca^{2+}] + [CaHCO_3^+] + [CaCO_3]$$
$$C_T = [CO_2] + [HCO_3^-] + [CO_3^{2-}] + [CaHCO_3^+] + [CaCO_3] \quad (38)$$

Combining these mass balances eliminates the ion-pair terms:

$$[Ca^{2+}] = [CO_2] + [HCO_3^-] + [CO_3^{2-}] \quad (39)$$

Combining Eq. (39) with the charge balance

$$[H^+] + 2[Ca^{2+}] + [CaHCO_3^+] = [HCO_3^-] + 2[CO_3^{2-}] + [OH^-] \quad (40)$$

to eliminate $[Ca^{2+}]$ gives a proton condition:

$$[HCO_3^-] + 2[CO_2] + [H^+] + [CaHCO_3^+] = [OH^-] + \cdots \quad (41)$$

At pH = 8 to 10, [H$^+$] and [CO$_2$] will be small, as will the ion pairs (at log $P = -3.5$ and pH = 8.3, [CaHCO$_3^+$] and [CaCO$_3$] are less than 1% of [HCO$_3^-$]—see Fig. 10.12), and these equations simplify to

$$[HCO_3^-] = [OH^-] + \cdots$$

$$\frac{[H^+][CO_3^{2-}]}{K_2} = \frac{K_w}{[H^+]} + \cdots \quad (42)$$

$$[CO_3^{2-}] = \frac{K_w K_2}{[H^+]^2} \quad (43)$$

The charge balance (Eq. 40 with [H$^+$] and [CaHCO$_3^+$] neglected) is

$$2[Ca^{2+}] = [HCO_3^-] + 2[CO_3^{2-}] + [OH^-] \quad (44)$$

Recall that $K_{so} = [Ca^{2+}][CO_3^{2-}]$

$$\frac{2K_{so}}{[CO_3^{2-}]} = \frac{[H^+][CO_3^{2-}]}{K_2} + 2[CO_3^{2-}] + \frac{K_w}{[H^+]} + \cdots \quad (45)$$

$$K_{so} = [CO_3^{2-}]^2 \left(1 + \frac{[H^+]}{2K_2}\right) + \frac{K_w[CO_3^{2-}]}{2[H^+]} + \cdots$$

Substitute for [CO$_3^{2-}$] from Eq. (43) to get

$$K_{so} = \frac{(K_w K_2)^2}{[H^+]^4}\left(1 + \frac{[H^+]}{2K_2}\right) + \frac{K_w^2 K_2}{2[H^+]^3} + \cdots$$

$$K_{so} = \frac{K_w^2 K_2}{[H^+]^3}\left(\frac{K_2}{[H^+]} + \frac{1}{2} + \frac{1}{2}\right) + \cdots$$

$$[H^+]^3 = \frac{K_w^2 K_2}{K_{so}}\left(1 + \frac{K_2}{[H^+]}\right) + \cdots \quad (46)$$

This is most easily solved by iteration. Neglecting the second term for a first approximation (using constants from Table 10.2) gives [H$^+$]^3 = 10$^{-29.84}$ or [H$^+$] = 10$^{-9.95}$. Using this value to estimate the second term gives a second approximation [H$^+$] = 10$^{-9.90}$. A third approximation does not change this value, which is marked on Fig. 10.14.

As a final check on the assumptions, calculate all the various species concentrations using the above equations and the equilibria in Table 10.2 (p. 381):

$$[H^+] = 10^{-9.90}$$
$$[OH^-] = K_w/[H^+] = 10^{-4.10}$$
$$[HCO_3^-] = [OH^-] = 10^{-4.10}$$
$$[CO_3^{2-}] = 10^{-4.53} \quad \text{from Eq. (42)}$$
$$[Ca^{2+}] = K_{so}/[CO_3^{2-}] = 10^{-3.95}$$
$$[CaCO_3] = 10^{+3.15} \; K_{so} = 10^{-5.33}$$
$$[CaHCO_3^+] = 10^{1.26} \, [Ca^{2+}][HCO_3^-] = 10^{-6.79}$$
$$[CaOH^+] = 10^{+1.3} \, [Ca^{2+}][OH^-] = 10^{-6.75}$$
$$[CO_2] = [H^+][HCO_3^-]/K_2 = 10^{-7.64}$$
$$P_{CO_2} = [CO_2]/K_H = 10^{-6.18} \text{ atm}$$

The assumptions made above, that $[H^+]$, $[CO_2]$, and the ion pairs can be neglected, is confirmed. $I = 3 \times 10^{-4}$ and $\gamma_+ = 0.98$.

Higher pH. To obtain pH values higher than 9.9 (and hence P_{CO_2} lower than $10^{-6.2}$), strong base must be added to the system; this decreases $[CO_3^{2-}]$ and increases $[Ca^{2+}]$. At sufficiently high pH (about 12.5), solid $Ca(OH)_2$ precipitates.

The equations for the line on Fig. 10.14 labeled "Full Chg Bal" are closely related to those used in the previous section. The charge balance, with ion pairs neglected, is Eq. (44). Substitute for $[HCO_3^-]$ and $[CO_3^{2-}]$ in terms of P_{CO_2} and $[H^+]$, and use $[Ca^{2+}] = K_{so}/[CO_3^{2-}]$ to get

$$\frac{2K_{so}[H^+]^2}{P_{CO_2}K_HK_1K_2} = \frac{P_{CO_2}K_HK_1}{[H^+]} + \frac{2P_{CO_2}K_HK_1K_2}{[H^+]^2} + \frac{K_w}{[H^+]} - [H^+] \qquad (47)$$

This can be rearranged to give a quadratic equation for P_{CO_2}:

$$P_{CO_2}^2 \left(\frac{K_HK_1}{[H^+]} + \frac{2K_HK_1K_2}{[H^+]^2} \right) + P_{CO_2} \left(\frac{K_w}{[H^+]} - [H^+] \right) - \frac{2K_{so}[H^+]^2}{K_HK_1K_2} = 0 \qquad (48)$$

An iterative solution is also possible, with the previous value of P_{CO_2} being used to evaluate the linear term. The results of this are plotted in Fig. 10.14. Note that when pH is larger than about 9, the quadratic solution is more direct and accurate than the iterative one. The pH obtained from the approximate Eq. (37) p. 385, is always larger than the quadratic solution; the difference is as large as 1 pH unit for pH = 12.

Including $[CaOH^+]$ in the charge balance introduces another term in Eq. (47) and produces a larger value of P_{CO_2} at high pH

BUFFER CAPACITY

$$\frac{K_{so}[H^+]^2}{P_{CO_2}K_HK_1K_2}\left(2 + \frac{K_{CaOH}K_w}{[H^+]}\right) = \frac{P_{CO_2}K_HK_1}{[H^+]} + \frac{2P_{CO_2}K_HK_1K_2}{[H^+]^2} + \frac{K_w}{[H^+]} - [H^+]$$

BUFFER CAPACITY

The buffering action of the carbon dioxide–carbonate system is very important in many practical applications: It is the principal acid–base system in seawater and in most fresh waters; it is a major factor in regulating the partial pressure of carbon dioxide in the atmosphere, and in regulating the acid–base balance of physiological systems. In this chapter, the general principles are set out; later, some detailed examples are given.

The buffer index of an acid–base system was defined in Chapter 4 (pp. 133–134) to be

$$\beta = \left(\frac{\partial C_b}{\partial pH}\right) \tag{49}$$

This definition applies to one-phase solutions; the constraint on this derivative is that the total concentration of each acid–base pair should be held constant while base (C_b) is added. Equation (49) also applies to a carbon dioxide–carbonate solution in a closed container, if C_T is held constant while base is added.

Types of Buffer Index

However, there are other variations on Eq. (49). If the solution is in equilibrium with gas-phase CO_2, a different constraint, constant P_{CO_2}, is more appropriate. If the solution is in equilibrium with solid $CaCO_3$, this provides an additional constraint, and can greatly increase the buffer capacity of the system. Here are the various buffer indices that will be developed in this section:

$$\beta_C = \left(\frac{\partial C_b}{\partial pH}\right)_{C_T}, \quad CO_2(\text{sol'n}) + H_2O, \text{ closed}$$

$$\beta_P = \left(\frac{\partial C_b}{\partial pH}\right)_{P_{CO_2}}, \quad CO_2(\text{gas phase}) + H_2O$$

$$\beta_{P,s} = \left(\frac{\partial C_b}{\partial pH}\right)_{P,s}, \quad CO_2(\text{gas}) + H_2O + CaCO_3 \text{ (solid)}$$

$$\beta_{D,s} = \left(\frac{\partial C_b}{\partial pH}\right)_{D,s}, \quad CO_2(\text{sol'n}) + H_2O + CaCO_3 \text{ (solid), closed}$$

The last derivative cannot be at constant C_T, since dissolution or precipitation of $CaCO_3$ changes C_T, but the quantity $D = 2(C_T - [Ca^{2+}])$, "the difference between carbonate and calcium", remains invariant if $CaCO_3$ dissolves or precipitates, and so can be held constant for this system. See p. 394.

Homogeneous Solution in Closed Systems

Evaluating the buffer indices in terms of solution composition proceeds as in Chapters 4 and 5: The charge balance, with strong base concentration C_b represented by $[Na^+]$ or alkalinity, is differentiated with respect to $[H^+]$, and that function is converted to a derivative with respect to pH: $d[H^+] = -2.303[H^+]\, dpH$ (Eq. 80, p. 134).

As above, the carbonate alkalinity is defined to be

$$A_C = [HCO_3^-] + 2\,[CO_3^{2-}] + [OH^-] - [H^+]$$

At constant C_T, with neither gas phase nor solid phase, Eqs. (27) and (28) (p. 380) yield

$$A_C = C_T \frac{K_{a1}[H^+] + 2K_{a1}K_{a2}}{[H^+]^2 + K_{a1}[H^+] + K_{a1}K_{a2}} + \frac{K_w}{[H^+]} - [H^+] \quad (50)$$

Take the derivative of each term with respect to $[H^+]$, holding C_T and the equilibrium constants constant, and obtain:

$$\left(\frac{\partial A_C}{\partial [H^+]}\right)_{C_T} = -C_T K_{a1}\frac{[H^+]^2 + 4K_{a2}[H^+] + K_{a1}K_{a2}}{([H^+]^2 + K_{a1}[H^+] + K_{a1}K_{a2})^2} - \frac{K_w}{[H^+]^2} - 1 \quad (51)$$

This expression is general for any diprotic acid (See Chap. 5, pp. 190–193), but because of the numerical values of K_{a1} and K_{a2} for CO_2, the middle term of the numerator $(4K_{a2}[H^+])$ is less than 1% of the other terms for any pH from 0 to 14 and is normally omitted.[27]

Converting the derivative with respect to $[H^+]$ to a derivative with respect to pH:

$$\beta_C = -2.303[H^+]\left(\frac{\partial A_C}{\partial [H^+]}\right)_{C_T}$$

$$= 2.303\left(C_T \frac{[H^+]K_{a1}([H^+]^2 + K_{a1}K_{a2})}{([H^+]^2 + [H^+]K_{a1} + K_{a1}K_{a2})^2} + \frac{K_w}{[H^+]} + [H^+]\right) \quad (52)$$

and identifying the various terms in Eq. (52) with the concentrations as given by Eqs. (26)–(28)

$$\beta_C = 2.303[HCO_3^-]\frac{[CO_3^{2-}]+[CO_2]}{C_T} + 2.303\,([H^+]+[OH^-]) \qquad (53)$$

In Fig. 10.15, the curve of β_C is overlaid on a logarithmic diagram similar to Fig. 10.9. Note that at sufficiently low pH, β_C is determined by $[H^+]$, and at sufficiently high pH, by $[OH^-]$. The minima in β_C correspond to the "CO_2 end point" (pH ≈ 4.5) and the "HCO_3^- end point" (pH ≈ 8.3) in the alkalinity titration. The maximum occurs when pH = pK_{a1} = 6.3, or $[CO_2] = [HCO_3^-]$. Note also that there is no minimum at the "CO_3^{2-} end point" (pH ≈ 10.5), in agreement with the observation that there is no inflection in the titration curve at that point.

Solution with Gas phase at Constant P_{CO_2}

The derivation of the buffer index at constant P_{CO_2} is a little simpler than at constant C_T. Make use of Eq. (22), p. 375 to evaluate the carbonate alkalinity in terms of P_{CO_2} and pH.

$$A_C = K_H P_{CO_2}\left(\frac{K_{a1}}{[H^+]} + 2\frac{K_{a1}K_{a2}}{[H^+]^2}\right) + \left(\frac{K_w}{[H^+]} - [H^+]\right) \qquad (54)$$

Take the derivative with respect to $[H^+]$, holding P_{CO_2} and the equilibrium constants constant.

$$\left(\frac{\partial A_C}{\partial [H^+]}\right)_{P_{CO_2}} = K_H P_{CO_2}\left(-\frac{K_{a1}}{[H^+]^2} - \frac{4K_{a1}K_{a2}}{[H^+]^3}\right) + \left(-\frac{K_w}{[H^+]^2} - 1\right) \qquad (55)$$

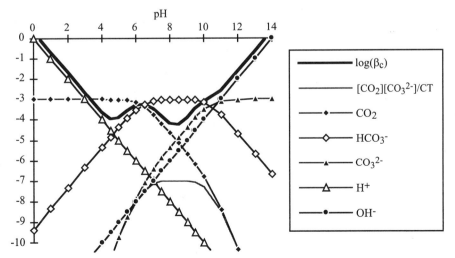

FIGURE 10.15. Logarithmic diagram for CO_2 at total concentration 10^{-3} M in a closed system. The bold line represents the buffer index as given by Eq. (52) or (53).

$$\beta_P = 2.303([HCO_3^-] + 4[CO_3^{2-}] + [OH^-] + [H^+]) \quad (56)$$

Solution with Solid $CaCO_3$ and Gas Phase at Constant P_{CO_2}

The buffer capacity in the presence of solid $CaCO_3$ and gaseous CO_2 can be obtained with a slight modification of Eq. (56). Add terms for $2[Ca^{2+}]$ and its ion pairs to the charge balance, express them in terms of P_{CO_2} and $[H^+]$ in equilibrium with $CaCO_3(s)$ (see Eq. 30), and perform the same sort of manipulation as above to obtain:

$$\beta_{P,s} = 2.303([HCO_3^-] + 4[CO_3^{2-}] + [OH^-] + [H^+] + 4[Ca^{2+}]$$
$$+ [CaHCO_3^+] + [CaOH^+]) \quad (57)$$

This curve, along with β_P, is plotted in Fig. 10.16. Comparing the situation with and without solid $CaCO_3$, you can see that the presence of the solid greatly restricts any change toward lower pH.

Buffer Index of Closed System with CO_2 and $CaCO_3$

Finally, the buffer index for a closed system in equilibrium with solid $CaCO_3$ requires considerable manipulation and does not yield simple and elegant functions as were obtained above.[28] For the following derivation the ion pairs $[CaHCO_3^+]$, $[CaCO_3]$, and $[CaOH^+]$ will be omitted, since they are normally small compared to the other concentrations (see Figs. 10.10–10.13).[29]

The procedure begins with the definition of D, the "difference between carbonate and calcium." Note $D = 0$ for $CaCO_3$ with no excess Ca^{2+} or CO_2.

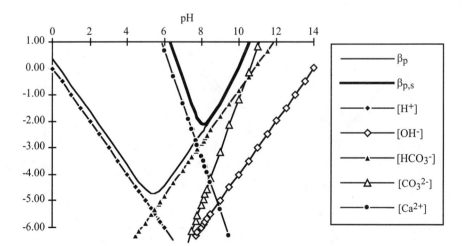

FIGURE 10.16. Buffer index of solutions in equilibrium with gaseous $CO_2(\beta_P)$ (Eq. 56) and also in equilibrium with solid $CaCO_3(\beta_{p,s})$ (Eq. 57). $P_{CO_2} = 0.001$ atm. Note that the presence of $CaCO_3$ greatly increases the buffer index in the lower pH range.

$$D = 2(C_T - [Ca^{2+}])$$

and the charge balance

$$[Alk] + [H^+] + 2[Ca^{2+}] = [HCO_3^-] + 2[CO_3^{2-}] + [OH^-]$$

Substitute for $[Ca^{2+}]$ in terms of D, and expand $C_T = [CO_2] + [HCO_3^-] + [CO_3^{2-}]$ to get

$$D = [Alk] + 2[CO_2] + [HCO_3^-] + [H^+] - [OH^-] \tag{58}$$

It is convenient to define the functions

$$x = 2[CO_2] + [HCO_3^-] \tag{59}$$

$$y = [H^+]^2 + K_{a1}[H^+] + K_{a1}K_{a2} \tag{60}$$

$$z = 2[H^+]^2 + K_{a1}[H^+] \tag{61}$$

Note that $x = C_T z/y$, and hence

$$C_T = \frac{xy}{z} \tag{62}$$

In equilibrium with $CaCO_3$, $[Ca^{2+}] = K_{so}/[CO_3^{2-}]$, hence

$$[Ca^{2+}] = \frac{K_{so} y}{C_T K_{a1} K_{a2}} \tag{63}$$

$$D = \frac{2xy}{z} - \frac{2zK_{so}}{xK_{a1}K_{a2}} \tag{64}$$

Equation (64) is a quadratic in x, which has the solution

$$x = \frac{zQ}{4y} \tag{65}$$

where

$$Q = \left(\frac{z}{4y}\right)\left(D + \sqrt{D^2 + \frac{16K_{so}y}{K_{a1}K_{a2}}}\right) \tag{66}$$

Equations (59)–(66) can be substituted in the modified charge balance (Eq. 58) to obtain [Alk] as a function of $[H^+]$ for a given value of D—this is the alkalinity titration curve in the presence of solid $CaCO_3$ in a closed system.

The variation of acidity and alkalinity with pH is shown in Fig. 10.17(a) as a "titration curve" of pH versus [Alk] at constant D, and in Fig. 10.17(b) as log of alkalinity or acidity versus pH at constant D. It is the derivative of these functions that gives the buffer index $\beta_{D,s}$. Note that the point of zero alkalinity is significantly

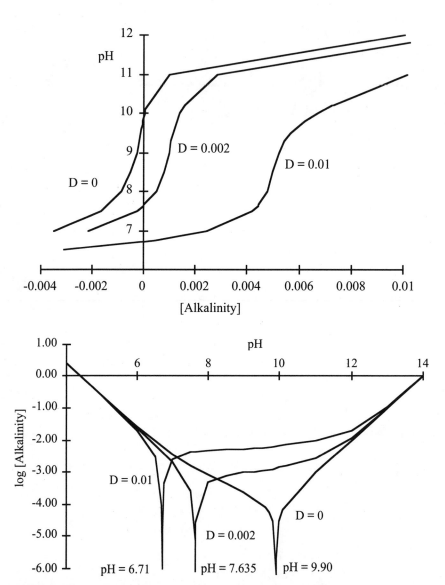

FIGURE 10.17. Acidity/alkalinity of carbonate solutions in equilibrium with solid $CaCO_3$. (a) "Titration curve"—pH as a function of alkalinity. (b) log of acidity/alkalinity versus pH. The point of zero alkalinity changes from pH = 4.5 in solutions without solid, to pH = 6.7 at D = 0.01, to pH = 7.6 at D = 0.002, to pH = 9.9 at D = 0.

higher (pH = 9.9 at D = 0, pH = 7.6 at D = 0.002, pH = 6.7 at D = 0.01) than the usual end point (pH = 4.5) for carbonate titrations without solid $CaCO_3$.

The derivative of the titration curve pH versus [Alk], with respect to pH, maintaining constant D and equilibrium with solid $CaCO_3$, gives the desired buffer index:

$$\beta_{D,s} = \left(\frac{\partial[\text{Alk}]}{\partial \text{pH}}\right)_{D,s} = -2.303[\text{H}^+]\left(\frac{\partial[\text{Alk}]}{\partial[\text{H}^+]}\right)_{D,s} \tag{67}$$

$$\beta_{D,s} = 2.303\left([\text{H}^+]\left(\frac{\partial x}{\partial[\text{H}^+]}\right)_{D,s} + [\text{H}^+] + [\text{OH}^-]\right) \tag{68}$$

The algebraic derivative involves the simultaneous solution of the following equations before substitution in Eq. (67). These functions are obtained directly by differentiation of Eqs. (60)–(66):

$$\frac{dx}{d[\text{H}^+]} = \frac{z}{4y}\left(\frac{dQ}{d[\text{H}^+]}\right) + \frac{Q}{4y}\left(\frac{dz}{d[\text{H}^+]}\right) - \frac{zQ}{4y^2}\left(\frac{dy}{d[\text{H}^+]}\right) \tag{69}$$

$$\frac{d[\text{Ca}^{2+}]}{d[\text{H}^+]} = \frac{K_{so}}{K_{a1}K_{a2}}\left[\frac{1}{x}\left(\frac{dz}{d[\text{H}^+]}\right) - \frac{z}{x^2}\left(\frac{dx}{d[\text{H}^+]}\right)\right] \tag{70}$$

$$\frac{dQ}{d[\text{H}^+]} = \frac{1}{2}\left(D^2 + \frac{16K_{so}}{K_{a1}K_{a2}}y\right)^{-1/2}\frac{16K_{so}}{K_{a1}K_{a2}}\left(\frac{dy}{d[\text{H}^+]}\right) \tag{71}$$

$$\frac{dz}{d[\text{H}^+]} = 4[\text{H}^+] + K_{a1} \tag{72}$$

$$\frac{dy}{d[\text{H}^+]} = 2[\text{H}^+] + K_{a1} \tag{73}$$

The algebraic solution of these equations yields

$$[\text{H}^+]\frac{dx}{d[\text{H}^+]} = \frac{Q}{4y^2}(K_{a1}[\text{H}^+])(K_{a1}K_{a2} + 4K_{a2}[\text{H}^+] + [\text{H}^+]^2)$$

$$+ \frac{(K_{a1}[\text{H}^+]) + 2[\text{H}^+]^2)}{y^2}\left(\frac{2K_{so}y}{K_{a1}K_{a2}}\right)\left(D^2 + \frac{16K_{so}y}{K_{a1}K_{a2}}\right)^{-1/2} \tag{74}$$

These algebraic combinations can be expressed in terms of species concentrations; for example, $C_T K_{a1}[\text{H}^+]/y = [\text{HCO}_3^-]$ or $K_{so}y/(C_T K_{a1} K_{a2}) = [\text{Ca}^{2+}]$. The result is

$$[H^+]\frac{dx}{d[H^+]} = \frac{D+\sqrt{D^2+16C_T[Ca^{2+}]}}{4C_T^2}([HCO_3^-]([CO_2]+[CO_3^{2-}])+4[CO_2][CO_3^{2-}])$$

$$+\frac{([HCO_3^-]+2[CO_2])^2}{C_T}\left(\frac{2[Ca^{2+}]}{\sqrt{D^2+16C_T[Ca^{2+}]}}\right) \quad (75)$$

This algebraic answer agrees exactly with the results of differentiating Eqs. (65) and (66) numerically. Figure 10.18 shows the results of this calculation (with $\Delta pH = 0.0001$). There are only small differences between $D = 0.002$ and 0. For $D = 0.01$, $\beta_{D,s}$ is a little higher in the pH range from 9 to 11.

Note that if $[Ca^{2+}] = 0$, $D = 2 C_T$, and the term $4[CO_2][CO_3^{2-}]$ is negligible; then $\beta_{D,s}$ from Eq. (75) reduces to the equation for β_C, as given by Eq. (53), p. 393.

A very rough approximation for $\beta_{D,s}$ can be obtained as follows: The lower-pH branch of the curve is a little above the curve for $[Ca^{2+}]$ and the high-pH branch is a little above the curve for $[OH^-]$—thus $\beta_{D,s} \approx 2.3 ([Ca^{2+}] + [OH^-])$ to within a factor of ±2 for a wide range of pH and D.

A magnified view of the difference between the rough formula and the numerically differentiated alkalinity is shown in Fig. 10.19. Although significant errors are introduced by using the rough approximation, they generally do not exceed a factor of ±2 or ±3, which can be adequate for many purposes.

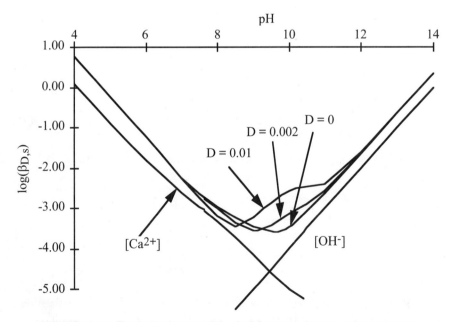

FIGURE 10.18. The buffer index of CO_2–$CaCO_3(s)$ solutions in a closed vessel.

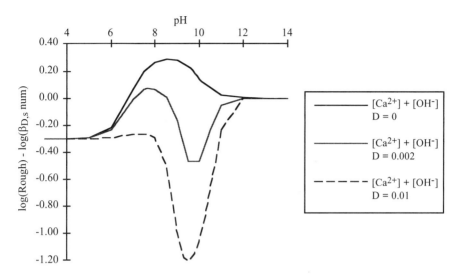

FIGURE 10.19. Difference between the rough approximation log $\beta_{D,s} \approx 2.3$ ([Ca^{2+}] + [OH^-]) and the numerically differentiated log $\beta_{D,s}$ for $D = 0$, 0.002 and 0.01.

Correction of Previously Published Equations

I am ashamed to admit that the equations for $\beta_{D,s}$ in both editions of *Carbon Dioxide Equilibria*[30] are incorrect. The algebraic solution (Eqs. 4.70 and 4.49, p. 99) is almost the same as Eq. (74) here, except that it contains a minus sign between the major terms, where there should be a plus sign.

The equation in terms of species concentrations (Eqs. 4.71 and 3.62, p. 100) is much more severely in error, as comparison with Eq. (75) here will show. Both the square root terms involving D, etc., were omitted in *Carbon Dioxide Equilibria*. Even though $\beta_{D,s}$ reduces to β_C when [Ca^{2+}] = 0, the remainder of the equation is incomplete. A graphical comparison is shown in Fig. 10.20.

Summary of Buffer Capacity Results

As we saw in the case of constant partial pressure, the presence of solid $CaCO_3$ greatly increases the buffering action of carbonate solutions, especially at low pH. Compare the following:

The system with $C_T = 0.001$ M and neither gas nor solid phase has a minimum buffer index of $10^{-3.5}$ to 10^{-4} over the pH range from 4 to 9 (Fig. 10.15).

The system with gas phase $P_{CO_2} = 0.001$ atm alone (Fig. 10.16) has a minimum buffer index of about 10^{-5} at pH = 5.

The system with solid but without gas phase ($D = 0.002$) (Fig. 10.18) has a minimum buffer index of about $10^{-3.5}$ at pH = 9.

The system with both gas phase ($P_{CO_2} = 0.001$ atm) and solid $CaCO_3$ (Fig. 10.16) has a minimum buffer index of 10^{-2} at pH = 8.

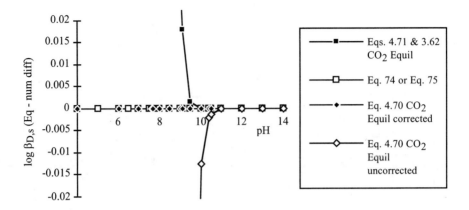

FIGURE 10.20. Comparison of Eq. 74 (or Eq. 75) with equations for $\beta_{D,s}$ published previously. Eq. 4.70 is algebraic, corresponding to Eq. 74; Eqs. 4.71 and 3.62 are in terms of species concentrations.

Thus the combination of solid and gas phase is in general more effective at buffering the solution than either one alone.

INTRODUCTION TO CASE STUDIES

In the second half of this chapter, I will discuss several topics of practical importance. These are more complicated and detailed than "examples"; so I have called them "case studies."

The first topic is the variety of forms for the equilibrium constants of the carbon dioxide–carbonate system. As the number of components in a solution increases, and as the concentration increases, ion pairs become more important. This section is an elaboration of the earlier discussion of ionic-strength dependence (pp. 369–372) to include the issues raised by the presence of ion pairs. At least six distinct forms have appeared in the literature, and I hope a side-by-side comparison will make them less confusing.

Second is a discussion of how carbon dioxide equilibria can be applied to solutions of physiological importance. The simplest are intravenous salt solutions such as Ringer's solution and Normosol; more complicated examples of chemical models are interstitial fluids, blood plasma, and red blood cell intercellular fluids.

Third, carbon dioxide equilibria are applied to test whether an unusually accurate alkalinity tiration can indeed be represented by the known composition and equilibria of seawater. A related problem is calculating the buffer factor, which relates changes in atmospheric carbon dioxide to changes in alkalinity and total dissolved carbonate in seawater—an important issue in understanding global climate change.

Proceeding along these lines, a general ion-pair model is developed and applied to find the saturation index for $CaCO_3$ in seawater, and also in mixtures of seawater and river water. It is used to show how mixing two supersaturated waters can yield a range of undersaturated waters—an important process in producing the fissures and caves of Karst topography.

SIX FORMS FOR THE EQUILIBRIUM CONSTANT[41]

In most of the derivations above, the ions were assumed to be individual ("free") charged species. In a few cases, however, additional equilibria were introduced to account for ion pairs such as $CaHCO_3^+$. As the number of components in the solution increases, and as the concentration increases, ion pairs become more important, and as a result several additional forms for the acid–base equilibria of CO_2 have been developed.

The following concentration equilibria were introduced on pp. 367–372. Although the distinction between free and total ion concentrations is not always explicit, the basic equilibria apply to the free ion concentrations:

$$[H^+]_f [HCO_3^-]_f = K_{a1} [CO_2]_f$$
$$[H^+]_f [CO_3^{2-}]_f = K_{a2} [HCO_3^-]_f$$

K_{a1} and K_{a2} depend on ionic strength as well as temperature. Activity coefficients, expressing the electrostatic interactions between ions, as well as a small salting-out effect, were introduced to express the ionic strength dependence:

$$[H^+]_f \gamma_+ [HCO_3^-]_f \gamma_- = K_{a1}^\circ [CO_2]_f \gamma_0$$
$$[H^+]_f \gamma_+ [CO_3^{2-}]_f \gamma_= = K_{a2}^\circ [HCO_3^-]_f \gamma_-$$

Activity coefficients at concentrations below 0.5 M can be estimated using the Davies equation (p. 49), which accounts approximately for the electrostatic effects (first term in brackets) and salting-out (second term).

$$\log(\gamma_z) = -0.5z^2 \left(\frac{\sqrt{I}}{1+\sqrt{I}} - 0.2I \right) \quad (76)$$

where I is the ionic strength (p. 45) and z is the charge on the ion.

You will recall that the activity coefficient of uncharged CO_2 (due to salting out, p. 367) was simply

$$\log \gamma_0 = b I \quad (77)$$

where b is approximately 0.1. A further simplification results from the absence of any complexes between CO_2 and the other components of physiological fluids or seawater, so that $[CO_2]_f = p[CO_2]_T$.

These examples will be presented as variations on the equilibrium K_{a1}. Small modifications will give the corresponding expressions for K_{a2}. The numerical value given with each equation corresponds to 25°C, at either $I = 0$ or 0.7 (seawater).

1. Activity of free ions (see Table 10.1):

$$K_{a1}^\circ = \frac{[H^+]_f \gamma_+ [HCO_3^-]_f \gamma_-}{[CO_2] \gamma_o} = 10^{-6.363} \tag{78}$$

K_{a1}° depends on temperature and to some extent on total pressure.

2. Concentration of free ions (recall $\gamma_+ = \gamma_- = 10^{-0.16}$ and $\gamma_o = 10^{+0.07}$ at $I = 0.7$):

$$K_{a1} = \frac{[H^+]_f [HCO_3^-]_f}{[CO_2]} = 10^{-5.97} \tag{79}$$

K_{a1} depends on ionic strength and temperature.

3. Empirical pH scale (NBS[42]) and free ions:

$$K'_{1(f)} = \frac{10^{-pH}[HCO_3^-]_f}{[CO_2]} = 10^{-6.08} \tag{80}$$

$K'_{1(f)}$ depends on ionic strength, temperature, and the method by which the pH scale is calibrated.[43]

4. Empirical pH scale (i.e., NBS) and total ions[44]:

$$K'_{1(T)} = \frac{10^{-pH}[HCO_3^-]_T}{[CO_2]} = 10^{-6.005} \tag{81}$$

Here the total concentration including ion pairs containing HCO_3^- (but not CO_3^{2-}, which is taken care of in the K_2 equilibrium) is used, and $K'_{1(T)}$ depends on ionic strength, temperature, pH method, and the concentrations of other (positive) ions in solution. For example, in seawater:

$$[HCO_3^-]_T = [HCO_3^-]_f + [NaHCO_3] + [MgHCO_3^+] + [CaHCO_3^+] + \cdots$$
$$= [HCO_3^-]_f (1 + K_{NaHCO_3}[Na^+] + K_{MgHCO_3}[Mg^{2+}] + K_{CaHCO_3}[Ca^{2+}] + \cdots)$$
$$= [HCO_3^-]_f \, \alpha_{HCO_3} \tag{82}$$

where $K_{MX} = [MX]/([M][X])$ is the formation constant for the ion pair MX (p. 435). For standard seawater, $[HCO_3^-]_f = 1.39$ mM, $[HCO_3^-]_T = 1.78$ mM, and $\alpha_{HCO_3} = 1.20 = 10^{+0.08}$.

5. To avoid the uncertainties in the pH scale, a number of workers have preferred to use equilibrium constants based on free H^+ concentration. Ideally, a cell

SIX FORMS FOR THE EQUILIBRIUM CONSTANT

without liquid junction (See Harned cell, Chapter 2) is calibrated using an electrolyte of accurately known activity, such as HCl. Modifying Eq. 79 to use the total concentration of HCO_3^- gives[45]

$$K_{a1(\text{free H})} = \frac{[H^+]_f[HCO_3^-]_T}{[CO_2]} = 10^{-6.00} \tag{83}$$

6. However, for titration in seawater, a convenient calibration point for the pH scale is when excess acid has been added (see discussion later in this chapter). This excess acid is not only free H^+, but also the species resulting from the reaction of H^+ with anions such as SO_4^{2-} and F^-. In this case, the acidity scale that results, called the "seawater scale" (SWS), is neither the NBS pH nor free $[H^+]$, but a hybrid[46]:

$$K_{1(\text{SWS})} = \frac{[H^+]_T[HCO_3^-]_T}{[CO_2]} = 10^{-5.83} \text{ to } 10^{-5.85} \tag{84}$$

$$[H^+]_T = [H^+]_f + [HSO_4^-] + [HF] + \cdots \tag{85}$$

$$[H^+]_T = [H^+]_f\left(1 + \frac{[SO_4^{2-}]_T}{[H^+]_f + K_{a(HSO_4)}} + \frac{[F^-]_T}{[H^+]_f + K_{a(HF)}} + \cdots\right) = [H^+]_f \alpha_H$$

α_H is independent of $[H^+]_f$ provided $[H^+]_f$ is small compared to $K_{a(HSO_4)}$ and $K_{a(HF)}$. For 35% seawater at 25°C, $pK_{a(HSO_4)} = 0.983$[47], $pK_{a(HF)} = 2.63$[48], $[SO_4]_T = 0.02824$, and $[F]_T = 7 \times 10^{-5}$, $\log \alpha_H = 0.115$.[49] This produces an equilibrium constant that is still different from the others.

Of these variations on the theme of pK_1, the most accurate are those that can be measured directly with the least ambiguity. The fifth type ($pK_{a1(\text{free H})}$) can be measured using a cell without liquid junction. Since it uses total concentrations, it does not rely on estimates of the ion-pairing constants or on empirical standards such as the NBS scale. Measurements at high ionic strength can be extrapolated to zero ionic strength to obtain pK_{a1}°.[50]

The next most reliable constant is the sixth type ($K_{a1(\text{SWS})}$), calibrated in a constant ionic medium at pH < 4.5. Although pH electrode setups employing liquid junction potentials are normally used for alkalinity titrations, all the factors $[H^+]_T$, $[HCO_3^-]_T$, and $[CO_2]$ can be determined independently, and therefore it is possible to measure $K_{a1(\text{SWS})}$ in a cell without liquid junction.

The second type (pK_{a1}) is almost identical with $pK_{a1(f)}$ at low ionic strengths or in weakly complexing media such as NaCl, but at high ionic strengths or in multicomponent solutions, the relationship between $[HCO_3^-]_T$ and $[HCO_3^-]_f$ requires additional data and introduces additional uncertainty.

Finally, the constants that employ the NBS pH scale, the third and fourth types, ($K'_{(f)}$ and $K'_{1(T)}$), while easy to use, suffer from uncertainty because of the changes in liquid junction potential between the standard solutions at $I = 0.05$ and the sample at considerably higher ionic strength.

Of all these choices, the highest pK_1 is 6.36 at $I = 0$ and the lowest is 5.85 (SWS), so that for rough calculations, an adjustment of the $I = 0$ value to $I = 0.7$ by means of the Davies equation gives a representative value ($pK_{a1} = 5.97$).

Analogous equilibrium expressions can be developed for K_2 and K_w, for example,

$$K_{2(SWS)} = \left(\frac{[H^+]_T[CO_3^{2-}]_T}{[HCO_3^-]_T}\right) = 10^{-8.90} \text{ to } 10^{-8.92} \qquad (86)$$

$$K_{w(SWS)} = [H^+]_T[OH^-]_T = 10^{-13.23} \qquad (87)$$

These values,[51] which include both the electrostatic ionic strength effects and the rather substantial ion pairs between the seawater cations and $[CO_3^{2-}]$ or $[OH^-]$, may be compared with the $I = 0$ values (Table 10.1)

$pK_{a2}^\circ = 10.33$, $pK_{2(SWS)}$ is 1.43 log units lower

$pK_w^\circ = 14.00$, $pK_{w(SWS)}$ is 0.77 log units lower

A summary of these variations on 10 relevant equilibria will be found in Table 10.13, p. 425.

PHYSIOLOGICAL FLUIDS

Physiological fluids such as blood plasma and the synthetic saline solutions used in intravenous therapy are closely related to seawater in their composition, being primarily NaCl, but are more dilute. In Table 10.3 four physiological fluids are listed: interstitial fluid, ISF; blood plasma; red blood cell intracellular fluid (RBC-ICF); and general intracellular fluid.[52] Also listed are Ringer's solution,[53] Ringer's solution to which $NaHCO_3$ has been added, and Normosol-R,[54] to which $NaHCO_3$ has been added. Ringer's and Normosol are used for intravenous therapy. The last fluid is seawater that has been diluted to 8‰ or ionic strength 0.16 to match the others.

Stable physiological activity occurs in an extremely narrow range of pH. An arterial blood pH less than 7.36 is considered abnormal and is called "acidemia"; pH greater than 7.44 is called "alkalemia." The more general terms "acidosis" and "alkalosis" refer to exposure to an acid load or alkaline load; these can cancel each other under the proper circumstances. The body responds to alteration of pH either by buffering or by excretion from the lungs (CO_2) or kidneys (HCO_3^-). Immediate buffering of an acid or alkaline load depends on the extracellular bicarbonate system and intracellular phosphate and protein—a total capacity of 12 to 15 meq/kg body weight. The most important excretion of acid occurs through the lungs—13,000 meq/day of CO_2; another important pathway is the excretion of nonvolatile acids (HSO_4^-, NH_4^+, uric acid), from the kidney—about 70 meq/day.[55]

TABLE 10.3. Ionic composition (millimole/L) of physiological fluids and seawater

Ion	ISF (Stewart p. 136)	Plasma	RBC-ICF	Gen ICF	Ringer	Ringer +NaHCO$_3$	Normosol +NaHCO$_3$	SW[56] salinity 8
Na$^+$	137	143	19	10	147.2	163.9	156.9	106.4
K$^+$	3	4	95	155	4	3.9	4.9	2.31
Mg^{2+}	4	4	5	10	0		1.4	11.2
Ca^{2+}	2	2	0	0	2.2	2.2	0	2.14
Cl$^-$	111	107	52	10	155.7	152.6	96	124.8
HCO$_3^-$	31	25	15	12	0	19.6	19.6	0.691
CO$_3^{2-}$	.045	.037	.014	.0067	0	0.1	0.1	0.0049
CO$_2$	2.3	1.8	1.8	2.3	varies	1.5	1.5	0.0106
SO$_4^{2-}$								4.56
Br$^-$								0.19
B(OH)$_4^-$								0.008
A^- (anion gap)	0	17	42	118	0	0	0	0
HA	0	3	18	82	0	0	0	0
Acetate							26.5	
Gluconate							22.6	
HA_T	0	0.02	0.06	0.2	0	0	0.0491	
pK_{a2}[57]		6.70	6.82	6.82			4.62 (acetate)	
							3.82 (gluconate)	
Strong anion equivalents	1	1	10	35				
Alk (carb)	31	25	15	12	0	19.6	19.6	0.79 incl. ip.
Alk (str ion)	40	51	62	140	0	19.6	68.7	0.50
SID (Stewart)	31	42	57	130	0			
Chg bal	9	9	5	10	0	−0.1	0	0.40
Buffer cap[58]	5.3	10.1	14.4	116.6	<0.1	3.7	3.8	0.4
Salinity (‰)								8.04
Ion str (M)	0.154	0.16	0.127	0.19	0.158	0.1744	0.1661	0.153
pH	7.4	7.4	7.2	7	5.0–7.5	7.4	7.4	8.00
P_{CO_2} (mm)	50	40	40	50	varies	33.65	34.24	0.24 ($10^{-3.5}$ atm)

Stewart[52] represents all the various organic acids and bases in intracellular fluids by a single monoprotic acid HA, with a pK_a of 6.7 to 6.82. Neither ISF nor Ringer's solution has any significant organic acid or base content; and although Normosol R has small amounts of acetate and gluconate, these are essentially fully ionized at pH = 7.4.

Charge Balance and "Strong Ion Difference"

In the older physiological literature, the charge balance is sometimes represented by a "Gamblegram"[59] (Fig. 10.21). This diagram shows the sum of the positive ion equivalents (i.e., [Na$^+$] or 2[Ca^{2+}]) in one column and the negative ion equivalents (i.e., [Cl$^-$] or 2[SO$_4^{2-}$]) in the other. The "strong ion difference" (SID) is the difference between nonreactive positive ions and nonreactive negative ions

$$\text{SID} = ([\text{Na}^+] + [\text{K}^+] + 2[\text{Ca}^{2+}] + 2[\text{Mg}^{2+}] + \cdots) - ([\text{Cl}^-] + 2[\text{SO}_4^{2-}] + \cdots)$$

The charge balance includes all ions, including those entering into acid–base reactions:

$$[\text{Na}^+] + [\text{K}^+] + 2[\text{Ca}^{2+}] + 2[\text{Mg}^{2+}] + [\text{H}^+] + \cdots$$
$$= [\text{HCO}_3^-] + 2[\text{CO}_3^{2-}] + [A^-] + [\text{OH}^-] + [\text{Cl}^-] + 2[\text{SO}_4^{2-}] + \cdots$$

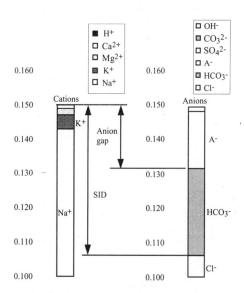

FIGURE 10.21. Gamblegram showing charge balance in isolated plasma. Numerical values are given in the second column of Table 11.1. The strong ion difference (SID) is the sum all nonreactive cation equivalents less nonreactive anion equivalents, and by the charge balance is equal to the alkalinity, that is, SID = [HCO$_3^-$] + [A$^-$] + $\cdots$.

and hence

$$\text{SID} = [\text{HCO}_3^-] + 2[\text{CO}_3^{2-}] + [A^-] + [\text{OH}^-] - [\text{H}^+] + \cdots \quad (88)$$

where "$+ \cdots$" represents all the terms small compared to the ones listed. You may recognize SID as the alkalinity (Eq. (22), p. 375 and Eq. (23), p. 379).

In Fig. 10.21, only the upper portion of the column representing these totals is shown. Almost all the positive ions are Na^+, and almost all the nonreactive negative ions are Cl^-. Reactive ions are primarily HCO_3^- and A^-. The SID is shown as a "gap" between the positive ions and the nonreactive negative ions. Ion pairs, particulary Na^+, Ca^{2+}, and Mg^{2+} with HCO_3^- and CO_3^{2-}, are minor components (see pp. 409–411). While $[\text{H}^+]$, $[\text{OH}^-]$, and $[\text{CO}_3^{2-}]$ should not be forgotten, they are too small to show on the Gamblegram.

The quantity $[A^-]$, the difference between the predominant extracellular cations and the predominant extracellular anions, is called the *anion gap*. In plasma and ISF, it is normally[60] less than 12 to 14 meq/L (in Table 10.3, 0 to 17 compared to > 100 meq/L for Na^+ and Cl^-) and consists of phosphate, sulfate, protein, and other endogenously produced or exogenously administered anions.[60]

Low arterial pH (acidosis) combined with a widened anion gap (increased $[A^-]$) has only a few causes: (1) ingestion of toxic chemicals such as salicylates, methanol, ethylene glycol; (2) acid retention as in duremia, diacetic ketoacidosis, lactic acidosis. Metabolic acidosis can also occur without changing the anion gap if bicarbonate is lost via the kidney (renal tubular acidosis) or gastrointestinal tract (as in diarrhea) but replaced by chloride. Alkalosis generally results from vomiting or the use of diuretics. Chloride is depleted, and the kidney attempts to maintain the plasma volume by reabsorbing sodium, exchanging hydrogen or potassium ions for sodium.[60]

The point of the Gamblegram is, first of all, to show that there are no "salts" in the sense of stoichiometric combinations such as NaCl; second to show that essentially all the reactive ions are negative; and third, to illustrate that the charge balance gives two independent measurements of the SID, provided all ions are taken into account. In Table 10.3, the SID is calculated both from the strong ion concentrations and from the weak acid concentrations.

For plasma, the sum of reactive anions gives 0.042 eq/L, of which 0.025 is carbonate alkalinity, and the difference between strong cations and strong anions is 0.051. The difference bewteen the two estimates, $0.051 - 0.042 = 0.009$, reflects uncertainties in the values composing the sums (this is the line in Table 10.3 labeled "Chg bal").

Models Relating pH to P_{CO_2}

The most important equilibrium calculations on physiological fluids relate pH to P_{CO_2} at constant alkalinity. These can be approached in three stages:

- A simplified model where $[\text{HCO}_3^-]$ and a single $[A^-]$ are dominant species;

- A more complex model including $[CO_3^{2-}]$, $[OH^-]$, $[H^+]$, and several $[A^-]$;
- A still more complex model including ion pairs.

The equilibrium constants for the carbonate system, adjusted to 37°C and $I = 0.16$, are given in Table 10.4. The equilibria and mass balances were presented at the beginning of this section.
For the simplified model,

$$\text{SID} = [HCO_3^-] + [A^-] + \cdots = \frac{K_H P_{CO_2} K_1'}{10^{-pH}} + \frac{K_a [HA]_T}{10^{-pH} + K_a} + \cdots \qquad (89)$$

Figure 10.22 shows the type of function obtained from Eq. (89) or (91).

If the total weak acid $[HA]_t$ is small, the second term of Eq. (89) is small and $\log P_{CO_2}$ is a linear function of pH with slope -1:

$$\log P_{CO_2} = \log(\text{SID}) + pK_H + pK_1' - pH + \cdots \qquad (90)$$

See the curves in Fig. 10.22 for ISF, Ringer's, Nomosol, and seawater, which have slope $d \log P_{CO_2}/d\text{pH} = -1.00$. If the second term of Eq. (89) is large compared to the first term, pH is independent of P_{CO_2}; see ICF curves in Fig. 10.22 at low P_{CO_2}. The calculations are easiest if values are chosen for pH, and P_{CO_2} is calculated from these:

$$P_{CO_2} = \frac{10^{-pH}}{K_H K_1'} \left(\text{SID} - \frac{K_a [HA]_T}{10^{-pH} + K_a} \right) \cdots \qquad (91)$$

Plasma, with a relatively small weak acid content, has a slope somewhat more negative than -1. If the weak acid content is large compared to the bicarbonate content, a maximum pH is reached at low P_{CO_2} (see curves for RBC-ICF and Gen ICF in Fig. 10.22).

Extending Eq. 89 to include additional weak acid–base systems can easily be accomplished by adding similar independent terms, as in Eq. (92). Including $[CO_3^{2-}]$, $[OH^-]$, and $[H^+]$, while more rigorous, has little effect on the results at the

TABLE 10.4. Equilibria and log K for free ions[61] at 37°C, $I = 0.16$

$\log \gamma_+ = -0.133$[62]	pK
$[CO_2] = K_H P_{CO_2}$	1.607
$[HCO_3^-] \, 10^{-pH} = K_1' \, [CO_2]$	6.163
$[CO_3^{2-}] \, 10^{-pH} = K_2' \, [HCO_3^-]$	9.930
$[OH^-] \, 10^{-pH} = K_w'$	13.486
$[A^-] \, 10^{-pH} = K_a'$	approx. 6.7

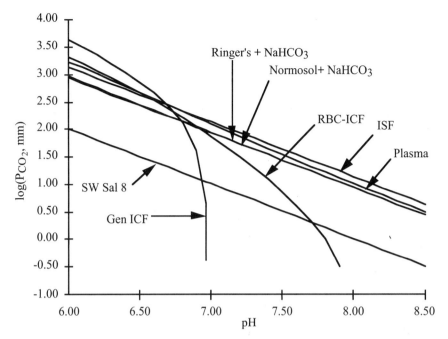

FIGURE 10.22. $\log(P_{CO_2})$ as a function of pH for the solutions listed in Table 10.3, as calculated from Eq. 89 or 91. The slope for seawater is the same as for the others (≈ -1), but P_{CO_2} is much smaller because seawater alkalinity (0.002) is ten times smaller than the physiological solutions (0.02). The solutions with high organic acid content (RBC-ICF and Gen ICF) approach a maximum pH at low P_{CO_2}.

resolution of Fig. 10.22. However, they make a difference on an expanded scale (see Fig. 10.23):

$$\text{SID} = [\text{HCO}_3^-] + \sum_i [A^-]_i + 2[\text{CO}_3^{2-}] + [\text{OH}^-] - [\text{H}^+] + \cdots \quad (92)$$

$$= \frac{K_H P_{CO_2} K_1'}{10^{-\text{pH}}} + \left(\sum_i \frac{K_{ai}[HA_i]_T}{10^{-\text{pH}} + K_{ai}} \right) + \frac{2 K_H P_{CO_2} K_1' K_2'}{10^{-\text{pH}}} + \frac{K_w'}{10^{-\text{pH}}} - \frac{10^{-\text{pH}}}{\gamma_+} + \cdots$$

Ion-Pair Models

Including ion pairs is more complicated, and there is some debate about how much influence they have in physiological solutions.[63] The calculations, if they are to be done, are best done iteratively. Table 10.5 gives equilibrium constants for the ion pairs likely to be found in Normosol-R, adjusted to $I = 0.16$ and 37°C. Other solutions in common use might contain calcium, lactate, and other materials.

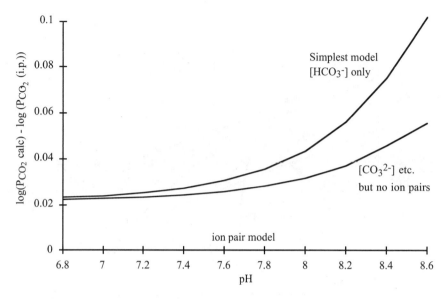

FIGURE 10.23. Comparison of three approximations to the pH–P_{CO_2} relationship for Normosol-R. The horizontal axis corresponds to the ion-pair model; the upper curve to the simplest approximation (Eq. 90) which includes only [HCO_3^-]. The lower curve was calculated using [CO_3^{2-}], [OH^-], etc. as well (Eq. 92).

For example, $[Na^+]_T$ is the known total (analytical) concentration of Na^+ species, and $[Na^+]_f$ is the concentration of free Na^+.

The calculation procedure in essence is this: Obtain a preliminary set of concentrations from the solution composition and the simplified model (Eq. 14 or 16). These are then used to calculate the ion-pair concentrations used to adjust the mass balances.

$$[Na^+]_T = [Na^+]_f + [NaHCO_3] + [NaCO_3^-] + [NaAc] + \cdots$$
$$= [Na+]_f(1 + K_{Na}[HCO_3^-] + K_{NaC}[CO_3^{2-}] + K_{NaAc}[Ac^-] + \cdots)$$
$$= \alpha_{Na}[Na^+]_f \tag{93}$$

TABLE 10.5. Equilibria and log K (free ions) at 37°C, I = 0.16 for an ion-pair model of Normosol-R[64]

$10^{-pH}[Ac^-] = K'_{aAc}[\text{H acetate}]$	−4.623
$10^{-pH}[G^-] = K'_{aG}[\text{H gluconate}]$	−3.38
$[NaHCO_3^\circ] = K_{Na}[Na^+][HCO_3^-]$	−0.52
$[MgHCO_3^+] = K_{Mg}[Mg^{2+}][HCO_3^-]$	+0.62
$[NaCO_3^-] = K_{NaC}[Na^+][CO_3^{2-}]$	+0.75
$[MgCO_3^\circ] = K_{MgC}[Mg^{2+}][CO_3^{2-}]$	+1.87
$[MgOH^+] = K_{MgOH}[Mg^{2+}][OH^-]$	+2.19
$[\text{Na acetate}] = K_{NaAc}[Na^+][Ac^-]$	−0.44
$[\text{Mg acetate}] = K_{MgAc}[Mg^{2+}][Ac^-]$	+0.78
$[\text{Mg gluconate}] = K_{MgG}[Mg^{2+}][G^-]$	+0.68

The equation $[Na^+]_T$ (calc) $- [Na^+]_T$ (obs) $= 0$ is then solved using Newton's method, which yields the formula[66]

$$[Na^+]_{f,new} = [Na^+]_{f,o}[Na^+]_T \frac{(obs)}{[Na^+]_T \text{ (calc)}} \qquad (94)$$

A similar procedure is carried out for $[Mg^{2+}]$, $[HCO_3^-]$, $[CO_3^{2-}]$, $[Ac^-]$, and $[G^-]$; then $[Na^+]_{f,new}$ is substituted for $[Na^+]_{f,o}$, and the calculation repeated until the values remain unchanged within a reasonable error limit. Some sample results are given in Table 10.6. The strongest ion pairs are NaAc, $NaHCO_3$, $NaCO_3^-$, $MgAc^+$, and MgG^+.

For physiological solutions, which are relatively dilute and have few strong ion pairs, the convergence is rapid. For more concentrated solutions such as seawater and brines, good convergence may be more difficult to obtain (pp. 434–448).

Figure 10.23 gives an example of how including ion pairs in the model affects the relationship between pH and P_{CO_2}. As mentioned above, acetate and gluconate, which are almost fully dissociated in this pH range, were included with the strong anions for simplicity (see Table 10.6; compare HAc to Ac^- and HG to G^-). In Fig.

TABLE 10.6. pH–P_{CO_2} calc. for Normosol-1; mass balance results at pH = 8.0

Mass balance on Na^+ (mM)		Mass balance on Mg^{2+} (mM)	
$[Na^+]_f$	154.50	$[Mg^{2+}]_f$	1.07
[NaAc]	1.40	$[MgCO_3]$	0.02
$[NaHCO_3]$	0.83	$[MgHCO_3^+]$	0.08
$[NaCO_3^-]$	0.21	$[MgAc^+]$	0.16
$[Na^+]_T$(obs)	156.94	$[MgG^+]$	0.12
		$[MgOH^+]$	7.3E-04
		$[Mg^{2+}]_T$(obs)	1.45

Mass balance on acetate (mM)		Mass balance on gluconate (mM)	
$[Ac^-]$	24.96	$[G^-]$–f	22.445
[HAc]	0.01	[HG]	0.001
[NaAc]	1.40	[NaG]	0.000
[MgAc]	0.16	$[MgG^+]$	0.115
$[Ac^-]_T$(obs)	26.53	$[G]_T$(obs)	22.561

Mass balance on CO_3^{2-} (mM)[65]			
$[CO_3^{2-}]_f$	0.246		
$[NaCO_3]$	0.213		
$[MgCO_3]$	0.020		
$[HCO_3^-]_f$	17.743		
$[NaHCO_3]$	0.828		
$[CaHCO_3^+]$	0.079		
$[CO_3^{2-} + HCO_3^-]_T$	19.128	May vary with pH due to CO_2 gain or loss	
Compare C_{NaHCO_3}	19.608		

10.23, the difference is displayed between the ion-pair model (horizontal axis), an intermediate model, and the simplest model

$$P_{CO_2}(mm) = 760\left(\frac{10^{-pH}[HCO_3^-]}{K_H K_1'} + \cdots\right) \quad (95)$$

All are plotted as a function of pH. The simplest model (Eq. 90 or 95) is represented by the upper curve. The lower curve, labeled "no ion pairs" is obtained using all the terms of Eq. 92, particularly $[CO_3^{2-}]$.

The most important conclusion is that P_{CO_2} calculated from the ion-pair model is always lower than either of the other approximations; hence the approximations tend to overestimate P_{CO_2} at all pH values. In the pH range from 6.8 to 7.4, the error is small: The ion-pair model gives P_{CO_2} values about 5% lower (log difference = 0.02) than either of the simpler models. At higher pH, however, the difference is much greater: 0.1 log units at pH = 8.6 corresponds to 25% difference in P_{CO_2}.

Figures 10.24 and 10.25 show how the equilibrium model fits some experimental data.[67] These experiments were performed with Normosol-R containing 1 meq of phosphate (to adjust its original pH to 7.4) and approximately 0.02 M $NaHCO_3$. This is the same solution as was used in the calculation example above. Seven gas mixtures containing from 0.5% to 7% CO_2 in nitrogen were bubbled through the solution, and the pH was measured several times. This experiment was performed six times, and the combined raw data are plotted on Fig. 10.24. A least-squares regres-

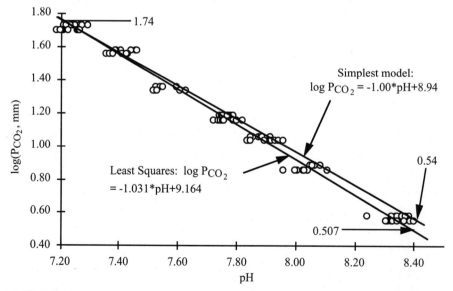

FIGURE 10.24. Experimental P_{CO_2} (mm) versus pH for Normosol-R containing 0.02 M $NaHCO_3$. These are the combined data of 6 tests using 7 different CO_2–N_2 mixtures equilibrated with the solution. Wong et al.[67]

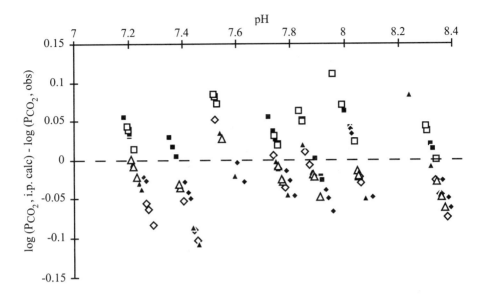

FIGURE 10.25. Comparison, on an expanded scale, of the experimental data[67] from Fig. 10.24 with P_{CO_2} values calculated using the ion-pairing model.

sion gave a slope of $d \log(P_{CO_2})/d\text{pH} = -1.031$, a little steeper than the -1.00 slope predicted by Eq. (90) or (95).

For each of the experimental points, a P_{CO_2} value was obtained using the ion-pair model, and Fig. 10.25 is a comparison of these calculated values with the observed values. The results cluster around the axis with a range of ±0.1, corresponding to about ±25% maximum error in an individual measurement. Although within each group (corresponding to a single P_{CO_2}) the difference (calc–obs) appears to decrease with increasing pH, there is no particular trend with time or test number.

The mean of all 122 calc–obs values is -0.0036 (the geometric mean of calculated P_{CO_2} is about 0.8% lower than the mean of observed P_{CO_2}). The standard deviation of a single measurement is 0.0440 (±10%), and the standard deviation of the mean is $\text{SD}/\sqrt{n} = 0.0040$ (±1%). Hence the mean is zero within experimental error.

Precipitation of CaCO₃ from Lactated Ringer's Solution

Another of the important synthetic solutions used in intravenous therapy is Ringer's solution, which contains calcium instead of magnesium (Table 10.7). A modification of this formula is "lactated Ringer's solution," in which lactic acid is added as a surrogate for the natural organic acids buffering physiological fluids. Its pH is adjusted to approximately 6.2. However, when additional buffering against acidosis is required, sodium bicarbonate is added (e.g., 10 mL of 1 M NaHCO₃ to 500 mL

TABLE 10.7. Composition of Lactated Ringer's Solution

	Lact. Ringer's	Lact. Ring. + NaHCO$_3$
Strong acid added[68]	1.36	1.1 mmol/L
pH	6.20 (adjusted)	7.5 to 8 (depends on P_{CO_2})
[Na$^+$]$_T$	130.33	147.38 mmol/L
[K$^+$]	4.02	3.94 mmol/L
[Ca^{2+}]$_T$	1.36	1.33 mmol/L
[Cl$^-$]	108.14	106.02 mmol/L
[Lactate$^-$]$_T$	27.57	27.03 mmol/L
[HCO$_3^-$]$_T$	small and variable	19.32 mmol/L approx[69]
[CO$_3^{2-}$]$_T$		0.26 mmol/L approx
Charge bal	0.00	0.00 mmol/L
Ionic Strength	138.43	155.58 mmol/LL

of lactated Ringer's). The increase in pH to 7.5 or 8 can cause precipitation of calcium carbonate, which becomes more severe as the mixture is stored for some time and CaCO$_3$(s) nucleates.

Proceeding as we did for Normosol in the previous example, mass balances are set up and the ion-pair concentrations obtained by iteration—beginning with the known total concentrations and using these to evaluate the ion-pair equilibria (see Eq. 93 and 94).

$$[Na^+]_T = [Na^+] + [NaHCO_3^\circ] + [NaCO_3^-] + \cdots$$

$$[Ca^{2+}]_T = [Ca^{2+}] + [CaHCO_3^+] + [CaCO_3^\circ] + [CaOH^+] + [Ca\ lactate] + \cdots$$

$$C_T = [CO_2] + [HCO_3^-] + [CO_3^{2-}] + [NaHCO_3^\circ] + [CaHCO_3^+] + [CaCO_3^\circ] + \cdots$$

$$[L]_T = [L^-] + [HL] + [CaL^+] + \cdots \tag{96}$$

Table 10.8 gives some equilibrium constants needed in addition to those in Table 10.4. The results of such a calculation are given in Table 10.9 and Fig. 10.26.

At every point in these calculations, it is possible to obtain a saturation index for CaCO$_3$ (Table 10.10):

$$SI = \log\left(\frac{[Ca^{2+}]_f[CO_3^{2-}]_f}{K_{so}}\right) \tag{97}$$

The saturation index calculated from the ion-pair model is substantially lower than that calculated from model without ion pairs. Since CaCO$_3$ will not precipitate if SI < 0, the ion-pair model predicts that precipitation will not occur at pH = 7 or below (see Table 10.10).

However, if SI > 0, CaCO$_3$ can precipitate. This may be slow, requiring nucleation and some time for crystal growth. For example, surface seawater in many oceanic areas is 4 times oversaturated (SI = +0.6; see p. 442), but there the formation

TABLE 10.8. Ion-pair constants used in model of lactated Ringer's solution[70]

Equilibrium expression	log K
$10^{-pH}[\text{lactate}^-] = K'_{aL}[\text{H lactate}]$	−3.721
$[\text{NaHCO}_3^\circ] = K_{\text{Na}}[\text{Na}^+][\text{HCO}_3^-]$	−0.52
$[\text{CaHCO}_3^+] = K_{\text{Ca}}[\text{Ca}^{2+}][\text{HCO}_3^-]$	+0.87
$[\text{NaCO}_3^-] = K_{\text{NaC}}[\text{Na}^+][\text{CO}_3^{2-}]$	+0.75
$[\text{CaCO}_3^\circ] = K_{\text{CaC}}[\text{Ca}^{2+}][\text{CO}_3^{2-}]$	+2.10
$[\text{CaOH}^+] = K_{\text{CaOH}}[\text{Ca}^{2+}][\text{OH}^-]$	+0.91
$[\text{Ca lactate}] = K_{\text{CaL}}[\text{Ca}^{2+}][L^-]$	+0.95

TABLE 10.9. Mass balances of ion-pair model for lactated Ringer's with NaHCO$_3$ added; pH = 8.0, P_{CO_2} = 8.33 mm Hg

Mass balance on Na$^+$ (mM)		Mass balance on Ca^{2+} (mM)	
$[\text{Na}^+]_f$	146.40	$[\text{Ca}^{2+}]_f$	0.948
$[\text{NaHCO}_3]$	0.79	$[\text{CaCO}_3]$	0.029
$[\text{NaCO}_3^-]$	0.20	$[\text{CaHCO}_3^+]$	0.125
$[\text{Na}^+]_T$	147.38	$[\text{Ca}L^+]$	0.232
		$[\text{CaOH}^+]$	1.11E-05
		$[\text{Ca}^{2+}]_T$	1.334

Mass balance on lactate (mM)		Mass balance on CO$_3^{2-}$ (mM)	
$[L^-]_f$	27.43	$[\text{CO}_3^{2-}]_f$	0.24
$[HL]$	0.00	$[\text{NaCO}_3]$	0.20
$[NaL]$	No data	$[\text{CaCO}_3]$	0.03
$[\text{Ca}L^+]$	0.23	$[\text{HCO}_3^-]_f$	17.76
$[L]_T$	27.66	$[\text{NaHCO}_3]$	0.79
		$[\text{CaHCO}_3^+]$	0.13
		$[\text{CO}_3 + \text{HCO}_3]_T$	19.14[71]

TABLE 10.10. Variation of saturation index in lactated Ringer's solution, calculated using pK_{so} = 7.54 at 37°C, I = 0.16[72]

pH	SI with ion pairs	SI without ion pairs
6.8	−0.27	0.16
7.0	−0.07	0.362
7.4	0.33	0.76
7.8	0.71	1.15
8.0	0.90	1.34
8.2	1.09	1.53
8.6	1.43	1.88

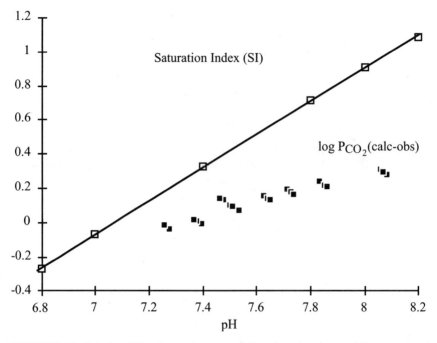

FIGURE 10.26. Calculated P_{CO_2} (assuming no precipitate forms)—observed P_{CO_2} as a function of pH in lactated Ringer's solution containing added $NaHCO_3$.[73] The saturation index for $CaCO_3$ is shown for comparison. As pH increases, the saturation index increases, and the observed P_{CO_2} is less than the calculated value.

of solid $CaCO_3$ is not normally spontaneous, and is typically biologically mediated by mollusks, forams, calcareous algae, and other organisms.

In Fig. 10.26, the observed[73] P_{CO_2} over lactated Ringer's solution becomes lower than the value calculated from the ion-pair model (which does not include precipitation in its mass balances), implying that the dissolved carbonate is decreasing due to precipitation of $CaCO_3$. Such precipitation is observed in the laboratory as turbidity in the solution. The deviation of P_{CO_2} from the calculated value exceeds the experimental error when the saturation index exceeds about 0.3.

CARBON DIOXIDE EQUILIBRIA IN SEAWATER

For a medium of nearly constant composition, such as seawater, the equilibria can usefully be expressed in terms of total concentrations, including the various ion paris. Seawater contains relatively high concentrations of $[Na^+]$, $[Mg^{2+}]$, and $[Ca^{2+}]$ as well as lower concentrations of the carbonate species, as summarized in Table 10.11.

Recall from the section on "Six Forms for the Equilibrium Constant" (p. 401) that one possible choice for the carbonate equilibrium expressions is K'_1 is $K'_{1(T)}$, etc. defined by Eq. (81). This expression and the analogous one for K_2 are:

CARBON DIOXIDE EQUILIBRIA IN SEAWATER

TABLE 10.11. Nominal composition of seawater of 35‰ salinity

Ion	mole/kg[74]	Ion	mole/kg
Na^+	0.4690	Cl^-	0.5459
Mg^{2+}	0.0528	SO_4^{2-}	0.0282
Ca^{2+}	0.0103	Br^-	0.00084
K^+	0.0102	Total carbonate	0.00217
Alkalinity	0.00236	Total borate	0.000397

$$10^{-pH} [HCO_3^-]_T = K'_{1(T)} [CO_2] \qquad (98)$$

$$10^{-pH} [CO_3^{2-}]_T = K'_{2(T)} = [HCO_3^-]_T \qquad (99)$$

where

$$[HCO_3^-]_T = [HCO_3^-]_f + [NaHCO_3] + [MgHCO_3^+] + [CaHCO_3^+] + \cdots$$
$$= [HCO_3^-]_f (1 + K_{NaHCO_3}[Na^+]_f + K_{MgHCO_3}[Mg^{2+}]_f + K_{CaHCO_3}[Ca^{2+}]_f + \cdots$$
$$= [HCO_3^-]_f \alpha_{HCO_3} \qquad (100)$$

Note that in this model there are no specific interactions between the major ions and uncharged CO_2. For the seawater scale (SWS), 10^{-pH} is replaced by $[H^+]_T$. Similarly,

$$[CO_3^{2-}]_T = [CO_3^{2-}]_f + [NaCO_3^-] + [MgCO_3] + [CaCO_3] + \cdots$$
$$= [CO_3^{2-}]_f (1 + K_{NaCO_3}[Na^+]_f + K_{MgCO_3}[Mg^{2+}]_f + K_{CaCO_3}[Ca^{2+}]_f + \cdots)$$
$$= [CO_3^{2-}]_f \alpha_{CO_3} \qquad (101)$$

In seawater with salinity in the range 5‰ to 45‰, the ratio of the total concentrations of major ions is nearly constant; α_{HCO_3} and α_{CO_3} are also nearly constant (at constant salinity), and the equilibria of Eqs. (98) and (99) can be related to the zero-ionic-strength constants and the conventional activity coefficients by Eq. 78 (p. 402) and its analogue for the second acidity constant:

$$10^{-pH} [HCO_3^-]_f \gamma_- = K°_{a1}[CO_2]\gamma_0 \qquad (102)$$

$$10^{-pH} [CO_3^{2-}] \gamma_= = K°_{a2} [HCO_3^-]_f \gamma_- \qquad (103)$$

Substituting Eqs. (102) and (103) in Eqs. (98) and (99) gives:

$$K'_1 \equiv K'_{1(T)} = \frac{K°_{a1} \alpha_{HCO_3} \gamma_0}{\gamma_-} \qquad (104)$$

$$K'_2 \equiv K'_{2(T)} = \frac{K°_{a2} \alpha_{CO_3} \gamma_-}{\alpha_{HCO_3} \gamma_=} \qquad (105)$$

K_1' and K_2' as defined here depend on the nature of the pH scale, in particular the changes in liquid junction and asymmetry potentials between the pH standard and the sample, and on the assumptions used to calculate γ (see Chapter 2, pp. 45–49). So long as the ionic medium remains essentially constant in composition, whether $[H^+]_f$ (see Eq. 83) or $[H^+]_T$ (see Eq. 84, p. 403) is substituted for 10^{-pH}, the constants are thermodynamically rigorous because α_{HCO_3} and α_{CO_3} are also nearly constant, as shown in Eqs. (100) and (101).

Ion-pairing constants appropriate to seawater will be given in Table 10.16, later in this chapter. As you saw in Chapter 9, and in Tables 10.5 and 10.8, they vary from about 1 to about 10^3, and of course vary with ionic strength. The factors α_{HCO_3} and α_{CO_3} are positive numbers, and enhance the normal electrostatic interaction effect expressed by $\gamma_=$ and γ_-.

Recently published values[75] for $K_{a1(\text{free H})}$ (defined by Eq. 83) in seawater

$$K_{a1(\text{free H})} \equiv \frac{[H^+]_f[HCO_3^-]_T}{[CO_2]} = \frac{10^{-pH}[HCO_3^-]_T}{\gamma_+[CO_2]} = \frac{K_1'}{\gamma_+} \quad (106)$$

were converted to K_1' using the Davies equation to calculate γ_+, and are plotted in Fig. 10.27 along with the predictions of the Davies equation for a hypothetical solution without ion pairs. Note that $pK_1' = 6.00$ in standard 35‰ seawater ($I = 0.7$). In a noncomplexing medium of the same ionic strength, $pK_1' = 6.12$.

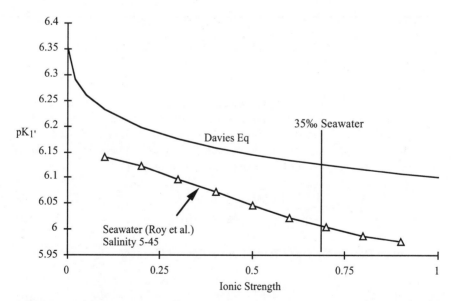

FIGURE 10.27. First acidity constant for CO_2 as a function of ionic strength. The upper curve is for pK_1' in noncomplexing solutions as predicted by the Davies equation; the lower curve was obtained from the experimental data for seawater[75] (bottom curve), with γ_+ for H^+ calculated from the Davies equation.

CARBON DIOXIDE EQUILIBRIA IN SEAWATER

The "oceanographic" second acidity constant K_2' is shown in Figs. 10.28 and 10.29. Four independent studies in seawater[76] agree well with each other, but all are much lower than the prediction of the Davies equation for noncomplexing media, because of the ion-pair formation between HCO_3^- and CO_3^{2-} and the cations Na^+, Mg^{2+}, and Ca^{2+}. So long as the ionic strength (salinity) and the relative composition of ions remain the same, these constants may be used as empirical data without concern for the details of ion pairing.

Numerical values for pK_1, pK_2, and pK_{so} at various salinities are given in Table 10.12 and plotted in Figs. 10.28–10.30 (pp. 419–421).

Alkalinity Titration in Seawater

As mentioned earlier in this chapter, the empirical definition of alkalinity is "the amount of strong acid required to bring pH from its initial value to a value corresponding to the CO_2 end point." The principles have been given in Chapters 4 and 5, and are applied here to seawater, a mixture of monoprotic and diprotic acids at ionic strength 0.7 (35‰), with the objective of testing how accurately an equilibrium model can fit such experimental data.

The first approximation consists only of the carbonate system and the borate system. More elaborate models include the acid–base systems of sulfate, fluoride, phosphate, silicate, and possible trace organics.[78] The initial pH of seawater is normally about 8, and the end point of the titration is normally around 4.5 (Fig. 10.31).

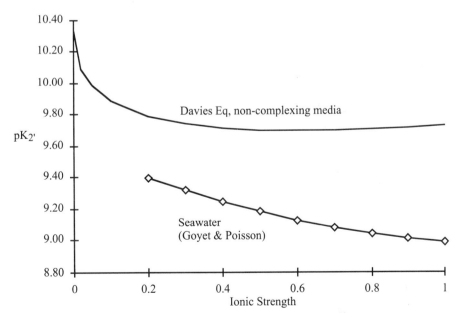

FIGURE 10.28. Second acidity constant K_2'. Measurements in seawater[76] agree well with each other (see Fig. 10.29), but are much lower than the prediction for noncomplexing media.

TABLE 10.12. Equilibrium constants for carbonate system in seawater of varying salinity[77]

I	Sal ‰	pK_1'	pK_1(SWS)	pK_2'	pK_2(SWS)	$pK_{so}'(T)$
0.1	5	6.140	6.035	9.47	9.62	7.31±0.01
0.2	10	6.123	5.993	9.41	9.40	
0.3	15	6.098	5.955	9.33	9.25	6.82±0.01
0.4	20	6.072	5.921	9.26	9.12	
0.5	25	6.047	5.891	9.19	9.06	6.56±0.02
0.6	30	6.027	5.866	9.13	9.00	
0.7	35	6.005	5.846	9.08	8.95	6.37±0.02
0.8	40	5.988	5.829	9.04	8.90	
0.9	45	5.975	5.817	9.00	8.86	6.25±0.04

As in Chapter 4 (pp. 136–137), derivation of the titration equation begins with the charge balance. [H$^+$] must be included because it is important in the region after the end point. On the other hand, because of the limited pH range, [OH$^-$] and its ion pairs with [Ca^{2+}] and [Mg^{2+}] are negligible. The species include ion pairs between the various seawater ions.

$$[H^+] + [Na^+] + [K^+] + 2\,[Mg^{2+}] + [MgHCO_3^+] + 2\,[Ca^{2+}] + [CaHCO_3^+] \quad (107)$$
$$= [HCO_3^-] + 2\,[CO_3^{2-}] + [NaCO_3^-] + [B(OH)_4^-] + [Cl^-]$$
$$+ 2[SO_4^{2-}] + [NaSO_4^-] + [HSO_4^-] + [OH^-] + [F^-] + \cdots$$

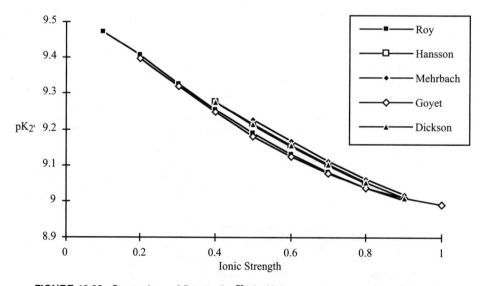

FIGURE 10.29. Comparison of five studies[76] of pK_2' in seawater on an expanded scale.

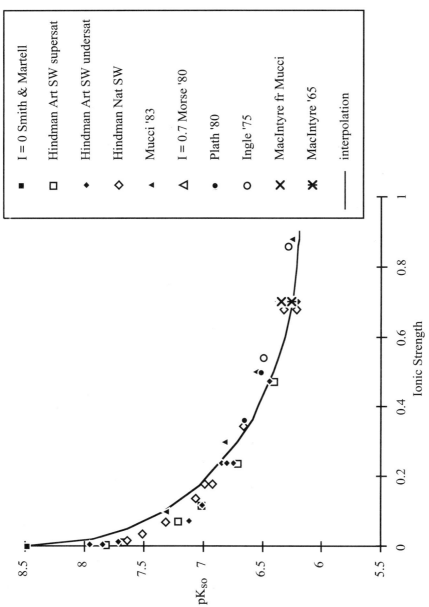

FIGURE 10.30. Ionic strength dependence of pK_{so} at 25°C in natural and artificial seawater from various workers[109]. The interpolation function is $pK_{so} = 8.48 - 0.51(8)(\sqrt{I}/(1 + 0.33a\sqrt{I}) - bI) + dI^2$ with $a = 5$, $b = 0.2$ and $d = 0.4$. Even with three adjustable parameters, it does not fit as well below $I = 0.2$ as one would like.

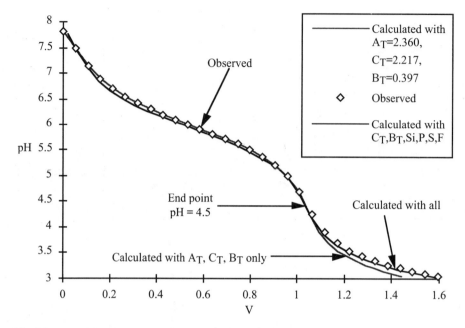

FIGURE 10.31. Seawater from 146 m in the Indian Ocean, Sample No. 306, titrated with 0.250 mol/kg HCl. Data from A. L. Bradshaw et al.[81] Calculated values from J. N. Butler[80] as described in Tables 10.13–10.15 (pp. 425–427). Deviations of calculated from observed values are plotted on an expanded scale in Fig. 10.33.

The mass balances on the ion pairs are:

$$[Na^+]_T = [Na^+] + [NaSO_4^-] + [NaCO_3^-] + [NaHCO_3] + \cdots \quad (108)$$

$$[K^+]_T = [K^+] + [KSO_4^-] + [KCO_3^-] + [KHCO_3] + \cdots \quad (109)$$

$$[Mg^{2+}]_T = [Mg^{2+}] + [MgSO_4] + [MgCO_3] + [MgHCO_3^+] + \cdots \quad (110)$$

$$[Ca^{2+}]_T = [Ca^{2+}] + [CaSO_4] + [CaCO_3] + [CaHCO_3^+] + \cdots \quad (111)$$

$$[HCO_3^-]_T = [HCO_3^-] + [NaHCO_3] + [KHCO_3] + [MgHCO_3^+] + [CaHCO_3^+] + \cdots \quad (112)$$

$$[CO_3^{2-}]_T = [CO_3^{2-}] + [NaCO_3^-] + [KCO_3^-] + [MgCO_3] + [CaCO_3] + \cdots \quad (113)$$

$$[SO_4^{2-}]_T = [SO_4^{2-}] + [HSO_4^-] + [NaSO_4^-] + [KSO_4^-] + [MgSO_4] + [CaSO_4] + \cdots \quad (114)$$

$$[B]_T = [B(OH)_4^-] + [NaB(OH)_4] + [B(OH)_3] + \cdots \quad (115)$$

$$[F^-]_T = [F^-] + [HF] \quad (116)$$

CARBON DIOXIDE EQUILIBRIA IN SEAWATER

The only chloride species is Cl^-. Since these data were obtained in a standard seawater, it is possible to use equilibrium constants containing the total concentrations instead of free ionic concentrations.

Substituting Eqs. 108–116 for the major ion concentrations in Eq. 107, and adding appropriate terms to each side, yields a charge balance on the total ionic concentrations:

$$[H^+] + [Na^+]_T + 2\,[Mg^{2+}]_T + 2\,[Ca^{2+}]_T$$
$$= [HCO_3^-]_T + 2\,[CO_3^{2-}]_T$$
$$+ [B(OH)_4^-] + [Cl^-] + 2\,[SO_4^{2-}]_T$$
$$- [HSO_4^-] + [F^-] + [OH^-]_T \cdots \qquad (117)$$

The charge balance can be divided into two groups, each equal to the alkalinity:

$$A_T = [HCO_3^-]_T + 2\,[CO_3^{2-}]_T + [B(OH)_4^-] - [H^+]$$
$$- [HSO_4^-] - [HF] + [OH^-]_T + \cdots \qquad (118)$$

$$A_T = [Na^+]_T + 2\,[Mg^{2+}]_T + 2\,[Ca^{2+}]_T - [Cl^-] - 2\,[SO_4^{2-}]_T - [F^-]_T + \cdots \qquad (119)$$

The equilibria are as derived above. Recall that K_1' stands for $K_{1(T)}'$, defined by Eq. 81 (p. 402). The values given are for 25°C, 35‰ seawater:

$$K_1' = \frac{10^{-pH}[HCO_3^-]_T}{[CO_2]} = 10^{-6.00} \qquad (120)$$

$$K_2' = \frac{10^{-pH}[CO_3^{2-}]_T}{[HCO_3^-]_T} = 10^{-9.12} \qquad (121)$$

$$K_B' = \frac{10^{-pH}[B(OH)_4^-]}{[B(OH)_3]} = 10^{-8.71} \qquad (122)$$

$$K_{HSO_4}' = \frac{10^{-pH}[SO_4^{2-}]}{[HSO_4^-]} = 10^{-1.11} \qquad (123)$$

$$K_F' = \frac{10^{-pH}[F^-]}{[HF]} = 10^{-2.76} \qquad (124)$$

The titrant, HCl, and the materials being titrated, are diluted as the titrant is added (see pp. 75–76 and 136–141). If V mL of HCl are added to V_0 mL of seawater, the charge balance, including the dilution factors, is obtained from Eq. (117):

$$\frac{A_T V_o}{V + V_o} = [HCO_3^-]_T + 2[CO_3^{2-}]_T + [B(OH)_4^-] - [HSO_4^-] - [F^-]$$

$$+ [OH^-] - [H^+] + \frac{C_{HCl} V}{V + V_o} \tag{125}$$

In this equation, the nonreactive ions (including the original [Cl$^-$] in the seawater) are in the A_T term on the left-hand side, and the excess [Cl$^-$] from the titrant is represented by the last term on the right-hand side. Using the mass balances (Eqs. 108–116) and equilibria (Eqs. 120–124), Eq. (125) becomes:

$$A_T V_o = \frac{C_T V_o (10^{-pH} K_1' + 2 K_1' K_2')}{10^{-2pH} + 10^{-pH} K_1' + K_1' K_2'} + \frac{B_T V_o K_B'}{10^{-pH} + K_B'} - \frac{C_S V_o K_S'}{10^{-pH} + K_S'} - \frac{C_F V_o K_F'}{10^{-pH} + K_F'}$$

$$- \left(\frac{10^{-pH}}{\gamma_+} - \frac{K_w'}{10^{-pH}} \right) (V + V_o) + C_{HCl} V \tag{126}$$

On rearranging, Eq. (126) becomes

$$\frac{V}{V_o} = \frac{A_T - C_T F - B_T F_B + C_S F_S + C_F F_F + G}{C_{HCl} - G} \tag{127}$$

where $C_S = [SO_4^{2-}]_T$ (Eq. 114), $C_F = [F^-] + [HF]$, and the carbonate pH function F (don't confuse it with fluoride ion!) and its analogues are defined by

$$F = \frac{10^{-pH} K_1' + 2 K_1' K_2'}{10^{-2pH} + 10^{-pH} K_1' + K_1' K_2'} \tag{128}$$

$$F_B = \frac{K_B'}{10^{-pH} + K_B'} \tag{129}$$

$$F_S = \frac{K_S'}{10^{-pH} + K_S'} \tag{130}$$

$$F_F = \frac{K_F'}{10^{-pH} + K_F'} \tag{131}$$

$$G = [H^+] - [OH^-] = \frac{10^{-pH}}{\gamma_+} - \frac{K_w'}{10^{-pH}} \tag{132}$$

A summary of equilibrium constants on various scales (pp. 401–403) is given in Table 10.13.

TABLE 10.13. Acid–base equilibrium constants for seawater (35‰)

$-\log K$	K_a ($I = 0.7$) (Eq. 4)[79]	Free H (Eq. 83)	Free H (lit[80])	NBS (Eq. 81) (lit[78])	$K'_{(f)}$ (Eq. 80)	SWS (Eq. 87) (lit[78])
K_{a1}	5.97	5.972	6.000	6.000	6.14	5.846
K_{a2}	9.70	9.022	9.086	9.115	9.86	8.945
K_b	8.85	8.723	8.75	8.71	9.01	8.598
K_w	13.68	13.351	13.365	13.6	13.84	13.226
K_{Si}	9.52	9.325	9.50	9.4	9.68	
K_{HPO_4}	8.47	9.075			8.63	8.71
$K_{H_2PO_4}$	6.06	6.075			6.22	5.978
$K_{H_3PO_4}$	1.92	1.875			2.08	1.576
K_{HSO_4}	1.36	1.108	0.983	1.1	1.52	0.983
K_F	2.78	2.755	2.63		2.94	2.63

The second column of Table 10.13 lists concentration equilibrium constants adjusted from $I = 0$ to 0.7 with the Davies equation. The third and fourth columns give constants based on the "free H" scale using [H⁺] with total concentrations of the other species (Eq. 83). The "NBS" scale or $K'_{(T)}$ (Eq. 81) in column 5 reports empirical results. The $K'_{(f)}$ or "pH & free ion" scale (Eq. 80) of column 6 uses $10^{-pH} = [H^+]\gamma_+$ and free concentrations of the other ions, and differs from the concentration scale of Column 2 only by $\log \gamma_+ = -0.16$. It is used for the ion-pair model of Table 10.15. The final column gives the SWS scale, which differs from the "free H" scale by inclusion of $[HSO_4^-] = C_s F_s$ and $[HF] = C_F F_F$ in $[H^+]_T$ (the SWS scale differs from the "free H" scale by the factor $\log \alpha_H = 0.115$. See the discussion of Eqs. (84) and (85) on p. 403).

If the "free H" scale is used (as in the equations above and Table 10.15), $[HSO_4^-]$ and $[HF]$ are included explicitly in the titration equation. If $[H^+]_T$, which already includes $[HSO_4^-]$ and $[HF]$, is used in the equilibria (SWS scale), these terms are omitted from the titration equation.[82]

Equations 127–132 define a curve of pH versus V that can be compared with the experimental values shown in Fig. 10.31 and Table 10.14. The equilibrium constants used for that calculation are listed separately in Table 10.13, Col. 6. In this comparison, several parameters (especially pK_{a2}) are not known accurately enough, but can be refined by adjustment to give a good fit of calculated to observed data. This method has been used to interpret data obtained at 909 stations by the GEOSECS global oceanographic program of the 1970s.[83]

The first parameter obtained from these data is A_T (Fig. 10.32). Near the end point at pH = 4.5, this is not very sensitive to values of the other parameters (compare with Fig. 10.33). However, the agreement of calculated and observed values is rather sensitive to the term in sulfate.[84] This may be partly compensated by adjustment of the pH scale (i.e., E_o or γ_+ as shown in Figs. 3.13–3.15, pp. 87–89). Adjusting these parameters is equivalent to adjusting the slope of Gran's plot (Fig. 3.12, p. 85).

TABLE 10.14. Data from Fig. 10.31 compared with calculated values[81]

V(ml)	pH	A (mM)	C_t (mM)	calc–obs (μmole/kg)
0.000	7.810	2.364	2.214	−3.96
0.053	7.485	2.361	2.217	−0.76
0.107	7.135	2.357	2.220	2.59
0.160	6.880	2.359	2.219	1.14
0.213	6.691	2.360	2.218	0.46
0.266	6.540	2.359	2.218	0.65
0.320	6.414	2.360	2.218	0.14
0.373	6.301	2.359	2.218	0.64
0.426	6.198	2.359	2.220	1.38
0.479	6.101	2.359	2.219	0.75
0.532	6.006	2.358	2.221	2.06
0.585	5.914	2.359	2.220	1.32
0.638	5.820	2.359	2.221	1.51
0.691	5.722	2.358	2.222	1.77
0.744	5.617	2.358	2.224	2.01
0.798	5.503	2.358	2.224	1.66
0.851	5.371	2.359	2.225	1.46
0.904	5.210	2.358	2.230	1.87
0.957	5.000	2.359	2.229	1.13
1.010	4.688	2.360	2.218	0.04
1.063	4.250	2.361	2.156	−1.14
1.117	3.902	2.361	2.057	−1.35
1.170	3.686	2.361	2.089	−0.66
1.223	3.538	2.360	2.178	−0.15
1.276	3.426	2.360	2.228	0.03
1.330	3.338	2.360	2.208	−0.02
1.383	3.263	2.359	2.581	0.71
1.436	3.200	2.360	2.006	−0.36
1.489	3.145	2.360	2.437	0.33
1.542	3.096	2.360	2.215	0.00
1.595	3.052	2.361	1.769	−0.54

The next parameter is C_T, which is most precisely obtained from data in the pH range from 6 to 8 (Fig. 10.32). Borate provides a significant contribution to the titration curve equation in this pH range (Eqs. 118 and 122) (see Fig. 10.33), and is normally taken to be proportional to salinity (0.42 mmol/kg at 35% salinity). Omission of borate produces large deviations of calculated from observed values. Phosphate makes a minor contribution (Fig. 10.33), but the effect of silicate is smaller than the experimental error. Finally, small adjustments in K_1' may give a superior fit, but of course the values used should not be outside the range of the best literature data obtained by independent measurements. One weakness of this technique is that if the potential of the glass electrode cell does not have slope precisely RT/F, systematic errors in C_T will result.

TABLE 10.15. Parameters used in calculating values given in Table 10.14.

The following equations were used to calculate values in Table 10.14:

$$A_T = 1000[(V/V_o)(C_a - G) - G] + C_T F + B_T F_B + C_{Si} F_{Si} - C_S F_S - C_F F_F + C_p F_p$$

$$C_T = 1000[0.001(A_T - B_T F_B - C_{Si} F_{Si} - C_p F_p + C_S F_S + C_F F_F) + G - V(C_a - G)/V_o]/F$$

$$\text{calc–obs} = 1000000 C_a \{[0.001(A_T - C_T F - B_T F_B - C_{Si} F_{Si} - C_p F_p + C_F F_F + C_S F_S) + G]/(C_a - G) - V/V_o\}$$

An estimated value of C_T is used to calculate A_T; the A_T values from the lowest pH region provide an improved estimate, which in turn is used to calculate a better value for C_T. This calculation converges after several iterations. The terms F_s (sulfate, Eq. 130), F_f (fluoride, Eq. 131), F_p (phosphate) and F_{Si} (silicate) are similar in form to F and F_b defined in Eqs. 128 and 129. All are small compared to the main terms.

The following parameters used in the calculations:

Salinity	35.024	
C_a	0.250	
V_o	110.741	
E_o	416.55 mV (this value was fitted to the low-pH EMF data. Bradshaw et al. gave 408.8)	
γ_+	1.000 (adjusting E_o is equivalent to adjusting γ_+; hence this was arbitrarily set = 1)	
A_t	2.360 mmole/kg (initial estimate)	2.359 is average of several approx to A
C_t	2.217 mmole/kg (initial estimate)	2.220 is average of several approx to C_T
C_b	0.397 mmole/kg as cited by Skirrow (1975)	
Silicate	0.021 mmole/kg (Bradshaw et al. data)	
Phosphate	0.002 mmole/kg (Bradshaw et al. data)	
Sulfate	28.219 mmole/kg Hansson (1973)	= 28.2 Sal/35
Fluoride	0.070 mmole/kg Hansson (1973)	= 0.07 Sal/35

Equilibrium constants are given under $K'_{(f)}$ in Table 10.13.

Effect of Increased Atmospheric CO$_2$ on the Oceans

Burning of coal and other human activities have caused a significant increase in atmospheric carbon dioxide, from less than 300 ppm in preindustrial times to 350 ppm in the present. This increase is expected to continue for at least another century before other energy sources are fully developed. Much of this additional carbon dioxide will find its way into the water and sediments of the deep oceans.[85]

The most rapid process by which the oceans remove CO_2 from the atmosphere is homogeneous dissolution and reaction with CO_3^{2-} to produce HCO_3^-. These chemical reactions proceed in the laboratory as fast as solutions can be mixed (see Chapter 1); the upper 100 m mixed layer of the ocean can reach equilibrium with the atmosphere in about 1 year. The calculations below will use the equilibria you are already familiar with to evaluate this process quantitatively.

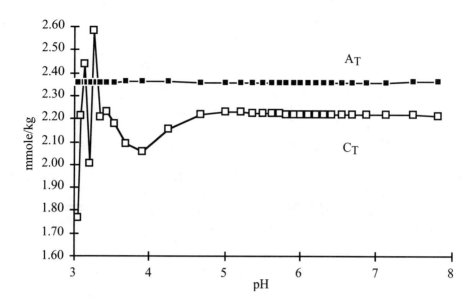

FIGURE 10.32. Data from Table 10.14. A_T is calculated as a function of pH using C_T = 2.22 mmole/kg, and C_T is calculated as a function of pH using A_T = 2.36 mmole/kg. The most precise values of A_T are obtained from the pH range 3 to 5; the most precise values of C_T are obtained from pH 6 to 8. Equations are given in Table 10.15, and equilibrium constants in Table 10.13.

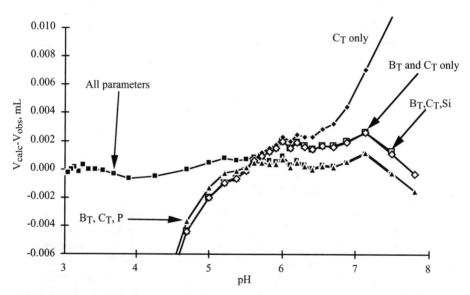

FIGURE 10.33. $V_{calc} - V_{obs}$ versus pH. These are the same data as were presented in Figs. 10.31 and 10.32, but displayed on a highly expanded scale. With all parameters (Table 10.15) the deviations are within ±0.001 mL (±2 μmole/kg in Table 10.14). When sulfate and fluoride are omitted, large negative deviations occur for pH < 5 (curves marked B_t, C_t, P). When borate is omitted, there are large positive deviations for pH > 7. The difference between "B_t and C_t only" and "all parameters" above pH = 5.5 reflects primarily the contribution of phosphate. The effect of silicate is barely noticeable.

CARBON DIOXIDE EQUILIBRIA IN SEAWATER

Additional processes for CO_2 removal include incresed weathering of carbonate and silicate rocks, dissolution of $CaCO_3$ particles and sediments in the oceans, and increased photosynthesis in areas where nutrients are in excess and the fixed carbon can be buried for geologic time. The rates and extent of these more complicated processes, as well as the effect of increased CO_2 on climate, are the subject of hundreds of books and thousands of journal articles. A critical review is beyond the scope of this book.[86]

The topic that will be considered here is how much CO_2 can be absorbed by homogeneous reaction with the mixed layer of seawater. One possible measure of this is the Revelle buffer factor[87]

$$B = \left(\frac{\partial \log(P_{CO_2})}{\partial \log(C_T)}\right)_A = \frac{C_T}{P_{CO_2}}\left(\frac{\partial(P_{CO_2})}{\partial(C_T)}\right)_A \qquad (133)$$

which is often used. The derivative of CO_2 partial pressure with respect to total carbonate C_T is taken at constant alkalinity A to reflect the fact that open ocean surface waters are mostly supersaturated with $CaCO_3$ and have few sources or sinks of alkalinity that might provide additional buffering capacity. However, these cannot be ruled out completely,[88] and so B as defined above represents a lower limit to the buffer capacity of the open ocean.

The basic equations have been developed earlier (Equations 1, 4, 5, pp. 367–368; Eqs. 21, 22, p. 375). The equilibria are expressed in terms of total concentrations (with the subscript T omitted) and borate is omitted for simplicity; P is an abbreviation for P_{CO_2}.

$$C_T = [CO_2] + [HCO_3^-] + [CO_3^{2-}] = K_H P\left(1 + \frac{K_{a1}}{[H^+]} + \frac{K_{a1}K_{a2}}{[H^+]^2}\right) \qquad (134)$$

$$A = [HCO_3^-] + 2[CO_3^{2-}] + [OH^-] - [H^+]$$

$$A = K_H P\left(\frac{K_{a1}}{[H^+]} + \frac{2K_{a1}K_{a2}}{[H^+]^2} + \frac{K_w}{[H^+]} - [H^+]\right) \qquad (135)$$

You can solve Eq. (135) for P in terms of A and $[H^+]$, substitute this in Eq. (134) to obtain C_T as a function of A and $[H^+]$, and from these two equations compute B as defined in Eq. (133) from

$$B = \frac{C_T\left(\frac{\partial P}{\partial [H^+]}\right)_A}{P\left(\frac{\partial C_T}{\partial [H^+]}\right)_A} \qquad (136)$$

Differentiate Eq. (134) with respect to [H⁺], holding A constant:

$$\left(\frac{\partial C_T}{\partial [H^+]}\right)_A = K_H P\left(-\frac{K_{a1}}{[H^+]^2} - \frac{2K_{a1}K_{a2}}{[H^+]^3}\right) + K_H\left(1 + \frac{K_{a1}}{[H^+]} + \frac{K_{a1}K_{a2}}{[H^+]^2}\right)\left(\frac{\partial P}{\partial [H^+]}\right)_A$$

and substitute back for the cominations of K and [H⁺] in terms of concentrations. This gives the denominator of Eq. (136) and involves no approximations yet:

$$\left(\frac{\partial C_T}{\partial [H^+]}\right)_A = -\frac{[HCO_3^-] + 2[CO_3^{2-}]}{[H^+]} + \frac{[CO_2] + [HCO_3^-] + [CO_3^{2-}]}{P}\left(\frac{\partial P}{\partial [H^+]}\right)_A \quad (137)$$

To get the term $(\partial P/\partial [H^+])_A$, required for the last part of Eq. (137) and for the numerator of Eq. (136), take the derivative of Eq. (135) holding A constant.

$$\left(\frac{\partial A}{\partial [H^+]}\right)_A = 0 = K_H P\left(-\frac{K_{a1}}{[H^+]^2} - \frac{4(2)K_{a1}K_{a2}}{[H^+]^3}\right) - \frac{K_w}{[H^+]^2}$$

$$-1 + K_H\left(1 + \frac{K_{a1}}{[H^+]} + \frac{2K_{a1}K_{a2}}{[H^+]^2}\right)\left(\frac{\partial P}{\partial [H^+]}\right)_A$$

Identify products as concentrations:

$$\frac{[HCO_3^-] + 2[CO_3^{2-}]}{P}\left(\frac{\partial P}{\partial [H^+]}\right)_A = \frac{[HCO_3^-] + 4[CO_3^{2-}] + [OH^-] + [H^+]}{[H^+]}$$

$$\left(\frac{\partial P}{\partial [H^+]}\right)_A = \frac{P}{[H^+]}\frac{[HCO_3^-] + 4[CO_3^{2-}] + [OH^-] + [H^+]}{[HCO_3^-] + 2[CO_3^{2-}]} \quad (138)$$

Substitute this in Eq. (137) to get

$$\left(\frac{\partial C_T}{\partial [H^+]}\right)_A = \frac{1}{[H^+]}\Bigg(-[HCO_3^-] + 2[CO_3^{2-}]$$

$$+ C_T \frac{[HCO_3^-] + 4[CO_3^{2-}] + [OH^-] + [H^+]}{[HCO_3^-] + 2[CO_3^{2-}]}\Bigg)$$

Multiplying out this expression, setting $C_T = [CO_2] + [HCO_3^-] + [CO_3^{2-}]$, you will find that all negtive terms are canceled by similar positive terms:

CARBON DIOXIDE EQUILIBRIA IN SEAWATER

$$\left(\frac{\partial C_T}{\partial [H^+]}\right)_A = \frac{[HCO_3^-]([CO_2]+[CO_3^{2-}]) + 4[CO_2][CO_3^{2-}] + C_T([OH^-]+[H^+])}{[H^+]([HCO_3^-] + 2[CO_3^{2-}])}$$

By taking only the largest terms, you can eliminate all but the first two terms of the numerator and the first of the denominator; $[HCO_3^-]$ cancels to give

$$\left(\frac{\partial C_T}{\partial [H^+]}\right)_A = \frac{[CO_2]+[CO_3^{2-}]}{[H^+]} + \cdots \tag{139}$$

When Eq. 138 and 139 are substituted in Eq. 136 (similar approximations hold), the final result is

$$B = \left(\frac{C_T}{P}\right)\left(\frac{P}{[H^+]}\right)\left(\frac{[H^+]}{[CO_2]+[CO_3^{2-}]}\right) = \frac{C_T}{[CO_2]+[CO_3^{2-}]} \tag{140}$$

Including the contribution of borate makes a significant difference[89]

$$B = C_T\left([CO_2]+[CO_3^{2-}] + \frac{\gamma[HCO_3^-] - 4[CO_3^{2-}]^2}{[HCO_3^-] + 4[CO_3^{2-}] + \gamma}\right)^{-1} \tag{141}$$

where

$$\gamma = \frac{C_B K_B' 10^{-pH}}{(K_B' + 10^{-pH})^2} \tag{142}$$

Here C_B is the borate concentration of seawater, and K_B' is the acidity constant for borate.

In Fig. 10.34, the calculated curve (Eq. 141) is compared with the GEOSECS field data from both Pacific and Atlantic Oceans. Agreement is excellent and within the uncertainties introduced by natural variations and by errors in the measurements and equilibrium constants. A value of about 10 for B tells you that if surface alkalinity remains constant, a change of 10% in the atmospheric partial pressure of CO_2 is required to produce a 1% change in the total CO_2 content of seawater in equilibrium with that atmosphere.

Although the time scale for mixing of the deep ocean with its surface waters is 1000–2000 years, the top few hundred meters (the mixed layer) mixes on a time scale of less than a year. How much of the recent increase in atmospheric CO_2 might have been taken up by this mixed layer? Broecker et al.[90] estimated the preindustrial atmosphere (290 ppm CO_2) to have approximately 145 moles of CO_2 above each square meter of ocean surface, and the mixed layer of the ocean to contain about 900 mole/m² of total carbonate.

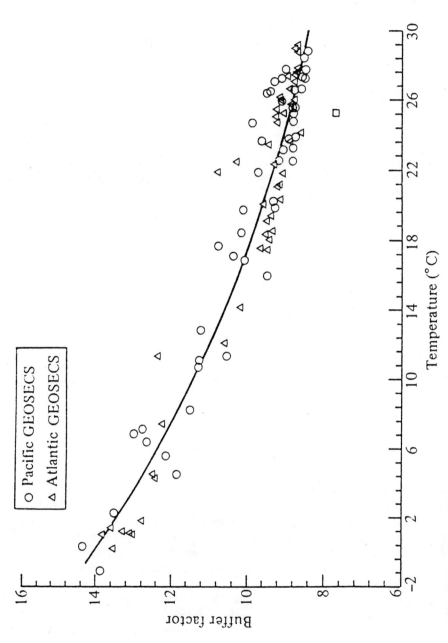

FIGURE 10.34. Homogeneous buffer factor as a function of temperature Reprinted with permission from Sundquist, Plummer, and Wigley, *Science* **204**:1203–5, 1979. Copyright 1979 American Association for the Advancement of Science.

Thus a 10% increase in atmospheric CO_2 (as occurred between 1960 and 1990, about 30 ppm or 15 moles/m^2) would produce an increase of about 1% in total dissolved carbonate (9 mole/m^2). By this mechanism, therefore, the oceans can absorb, within a year or two, about half the increase in atmospheric CO_2 (9 out of 15 mole/m^2). However, attempts to measure the predicted lag (the oceans should have an equilibrium P_{CO_2}, which is 1 to 3 ppm less than the atmosphere) have tended to be obscured by natural variations.

One interesting (and fortunate) phenomenon is predicted by Eq. 140 or 141: As atmospheric carbon dioxide dissolves and increases C_T, the buffer index increases to a maximum, about twice its present value.

The calculation begins with a choice for A, then for each value of P_{CO_2}, a value of pH is obtained by solving Eq. (135) (using Newton's method with a starting value for pH approximately 8). $[CO_2]$, $[HCO_3^-]$, and $[CO_3^{2-}]$ in Eq. (140) or (141) can be expressed in terms of $[H^+]$ and K_{a1} or 10^{-pH} and K_1'. This pH value together with the assumed P_{CO_2} value, is substituted in Eq. (134) to get C_T. Finally, B is calculated from Eq. (140) (or Eqs. 141 and 142). Figures 10.35–10.37 show pH, P_{CO_2}, and B, each as functions of C_T with A held constant at 0.0022 mole/kg. The equilibrium constants used in this calculation are given in Table 10.13 under "NBS."

The point labeled "present atm & sw" occurs at $A = 0.0022$, $P_{CO_2} = 0.00035$ atm, pH = 8.17, $C_T = 0.00200$, and $B = 9.09$, and falls to the left of the maximum at $B = 18.7$. That maximum occurs when $C_T = A = 0.0022$, pH = 7.55, and $P_{CO_2} = 0.0017$ atm. Thus atmospheric CO_2 could increase by a factor of five before the homogeneous buffer capacity began to decline.

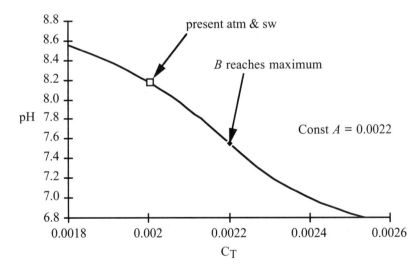

FIGURE 10.35. pH as a function of C_T at constant A.

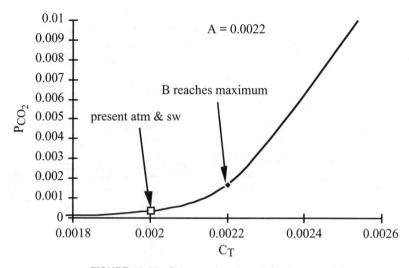

FIGURE 10.36. P_{CO_2} as a function of C_T at constant A.

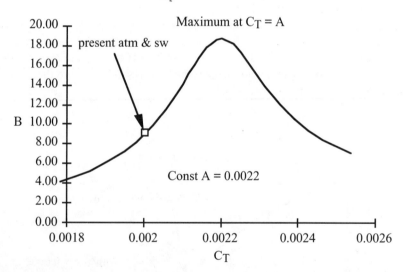

FIGURE 10.37. Homogeneous buffer index as a function of C_T at constant A. Note that as C_T increases from its present value, B also increases. The maximum $B = 18.7$ occurs when $C_T = A = 0.0022$.

Other Salinities: Ion-Pair Model for Seawater

The analysis above depends on having empirical equilibrium constants measured in a standard 35‰ seawater medium. If the system you are considering has a different salinity, there are some empirical data in the literature (Figs. 10.28–10.30, Table 10.12, pp. 420–421) covering the salinity range from 5 to 45‰.

But if you are dealing with a saline medium with different ionic composition from seawater, you must estimate the equilibrium constants in that medium. One approach is by applying ion-pairing equilibria and activity coefficients measured independently. The literature data in Table 10.16 are useful in calculations on the aqueous CO_2–carbonate system. These equilibria are especially important in seawater and brines, and significant in physiological and fresh-water systems. Unfortunately, few data are available at other ionic strengths and temperatures. Other theoretical approaches to high ionic strengths make use of empirical ionic interaction coefficients[92,110] or hydration theory.[93]

The principles used in developing an ion-pairing model were already presented in connection with physiological solutions (pp. 409–414). The seawater ion-pair model presented here also relies on a set of equilibria and mass balances.[94] The input data shown in Table 10.17 are in four groups. First are the ion-pairing constants, taken from Table 10.16. The activity coefficients are calculated from the Davies equation and are used to adjust the equilibrium constants to the ionic strength corresponding to the salinity; in this case $I = 0.7$ for 35‰. For example,

$$K^\circ_{NaSO_4} = \frac{[NaSO_4^-]\gamma_-}{[Na^+]\gamma_+[SO_4^{2-}]\gamma_=} = \frac{[NaSO_4^-]}{[Na^+][SO_4^{2-}](\gamma_+)^4} = \frac{K_{NaSO_4}}{(\gamma_+)^4} \quad (143)$$

where the activity coefficients have been combined using $\gamma_- = \gamma_+$ and $\gamma_= = (\gamma_+)^4$. Hence

$$\log(K_{NaSO_4}) = \log(K^\circ_{NaSO_4}) + 4 \log \gamma_+ \quad (144)$$

Similar equations apply to the other ion pairs.

TABLE 10.16. Ion-pairing constants for the seawater model at 25°C[91]

Cation	Anion	log K (I = 0) (Smith & Martell)	log K (I = 0) (Clegg & Whitfield)	(I = 0.7, expt)	(I = 0.7, Davies)
Ca^{2+}	CO_3^{2-}	3.15±0.07	3.15	2.21	1.89
Ca^{2+}	HCO_3^-	1.26±0.03	1.10	0.29	0.63
Ca^{2+}	SO_4^{2-}	2.31±0.01	2.31	1.03	1.05
Ca^{2+}	OH^-	1.3±0.1	1.15		0.67
Mg^{2+}	CO_3^{2-}	2.92±0.07	2.88	2.05	1.66
Mg^{2+}	HCO_3^-	1.01±0.06	1.07	0.21	0.38
Mg^{2+}	SO_4^{2-}	2.23±0.02	2.25	1.01	0.97
Mg^{2+}	OH^-	2.8	2.21		2.17
Na^+	CO_3^{2-}	1.27±0.3	0.7	0.63	0.64
Na^+	HCO_3^-	−0.25	−0.25	−0.55	−0.57
Na^+	SO_4^{2-}	0.70±0.05	0.82		0.07
Na^+	OH^-	0.1±0.3	−0.2		−0.22
K^+	SO_4^{2-}	0.85±0.1	0.85		0.22
K^+	CO_3^{2-}		0.7		0.07

TABLE 10.17. Seawater ion-pair model—inputs[96,a]

	Ion-pairing constants			Equilibrium consts			Total conc. (mM)		Initial free conc.	
	$\log K^\circ$	$\log K_{sw}$		pK°	$pK_{f(sw)}$					
NaSO$_4^-$	0.70	0.07	K_H	1.464	1.53	Cl$_T$	545.88	input	Na$_i$	469.2
NaCO$_3^-$	1.27	0.64	$K'_{a1(f)}$	6.363	6.14	Na$_T$	469.00	input	K$_i$	9.8
NaHCO$_3$	−0.25	−0.57	$K'_{a2(f)}$	10.329	9.86	K$_T$	10.21	input	Mg$_i$	46.3
NaOH	0.10	−0.22	K_w	13.997	13.84	Mg$_T$	52.80	input	Ca$_i$	9.2
KSO$_4^-$	0.85	0.22	$K_{so(f)}$	8.48	7.22	Ca$_T$	10.28	input	SO$_{4i}$	14.2
KCO$_3^-$	0.70	0.07	K_b	9.24	9.01	SO$_{4T}$	28.24	input	CO$_{3i}$	0.03
MgSO$_4$	2.23	0.97				B$_T$	0.476		HCO$_{3i}$	1.7
MgCO$_3$	2.92	1.66				P_{CO_2}	0.00035		OH$_i$	1.85E-03
MgHCO$_3^+$	1.01	0.38				pH	8.270			
MgOH$^+$	2.80	2.17				(adjust pH to make A = 2.20)				
CaSO$_4$	2.31	1.05								
CaCO$_3$	3.15	1.89								
CaHCO$_3^+$	1.26	0.63								
CaOH$^+$	1.30	0.67								

[a] Inputs: Ionic strength, 0.7; salinity, 35‰; temp., 25°C; $\log \gamma_\pm = -0.1578$.

The second group of input data are the acid–base equilibrium constants, listed in the fifth and sixth columns of Table 10.17. As with the ion-pairing equilibria, the zero-ionic-strength values have been adjusted to $I = 0.7$. For example,

$$K_2' = \frac{10^{-pH}[HCO_3^-]}{[CO_2]} = \left(\frac{[H^+]\gamma_+[HCO_3^-]\gamma_-}{[CO_2]\gamma_o}\right)\left(\frac{\gamma_o}{\gamma_-}\right) = K_{a2}^\circ \frac{\gamma_o}{\gamma_-} \quad (145)$$

Since $pK_{a2}^\circ = 6.363$ (Table 10.1); from the Davies Eq., $\log \gamma_- = \log \gamma_+ = -0.158$; and $\log \gamma_o = 0.1\ I = +0.070$,

$$pK_1' = pK_{a1}^\circ + \log \gamma_+ - 0.1\ I = 6.135$$

Similarly,

$$K_2' = \frac{10^{-pH}[CO_3^{2-}]}{[HCO_3^-]} = \left(\frac{[H^+]\gamma_+[CO_3^{2-}]\gamma_=}{[HCO_3^-]\gamma_-}\right)\left(\frac{\gamma_-}{\gamma_=}\right) = K_{a2}^\circ \frac{\gamma_-}{\gamma_=} \quad (146)$$

$$pK_2' = pK_{a2}^\circ + 3\log \gamma_+ = 10.329 + 3(-0.158) = 9.855$$

The values obtained in this way are listed in Table 10.13 (p. 425) in the column labeled "$K_{(f)}'$" (Eq. 80, p. 402). They differ from either the "free H" (Eq. 83) or "NBS" (Eq. 81) scales, which use total ionic concentrations, including ion pairs. An additional difference should be noted in that the pH scale used here is based on the Davies equation, and the NBS scale is based on an empirical standard.

Except for the adjustment to $I = 0.7$, this calculation is equivalent to one that uses concentrations on a free [H$^+$] scale, and eliminates the uncertainties due to liquid junction potentials that are inherent in the NBS pH scale.[95] The next group of data (Columns 7 and 8 in Table 10.17) include the composition of seawater in terms of total elemental concentrations, as well as the partial pressure of CO_2 and the pH. In this example, P_{CO_2} is set equal to the atmospheric average 3.5×10^{-4} atm. pH could be set equal to a measured value, but in this example, it was adjusted to make the total alkalinity $A = 2.200$ meq/L. This calculation will be described shortly.

The last group of data in Table 10.17 are starting values for the various concentrations. These are rather arbitrary, but of course the closer they are to the final answers, the faster will be the convergence of the iterative calculation.

The mass balances are computed in the form shown in Table 10.18.[97] The initial composition values are used to start the calculation (when the variable "init" = 0), but when the iterations are proceeding ("init" = 1), the values are taken from the results of the previous iteration. For example, if init = 0, free Na$^+$ (cell C25) is set equal to Na$_i$ = 469.2 mM (see top right of Table 10.17), but after init = 1, free Na$^+$ is set equal to the contents of cell B25, that is, the result of the first iteration. As you can see from the formula in cell B25, the iteration uses Newton's method in the form derived earlier in this chapter (Eq. 94, pp. 410–411).

TABLE 10.18. Seawater ion-pair model—Formulas[a]

	A	B	C	D	E			L
21	init =	1	(0=initialize,1=run: 10					
22	Ion			CO3=	HCO3-			
23		Free(out)	(mM)	=D24*D30/D29	=E24*E30/E29			
24			Free(in)	=IF(init=0,CO3i,D23)	=IF(init=0,HCO3i,E23)			
25	Na+	=C25*J25/I25	=IF(init=0,Nai,B25)	=$C25*$D$24*10^(NaCO3-3)	=$C25*$E$24*10^(NaHCO3-3)			
26	K+	=C26*J26/I26	=IF(init=0,Ki,B26)	=$C26*$D$24*10^(KCO3-3)	0			
27	Mg++	=C27*J27/I27	=IF(init=0,Mgi,B27)	=$C27*$D$24*10^(MgCO3-3)	=$C27*$E$24*10^(MgHCO3-3)			
28	Ca++	=C28*J28/I28	=IF(init=0,Cai,B28)	=$C28*$D$24*10^(CaCO3-3)	=$C28*$E$24*10^(CaHCO3-3)			
29			Tcalc	=SUM(D24:D28)	=SUM(E24:E28)			
30			Tobs	=CO3t	=HCO3t			
31			Δ	=D29-D30	=E29-E30			
32			% free	**=D24/D30**	=E24/E30			
33								

	F	G	H	I	J	K	L
22	Ion	SO4=	OH-				
23		=G24**G30/G29	=H24**H30/H29		Tcalc	Δ	% free
24		=IF(init=0,SO4i,G23)	=IF(init=0,OHi,H23)				
25	Na+	=$C25*$G$24*10^(NaSO4-3)	=$C25*$H$24*10^(NaOH-3)	=SUM(C25:H25)	=Nat	=I25-J25	=C25/J25
26	K+	=$C26*$G$24*10^(KSO4-3)	0	=SUM(C26:H26)	=Kt	=I26-J26	=C26/J26
27	Mg++	=$C27*$G$24*10^(MgSO4-3)	=$C27*$H$24*10^(MgOH-3)	=SUM(C27:H27)	=Mgt	=I27-J27	=C27/J27
28	Ca++	=$C28*$G$24*10^(CaSO4-3)	=$C28*$H$24*10^(CaOH-3)	=SUM(C28:H28)	=Cat	=I28-J28	=C28/J28
29		=SUM(G24:G28)	=SUM(H24:H28)				
30		=SO4t	=OHi				
31		0	0				
32		=G24/G30	**=H24/H30**				

[a] $Iy = \log \gamma_+$; MX stands for the ion pairing constant $\log(K_{MX}) = \log[MX]/([M][X])$; $[B(OH)_4] = 0.176 = B_T\, 10^{-pK_b}/(10^{-pH-Iy} + 10^{-pK_b})$

$$[HCO_3]_T = P_{CO_2} 10^{(-pKH-pK_1+pH+Iy)}(1000 + 10^{NaHCO_3}[Na^+] + 10^{MgCO_3}[Mg^{2+}] + 10^{CaCO_3}[Ca^{2+}])$$

$$[CO_3]_T = P_{CO_2} 10^{(-pK_H-pK_1-pK_2+2pH+2Iy)}(1 + 10^{NaCO_3}[Na^+] + 10^{MgCO_3}[Mg^{2+}] + 10^{CaCO_3}[Ca^{2+}])$$

$$A_T = [HCO_3]_T + 2[CO_3]_T + [OH]_T - [OH]_T + B_t 10^{-pKb}/(10^{-pH-Iy} + 10^{-pKb})$$

$$C_T = (10^{-pKH})P_{CO_2} + [HCO_3]_T + [CO_3]_T$$

CARBON DIOXIDE EQUILIBRIA IN SEAWATER   439

$$[Na^+]_{f,new} = [Na^+]_{f,o} \, [Na^+]_T \, (obs)/[Na^+]_T(calc) \qquad (94)$$

The calculated total $[Na^+]_T(calc)$ is the sum of C25 to H25 along row 25 (and found in cell I25—see second section of Table 10.18) and $[Na^+]_T(obs) = J25 = 545.88$ comes from the "total concentration" section of Table 10.17.

Analogous calculations are made for K^+, Mg^{2+}, and Ca^{2+} along rows 26, 27, and 28, and for CO_3^{2-}, HCO_3^-, SO_4^{2-}, and OH^- in columns D, E, G, and H (Column F is a repetition of the legend in Column A).

After 10 iterations, the mass balances are normally correct to within 10^{-8} mM and after 20 iterations to within 10^{-15}, the roundoff error of the calculations. Note that to do these iterations, it is normally necessary to choose "Iterate" (with Excel, this is under "Calculation" in the "Option" menu). When editing the spreadsheet, it is convenient to limit the calculation to one iteration; to obtain answers, the maximum number of iterations should be set equal to 10 or 20, and the maximum change to zero (or else the iteration may stop before the equations are balanced—see pp. 33–34).

Three results of the sample calculation are given in Table 10.19. For example, cell D24, at the intersection of the Na^+ row and the CO_3^{2-} column, shows the value of $[NaCO_3^-] = 0.0600$ mM. Other ion-pair concentrations are shown at the analogous intersections.

Along the last column and bottom row of the spreadsheet are displayed the percentages of each ion which are free. 86% to 98% of the cations are free, but for the anions, the percentages are much smaller, 17% for CO_3^{2-} and 12% for OH^-. The result is that processes such as the precipitation of carbonate and hydroxide minerals are greatly affected by the presence of these major ions.

Three direct tests can be made of the ion-pairing model. Each involves comparing an equilibrium constant calculated from the ion-pairing model with an experimental value from the literature. The lower part of Table 10.20 (Cells O12 to P13) give the values for pK_1' and pK_2' obtained from the literature and from the ion-pair model with the equilibrium constant values given in Table 10.17.

The ion-pair model value of $pK_1' = 6.03$ is an excellent agreement with the experimental value 6.01. Adjustment of $\log K_{NaHCO_3}$, from -0.25 to -0.10 (or $\log K_{CaHCO_3}$ from 1.26 to 1.6) could make the agreement exact.

$pK_2' = 9.28$ from the ion-pairing model, compared with the experimental value of 9.08, does not look so good, but in column P of Table 10.20, you will note that pK_2' can be brought down to 9.08 if $\log K_{NaCO_3}$ is increased from 1.27 to 1.64. Of course, one or more of the other constants could also be adjusted and $\log K_{NaCO_3}$ would not have to be changed so much. For example, in Table 10.20, adjustments of $\log K_{MgCO_3}$ from 2.92 to 3.28 (column S), or adjustment of $\log K_{CaCO_3}$ from 3.15 to 3.85 (column X), will accomplish the same result. Adjusting $\log K_{CaSO_4}$, on the other hand (columns U and V), does not help. Considering the stronger ionic strength dependence of pK_2' and its further dependence on $[CO_3^{2-}]$, most of which is bound up in pairs with Na^+, Mg^{2+}, and Ca^{2+}, this result is quite acceptable.

The final test might be the solubility product K_{so}. Here there is an extremely strong dependence on ionic strength ($\gamma_\pm^8$) or $8(-0.16) = -1.28$ log units at $I = 0.7$. In addition, both Ca^{2+} and CO_3^{2-} ions form strong complexes with the other ions.

TABLE 10.19. Seawater ion-pair—outputs (mM)[a]

	A	B	C	D	E	F	G	H	I	J	K	L
22	Ion			CO3=	HCO3-	Ion	SO4=	OH-				
23		Free(out)	(mM)	0.0299	1.3946		13.526	3.28E-04				
24			Free(in)	0.0299	1.3946		13.526	3.28E-04	Tcalc	Tobs	Δ	% free
25	Na+	461.45	461.45	0.0600	0.1750	Na+	7315	9.21E-05	469.00	469.00	0.0E+00	98.4%
26	K+	9.98	9.98	0.0003	0.0000	K+	0.224	0.00E+00	10.21	10.21	0	97.8%
27	Mg++	46.71	46.71	0.0635	0.1559	Mg++	5.868	2.26E-03	52.80	52.80	0.0E+00	88.5%
28	Ca++	8.626	8.626	0.0199	0.0512	Ca++	1.303	1.32E-05	10.000	10.000	1.8E-15	86.3%
29			Tcalc	0.1737	1.7766		28.236	2.69E-03				
30			Tobs	0.1737	1.7766		28.236	2.69E-03				
31			Δ	-2.8E-17	0.0E+00		0	0				
32			% free	17.2%	78.5%		47.9%	12.2%				

[a] pH, 8.270; [HCO3-]T, 1.777 mM; A, 2.200 mM; [CO$_3^{2-}$]T, 0.174 mM; C_T, 1.951 mM; [OH]T, 0.0027 mM; SI(f), 0.629.

TABLE 10.20. Sensitivity analysis of ion-pair constants[98]

M	N	O	P	Q	R	S	T	U	V	W	X	Y
	Davies	Lit val										
	logKsw	log K°	log K°	log K°	log K°	log K°	log K°	log K°	log K°	log K°	log K°	log K°
NaCO3	0.64	1.27	1.64	0.90	1.27	1.27	1.27	1.27	1.27	1.27	1.27	1.27
MgCO3	1.66	2.92			1.92	3.28	3.92	2.92	2.92	2.92	2.92	2.92
CaSO4	1.05	2.31						1.31	3.31	2.31	2.31	2.31
CaCO3	1.89	3.15								2.15	3.85	4.15
	IP model	Lit	IP model	IP model	IP model	IP model	IP model	IP model	IP model	IP model	IP model	IP model
pK1'(T)	6.03	6.01	6.03	6.03	6.03	6.03	6.03	6.03	6.04	6.03	6.03	6.03
pK2'(T)	9.28	9.08	9.08	9.40	9.50	9.08	8.59	9.27	9.30	9.34	9.08	8.92

Unfortunately, seawater is supersaturated with $CaCO_3$, and so all that can be compared is the saturation index, as calculated from the ion-pairing model (Eq. 147), with the saturation index calculated from total concentrations (see Eq. 148):

$$SI = \log\left(\frac{[Ca^{2+}]_f[CO_3^{2-}]_f}{K_{so}}\right) \qquad (147)$$

$$SI = \log\left(\frac{[Ca^{2+}]_T[CO_3^{2-}]_T}{K_{so(T)}}\right) \qquad (148)$$

When $pK_{so(f)} = 7.22$ (Table 10.17, p. 436) is used with the free-ion concentrations in Table 10.19 (p. 440), Eq. 147 gives SI = $\log[(8.63 \times 10^{-3})(0.0299 \times 10^{-3})] + 7.22$ = 0.63. If the total concentrations are used with $pK_{so(T)} = 6.36$,[99] Eq. 148 gives SI = 0.60, in excellent agreement.

Variations in the Model. In Table 12.6, a comparison is made among various models for ion pairing in seawater and various methods of calculation. To facilitate a direct comparison among cases, this model was run using $[H^+] = 6.31 \times 10^{-6}$ mM, which corresponds to pH = 8.20 + 0.16 = 8.36. The output for this calculation for a few ion pairs is given in Table 10.21. Other calculations summarized in that table include changing the solution composition to conform with Chapter 12, changing the constant term b in the Davies equation from 0.20 to 0.24 (to conform with the program MINEQL), calculating the ionic strength from the free-ion concentrations, and iterating that to convergence. This reduces the ionic strength from $I = 0.70$ to 0.646 and changes $\log \gamma_+$ from -0.158 to -0.145.

Finally, the activity coefficients of the neutral ion pairs is changed from $\log \gamma_o = 0$ (Table 10.17) to $\log \gamma_o = 0.1\ I$. For each of these calculations, minor adjustments in pH and P_{CO_2} were necessary to meet the constraints placed by Table 12.3: $[H^+] = 6.31 \times 10^{-6}$ mM and $C_T = 2.42$ mM. The last column of Table 10.21 corresponds to case 3b of Chapter 12.

Many of the output values are not sensitive to any of these changes. Of course $[CO_3^{2-}]$ and the carbonate ion pairs are sensitive to changes in pH. One way to see these variations is to compare the saturation index calculated from the free ionic concentrations with that calculated from the total ionic concentrations. In the second column of Table 10.21 (same as Table 10.19, pH = 8.27) S.I.(free) = 0.63 and S.I.(total) = 0.60. In column 3, the pH is changed to 8.36, but the composition is otherwise the same, and S.I.(free) = 0.81 and S.I.(total) = 0.78. Changing the composition to conform with Table 12.3 gives the same results. In each case the difference is 0.03.

Changing the coefficient b in the Davies equation also has a significant effect: All three of the last columns display a higher saturation index [S.I.(free) = 0.85 to 0.86] than the previous ones. The difference between S.I.(free) and S.I.(total) is also larger, 0.05 to 0.09. Adding $\log \gamma_o = 0.1\ I$ has a greater effect than iterating the ionic strength. In Chapter 12, further comparisons will be made.

TABLE 10.21. Effect of parameter change (each column contains all the changes in the previous ones)

Species	Table 10.19 comp⟶	Ch 12.6 comp⟶	$b = 0.24$	I by iteration	$\log \gamma_o = 0.1\ I$
pH	8.27	8.36	8.358	8.344	8.345	8.345
$P_{CO_2} \times 10^4$	3.50	3.50	3.47	3.61	3.59	3.67
Mg^{2+}	46.7	46.7	47.0	46.1	46.2	46.8
Ca^{2+}	8.63	8.61	8.78	8.58	8.61	8.74
SO_4^{2-}	13.5	13.5	13.5	12.21	12.34	12.81
$NaSO_4^-$	7.32	7.32	7.28	7.50	7.47	7.75
$MgSO_4$	5.87	5.86	5.89	6.76	6.67	6.05
$CaSO_4$	1.30	1.30	1.32	1.51	1.49	1.36
$MgCO_3$	0.0635	0.0961	0.0952	0.107	0.105	0.0929
CO_3^{2-}	0.0299	0.0453	0.0445	0.0393	0.0398	0.0402
$CaCO_3$	0.0199	0.0301	0.0302	0.0337	0.0333	0.0294
S.I. (free)	0.63	0.81	0.81	0.86	0.85	0.86
S.I. (total)	0.60	0.78	0.78	0.80	0.80	0.77

TABLE 10.22. Composition of a mixture of 1 part seawater and 3 parts river water

Conc (mM)	SW comp	RW comp	Mixture
Na	468.04	0.274	117.22
K	10.0	0.059	2.54
Mg	53.27	0.169	13.44
Ca	10.33	0.375	2.86
Cl	545.88	0.220	136.64
SO4	28.2	0.117	7.14
Borate	0.43	0.000	0.11
Alk	2.20	0.957	1.268
Ct	1.956	0.952	1.217
Ion str	696.88	2.077	175.78
Sal	35.00	0.104	8.83
pH	8.270	8.235	8.188
SI	0.645	−0.701	−0.045

A Mixture of Seawater and River Water. The ion-pair model becomes most useful when the solution is not one of the standard salinities or exactly the ionic composition of seawater. Here is an example where 1 part seawater is mixed with 3 parts river water (Table 11.20). The total ionic concentrations follow the standard mixing rule:

$$[C]_{\text{mixt}} = x_R[C]_R + x_S[C]_S \qquad (149)$$

where x_R is the fraction of river water and x_S is the mole fraction of seawater; $[C]_R$ and $[C]_S$ are the concentrations of a component in the river water and seawater, respectively. If the concentrations are in mol/L, the fractions are volume fractions; if the concentrations are in mol/kg, the fractions are based on the weight of water in each solution, which is approximately the weight fraction of solution.

Note that the materials to which this mixing rule applies are all "conservative"; that is, the same total mass of material is present in the final mixture as was present in the two initial solutions. On the other hand, variables like pH, $[CO_3^{2-}]$, or P_{CO_2}, which do not represent an unchanging total mass, are not conservative. Indeed, the nonlinear behavior of pH and saturation index as saline solutions are mixed with freshwater has been well documented.[100]

The layout of Table 10.23 contains the input and output information such as was presented in Tables 10.17 and 10.19. The formulas are the same as presented in Table 10.18. The big differences in the input data are in the salinity, and hence the ionic strength (0.177 instead of 0.7), and in the composition of the mixed solution (Table 10.22), for which all the species except alkalinity are considerably lower.

As before, pH was adjusted to bring the alkalinity to the input value (pH = 8.188 gives A = 1.270 mM). The iterative machinery then produces an output (Cells D24

TABLE 10.23. Results of ion-pairing model for mixtures of seawater and river water

	A	B	C	D	E	F	G	H	I	J	K	L
1	SEAWATER ION PAIR MODEL				Ka1 & Ka2 are free constants for Seawater							
2	Inputs	RW fract	0.75		Ion Str	0.177						
3	Salinity	8.828	Temp =	25°C	log γ	-0.13						
4	Ion-pairing constants			Equilibrium Consts.			Total Conc's (mM)				Initial Free Conc	
5		log K°	logKsw	pK°	pKf(sw		Clt	136.635	input	Nai	118.346	
6	NaSO4	0.70	0.18	KH	1.464	1.48	Nat	117.216	input	Ki	2.472	
7	NaCO3	1.27	0.75	Ka1(f)	6.363	6.22	Kt	2.544	input	Mgi	11.678	
8	NaHCO3	-0.25	-0.51	Ka2(f)	10.329	9.94	Mgt	13.444	input	Cai	2.321	
9	NaOH	0.1	-0.16	Kw	13.997	13.87	Cat	2.864	input	SO4i	3.582	
10	KSO4	0.85	0.33	Kso(f)	8.48	7.44	SO4t	7.138	input	CO3i	0.008	
11	KCO3	0.70	0.18	Kb	9.24	9.09	Bt	0.108	input	HCO3i	0.429	
12	MgSO4	2.23	1.19				A(target	1.268	input	OHi	2.09E-03	
13	MgCO3	2.92	1.88				PCO2	3.50E-04	input			
14	MgHCO3	1.01	0.49				pH	8.188	(pH adj to match A to A(target))			
15	MgOH	2.80	2.28						Outputs (mM)			
16	CaSO4	2.31	1.27					HCO3t	1.1797	sw calc*		
17	CaCO3	3.15	2.11					CO3t	0.0370	sw calc*		
18	CaHCO3	1.26	0.74					OHt	0.0021	sw calc*		
19	CaOH	1.30	0.78					A	1.268	sw calc*		
20								Ct	1.217	sw calc*		
21	Init =	1	(0=initialize,1=run; 10 iterations)									
22	Ion			CO3=	HCO3-	Ion	SO4=	OH-				
23		Free(out	(mM)	0.0127	1.0838		5.024	5.32E-04				
24			Free(in)	0.0127	1.0838		5.024	5.32E-04	Tcalc	Tobs	Δ	% free
25	Na+	116.29	116.29	0.0083	0.03889	Na+	0.882	4.27E-05	117.22	117.22	0.0E+00	99.2%
26	K+	2.52	2.52	4.8E-05	0.0000	K+	0.027	0.00E+00	2.54	2.54	0	98.9%
27	Mg++	12.43	12.43	0.0119	0.0415	Mg++	0.962	1.26E-03	13.44	13.44	1.8E-15	92.4%
28	Ca++	2.602	2.602	0.0042	0.0155	Ca++	0.242	2.63E-04	2.864	2.864	0.0E+00	90.8%
29				0.0370	1.1797		7.138	2.09E-03				
30			Tcalc	0.0370	1.1797		7.138	2.09E-03		SI(f) =	-0.045	
31			Tobs	0.0E+00	0.0E+00		0	0		pKso(T) =	6.92	
32			Δ	34.2%	91.9%		70.4%	25.4%		SI(T) =	-0.054	
			% free									

to H28) in terms of free-ion and ion-pair concentrations. Note that the cations are almost all free, but that CO_3^{2-} and OH^- both still exist largely as ion pairs.

The most interesting result is the saturation index (in the lower right corner, Cells J30 to K32), which is significantly lower than standard seawater; indeed it is negative, indicating undersaturation. From the ion-pair model, SI = −0.045; from the total ionic concentrations, together with an experimental value pK_{so} = 6.92 at I = 0.18,[101] SI = −0.054, in excellent agreement. An interpolated value from the more recent data of Mucci[102] gives pK_{so} = 7.03 at I = 0.18, yielding SI = +0.057.

Repetitive calculations such as those shown in Table 10.23 can yield the saturation index of any mixture of solutions containing calcium and carbonate. Figure 10.38 shows the saturation index for a mixture of seawater (left end member) with river water (right end member) of the composition given in Table 10.22, all under a constant CO_2 partial pressure of 3.5×10^{-4} atm. The seawater is supersaturated with $CaCO_3$, and the fresh water is undersaturated, but the mixtures tend to be more saturated than a linear interpolation would predict.

The curves of Fig. 10.39 represent the saturation index of a mixture of seawater ($[Ca^{2+}]_T$ = 10.33 mM, [Alk] = 2.20 mM, P_{CO_2} = 3.5×10^{-4} atm), and a fresh "hard" water similar to that in Table 10.22, but containing much higher calcium (5.5 mM) and alkalinity (11 meq/L), as well as being in equilibrium with a higher partial pressure (0.1 atm) of CO_2. This water is supersaturated with $CaCO_3$ and so is seawater, but a wide range of their mixtures are undersaturated.[103] In computing the saturation index of the mixtures, using a spreadsheet like Table 10.23, it is tempting to make a linear interpolation between the seawater and freshwater values: P_{CO_2} = $0.10x + 3.5 \times 10^{-4}(1-x)$. Although it might be possible to arrange a set of experi-

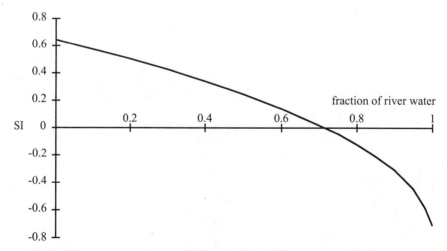

FIGURE 10.38. Saturation index for calcium carbonate in mixtures of seawater and a low-calcium, undersaturated fresh water under a constant atmospheric partial pressure of CO_2.

CARBON DIOXIDE EQUILIBRIA IN SEAWATER

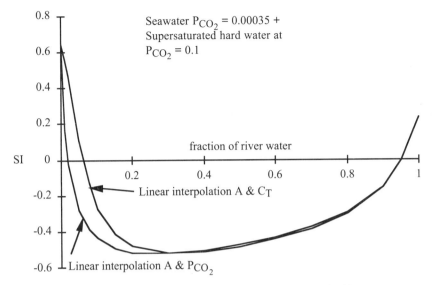

FIGURE 10.39. Saturation index of mixtures of seawater (left member) with a supersaturated hard fresh water (right member).

ments where P_{CO_2} was adjusted to conform to that formula for each mixture, a more plausible constraint is to take some of the seawater (with its appropriate pH and P_{CO_2}) and mix it with a quantity of the fresh water (with its appropriate pH and P_{CO_2}). The composition of the intermediate mixtures will be determined by the conservative quantities alkalinity and total carbonate, and the pH and P_{CO_2} of the mixtures will be determined by A and C_T:

$$A_{mix} = A_{RW} x + A_{SW} (1-x)$$
$$C_{T\ mix} = C_{T\ RW} x + C_{T\ SW} (1-x)$$

As before, pH is adjusted to match the computed value of A to A_{mix}. Likewise, $C_{T\ mix}$ is part of the input; and P_{CO_2}, instead of being part of the input, is computed from the equation for C_T, which is solved for P_{CO_2}:

$$C_T = K_H P_{CO_2} + [HCO_3^-]_T + [CO_3^{2-}]_T \qquad (150)$$

The rigorous calculation is the linear interpolation of A and C_T, but the less rigorous calculation that makes a linear interpolation of P_{CO_2} agrees well for $x > 0.2$. Another way of looking at this difference in constraints is in terms of P_{CO_2} itself. Figure 10.40 shows the difference between P_{CO_2} obtained in the course of a linear interpolation of A and C_T, and the P_{CO_2} obtained by direct linear interpolation. For $x < 0.3$, the first method gives a smaller value; for $x > 0.3$, it gives a larger value.

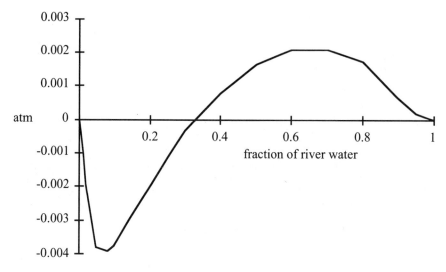

FIGURE 10.40. P_{CO_2} for the mixtures of seawater and river water (see Fig. 10.39) obtained from a linear interpolation of A and C_T, less the P_{CO_2} obtained by direct interpolation.

Since P_{CO_2} at the seawater end is low (0.00035 at $x = 0$), these deviations show up much more strongly in that region.

PROBLEMS[31]

1. What would be the pH of a soda water made by saturating distilled water with pure CO_2 at 1.0 atm pressure?

2. Find the amount of strong base required to change the pH of water saturated with CO_2 at 0.10 atm from its initial value of pH = 4.4 to 5.5, assuming equilibrium is maintained with the gas phase.

3. What is the alkalinity of a solution containing 0.01 M $NaHCO_3$ afer it has reached equilibrium with an atmoshere containing CO_2 at partial pressure 1.0 atm? At partial pressure 0.01 atm?

4. If NaOH is aded to water saturated with CO_2 at $10^{-3.5}$ atm and the solution is kept in equilibrium with the atmosphere, at what pH will the composition of Na_2CO_3 be reached?[32]

5. If NaOH is aded to water saturated with CO_2 at $10^{-3.5}$ atm and the solution is kept in a closed vessel while the alkali is added, at what pH will the composition of Na_2CO_3 be reached?[33]

6. If alkalinity is 1.0×10^{-3} equivalents/L, express this quantity in "ppm as $CaCO_3$."[34]

PROBLEMS

7. A water is analyzed and reported to have pH = 7.2, total hardness 175 ppm as $CaCO_3$, calcium hardness 160 ppm as $CaCO_3$, and alkalinity 165 ppm as $CaCO_3$. Find the molar concentration of $[Ca^{2+}]$, $[Mg^{2+}]$, $[HCO_3^-]$, $[CO_3^{2-}]$ and speculate about the remaining ions if the total dissolved solids are 280 mg/L.[35]

8. Acid rain is produced when acidic air pollutants (e.g., H_2SO_4 and HNO_3) are washed out of the atmosphere by rain. Compare the pH to be expected for rain in equilibrium with the average atmosphere ($P_{CO_2} = 10^{-3.5}$ atm)
   (a) If the rain is distilled water;
   (b) If the rain contains 10^{-5} M HNO_3 and 2×10^{-5} M H_2SO_4.

9. Recent measurements of the meltwater from high-altitude snow in Greenland and the Himalayas have given values near pH = 5.1 instead of the expected value of about 5.6. Use the equilibrium constants in Table 10.1 to calculate pH of water at 0°C in equilibrium with CO_2 at a partial pressure of $10^{-3.5}$ atm. What would the partial pressure of CO_2 have to be in order to reduce the pH to 5.1? What acidity would be required for the same reduction in pH at $P = 10^{-3.5}$ atm?[36]

10. Calculate the equivalence point pH for the titration of water with alkalinity 2×10^{-3} equivalents/L and pH = 7.5 with 0.10 M HCl.

11. A solution with alkalinity 2.5×10^{-3} and total carbonate 2.0×10^{-3} is titrated with 0.0100 M HCl solution. This titration is done in a sealed vessel so that no CO_2 escapes or is absorbed from the atmosphere.
    (a) Calculate the pH at the start of the titration.
    (b) Find the pH and partial pressure of CO_2 at the equivalence point.
    (c) Estimate the titration error if the end point is taken to be pH = 4.60.

12. An industrial waste from a metals industry contains approximately 5×10^{-3} M H_2SO_4. Before being discharged into a stream, the water is diluted with tap water in order to raise the pH to an acceptable level. The tap water has the following composition: pH = 6.5, alkalinity 2×10^{-3} equivalents/L, $[Ca^{2+}] = 2 \times 10^{-3}$ M, $[Mg^{2+}] = 10^{-4}$ M, $[Cl^-] = 10^{-3}$ M.
    (a) What is the initial pH of the waste?
    (b) What dilution is necessary to raise the pH to 4.5?

13. A solution contains 10^{-3} M $NaHCO_3$ and 10^{-3} M Na_2CO_3. Calculate its pH and its buffer capacity β_C.

14. Find the buffer index β_p of a solution containing CO_2 at a partial pressure of 10^{-2} atm in water with alkalinity 3×10^{-3} equivalents/L.

15. A well-known source of error in acid–base titrations is the possible presence of carbonate. If 10^{-3} M HCl is titrated with 10^{-3} M NaOH, the equivalence point would be expected to occur at pH = 7.00 (see Chapter 3). It is more likely that the solution has absorbed CO_2 from the atmosphere. If $[CO_2] = 10^{-5}$ M, find the titration error if the end point were taken at pH = 5.0, 6.0, and 7.0.

16. A classical method for measuring alkalinity[37] is to treat the sample with a measured excess of strong acid, driving off the CO_2 by purging it with N_2 or by boiling, and back-titrating with strong base to pH 6.0–7.0, where borate is negligibly ionized. Show that this method gives the correct result.

17. Culberson's modification of the Gripenberg method[38] consists of adding 30 mL of 0.01 M HCl to a 100 mL sample of seawater.[39] The sample is purged with CO_2-free, water-saturated air, and the final pH is measured. Show that the alkalinity is given by

$$A = \frac{1000}{V_s}VM - \frac{1000}{V}(V_s + V)\frac{10^{-pH}}{\gamma_H}$$

where V_s is the sample volume, V is the HCl volume, M the molar concentration of HCl, and γ_H is the activity coefficient of hydrogen ion.

18. Acid rain containing C_T mole/L of CO_2, C_S mole/L of H_2SO_4, C_A mole/L of HNO_3, and C_N mole/L of NH_4^+ is titrated with strong base of concentration C_b. Derive an equation for the titration curve and plot a sample curve for $C_T = C_S = C_A = C_N = 1.0 \times 10^{-5}$ M and $C_b = 0.001$ M. If the end point were determined by the Gran method, what acidity would be obtained? How could you improve on this result?

19. Show that the alkalinity end point determined by the Gran method in the presence of a noncomplexing ionic medium (e.g., 1.0 M NaCl) is the same as the equivalence point.

20. Protonation of sulfate in seawater distorts the shape of the Gran plot at low pH. Compare the Gran plot for 10^{-3} M HCO_3 titrated with 0.01 M HCl with and without a typical amount of sulfate found in seawater (0.028 M, $pK_a = 0.9$). Estimate the titration error introduced by the sulfate if the Gran plot is forced to fit a straight line from pH = 3 to 4.5.

21. In deriving the equation for the buffer index, the formula $d[H^+] = -2.303[H^+]\,d\text{pH}$ (Chapter 4, Eq. 76) is employed instead of the more accurate $[H^+] = -2.303[H^+][(d\text{pH} + d(\log \gamma_+))]$. Estimate the effect of neglecting the term $d(\log \gamma_+)$ on the buffer index if the pH of pure water is changed from 5.1 to 5.0 by addition of HCl.[40]

22. In deriving the equation for the buffer index β_C (Chapter 9, Eqs. 51–53) the term $4K_{a2}[H^+]$ was neglected. What error is introduced by this approximation in calculating the buffer index of 0.01 M $NaHCO_3$?

23. Find the partial pressue of CO_2 in equilibrium with solid $CaCO_3$ if the solution contains 1.0×10^{-3} M dissolved Ca^{2+} at pH = 8.0.

24. Acid rain (containing 10^{-5} M HNO_3 and 2×10^{-5} M H_2SO_4) in equilibrium with atmospheric CO_2 at $P = 10^{-3.5}$ atm falls on limestone and comes to equilibrium

PROBLEMS   451

with $CaCO_3$ and CO_2. What is the pH and alkalinity of the runoff water? How much would it differ if the rain contained only water and CO_2?

25. Rain falls on soil where all its oxygen content is converted to CO_2 by microbial processes, so that it is in equilibrium with $P_{CO_2} = 0.2$ atm. What is its alkalinity and pH? What effect would acidity in the rain have? Compare the composition of this runoff/leachate with the answer to Problem 24.

26. If a solution containing 10^{-3} M total calcium and pH = 7.5 is just barely saturated with $CaCO_3$,
    (a) Find the partial pressure of CO_2.
    (b) If this solution is placed in a closed vessel with excess $CaCO_3$, what pH and alkalinity result?

27. Derive the equation for $\beta_{P,s}$ (corresponding to Eq. 57 including ion pairs).

28. Derive the equations for $\beta_{D,s}$ (corresponding to Eqs. 68–75 including ion pairs).

29. Water is equilibrated with CO_2 at a partial presure of 10^{-5} atm. The vessel is then sealed and equilibrated with $CaCO_3$. Find the equilibrium pH and alkalinity. Evaluate D and $\beta_{D,s}$.

30. When samples of water are shipped from distant places, pH and P_{CO_2} may change during shipment, due to leaks and microbial action, which produces CO_2 and can change alkalinity. But $[Ca^{2+}]_T$ and alkalinity can be accurately determined and are less vulnerable. Some workers have estimated the original pH from

$$pH = pK_{a2}^\circ - pK_{so}^\circ - \log[Ca^{2+}]_T - \log(A) + 2.54\left(\frac{\sqrt{I}}{1+\sqrt{I}} - 0.2I\right)$$

Derive this equation. Improve on it if you can. What limitations does this method have? Distinguish limitations resulting from the basic model, those resulting from approximations, those resulting from experimental difficulties.

31. Using Eqs. (1)–(8) (pp. 367–368), verify the various equilibrium constants in Table 10.4 (p. 408).

32. The equilibrium constants for calcium–carbonate ion pairs at 25°C, $I = 0$ (Table 10.16) together with their temperature dependence[64] are:

$$[CaCO_3(aq)] = 10^{+3.15}[Ca^{2+}][CO_3^{2-}], \quad 25°C$$
$$\log K = +5.49 - 713/T$$
$$[CaHCO_3^+] = 10^{+1.26}[Ca^{2+}][HCO_3^-] \quad 25°C$$
$$\log K = -2.74 + 0.0133\, T$$

where T is in K. Check the consistency of the temperature dependence with the 25°C values. Including activity coefficients and ion pairs, calculate the pH and alkalinity of water saturated with calcite $CaCO_3(s)$ and with CO_2 at $P = 1$ atm, at 0 and 50°C. See Table 10.1 for temperature dependence of other constants. (Note that the Debye–Hückel coefficients also have a small temperature dependence—see Chapter 2, p. 45 and footnote 62 of this chapter).

33. Garrels and Christ[65] say "the role of CO_2 in rainwater probably has been overrated, whereas the effects of hydrolysis and of CO_2 in the soil atmosphere have been underrated." Show this by comparing the solubility of $CaCO_3$ in pure H_2O with the case where the water is first equilibrated with soil-derived CO_2 at P atm, where $10^{-3.5} < P < 0.2$. What effect do ion pairs have on these conclusions?

34. The reaction of dolomite $MgCa(CO_3)_2(s)$ with CO_2 in river water or groundwater is similar to that of calcite, and its solubility is controlled by a solubility product of the form[66]

$$K_D^\circ = [Mg^{2+}][Ca^{2+}][CO_3^{2-}]^2 = 10^{-17.0\pm0.02}$$

Calculate the equilibrium concentrations of Ca^{2+} and Mg^{2+} in equilibrium with dolomite and CO_2 at 10^{-2} atm. Estimate the error incurred by neglecting activity coefficients and ion pairs.

35. Water percolating through a rock system containing both calcite and dolomite may come into simultaneous equilibrium with both. Calculate the ratio $[Ca^{2+}]/[Mg^{2+}]$ at 25 and 0°C (see Footnote 18, p. 454 and Table 10.1, p. 368 for constants) for the CO_2 partial pressure range from 10^{-4} to 1 atm. Estimate the error incurred by neglecting activity coefficients and ion pairs.

36. How would the curves of P_{CO_2} versus pH for general intracellular fluid (Gen ICF, Table 10.3 and Fig. 10.22) and red blood cell intracellular fluid (RBC-ICF) change if $pK_a = 6.82$ were 6.52 or 7.12? How would the curves change if the HA listed in Table 11.1 consisted of a diprotic acid with $pK_{a1} = 6.52$ and $pK_{a2} = 7.12$?

37. Wimberley et al.[63] measured Na^+ and K^+ in 160 mM solutions using ion-selective electrodes and found that the cation activity decreased when Cl^- was replaced by HCO_3^-: 9.2 mM for Na^+ and 7.3 mM for K^+. Using the ion-pairing constants $\log K = -0.52$ for $NaHCO_3$ (Table 10.5, p. 410) and $\log K = -0.7$ for $KHCO_3$, and no ion pairing for chlorides, calculate what the result of the experiment would be. Assuming that the experimental results represented only the change in activity, calculate the ion-pairing constants. What other factors might be involved?

38. Calculate the concentration of $NaHCO_3$ and $KHCO_3$ ion pairs in plasma (Table 10.3, p. 405).

39. Using the equilibrium constants from Table 10.13 and the values of A_t and C_t

from Table 10.15, calculate the pH of seawater. Compare with the first entry in Table 10.14.

40. Calculate the equilibrium partial pressure of CO_2 from the data obtained in Problem 39.

41. If the atmospheric partial pressure of CO_2 were doubled (as might occur by 2100 if fossil fuel use continues unabated), and the alkalinity remained unchanged (as it would if the water sample were taken from the open ocean), find the new C_t and pH.

Additional problems will be found in *Carbon Dioxide Equilibria*, Chapter 5.

NOTES

1. Recall from Chapter 1 that activity is usually taken to be partial pressure of the gas in atmospheres. At high pressures, an activity (or "fugacity") coefficient may be required to correct by a few percent for nonideal behavior. See p. 12 and Note 16 in Chapter 1, also Note 4 below.
2. These data were selected from L. G. Sillén and A. E. Martell, *Stability Constants*, Special Publications No. 17 (1964) and 25 (1971), The Chemical Society, London, and recent literature. Detailed references are given in footnotes to the text below. Values at 25°C and various ionic strengths can be found in R. M. Smith and A. E. Martell, *Critical Stability Constants*, Vols. 1–6, New York. Plenum Press, 1974–1989.
3. Concentration in mol/L = (mol/kg)(kg/L); the density of dilute solutions varies from 0.99979 kg/L at 0°C to 0.99707 at 25°C to 0.95838 at 100°C. The corresponding log values are –0.000, –0.001, and –0.018, all of which are within the experimental error of the values given in Table 10.1. Individual sources and discussion of uncertainties are given below.
4. A. J. Ellis and R. M. Golding, *Am. J. Science* **261**:47–60, 1963. At high partial pressures of CO_2, a fugacity coefficient $\alpha_{CO_2} = f_{CO_2}/P_{CO_2}$ is introduced to express the nonideality of the gas phase.

$$[CO_2]\gamma_o = K_H^\circ P_{CO_2} \alpha_{CO_2}$$

α_{CO_2} depends only weakly on temperature and pressure. At pressures near atmospheric and temperatures near ambient, $\alpha_{CO_2} = 1.000$; in water α_{CO_2} varies from 0.985 at 177°C and 15.7 atm to 1.02 at 334°C and 59.2 atm. Detailed data will be found in the paper by Ellis and Golding.
5. Smith and Martell, Vol. 6, 1989.
6. $pK_H = pK_H^\circ + bI = 1.466 + (0.1)(0.7) = 1.54$
7. pK_H values for 0–50°C were taken from H. S. Harned and R. Davis, *J. Am. Chem. Soc.* **65**:2030, 1943. Those for 50–350°C from A. J. Ellis and R. M. Golding, *Am. J. Sci.* **261**:47, 1963, and H. C. Helgeson, *J. Phys. Chem.* **71**:3121, 1967. Smith and Martell (1986) give a critically evaluated $pK_H^\circ = 1.466 \pm 0.002$ at 25°C, $I = 0$.
8. K. F. Wissbrun, D. M. French, and A. Patterson, *J. Phys. Chem.* **58**:693, 1954; D. M. Kern, *J. Chem. Educ.* **37**:14, 1960.

9. Some textbooks refer to the sum $[H_2CO_3] + [CO_2]$ as $[H_2CO_3]$, and others as $[H_2CO_3^*]$ or $[CO_2^*]$. The first is inaccurate and the second is cumbersome. The most precise notation would be $[CO_2]_T$, indicating the total of all uncharged species containing CO_2, whatever they were.

10. The traditionally accepted data for pK_{a1}° were obtained by H. S. Harned and R. Davis, *J. Am. Chem. Soc.* **65**:2030, 1943, A. J. Ellis, *Am. J. Sci.* **257**:287, 1959, and evaluated by H. C. Helgeson, *J. Phys. Chem.* **71**:3121, 1967. The most recent published data, by R. N. Roy et al., *J. Chem. Thermo.* **14**:473–482, 1982 (NaCl); *ibid.* **15**:37–47, 1983 (KCl) are slightly higher than these:

    At 5°C, $I = 0$: 6.525 (Roy, extrap. from NaCl media), 6.512 ± 0.006 (Roy, extrap. from KCl media) rather than 6.517 from Harned and Davis;

    At 25°C, $I = 0$: 6.366 (NaCl), 6.360 ± 0.005 (KCl) rather than 6.352 ± 0.010 (Harned and Davis);

    At 45°C, $I = 0$: 6.292 (NaCl), 6.298 ± 0.009 (KCl) rather than 6.290 (Harned and Davis). Although these three data sets agree within Smith and Martell's (Vol. 6, 1989) estimated error of ±0.010, in Table 10.1, the older values have been adjusted slightly to agree better with Roy's recent measurements. The constant ΔH (for 25°C) given in the Smith and Martell tables gives a reliable prediction of the temperature dependence only between about 15 and 35°C (see Fig. 10.2). Outside this range, original data must be obtained from the literature.

11. Including the ion pairs in a total concentration is analogous to the formalism used for the conditional stability constants of metal–chelate complexes (Chapter 8, pp. 295–296).

12. Data for pK_{a2} were obtained by H. S. Harned and S. R. Scholes, *J. Amer. Chem. Soc.* **63**:1706, 1941, F. Cuta and F. Stráfelda, *Coll. Czech. Chem. Commun.* **48**:1308, 1954 and H. C. Helgeson, *J. Phys. Chem.* **71**, 3121, 1967. Smith and Martell list pK_{a2} = 10.320 ± 0.01 at 25°C, $I = 0$.

13. Data in Table 10.1 were interpolated from T. Ackermann, *Z. Elektrochem.* **62**:411, 1958, and H. G. Helgeson, *J. Phys. Chem.* **71**:3121, 1967. Other sources are tabulated by Sillén and Martell (1964 and 1971). The value pK_w° = 13.999 at 25°C was selected by Smith and Martell (1989).

14. Data for Table 10.1 at temperatures ≤ 100°C have been taken from C. L. Christ, P. B. Hostetler, and R. M. Siebert, *Jour. Research U.S. Geol. Survey* **2**:175–184, 1974, with a slight adjustment (−0.004) to agree with pK_{so}° = 8.48, the value selected for 25°C by Smith and Martell (Vol. 6, 1989). For higher temperatures the data of A. J. Ellis, *Am. J. Sci.* **261**:259–67, 1963, were used, with minor adjustments for ion pairs. Literature data for pK_{so} from over 20 investigators have been re-evaluated, using an ion-pair model, by Jan Vanderdeelen (University of Ghent, for IUPAC, 1980).

15. F. J. Millero, P. J. Milne, and V. L. Thurmond, *Geochim. et Cosmochim. Acta* **48**:1141–43, 1984.

16. K. S. Pitzer, *J. Phys. Chem.* **77**:268–77, 1973. See Chapter 12, pp. 492–494.

17. Smith and Martell, Vol. 6, 1989. Calcite containing several percent MgCo3 ("magnesian calcite") is also an important part of the magnetic process. See W. D. Bischoff, M. A. Bertram, F. T. Mackenzie, and F. Biship, *Carbonates and Evaporites* **8**:82–89 (1993); M. A. Bertram, F. T. Mackenzie, F. C. Bishop and W. D. Bischoff, *Amer. Mineralogist* **76**:1889–1896 (1991).

18. D. Langmuir, *Geochim. Cosmochim. Acta* **35**:1023–45, 1971. He also gives pK_{so} = 16.6 at 10°C and 16.9 at 20°C.

19. The typical atmospheric partial pressure of CO_2 has increased from a preindustrial value of less than 300 µatm to its 1996 value of 350 µatm or $10^{-3.45}$.

20. The result using constants for $I = 0$ is 2.08×10^{-3}; with constants adjusted to $I = 0.001$ by the Davies equation ($\log \gamma_+ = -0.0152$), the result is 2.01×10^{-3} atm.

21. See A. G. Dickson, *Marine Chemistry* **40**:49–63, 1992, for a historical review of the alkalinity concept and a critical analysis of current research on this topic.

22. J. N. Butler, *Carbon Dioxide Equilibria*, Addison-Wesley, 1982, Lewis, 1991, p. 52.

23. "The total alkalinity of a natural water is defined as the number of moles of hydrogen ion equivalent to the excess of proton acceptors ... over proton donors ... in one kilogram of sample" (proton acceptors are defined as bases formed from weak acids with $pK_a \geq 4.5$; proton donors are defined as acids with $pK_a \leq 4.5$ at 25°C and $I = 0$). A. G. Dickson, *Deep-Sea Res.* **28A**:609–23, 1981; J. N. Butler, *Mar. Chem.* **38**:251–82, 1992.

24. The superscript "ep" refers to the equivalence point, corrected for dilution. This is approximated by the empirical end point (see pp. 83–84, 139–140).

25. NH_3 and H_2S are the two most important exceptions.

26. The first five constants were taken from Table 10.1. The ion-pairing constants were selected from Smith and Martell (Vol. 6, 1989).

27. If the middle term is included, a term $(4)(2.303)[CO_2][CO_3^{2-}]/C_T$ is added to Eq. (53). This term is plotted in Fig. 10.15.

28. J. N. Butler, *Carbon Dioxide Equilibria and Their Applications*, Addison-Wesley, 1982; reprinted by Lewis, 1991; pp. 93–94 and 98–100.

29. The expansion of this derivation to include ion pairs is presented as problem 28.

30. J. N. Butler, *Carbon Dioxide Equilibria*, Addison-Wesley, 1982, and Lewis, 1991 (pp. 98–100). My feeble defense is that in 14 years no one else has pointed out these errors to me.

31. Additional problems will be found in J. N. Butler, *Carbon Dioxide Equilibria and Their Applications*, Addison-Wesley, 1982; Lewis, 1991.

32. Answer: Never. The proton condition for Na_2CO_3 is $[HCO_3^-] + \cdots = [OH^-] + \cdots$ and as long as the solution is in equilibrium with CO_2, both sides of this balance increase with increasing pH at the same slope.

33. In contrast, if C_T is fixed or limited, $[HCO_3^-]$ is limited but $[OH^-]$ increases monotonically with pH.

34. The engineering literature often expresses concentrations in terms of weight, not moles. This is awkward when a parameter such as alkalinity is involved, because it consists of a sum of several, often interconvertiible, species with different formula weights: $A = [HCO_3^-] + 2[CO_3^{2-}] + \cdots$. This problem is addressed by expressing alkalinity in terms of the equivalent weight of $CaCO_3$. Every mole of $CaCO_3$ that dissolves or precipitates changes the alkalinity by two equivalent units; hence $A = (100.09 \text{ g/mole})(0.5)(1.0 \times 10^{-3} \text{ mole/L})(10^{-3} \text{ mg/g}) = 50$ mg/L or 50 ppm by weight.

35. Any quantity that is related to alkalinity or calcium ion concentration can be expressed in terms of ppm $CaCO_3$ by using simple stoichiometry to determine how many moles of $CaCO_3$ would be required to produce the same (or in the case of acids, the opposite) effect on alkalinity. For this problem the numerical answers are $[Ca^{2+}] = 1.60 \times 10^{-3}$ M, $[Mg^{2+}] = 1.5 \times 10^{-4}$ M, $[HCO_3^-] = 3.30 \times 10^{-3}$ M, $[CO_3^{2-}] = 2.5 \times 10^{-6}$, and by difference $[Cl^-] = 1.95 \times 10^{-4}$. This last component could be any nonreactive anion; e.g., $[SO_4^{2-}] = 0.98 \times 10^{-4}$.

36. The calculated value is pH = 5.60. $P_{CO_2} = 10^{-2.5}$ atm would reduce this to 5.1. Acidity $10^{-5.15}$ would also reduce pH to 5.1.
37. Gripenberg, 5th Hydrol. Conf. Helsingfors Comm. 10B (1937); see J. P Riley and G. Skirrow, *Chemical Oceanography*, Vol. 2; Academic Press, NY, 1975, p. 27.
38. C. Culberson, R. M. Pytkowicz, and J. E. Hawley, *J. Mar. Res.* **28**: 15, 1970. See Riley and Skirrow, *op. cit.*, p. 28.
39. See later sections of this chapter for a full discussion of the equilibria in seawater. Constants including ion pairs will be found in Table 10.10.
40. dpH = 0.10. If there is no added salt, log $\gamma_+ = 1.43 \times 10^{-3}$ at $I = 10^{-5.1}$ and 1.61×10^{-3} at $I = 10^{-5.0}$, giving $d(\log \gamma_+) = 1.7 \times 10^{-4}$; the error is 0.17% of dpH.
41. See A. G. Dickson, *Geochim. Cosmochim. Acta* **48**:2299–308, 1984, for a clear and rigorous comparison of the NBS, SWS, and free H scales for pH in seawater.
42. U.S. National Bureau of Standards, now NIST, U.S. National Institute of Standards and Technology.
43. From Eq. 79 using the Davies equation ($\gamma_+ = 10^{-0.16}$ at $I = 0.7$) for $10^{-pH} = [H^+]\gamma_+$ you would get $K'_{1(f)} = 10^{-6.13}$. This is smaller by $10^{-0.05}$ times than the value ($10^{-6.08}$) given in Eq. 80. That value includes the systematic error in the liquid junction potential resulting from calibration in the NBS pH standards containing 0.05 M phosphate or borate, and measurement in seawater. It was obtained from the literature value quoted in Eq 81 ($10^{-6.00}$) for the NBS scale together with the ion-pairing fraction $\alpha_{HCO_3} = 10^{0.08}$ (Eq 82)
44. C. Mehrbach, C. H. Culberson, J. E. Hawley, and R. M. Pytkowicz, *Limnol. Oceanog.* **18**:897–907, 1973.
45. A. G. Dickson, *Deep-Sea Res.* **28A**:609–23, 1981.
46. A. G. Dickson, *Geochim. et Cosmochim. Acta* **48**:2299–308, 1984; C. Goyet and A. Poisson, *Deep-Sea Res.* **36**:1635–54, 1989; I. Hansson, *Deep-Sea Res.* **20**:461–91, 1973.
47. A. G. Dickson, *J. Chem. Thermodyn.* **22**:113–27, 1990.
48. A. G. Dickson and J. P. Riley, *Mar. Chem.* **7**:89–99, 1979.
49. J. N. Butler, *Mar. Chem.* **38**:251–82, 1992. Note log α is quite sensitive to pK_a for HSO_4^-; values in the literature vary from 0.115 to 0.139. Note also that the difference between the literature values quoted in Eqs. 83 and Eq 85 is 0.15 to 0.17. Applying log α_H = 0.115 to Eq 85 gives p$K_{a1(f)}$ = 5.95 to 5.97 for Eq 83. This is nearly in agreement with O. Johansson and M. Wedborg, *Oceanol. Acta* **5**:209–18, 1982, who obtained p$K_{a1(f)}$ = 5.986.
50. R. N. Roy et al., *Marine Chemistry*, **44**:249–67, 1993. See Chapter 11 for more detail.
51. The value for K_2 is from Goyet and Poisson, 1989, and for K_w is from Dickson and Riley, 1979, both cited above.
52. P. A. Stewart, *How to Understand Acid–Base*, Elsevier, New York, 1981, p. 136.
53. Ringer's solution is produced by McGaw, Inc., Irvine, CA 92714 (U.S. Pat. 4,803,102). This solution alone does not have a specified carbonate content and hence does not have a reproducible pH–P_{CO_2} relationship. A typical ratio for added bicarbonate is 500 ml Ringer's + 10 mL 1M $NaHCO_3$ (Astra Pharmceutical Products, Westborough, MA 01581).
54. Normosol-R is produced by Abbott Laboratories, North Chicago, IL 60064.

NOTES

55. M. C. Fishman, A. R. Hoffman, R. D. Klausner, M. S. Thaler, *Medicine*, 4th Ed., Lippincott-Raven, 1996, pp. 135–40.
56. Free-ion concentrations in seawater of 8% (nominally I = 0.16 based on total concentrations) were evaluated with an ion-pair model such as is described in Table 10.19, p. 440. Note $[CO_2] = K_H P_{CO_2}$ (mm)/760.
57. Stewart (Note 52) does not distinguish between concentrations and activities. His pK_a is $[H^+][A^-]/[HA]$. The constants in Table 10.3 for acetate and gluconate (components of Normosol R) are $pK' = 10^{-pH}[A^-]/[HA]$ estimated at $I = 0.16$ and 37°C. Later (Table 10.7, p. 414) we will consider "lactated Ringer's solution," which contains approximately 0.027 M lactate.
58. $\beta \approx 2.3([CO_2][HCO_3^-]/C_T + [HA][A^-]/[HA]_T)$. See Chaps. 4, 5, and Equation 53 above.
59. J. L. Gamble, *Chemical Anatomy, Physiology and Pathology of Extracellular Fluid. A lecture syllabus*, Dept. of Pediatrics, Harvard Medical School, Spaulding-Moss, Boston, 1941.
60. Fishman et al., *op. cit.*
61. Note that the carbonate equilibria given in Table 11.2 are based on free concentrations, not on total concentrations including ion pairs, as is done for seawater (see below). These constants were derived from $I = 0$ data: $pK_H = 1.63$, $pK_1 = 6.16$, $pK_2 = 9.86$.
62. R. A. Robinson and R. H. Stokes, *Electrolyte Solutions*, Butterworths, 1968, p. 468, gives the Debye–Hückel factor $A = 0.5115$ at 25°C and 0.5232 and 37°C. With $I = 0.16$ in the Davies equation

$$\log \gamma_z = -Az^2\left(\frac{\sqrt{I}}{1+\sqrt{I}} - 0.2I\right) = -0.1327z^2$$

63. See P. D. Wimberley, O. Siggaard-Andersen, N. Fogh-Andersen, and A. B. T. J. Boink, "Are sodium bicarbonate and potassium bicarbonate fully dissociated under physiological conditions?" *Scand. J. Clin. Lab. Invest.* **45**:7–10, 1985; H. Monoi, "Possible existence of ion pairs at the mouths of ion channels," *Biochim. Biophys. Acta* **693**:159–64, 1982; R. L. Coleman and C. C. Young, "Evidence for formation of bicarbonate complexes with Na^+ and K^+ under physiological conditions," *Clin. Chem.* **27**:1938, 1981; ibid. **28**:1705, 1982.
64. Selected from R. M. Smith and A. E. Martell, *Critical Stability Constants*, Vols. 1–6, Plenum Press, NY, 1974–1989. Constants were adjusted to $I = 0.16$ using the Davies equation.
65. See Table 11.2 for carbonate equilibrium constants.
66. See p. 27. If x_o is a first approximation to $f(x) = 0$, the second approximation is $x_1 = x_0 - f(x_o)/f'(x_o)$. Set $x = [Na^+]_f$, $f(x) = \alpha_{Na}[Na^+]_f - [Na^+]_T(obs)$,

$$f'(x) = \alpha_{Na} = \frac{[Na^+]_T(calc)}{[Na^+]_f}$$

to get Eq. 94.
67. D. Wong, D. Buse, and L. Savage, Via Medical Corp., San Diego CA. Unpublished data, 1996.

68. Add 1.36 mM strong acid to adjust the charge balance to zero at pH = 6.2 for lact. Ringer's, 1.1 mM to adjust chg. bal. = 0 at pH = 8.
69. $[HCO_3^-]_T$ and $[CO_3^{2-}]_T$ were calculated from $C_t = 19.608$, using $pK_1' = 6.165$ and $pK_2' = 9.865$ ($I = 0.15$, no ion pairs included).
70. Selected from R. M. Smith and A. E. Martell, *Critical Stability Constants*, Vols. 1–6, Plenum Press, NY, 1974–1989. Constants were adjusted to $I = 0.16$ using the Davies equation. There appear to be no data on sodium lactate ion-pair equilibria. By analogy with Na acetate (log $K = -0.44$; see Table 10.5), these ion pairs would be expected to be quite weak. See Table 11.2 for carbonate equilibrium constants.
71. May vary due to CO_2 gain or loss. Compare with 19.61 mM, the initial total concentration of $NaHCO_3$.
72. From Table 10.1, $pK_{so} = 8.60$ at 37°C, $I = 0$. At $I = 0.16$, $\log \gamma_+ = -0.133z^2$, and $pK_{so} = 7.54$ for free ions.
73. D. Wong, D. Buse, and L. Savage, Via Medical Corp., San Diego, CA. Unpublished data, 1996.
74. Following the conventions of oceanography, concentrations are given in moles per kg seawater. These values are related to concentrations in moles/L by the density of seawater (approx. 1.0025 kg/L), and hence are about 2.5% higher than concentrations in moles/L.
75. R. N. Roy et al., *Marine Chemistry*, **44**:249–67, 1993.
76. Roy et al., *Marine Chemistry* **44**:249–67, 1993; I. Hansson, *Deep-Sea Res.* **20**:461–78, 1973; C. Mehrbach et al., *Limnol. Oceanog.* **18**:897–907, 1973; C. Goyet and A. Poisson, *Deep-Sea Res.* **36**:6135–1654, 1989; A. G. Dickson and F. J. Millero, *Deep-Sea Res.* **34**:1733–43, 1987, is a review of the above.
77. $pK_1' = pK_{1(T)}'$ is defined by Eq. (81), p. 402; $pK_{1(SWS)}$ is defined by Eq. (84). pK_2' is defined by Eq. 105. Data for $pK_{a1(free\ H)}$ (Eq. 83) obtained by R. N. Roy et al., *Marine Chemistry* **44**:249–67, 1993, were converted to pK_1' using the Davies equation to evaluate γ_+. SWS values are from A. G. Dickson and F. J. Millero, *Deep-Sea Res.* **34**, 1733, 1987. Values for $pK_{so}(T) = [Ca^{2+}]_T[CO_3^{2-}]_T$ are from A. Mucci, *Am. J. Sci.* **283**:780–99, 1983; J. W. Morse, A. Mucci, and F. J. Millero, *Geochim. et Cosmochim. Acta* **44**:85–94, 1980, give $pK_{so} = 6.358 \pm 0.002$ at 35‰, in good agreement.
78. A. G. Dickson, *Marine Chemistry* **40**:49–63, 1992; J. N. Butler, *Marine Chemistry* **38**:251–82, 1992.
79. Free [H+] and free-ion constants from Table 10.1 and Smith and Martell; adj to $I = 0.7$ with the Davies Eq.
80. J. N. Butler, *Mar. Chem.* **38**:251–82, 1992, Table 10.1.
81. Data from A. L. Bradshaw, P. G. Brewer, D. K. Shafer, and R. T. Williams, *Earth & Planetary Sci. Letters* **55**:99–115, 1981. Sample 306, Station 449, 146 m depth (Indian Ocean).
82. A. G. Dickson notes this applies only for pH > 5. See A. G. Dickson, *Deep-Sea Res.* **28A**:609–23, 1981.
83. T. Takahashi et al., in B. Bolin (Ed.) *SCOPE 16: Carbon Cycle Modeling*, Wiley, NY, 1981, pp. 271–86; A. L. Bradshaw, P. G. Brewer, D. K. Shafer, and R. T. Williams, *Earth & Planetary Sci. Letters* **55**:99–115, 1981.
84. A. G. Dickson and J. P. Riley, *Mar. Chem.* **7**:89–99, 1979; A. G. Dickson, *J. Chem. Thermodyn.* **22**:113–27, 1990; J. N. Butler, *Mar. Chem.* **38**:251–82, 1992.

85. M. Heimann, Ed., *The Global Carbon Cycle*, Springer New York, 1993; I. G. Enting, K. R. Lassey, R. A. Houghton, *Projections of Future CO_2*, Australia: CSIRO, Division of Atmospheric Research technical paper no. 27, 1993; H. D. Holland, *Chemistry of the Atmospheres and Oceans*, Wiley, New York, 1978; N. R. Andersen and A. Malahoff, Eds., *The Fate of Fossil Fuel CO_2 in the Oceans*, Plenum, New York, 1977.

86. Two recent papers on the geochemistry of CO_2 in the open ocean are: C. L. Sabine, F. T. Mackenzie, C. Winn, and D. M. Karl, *Global Biogeochemical Cycles* **9**:637–51, 1995; N. R. Bates, A. F. Michaels, and A. H. Knap, *Marine Chem.* **51**:347–58, 1996.

87. R. Revelle and H. E. Suess, *Tellus* **9**:18–27, 1957; E. T. Sundquist, L. N. Plummer, and T. M. L. Wigley, *Science* **204**:1203–4, 1979.

88. However, a sudden dramatic decrease in alkalinity in the open ocean of the Sargasso Sea, apparently resulting from a bloom of coccolithophores (calcareous algae), has been observed by N. R. Bates, A. F. Michaels, and A. H. Knap, *Marine Chem.* **51**:347–58, 1996.

89. E. Sundquist, Thesis, Harvard University, 1979; Sundquist, Plummer, and Wigley, *Science* **204**:1203–5, 1979.

90. W. S. Broecker, T. Takahashi, H. J. Simpson, and T. H. Peng, *Science* **206**:409–18, 1979—look for more recent reference.

91. Selected from R. M. Smith and A. E. Martell, *Critical Stability Constants*, Vols. 1–6, Plenum Press, NY, 1974–1989. Additional values were taken from S. Clegg and M. Whitfield, Chapter 6, of *Activity Coefficients in Electrolyte Solutions*, K. S. Pitzer, Ed., CRC Press, 1991, p. 408. Original literature references are given there. Experimental values for $I = 0.7$ (or standard seawater), where available, are given in the fourth column. Constants in the last column were adjusted from $I = 0$ to 0.7 using the Davies equation

92. K. S. Pitzer, Ed., *Activity Coefficients in Electrolyte Solutions*, CRC Press, 1991. See especially Chapters 3 (by Pitzer) and 6 (by Clegg and Whitfield).

93. T. J. Wolery and K. Jackson, Chapter 2, *Chemical Modeling of Aqueous Systems II*, D. C. Melchior and R. L. Bassett, Eds., American Chemical Society Symposium Series No. 416, 1990; R. Heyrovska, *J. Electrochem Soc.* **143**:1789–93, 1996.

94. R. M. Garrels and M. E. Thompson, *Am. J. Sci.* **260**:57–66, 1962, developed the first ion-pairing model of seawater; see also W. S. Broecker and T. Takahashi, *J. Geopys. Res.* **71**:1575–602, 1966; M. Whitfield, *Mar. Chem.* **1**:251–66, 1973; J. N. Butler, *Carbon Dioxide Equilibria*, 1982 and 1991, pp. 123–28.

95. It would also be possible to carry out these calculations on the SWS scale (see Eqs. 84 and 85, p. 403). Compare the values in Table 10.17 with Table 10.13.

96. See also A. Dickson and M. Whitfield, An ion-association model for estimating acidity constants (at 25°C and 1 atm total pressure) in electrolyte mixtures related to seawater (ionic strength < 1 mol/kg H_2O), *Mar. Chem.* **10**:315–33, 1981.

97. This format for computing the ion-pairing model was developed by Jonathan Betts in 1990, while he was an undergraduate in the Marine Chemistry course at Harvard. The computations displayed here were performed using Excel 3.0 software, but this algorithm can easily be adapted to other spreadsheet applications by minor adjustments of the code.

98. The literature values for $pK'_{1(T)}$ and $pK'_{2(T)}$ were obtained from the concentration constraints of R. N. Roy et al., *Marine Chemistry* **44**:249–67, 1993, by using $\log \gamma_+ = -0.16$ from the Davies equation to convert $[H^+]$ to pH.

99. J. W. Morse, A. Mucci, and F. J. Millero, *Geochim. Cosmochim. Acta* **44**:85–94, 1980.
100. T. M. L. Wigley and L. N. Plummer, *Geochim. Cosmochim. Acta* **40**:989–95, 1976.
101. J. C. Hindman, *Properties of the System $CaCO_3$–CO_2–H_2O in Seawater and Sodium Chloride Solutions*. Thesis, Univ. California Los Angeles, 1943, 160 pp. Numerical data can be found in J. M. Edmond and J. M. T. M. Gieskes, *Geochim. Cosmochim. Acta* **34**:1261–91, 1976.
102. A. Mucci, *Am. J. Sci.* **283**:780–99, 1983.
103. T. M. L. Wigley and L. N. Plummer, *Geochim. Cosmochim. Acta* **40**:989–95, 1976.
104. R. L. Jacobson and D. Langmuir, *Geochim. Cosmochim. Acta* **38**:301–18, 1974.
105. R. M. Garrels and C. L. Christ, *Solutions, Minerals, and Equilibria*, Harper & Row, New York, 1965.
106. D. Langmuir, *Geochim. Cosmochim. Acta* **35**:1023–45, 1971. PK_D is 16.6 at 0°C, 16.7 at 10°C, 16.9 at 20°C, and 17.0 ± 0.02 at 25°C.
107. P. D. Wimberley, O. Siggaard-Andersen, N. Fogh-Andersen, and A. B. T. J. Boink, *Scand. J. Clin. Lab. Invest.* **45**:7–10, 1985.
108. Wimberley et al. calculate 1.2 mM $NaHCO_3$ and 0.003 mM $KHCO_3$, and note that $NaHCO_3$ accounts for about the same amount of CO_2 as the uncharged dissolved species. They suggest that there may be about 2 mM of unknown CO_2 species in normal plasma.
109. J. C. Hindman, *Properties of the System $CaCO_3$–CO_2–H_2O in Seawater and Sodium Chloride Solutions*. Thesis, Univ. California Los Angeles, 1943, 160 pp. Data were published in J. M. Edmond and J. M. T. M. Gieskes, *Geochim. Cosmochim. Acta* **34**:1261–91, 1976; A. Mucci, *Am. J. Sci.* **283**:780–99, 1983; J. W. Morse, A. Mucci and F. J. Millero, *Geochim. Cosmochim. Acta* **44**:85–94, 1980; D. C. Plath, K. S. Johnson, and R. M. Pytkowicz, *Marine Chem.* **10**:9–29, 1980; S. E. Ingle, *Marine Chem.* **3**:301–19, 1975; W. G. MacIntyre, *The Temperature Variation of the Solubility Product of Calcium Carbonate in Sea Water*. Canada Fisheries Research Board, Rept. Ser. No. 200, 153 pp. Data published by Mucci, *op. cit.*

    Mucci's data, arguably the best seawater solubility data for calcite, have been severely misquoted. Mucci's paper gives a polynomial that does not represent the data accurately. This polynomial was quoted by Clegg and Whitfield (in K. S. Pitzer, Ed., *Activity Coefficients in Electrolyte Solutions*, p. 367) with two typographical errors—making the data essentially unretrievable!
110. *Note added in proof*: The most recent ionic interaction model for seawater and other natural waters employs the Pitzer equations (pp. 492–494). F. J. Millero and R. N. Roy, "A Chemical Equilibrium Model for the Carbonate System in Natural Waters," *Croatia Chemica Acta* **70**:1–38 (1997).

11

pH IN BRINES[1]

Introduction
Definition of pH
pH in Brines: Limitations and Ambiguities
Glass Electrode with Liquid Junction Reference
    Calculating Liquid Junction Potentials
Hydrogen Electrode: The Harned Cell
Spectrophotometric Method
    Comparison of Potentiometric and Spectrophotometric Methods
Glass Electrode and Chloride Ion-Selective Electrode
    Evaluating HCl Activity
    Some Experimental Results
    Converting pHCl to pH
    Other Factors Affecting pH Determination in Brines
Conclusion
Comparison of Methods
Notes on Hydration Theory
Problems

INTRODUCTION

The pH scale as developed in Chapter 2 applies rigorously only to solutions dilute enough that the Debye–Hückel theory describes the activity coefficients within experimental error. For most systems, this means the ionic strength must be less than 0.1 M for accurate results. What is to be done with solutions of higher ionic strength—up to 6 M or more?

DEFINITION OF pH

In Chapter 2 pH was conceptually defined as the negative common logarithm of the hydrogen ion activity:

$$pH = -\log\{[H^+](\gamma_+)\} \quad (1)$$

At low ionic strengths, γ_+ approaches 1.0, by definition. At high ionic strengths, however, γ_+ depends not only on I but on the individual concentrations of other ions, particularly those of opposite charge.

Because it is not feasible to separate enough single ions from an ionic solution to make accurate analytical measurements, only combinations of activity coefficients corresponding to overall neutral charge can be measured experimentally. Thus the theories for activity coefficients can be empirically verified only for combinations such as

$\gamma_+\gamma_-$ (as NaCl),
$\gamma_+\gamma_-/\gamma_o$ (as K_a for HAc),
$(\gamma_+)^2\gamma_=$ (as Na_2SO_4),
or γ_+/γ_+ (as K_a for NH_4^+).

One approach to assigning activity coefficients to single ions relies on an arbitrary division of one of the experimentally determinable combinations. For example, MacInnes[2] assumed that γ_+ for K^+ and γ_- for Cl^- were equal (because the ionic mobilities of K^+ and Cl^- are nearly equal), and set both γ_+ and γ_- equal to $(\gamma_+\gamma_-)^{1/2}$ for KCl. This approach was used in the Garrels–Thompson model for seawater.[3]

Many theoretical and empirical expressions have been developed to calculate activity coefficients at higher ionic strengths. Some of the simpler ones such as the Debye-Hückel theory and the Davies equation, were described in Chapter 2 and have been used throughout this book.

pH in Brines: Limitations and Ambiguities

pH measurements are less certain in brines than in dilute solutions for several reasons.[4]

- The usual pH electrode–calomel reference electrode combination does not measure hydrogen ion activity (i.e., pH) alone, but rather measures a combination of hydrogen ion activity, counter-ion activity, membrane potentials, and liquid junction potentials.
- In the standard two-electrode or combination-electrode pH setup, liquid junction potentials arise between the calibration or test solution and the salt bridge leading to the reference electrode. When the calibration and test solutions are similar in composition, the liquid junction potentials are close in value, and errors are small. If the calibration and test solutions are very different in composition, however, the difference in liquid junction potentials can also be large—introducing an error equivalent to one or more pH units.

- The standard buffers developed by the National Bureau of Standards,[5] and widely available commercially, have much lower ionic strength (< 0.1 M) than brines (4–6 m), and standard buffers of composition similar to the brines are not available.
- Liquid junction potentials have traditionally been eliminated by using the a "Harned cell,[6]" a hydrogen gas/platinum electrode with a silver/silver chloride electrode (see Chapter 2, p. 53). The Harned cell has been used to measure the activity of HCl in a wide variety of multicomponent solutions, and is still the method of choice for the most accurate potentiometric measurements on acid–base systems.[7] A variant on this theme, using a glass electrode and a chloride ion-selective electrode, has been investigated, and some of the results are described below.
- However, even a cell without liquid junction measures the combination of hydrogen and chloride activities, not the hydrogen ion activity alone. To establish a pH scale, some theoretical assumptions must be made to evaluate the activity coefficient of at least one of the ions. For the NBS scale, the Bates–Guggenheim convention[8] for the activity coefficient of chloride ion was chosen:

$$\log(\gamma_{Cl}) = -\frac{A\sqrt{I}}{1 + 1.5\sqrt{I}} \quad (2)$$

but this convention was not intended to apply at ionic strengths greater than 0.1. Other possibilities for calculating activity coefficients theoretically include Guggenheim's[9] and Pitzer's[10] equations, which are extensions of the Debye–Hückel theory.

Hydration theory, although not so widely used, and not as accurate at low concentrations, can fit data at higher ionic strengths. Some notes on this approach are appended to this chapter. Data for NaCl,[11] and other 1:1 electrolytes,[12] have been fitted with this type of equation. Hydration theory has also been used with the computer program EQ3/6 (see Chapter 12) for binary salt mixtures[13] with only two adjustable parameters per component. See discussion at the end of this chapter (pp. 467–480).

- A related method uses spectrophotometric measurements of organic dyes, which are acid–base couples (see Chapter 4, pp. 126–130), and obtains pH from their optical absorbance. This method does not suffer from liquid junction potentials, but, like the others, requires extrathermodynamic assumptions, or a model of the activity coefficients of the ions involved; and to be generally applicable, needs standard buffers with composition similar to brines.
- If the high ionic strength is maintained with a relatively inert electrolyte, such as $NaClO_4$ or NaCl, a pH scale can be defined for that medium in terms of hydrogen ion concentration, by measuring the potential of an electrochemical cell (ideally, the Harned cell) after known additions of HCl or other strong acid

have been made. This scale[14] is $pm_H = -\log m_H$, which defines the reference state for activity to be that ionic medium, and defines the activity coefficient on that scale to be 1.00. The seawater scale (SWS) originated by Hansson (see Chapter 10, pp. 403–404) is an example. While such a scale is useful for evaluating acidity and complex formation constants, the relation between pm_H in one medium and pm_H in another medium involves the same problems as relating pm_H to pH (where the reference state is $I = 0$).

GLASS ELECTRODE WITH LIQUID JUNCTION REFERENCE

The advantages of this conventional and widely used method are that it is familiar to most chemists and engineers, equipment is inexpensive, and standard pH calibration solutions are readily available. There is plenty of evidence that the glass pH electrode is well behaved in electrolytes such as NaCl, even at high ionic strength.[15]

However, the primary source of uncertainty in the conventional glass–calomel pH system is the liquid junction potential between the sample and the calomel reference electrode. It is normally assumed that this liquid junction potential (included in $E°$) is the same in the standard and the test solution, but this assumption is less accurate as the composition of the sample becomes more different from that of the calibrating buffer.

Another major disadvantage for use in brines is the lack of readily available standard solutions to calibrate the electrode system in brinelike media. The National Bureau of Standards method uses the glass-calomel system, but relies on electrode calibration in buffers of ionic strength 0.05 to 0.1 m. Changes in liquid junction potential between these dilute buffers and a brine, and with variations in brine composition, are large and not easy to predict accurately.

For example, preliminary work by Felmy and co-workers[16] determined a correction factor for glass electrode pH measurements in brines, calibrating the pH meter and electrodes in National Bureau of Standards (NBS) buffers at pH = 4 (0.05 m potsssium acid phthalate) and 7 (0.025 m KH_2PO_4 and 0.025 m Na_2HPO_4). These electrodes are then used to measure potentials in brine to which aliquots of HCl were added to achieve apparent pH values in the range 4.5 to 2.5. Similarly, they titrated the brine with NaOH to pH 12, and the data (10^{+pH} vs. moles of NaOH per liter) were fitted to a straight line. The influence of the liquid junction depressed the apparent pH values from calculated values by approximately 1 unit (60 mV) in saturated NaCl at 25°C and by approximately 3 units (180 mV) in saturated $MgCl_2$.

Calculating Liquid Junction Potentials

Instead of ignoring E_J in the glass electrode/calomel electrode system, or assuming it does not change, can it be calculated from tabulated ionic mobilities? Not very accurately. Liquid junction potentials between most of the standard buffers and saturated KCl have been estimated[17] to be −1 to +3 mV, but for 1 M HCl the liquid junction potential is about +14 mV, and for 1 M NaOH or KOH, it is about −8 mV. At higher ionic strengths, the equations describing liquid junction potentials are more complicated and less reliable.

But it is still a reasonable hypothesis that the liquid junction potential would remain constant if the brine composition is also constant. In seawater, this approach has been most successful, with a precision of 0.001 pH units (and accuracy of 0.01 units) being attained.[18]

HYDROGEN ELECTRODE: THE HARNED CELL

This method, described in Chapter 2, is the primary standard for pH measurements. The most precise and accurate measurements (<0.0002 pH units) have been made with the Harned cell. This method essentially eliminates uncertainties from liquid junction potentials and membrane potentials, and provides an absolute calibration scale.

But it has some disadvantages. A hydrogen electrode requires a source of highly pure hydrogen gas, either from a tank or from an electrolysis cell. The sealed all-glass cells, platinum electrodes, and silver/silver chloride electrodes are usually fabricated by hand in the laboratory at considerable expense, and are not generally available commercially. Measurements of the highest quality require care and experience.[19]

SPECTROPHOTOMETRIC METHOD

The primary advantages of the spectrophotometric method are precision and infrequent calibration. As discussed in Chapter 4 (pp. 126–130), for samples that tend to fall in a narrow range of pH (as seawater and some brines do), and that can be conveniently brought to a spectrophotometer before their composition changes, the spectrophotometric method provides an accurate pH method with builtin calibration. The three independent parameters needed for calibration do not need to be remeasured frequently (as does the calibration of a pH meter). However, the acidity constant of the indicator depends on ionic strength and may depend strongly on the chemical composition of the ionic medium. The extinction coefficient ratios are less sensitive to the medium.

Byrne and Breland[20] reported measurements on 112 samples of seawater of varying salinity and pH, and found that 93% of these measurements were within 0.004 pH units of a best-fit quadratic function. Replicates agreed within 0.0005 pH units. Such precision is as good or better than has been achieved by any method,[21] but the absolute accuracy of the method was not addressed.[22]

Spectrophotometric techniques have also been suggested for measurement of pH in brines.[23] Values of $pm_H = -\log(m_H)$ were measured using cresol red in solutions up to 7 M ionic strength. pK_a decreased with ionic strength to 3 M in NaCl, and increased with ionic strength in both NaCl and $NaClO_4$ solutions above 3 M (Fig. 11.1). The spectrophotometric pm_H values varied from 8.0 to 9.0 and correlated well

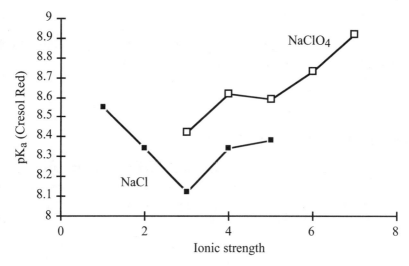

FIGURE 11.1. pK_a for the indicator Cresol Red in NaCl and $NaClO_4$ up to $I = 7$.

with glass–calomel electrode pH (NBS) measurements, but were about 0.8 pH units higher in 5 M $NaClO_4$.

The primary disadvantage is that a spectrophotometer tends to be larger, more expensive, and less portable than a pH meter. This means the sample must usually be brought to a central laboratory (although measurements have been done at sea), and changes in its composition during sampling, transit, and storage are more likely.

The other important disadvantage is that the range of pH that can be measured with a single indicator is narrow. The error in pH is least when the absorbances of the acidic and basic species are equal, at approximately pH = pK_a, and increases as pH becomes either larger or smaller.

In sufficiently acidic solutions, the concentration of added acid m_H can give an accurate value for [H⁺], provided any reactions with weak bases are accounted for (see above). But if an indicator with $pK_a = 8$ is measured at pH = 3, the absorbance of its basic form is too weak to measure accurately; at higher pH, the concentration of added acid is so small it cannot be accurately measured. To cover a wider pH range, a number of indicators can be calibrated independently and the pH scales so obtained reconciled where they do not agree exactly. Such a procedure was used by Byrne.[24] But this chain becomes less accurate as pH becomes higher. More commonly, the pH range covered is quite small. In his standardization of Tris (seawater) buffer, Byrne employed a pH range of only 8.199–8.271.

Indicator dye species, being acids or bases themselves, will enter into acid–base reactions. If the buffer capacity of the test solution is low, it is necessary to correct for reaction of the indicator species with other species, or a systematic error will result. While this is not difficult, it is not always done. More difficult is anticipating and eliminating acidic and basic impurities (such as CO_2), which can introduce large errors when the concentration of added acid or base is low.

Comparison of Potentiometric and Spectrophotometric Methods

Much of this discussion has already been presented in Chapter 4 (p. 129), but there are a few additional points to be made.

- The most precise and accurate measurements of hydrogen chloride activity pHCl, which is the basis for the pH scale, are obtained with the Harned cell, using hydrogen and silver chloride electrodes. This method is ten times as precise and 100 times as accurate as a glass electrode with calomel reference electrode.
- Careful spectrophotometric pH measurements are as precise and accurate as careful potentiometric measurements using a glass electrode.
- The spectrophotometric method does not confer any special advantage in measuring either activity or concentration of hydrogen ion in multicomponent solutions. Both the spectrophotometric and potentiometric methods inherently measure the thermodynamic activity of an ionized acid (i.e., HCl), which is translated into the activity of the hydrogen ion by conventional assumptions.
- It is possible, for either method, to arrange the calibration so that the hydrogen ion activity is equal to its molar concentration by employing a reference state based on an electrolyte of nearly constant composition. Two common ones are seawater (SWS or Hansson scale) and 1–5 M $NaClO_4$ (pc_H or pm_H scale)— a brine of consistent composition would be equally satisfactory.
- Both potentiometric and spectrophotometric methods typically rely on standard buffers or on extrapolation to infinite dilution to calibrate their pH scale. Of course, a nonstandard buffer used in the spectrophotometric method can be calibrated by measurements with a Harned cell. Either method can employ addition of a known reagent such as HCl to establish a concentration scale, provided there are no components that would react with the acid or base additive. This approach is more difficult and less accurate for indicators with pK_a in the range from 4 to 10, where other buffers must be developed for the medium under study. Trace impurities in the ionic medium can become critical factors.
- The influence of ionic medium composition on the spectrophotometric method has not been thoroughly investigated. As noted in Fig. 11.1, pK_a for cresol red may vary from 8.2 in 3 M NaCl to 8.92 in 7 M $NaClO_4$. Other ions also affect these values: pK_a = 8.55 for cresol red in 1.0 M NaCl, and pK_a = 7.76 in seawater extrapolated to 1.0 M ionic strength (50% salinity). This discrepancy may have to do with the presence of Mg^{2+}, SO_4^{2-}, or other components in seawater, but has not been quantitatively explored. This effect of other ions would be expected to be even more important in brines.
- A careful comparison of the spectrophotometric technique with a Harned cell, where both methods were standardized on the same pH scale in high-ionic-strength brines, would be valuable.[25] This experiment could be effectively done by examining the behavior of indicators in precisely the same solutions as the activity of HCl has been measured (see Table 12.2).

GLASS ELECTRODE AND CHLORIDE ION-SELECTIVE ELECTRODE

This method has the advantage of a cell without liquid junction. It employs commercially available glass electrodes and $AgCl/Ag_2S$ solid-state chloride ion-selective electrodes.[26] Measurements can be made in conventional laboratory apparatus, without the need for hydrogen gas. Unlike the glass–calomel electrode system, this system can be easily employed in solutions of high ionic strength.[27]

It has not been widely employed, however, partly because its advantages over the liquid-junction reference system have not been widely recognized, partly because chloride ion-selective electrodes have only recently been developed, and partly because standard solutions for calibration have not been readily available.

One important factor is that neither the glass electrode nor the chloride ion-selective electrode is calibrated on an absolute scale the way the hydrogen and silver chloride electrodes are. Therefore, each experimental setup must be calibrated with a solution of known HCl activity to evaluate the potentials of the internal reference electrodes and the asymmetry potentials at the electrode interfaces. This can be done by titrating the solution with concentrated HCl or NaOH to get a range of potentials, and correlating the potential with the concentration of added acid or base (as in the Gran method—see pp. 84–87).

Evaluating HCl Activity

The electrode system can also be calibrated using the Nernst equation together with buffers of known HCl activity:

$$E = E° + \frac{2.303RT}{F} \log(m_H m_{Cl} \gamma_\pm^2) = E° - 59.15 \text{ pHCl} \tag{3}$$

where the numerical factor 59.15 is in mV and corresponds to 25°C.

Since the activity of many solutions containing HCl and other electrolytes have already been measured accurately using the Harned cell, and these results are already in the literature,[28] preparation of standard buffers in the pH range 2–4 is primarily a matter of obtaining pure reagents and measuring them accurately. The activity coefficient $\gamma_\pm$ of HCl in Eq. (3) can be obtained from the experimental literature values $\gamma_\pm^°$ for pure HCl together with Harned's Rule (Table 11.1), or from a theoretical calculation[29] (see Table 11.2 below).

For example, at constant ionic strength $I = m_{HCl} + m_{NaCl}$, the activity coefficient of HCl in the mixture is given by a linear mixing rule.[30]

$$\log \gamma_\pm = \log \gamma_\pm^° - \alpha_{12} m_{NaCl} \tag{4}$$

The Harned Rule coefficient α_{12} for this system is 0.030 ± 0.001 from $I = 1$ to 6 (Table 11.1, Fig. 11.2). In particular, when HCl is only a trace amount, Eq. (4) becomes

$$\log \gamma_\pm^{tr} = \log \gamma_\pm^° - \alpha_{12} I \tag{5}$$

GLASS ELECTRODE AND CHLORIDE ION-SELECTIVE ELECTRODE

TABLE 11.1. Activity coefficients of pure HCl ($\log \gamma_\pm^\circ$), trace HCl in NaCl ($\log \gamma_\pm^{(tr)}$) and Harned Rule coefficient α_{12}

Ionic Str	$\log \gamma_\pm^\circ$	$\log \gamma_\pm^{(tr)}$	α_{12}	Ref.[31]
0.1	−0.099	−0.103	0.042	(Harned)
0.100	−0.099	−0.103	0.0400	(Macaskill)
0.381	−0.122	−0.135	0.0353	(Macaskill)
0.5	−0.121	−0.140	0.0375	(Harned)
0.673	−0.113	−0.134	0.0305	(Macaskill)
0.872	−0.101	−0.128	0.0306	(Macaskill)
1.0	−0.092	−0.124	0.0324	(Harned)
1.5	−0.048	−0.095	0.0316	(Harned)
2.0	0.004	−0.058	0.0308	(Harned)
3.0	0.119	0.027	0.0306	(Harned)
4.0	0.246	0.128	0.0296	(Harned)
5.0	0.375	0.227	0.0297	(Harned)
6.0	0.508	0.336	0.0287	(Harned)

TABLE 11.2. Comparison of pHCl based on experimental data of Harned[33] with values calculated from the Pitzer equations by the program EQ3[34]

I	m_{HCl}	$\log \gamma_\pm$ Harned	$\log m_H$	$\log m_{Cl}$	pHCl Harned	pHCl EQ3
4.00	9.98×10^{-4}	0.128	−3.008	0.602	2.150	2.154
5.00	9.93×10^{-4}	0.227	−3.003	0.703	1.846	1.859
6.00	1.001×10^{-3}	0.336	−3.000	0.778	1.550	1.575

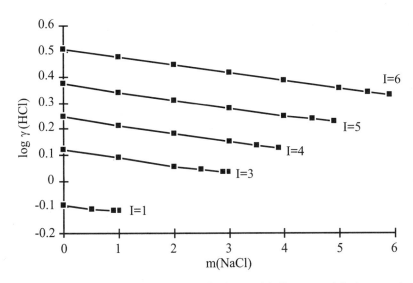

FIGURE 11.2. Activity coefficient of HCl in NaCl. The straight lines at each ionic strength correspond to Harned's Rule (Eq. 4) with α_{12} = 0.03. See Table 11.1.

Example 1. In Table 11.1 and Fig. 11.3, $\log \gamma_{\pm}^{\circ}$ and $\log \gamma_{\pm}^{tr}$ are compared as a function of ionic strength. The numerical values of $\log \gamma_{\pm}^{tr}$ and the corresponding values of pHCl are given in Table 12.2 for three solutions of $I = 4, 5,$ and 6 containing 10^{-3} M HCl.

The pHCl values calculated from Harned's data for $\gamma_{\pm}^{tr}$ are compared in Table 11.2 with pHCl obtained from Pitzer's equations as implemented by the program EQ3.[32] The EQ3 values are approximately 0.004 to 0.025 higher than Harned's experimental values. Since m_{Cl} and m_H are accurately known, this difference corresponds to an excellent agreement within 0.002 to 0.01 in $\log \gamma_{\pm}^{tr}$.

Some Experimental Results

Figure 11.4 shows an example of the correlation between potential and activity of HCl in 4 M NaCl obtained by Yelton et al.[35] They used four buffers:

1. 0.001 m HCl ($pm_H = 3$, pHCl = 2.154);
2. 0.0500 m H_3BO_3 + 0.00766 m NaOH ($pm_H = 8$, pHCl = 7.151);
3. 0.0500 m H_3BO_3 + 0.02516 m NaOH ($pm_H = 9$, pHCl = 8.153);
4. 0.007128 m NaOH ($pm_H = 12$, pHCl = 11.154).

The Nernst equation seems to be obeyed within the resolution of the plot, but not exactly: A least-squares fit to these data yields a slope of 58.71, not 59.15 mV.

Note that for buffer 1, $pm_H = -\log m_{HCl} = 3.00$, and for buffer 4, pm_H yields $-\log[H^+][OH^-] = pK_w = 14.15$. Compare this with the experimental $pK_w = 14.18$ at

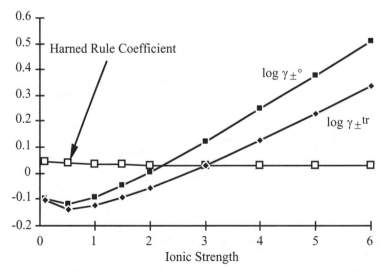

FIGURE 11.3. Activity coefficient of pure HCl ($\log \gamma_{\pm}^{\circ}$), activity coefficient of trace HCl in NaCl ($\log \gamma_{\pm}^{tr}$) and Harned rule coefficient α_{12} as a function of ionic strength.

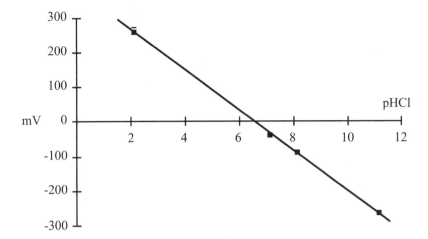

FIGURE 11.4. Potential of a glass electrode–chloride ion-selective electrode system in 4 m NaCl. The activity of HCl is expressed as pHCl = $-\log m_H - \log m_{Cl} - 2 \log \gamma_\pm$.

$I = 3.0$ in Table 3.1 (p. 67). Extrapolation by the Davies equation (Fig. 3.2, p. 68) to $I = 4$ gives $pK_w = 14.13$.[36] The two borate buffer concentrations can be translated in pK_a values for boric acid. For Buffer 3:

$$pK_a = -\log \frac{[H^+][B^-]}{[HB]} = -\log \frac{10^{-9.00}(0.02516)}{0.0500 - 0.02516} = 8.994$$

This agrees with Smith and Martell's value at $I = 3$, $pK_a = 8.97$. There are apparently no experimental data at higher ionic strengths.[37]

Another approach is to use the HCl data (Buffer 1) to determine $E°$, and to fix the slope at the Nernst value of 59.15 mV. The results of such a calculation are shown in Fig. 11.5. The deviations are smallest at $I = 4$, pHCl = 2 and 8, and the least discrepancy appears to be between the first three buffers. For the whole set, the standard deviation of a measurement is 6–8 mV (0.1 pH units), and the standard deviation of the mean of 48 values is 0.8 mV (0.01 pH units). While this is not anywhere near the precision of the Harned cell (±0.01 to 0.1 mV), it is adequate for many purposes.

The highest pH (Buffer 4) shows a consistent positive deviation from the Nernst equation, which corresponds to a lower pH than expected (compare with Fig. 11.4). Since the vessels were open, this discrepancy may have resulted from reaction between NaOH in the buffer and CO_2 in the laboratory atmosphere.

Barker[38] also made measurements, using a glass electrode and chloride ion-selective electrode, in a number of buffers containing 6 m NaCl at pH values from 2 to 12. Within a set of measurements made with a single buffer, a standard deviation of 0.6 mV (0.01 pH units) was common.

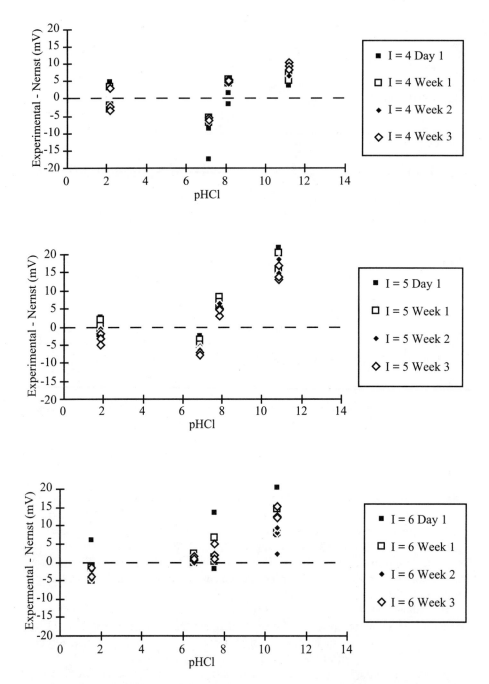

FIGURE 11.5. Deviation of experimental values from Nernst extrapolation for *I* = 4, 5, and 6.

In contrast, the difference of some buffers from the Nernst equation was ±5–35 mV—ten or more times as great as the precision within a set. This is probably due to impurities in the components of the buffer solution. For example, the NaCl Barker used appears to have had a basic impurity that biased the pH = 4 buffer to higher pH than intended, and the pH = 12 buffer appears to have decreased in pH because of exposure to atmospheric CO_2.

Converting pHCl to pH

The final stage of these calculations is to use the known m_{Cl} of the solution with a theoretical value for γ_{Cl} to obtain a value for hydrogen ion activity and hence pH as it is usually conceived. From Eq. (3)

$$pHCl = -\log(m_H m_{Cl} \gamma_\pm^2)$$

Since

$$\gamma_\pm^2 = \gamma_H \gamma_{Cl}$$

$$pH = -\log m_H - \log \gamma_H = pHCl + \log m_{Cl} + \log \gamma_{Cl}$$

Example 2. The first buffer in Table 11.2 at $I = 4$ has $\log m_H = -3.008$, $\log m_{Cl} = 0.602$, and $\log \gamma_\pm = 0.128$ from Harned's data,[39] hence pHCl = 2.150 (from Harned's measurements) or 2.154 (from EQ3).

The activity coefficient of the individual chloride ion could be estimated from the Davies equation to be $\log \gamma_{Cl} = +0.068$ at $I = 4$, and the empirical measurement of pHCl thereby converted to pH = 2.154 + 0.602 + 0.068 = 2.824. The EQ3 program[40] calculates γ_{Cl} under the assumption that $\gamma_H = 1.000$ (the pm_H scale) and yields pH = 3.000.

But it must be borne in mind that the value used for γ_{Cl} is arbitrary. A number of choices are outlined in Table 11.3.

The Bates–Guggenheim convention (Eq. 2, p. 463) was used to calculate γ_{Cl} in the establishment of the NBS pH scale.[42] The mean activity coefficient of KCl was MacInnes' choice for dividing the mean activity of positive and negative ions, because K$^+$ and Cl$^-$ have nearly equal mobilities (p. 462).[43] Knauss et al. proposed

TABLE 11.3. Some choices for γ_{Cl} and the resulting pH values; 4 m NaCl containing 10^{-3} m HCl

pHCl	$\log \gamma_{Cl}$	pH	Method
2.154	−0.342	2.414	Debye–Hückel, $a = 3$, $I = 4$
2.154	−0.255	2.501	Bates–Guggenheim, $I = 4$
2.154	−0.239	2.517	mean act. coeff. of 4 m KCl
2.154	+0.068	2.824	Davies,[41] $I = 4$
2.154	+0.244	3.000	pmH, EQ3, or $\gamma_+ = 1$

the pm_H (pm_H) scale as being the simplest conceptually. Clearly, a range of options is available that covers more than half of pH unit for the same solution, but there is as yet no agreement on which to use.

It must be emphasized that separating pH from pCl by a calculation does not predict the result that would be obtained with a glass–calomel combination pH electrode. For example, Barker[44] found that in one buffer the pH measured with a conventional combination electrode was 6.22, and pH calculated from measurements with the glass electrode–chloride ion electrode system (calibrated with EQ3 calculations) was 7.17. This difference is nearly a full pH unit. However, the establishment of a self-consistent pH scale in a particular brine makes it feasible to calibrate a glass–calomel system with liquid junction, and hence to be fairly confident of ordinary pH measurements in solutions of similar ionic composition.

Overall, these experiments are encouraging, and tell us that relatively precise activity data can be obtained in brines with commercially available equipment, and that an empirical pHCl scale can be readily established. However, consistent conversion of pHCl to pH requires an arbitrary decision as to how to evaluate $\log \gamma_{Cl}$.

Other Factors Affecting pH Determination in Brines

Calibration of pH systems in brines by titration with HCl can be tricky if they contain weak bases. For example, if substantial Na_2SO_4 is present, formation of HSO_4^- will distort the shape of the Gran plot. Similar effects occur in the presence of other weak bases such as F^-, $H_2PO_4^-$, or anions of organic acids. Such problems can in principle be compensated by adding an acid–base equilibrium term.

Example 3. HCl + Na_2SO_4 (compare with Chapter 4, pp. 119–120):

$$[H^+] + [Na^+] = [HSO_4^-] + 2\,[SO_4^{2-}] + [Cl^-]$$

$$[H^+] = \frac{C_{SO_4}[H^+]}{[H^+] + K_a} + \frac{2C_{SO_4}K_a}{[H^+] + K_a} + C_{HCl} - \frac{K_w}{[H^+]}$$

Neglecting [OH$^-$], this gives a quadratic in [H$^+$]:

$$[H^+]^2 + [H^+]\,(K_a + C_{SO_4} - C_{HCl}) - K_a C_{HCl} = 0$$

This function is plotted for several values of C_{SO_4} at $I = 3$ ($pK_a = 0.91$) in Fig. 11.6.

But such a correction is not always this easy. For example, K_a for HSO_4^- varies strongly with ionic strength (Table 11.4 and Fig. 11.7), and with electrolyte composition because of the formation of ion pairs.[45]

Similarly, addition of NaOH to brines containing Mg, Fe, Al, etc. will precipitate hydroxides of these elements; the "NaOH added" will be mostly removed by the precipitation reaction, and hence a much larger amount of NaOH must be added

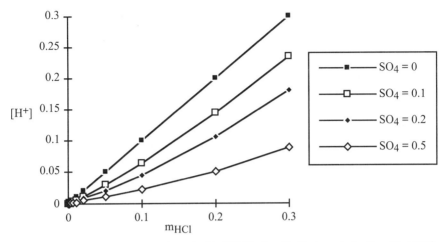

FIGURE 11.6. [H$^+$] is a linear function of added HCl in 3 m NaCl alone, but with added SO$_4^{2-}$ the function is a curve with slope substantially less than 1.0.

than if the cations were inert like Na$^+$. The correlation of 10^{+pH} with m_{NaOH}, which is linear in pure water or NaCl, will be greatly distorted (Fig. 11.8).[52]

Recently, pH measurements of HCl–NaCl ($m = 0.57$) were made under supercritical conditions (400°C and 40 megapascals) using the cell

$$\mathrm{Ag/AgCl/Cl^-, H^+, H_2O/YSZ/HgO/Hg}$$

where YSZ stands for a zirconia (ZrO$_2$) membrane stabilized with 9% yttrium oxide.[53] A precision of ±5 mV or 0.05 pH units was obtained. Such techniques are promising.

TABLE 11.4. pK_a of HSO$_4^-$ at 25°C[46]

I	pK_a	Ref.
0	1.987	Lietzke et al.[47]
0	1.98±0.01	Pitzer et al.[48]
0	1.99±0.02	Young et al.[49]
0	1.99±0.01	Smith and Martell, Vol. 6 (1989)
0	1.98±0.02	Mussini et al.[50]
0	1.96±0.02	Dickson et al.[51]
0.1	1.55±0.05	Dickson et al.[51]
0.5	1.32±0.06	Smith and Martell, Vol. 6
1.0	1.10±0.08	Smith and Martell, Vol. 6
2.0	1.01±0.07	Smith and Martell, Vol. 6
3.0	0.91±0.02	Smith and Martell, Vol. 6
5.0	1.53	Smith and Martell, Vol. 6

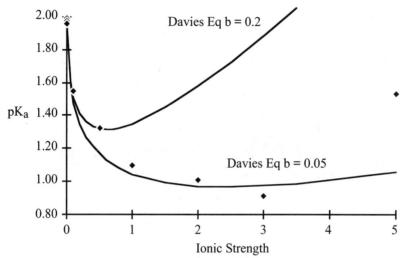

FIGURE 11.7. pK_a for HSO_4^- as a function of ionic strength. The normal Davies equation fits only the range for $I < 0.5$; but if b is adjusted to 0.05, the fit is adequate to $I = 3$. See Table 11.4 for references.

Conclusion

The glass electrode–chloride ion-selective electrode cell is currently not nearly as precise or accurate as the Harned cell or the spectrophotometric method. However, it is much more portable and flexible and could be a valuable addition to the arsenal of activity-measuring techniques. It needs to be investigated more accurately,

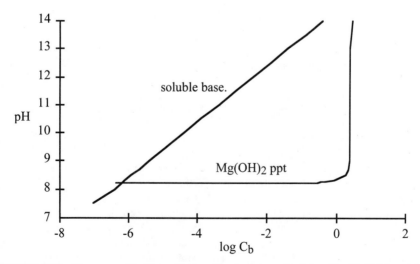

FIGURE 11.8. pH versus added base for the situation where the base (like NaOH) is soluble at all pH, and for the situation where $Mg(OH)_2$ precipitates.

COMPARISON OF METHODS

Here is a brief comparison of the precision (first number) and accuracy of the various methods under consideration:

Method	mV	pH units
Harned cell	0.01–0.1	0.0002–0.002
Spectrophotometer	N/A	0.001–0.004
Glass/calomel	0.05–0.5	0.001–0.01
Glass/Cl ion	< 5	< 0.1

The best values of the Harned cell data show reproducibility within a single data set of as good as ±0.01 mV, but agreement of the standard potential for Ag/AgCl among workers in different places and times is more like ±0.1 mV.[54]

As mentioned above, the pH in seawater, obtained by the cresol red (or m-cresol purple) spectrophotometric method, can be replicated to within 0.001 units, and a large data set can have a precision of ±0.004.

Careful glass electrode measurements can achieve a precision of ±0.05 mV,[55] corresponding 0.001 pH units. An estimate of the absolute accuracy of ±0.5 mV (±0.001 pH units) comes from a comparison of the activity coefficient of NaCl, measured by four different investigators, employing a glass electrode, sodium amalgam electrode, and isopiestic methods.[56]

Measurements with the glass electrode–chloride ion selective electrode are still in a preliminary stage. It is possible that this method will yield much more accurate results as the technique is developed.

NOTES ON HYDRATION THEORY[57]

The activity of water a_w in a salt solution is related to the osmotic coefficient ϕ by

$$\ln(a_w) = \frac{-\nu \phi m}{55.51} \qquad (6)$$

where ν is the number of ions produced by dissolving one mole of salt (2 for NaCl), m is the total number of moles of salt per kg of water, and 55.51 is the number of moles of water in 1 kg. If the salt is incompletely dissociated, with degree of dissociation α, the total number of moles of solute is im where $i = 1 + \alpha$ is the van't Hoff factor, and

$$v\phi = 55.51 \left(\frac{i}{55.51 - mn_b} \right) \tag{7}$$

where n_b is "hydration number in the bulk". The water activity is expressed as the mole fraction of "free water" in terms of n_s (the "hydration number in the surface"):

$$a_w = \frac{55.51 - mn_s}{55.51 - mn_s + im} \tag{8}$$

Combining Eqs. (7) and (8) to eliminate i gives an equation of the form

$$\frac{55.51 - mn_s}{55.51 - mn_b} + \frac{a_w \ln(a_w)}{1 - a_w} = 0 \tag{9}$$

Given values for m, n_s, and n_b, a_w (and hence from Eq. 6, ϕ) can be obtained from the combined nonlinear equation by iteration using Newton's method. An example is given in Fig. 11.9, where a good fit to the experimental data is obtained for $m > 0.5$.

Activity coefficients can be obtained from ϕ by the standard integration method[58]

$$-\ln(\gamma) = (1 - \phi) + \int_0^m (1 - \phi) \, d \ln(m) \tag{10}$$

This integral was evaluated for ϕ calculated from Eqs. (6)–(9), and is compared in Fig. 11.10 with the experimental values for NaCl tabulated by Robinson and Stokes.[59]

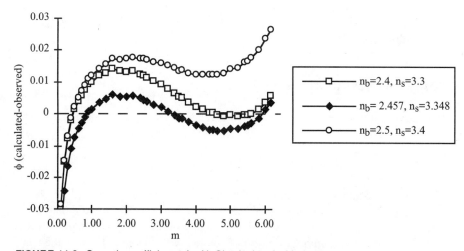

FIGURE 11.9. Osmotic coefficient ϕ for NaCl calculated with three sets of the parameters n_b and n_s compared with experimental values[65] (horizontal axis). The lowest curve $n_b = 2.457$ and $n_s = 3.348$ corresponds to the best fit made by Heyrovska (Table 1, Ref. 60.)

NOTES ON HYDRATION THEORY

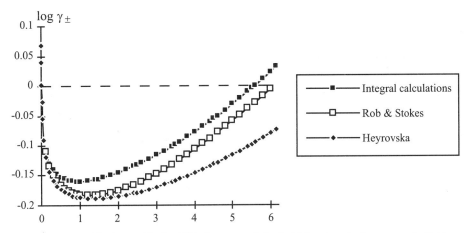

FIGURE 11.10. The integral in Eq. (10) gives unrealistically high values for low concentrations; hence the Davies equation was used to fix the value of $\gamma_\pm$ at $m = 0.10$. Even so, the integral calculations give higher values than the experimental rresults. The curves are nearly parallel, so that a better fit could be obtained for interpolation at high concentrations if the integral was fixed to agree with experiment at $I = 1.0$ or higher. (Compare with Fig. 3, Ref. 60.)

Heyrovska proceeded by calculating a degree of dissociation $\alpha = i - 1$ (Eq. 6 of Ref. 60)

$$\alpha = \left(\frac{55.51 - mn_s}{m}\right)\left(\frac{1 - a_w}{a_w}\right) - 1 \tag{11}$$

where a_w is related to ϕ by Eq. (6). For Fig. 12.11, a_w was calculated from the experimental value of ϕ and also from the values of ϕ displayed in Fig. 12.9. The

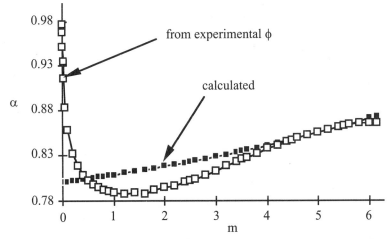

FIGURE 11.11. The degree of dissociation α of NaCl calculated from experimental osmotic coefficients and from Eq. (11). a_w was calculated from the experimental values of ϕ by Eq. (6) and also from the values of ϕ displayed in Fig. 11.9. (See Table 1, Ref. 60 for data.)

magnitude of α obtained by the 2 methods is similar to $m > 0.5$, but the calculated value does not approach 1 as $m \to 0$, as would be expected. The expression obtained by Heyrovska[60,61] was

$$\log(m\gamma_\pm) = \delta_A \log\left(\frac{\alpha m}{(55.51 - mn_h)r_o}\right) \qquad (12)$$

Here δ_A is an empirical "solvent-solute polarization" parameter (0.947 for NaCl in water at 25°C), n_h is a hydration number (n_s or n_b from Eqs. 7 and 8), and r_o is another empirical constant, which adjusts the zero of the log $\gamma_\pm$ scale: "equal to $\alpha m/(55.51 - mn_h)$ when $m\gamma_\pm = 1$." Heyrovska obtained the value $r_o = 0.0233$.[62] The results of Eq. (12) (within $n_h = n_s = 3.36$) are displayed as the third curve in Fig. 12.10. It gives better agreement with the experimental values than the integral when $m > 3$, but poor agreement for $m < 3$. At concentrations below $m = 0.01$, this method gives unrealistic positive values for log $\gamma_\pm$.

These hydration theory approaches to predicting osmotic or activity coefficients all require at least three empirical parameters. Even though the theory limits their range, this is a limitation. The equations developed here fail to agree with experiment for $m < 0.5$. At low concentrations, the Debye–Hückel equations and its extensions are still the most accurate predictors of osmotic and activity coefficients.[63]

PROBLEMS

1. Calculate the $pa_H\gamma_{Cl}$ for several HCl–NaCl mixtures using data in Table 11.1.

2. Calculate the pH of the above using several formulas for log γ_{Cl} (Table 11.3).

3. The following is the composition of WIPP Brine A, from the Waste Isolation Pilot Plant in Carlsbad NM[64]:

Component	mg/L
Na	42,000
Mg	30,000
K	35,000
Ca	600
B	220
Cl	190,000
SO$_4$	3500
HCO$_3$	700
pHCl	8.34
pH	9.91

Test the charge balance (in moles/L) and add an unknown cation or anion to make up the difference. Calculate the ionic strength. Estimate the pH based on the borate and carbonate buffering systems, using extrapolated acidity constants from Figs. 10.1 and 10.3 for carbonate, and the following data for borate at 25°C from Smith and Martell:

I	pK_a
0	9.236 ± 0.001
0.1	8.97 ± 0.02
0.7	8.85
2.0	8.94 ± 0.07
3.0	8.97

Look up data on activity coefficients for Na,Mg,K–Cl mixtures. See K. S. Pitzer,[10] (p. 172).

4. Compare the predictions of hydration theory, Debye–Hückel thoery, and the Pitzer equations[29] (Chap. 12, pp. 492–494) for aqueous NaCl and other electrolytes. Data can be found in Appendix 8.10 of Robinson and Stokes, *Electrolyte Solutions*.[59] Pay particular attention to the number of adjustable parameters required to get a good fit.

NOTES

1. This work was supported in part by Sandia National Laboratories, Albuquerque, NM, under the supervision of Dr. Lawrence H. Brush.
2. D. A. MacInnes, *J. Amer. Chem. Soc.* **41**:1086, 1919; *Principles of Electrochemistry* (reprint), Dover, 1961.
3. R. M. Garrels and M. E. Thompson, *Am. J. Sci.* **260**:57–66, 1962.
4. See A. G. Dickson, *Geochim. Cosmochim. Acta* **48**:2299–308, 1984, for a critical review of pH scales used in saline media such as seawater.
5. R. G. Bates, *Determination of pH; Theory and Practice*, Wiley, 1973.
6. H. S. Harned and R. W. Ehlers, The thermodynamics of aqueous hydrochloric acid solutions from electromotive forces. *J. Am. Chem. Soc.* **55**:2179, 1933.
7. J. N. Butler and R. N. Roy, "Potentiometric Methods," in *Activity Coefficients in Electrolyte Solutions*, K. S. Pitzer, Ed., CRC Press, 1991.
8. R. G. Bates and E. A. Guggenheim, *Pure Appl. Chem.* **1**:163, 1960.
9. E. A. Guggenheim, *Thermodynamics*, 4th ed., North Holland, Amsterdam, 1959.
10. K. S. Pitzer, Ed., *Activity Coefficients in Electrolyte Solutions*, CRC Press, 1991, Chapters 3 and 6; C. E. Harvie, N. Møller, and J. H. Weare, *Geochim. Cosmochim. Acta* **48**:723–51, 1984.
11. R. Heyrovska, *J. Electrochem. Soc.* **143**:1789–93, 1996.
12. R. Heyrovska, *Chem. Listy*, in press, 1996; *Croat. Chem. Acta*, **70**:39–54, 1997.
13. K. J. Jackson and T. J. Wolery, "Extension of the EQ3/6 computer codes to geochemical modeling of brines," *Mat. Res. Soc. Symp. Proc.* **44**:507–14, 1985.

14. pmH, sometimes called "rational pH," means $-\log m_H$ (where m_H is in mol/kg), a notation suggested by T. J. Wolery and R. E. Mesmer (private communications). In concentrated solutions it is important to distinguish beween pm_H and $pc_H = -\log[H^+]$ (where $[H^+]$ is in mol/L).

15. J. N. Butler and R. N. Roy, "Potentiometric Methods," in *Activity Coefficients in Electrolyte Solutions*, K. S. Pitzer, Ed., CRC Press, 1991.

16. A. R. Felmy, Battelle Pacific Northwest Lab, Richland, WA 99352. Report TIP-WIPP-ETG-10. The temperature was 25 ± 0.1°C, and pH measurements are taken with a precision of 0.005 units (0.3 mV). The pH with which the experiments were compared was calculated using the EQ3 program of Wolery and Jackson (*op. cit.*).

17. R. G. Bates, *Determination of pH*, Wiley, 1973, p. 38.

18. C. H. Culberson, "Direct potentiometry," in *Marine Electrochemistry*, M. Whitfield and D. Jagner, Eds., Wiley, 1981; J. N. Butler, *Mar. Chem.* **38**:251–82, 1992; A. G. Dickson, *Mar. Chem.* **44**:131–42, 1993.

19. D. J. G. Ives and G. J. Janz, *Reference Electrodes*, Academic Press, 1961; R. N. Roy, private communication, 1990.

20. R. H. Byrne, *Anal. Chem.* **59**:1479–81; R. H. Byrne, G. Robert-Baldo, S. W. Thompson, and C. T. A. Chen, *Deep-Sea Research* **35**:1405–10, 1988; R. H. Byrne and J. A. Breland, *Deep-Sea Research* **36**:803–10, 1989.

21. J. N. Butler and R. N. Roy, *op. cit.*

22. A. Dickson, private communication, suggests approximately ±0.005.

23. M. Solache-Rios and G. R. Choppin, *Determination of the pcH in Highly Saline Waters using Cresol Red*, Report SAND91-7068J, 1991. Sandia National Laboratories, Albuquerque NM.

24. R. H. Byrne, *op. cit.*

25. Such experiments are underway for seawater (A. G. Dickson, private communication).

26. J. W. Ross, in *Ion-Selective Electrodes*, Ed., R. A. Durst, National Bureau of Standards, Special Publication 314, 1969; K. Knauss, T. J. Wolery, and K. J. Jackson, *Geochim. Cosmochim. Acta* **54**:1519, 1990.

27. Although the equilibrium solubility of AgCl in 4–6 M NaCl is as high as 0.01 m (see Fig. 7.14), Knauss et al. (*op. cit.*) reported that the dissolution of the electrode material did not appear to introduce any obvious errors except in the case of bromide.

28. J. N. Butler and R. N. Roy, *op. cit.*; R. A. Robinson and R. H. Stokes, *Electrolyte Solutions*, Butterworths, 1965; J. B. Macaskill, R. A. Robinson, and R. G. Bates, *J. Solution Chem.* **6**:385–92, 1977; R. A. Robinson, *J. Solution Chem.* **9**:449–54, 1980. More recent work has reported more accurate values for HCl activity in NaCl at temperatures other than 25°C, but α_{12} is still 0.03.

29. K. S. Pitzer, *Electrolyte Solutions*, CRC Press, 1991, pp. 75–153. More details are given in Chapter 13. Pitzer's equations have been implemented by T. J. Wolery, *EQ3NR, a Computer Program for Geochemical Aqueous Speciation Solubility Calculations*, UCRL-MA-110662-PT-III, Lawrence Livermore National Laboratory, Livermore, CA, 1992.

30. H. S. Harned, 1935, *op. cit.*

31. H. S. Harned, 1935, *op. cit.*; J. B. Macaskill, R. A. Robinson, and R. G. Bates, *op. cit.*

32. T. J. Wolery, *EQ3NR, op. cit.*, see Chap. 12.

33. H. S. Harned, *J. Am. Chem. Soc.* **57**:1865–73, 1935.
34. W. G. Yelton, S. Free, and L. R. Montano, unpublished data, Sandia National Laboratories, Albuquerque, NM, 1995.
35. W. G. Yelton, S. Free, and L. R. Montano, unpublished data, Sandia National Laboratories, Albuquerque, NM, 1995. They also made measurements at $I = 5$ and 6. See Fig. 11.5 for some samples.
36. Experimental values at $I = 4$ are not listed in Smith and Martell. Sillén and Martell (Spec. Publ. 17) cite $pK_w = 14.80$ in 4 M $NaClO_4$ from the thesis of A. Kleiber at the University of Strasbourg, 1957; in Spec. Publ. 25, they cite $pK_w = 14.42$ in 3.8 M $NaClO_4$ from R. Fischer and J. Byé, *Bull. Soc. Chim. France*, 2920, 1964. Both these values are higher than the Davies equation or the calculations of Yelton et al. There may also be differences between $NaClO_4$ and NaCl media.
37. The concentrations for Buffer 2 yield $pK_a = 8.7425$, significantly lower than Buffer 3 or the Smith and Martell values. The NaOH concentrations used in these buffers were obtained from an EQ3 calculation, and this may be the source of the discrepancy.
38. G. Barker, unpublished data, Sandia National Laboratories, Albuquerque, NM, 1991.
39. Harned, 1935, *op. cit.* The same values are listed in Robinson and Stokes, Appendix 8.10.
40. Yelton et al., *op. cit.*
41. The Davies equation would give directly $\log \gamma_H = +0.068$, hence pH $= -\log m_H - \log \gamma_+$ $= -(-3.008) - 0.068 = 2.940$; if the Davies value for γ_{Cl} is used with $\gamma_\pm$, this yields $\log \gamma_H = 2(0.128) - 0.068 = 0.188$, and hence pH $= 2.824$.
42. R. G. Bates and E. A. Guggenheim, *Pure Appl. Chem.* **1**:163, 1960.
43. D. A. MacInnes, *J. Amer. Chem. Soc.* **41**:1086, 1919.
44. Barker, *op. cit.*
45. H. F. Holmes and R. E. Mesmer, *J. Chem. Thermodynamics* **28**:67–81, 1996.
46. Note that while four papers from 1961 to 1989 gave values between 1.96 and 2.01, the most recent study (Dickson et al.) recalculated these data to give 1.96 ± 0.02.
47. M. H. Lietzke, R. W. Stoughton, and T. F. Young, *J. Phys. Chem.* **65**:2247, 1961.
48. K. S. Pitzer, R. N. Roy, and L. F. Silver, *J. Am. Chem. Soc.* **99**:4930, 1977.
49. T. F. Young, C. R. Singleterry, and I. M. Klotz, *J. Phys. Chem.* **82**:671, 1978.
50. P. R. Mussini, P. Longhi, T. Mussini, and S. Rondinini, *J. Chem. Thermodyn.* 625, 1989.
51. A. G. Dickson, D. J. Wesolowski, D. A. Palmer, and R. E. Mesmer, *J. Phys. Chem.* **94**:7978–85, 1990. This paper contains a critical review of the literature as well as data at temperatues to 250°C. A. G. Dickson, *J. Chem. Thermodyn.* **22**(2):113–27, 1990, measured accurate acidity constants for HSO_4^- in seawater up to 45% ($I = 0.9$ m) and discusses the problems with evaluation of a relatively large K_a in multicomponent media of high ionic strength.
52. The charge balance for addition of NaOH (C_b molar) to $MgCl_2$ (C_{Mg} molar) is

$$[H^+] + [Na^+] + [Mg^{2+}] = [Cl^-] + [OH^-]$$

$$C_b = 2 C_{Mg} - K_{s0}/[OH^-]^2 - [H^+] + [OH^-]$$

Values are chosen for pH, hence for $[H^+]$, $[OH^-]$ is calculated from $K_w/[H^+]$ and C_b calculated from the above equation. The result is plotted as pH vs C_b in Fig. 12.8.

53. K. Ding and W. E. Seyfried, *Science* **272**:1634–36, 1996.
54. J. N. Butler and R. N. Roy, "Experimental methods: Potentiometric," in *Activity Coefficients in Electrolyte Solutions*, K. S. Pitzer, Ed., CRC Press, 1991, Table 1, p. 159. For example:
    R. N. Roy, S. A. Rice, K. M. Vogel, L. N. Roy, and F. J. Millero, *J. Phys. Chem.* **94**:7706, 1990, 222.58 mV;
    R. G. Bates and J. B. Macaskill, *Pure Appl. Chem.* **50**:1701, 1978 found $E° = 222.42$ mV;
    R. S. Greeley, W. T. Smith, R. W. Stoughton, and M. H. Lietzke, *J. Phys. Chem.* **64**:652–57, 1960, 222.33 mV;
    R. G. Bates and V. E. Bower, *J. Res. Nat. Bureau Standards* **53**:283–90, 1954, 222.34 ± 0.02 mV at 25°C.
55. R. D. Lanier, *J. Phys. Chem.* **69**:2697, 1965; *ibid. J. Phys. Chem.* **69**:3992, 1965.
56. J. N. Butler and R. N. Roy, "Experimental methods: Potentiometric," in *Activity Coefficients in Electrolyte Solutions*, K. S. Pitzer, Ed., CRC Press, 1991, p. 170. See p. 165 for references to glass pH electrode.
57. R. Heyrovska, *J. Electrochem Soc.* **143**:1789–93, 1996; private communication, 1996.
58. R. A. Robinson and R. H. Stokes, *Electrolyte Solutions*, Butterworths, 1965, p. 180.
59. R. A. Robinson and R. H. Stokes, *Electrolyte Solutions*, Butterworths, 1965, App. 8.3, p. 476.
60. R. Heyrovska, *J. Electrochem. Soc.* **143**:1789–93, 1996.
61. R. Heyrovska, *J. Electrochem. Soc.* **144**:2380–2834, 1997.
62. R. Heyrovska, *Croatia Chem. Acta* **70**:39–54 (1997). Note from Fig. 12.10 that when m is approximately $1/\gamma = 10^{0.18} = 1.5$, $m\gamma_\pm = 1$ at $m = 1.5$; From Fig. 12.11 at $m = 1.5$, $\alpha = 0.788$, and with $n_h = n_s = 3.36$, the expression quoted above yields $r_o = 0.0233$.
63. T. J. Wolery and K. J. Jackson, *Proc. 7th Intern Symp on Water–Rock Interaction*, A. A. Balkema, Rotterdam, pp. 189–93, 1992; T. J. Wolery and K. J. Jackson, "Activity coefficients in aqueous salt solutions," Chapter 2 of *Chemical Modelling of Aqueous Systems II*, D. C. Melchior and R. L. Bassett, Eds., American Chemical Society Symposium Series No. 416, ACS, Washington, DC, 1990. These papers reported on the effect of applying a hydration correction to a model with both a Debye-Hückel term and a hard-core repulsion term. Fitting exercises with various pure aqueous electrolytes yielded reasonable fits, but with near-zero hydration numbers and with ion size parameters smaller than those obtained by Stokes and Robinson (*J. Am. Chem. Soc.* **70**:1870–78, 1948). Since there must be hard-core repulsion, Wolery (private communication) concluded that the concept of a hydrated ion as a thermodynamic species as envisaged by Stokes and Robinson was wrong.
64. S. P. Pednekar, Battelle Pacific Northwest Laboratories, Richland, Washington, private communication, 1994.
65. W. J. Hamer and Y. C. Wu, *J. Phys. Chem. Ref. Data* **1**:1067, 1972; Robinson and Stokes, App. 8.3.

12

AUTOMATED COMPUTATION METHODS

David R. Cogley

Introduction
  Historical/Computational Context
Computational Accuracy
  Measured Concentration Values
  Disequilibrium
Ion Interaction versus Ion Association Approaches
  Modified, Extended Debye–Hückel Approach
  Davies Equation
  Pitzer's Ion Interaction Approach
  Formation Constants
  Conversions between Molal and Molar Scales
  Extrapolation
Setting Up the Computation
  The Model
  Conditions
  Component Concentrations
  Reaction Stoichiometries
  Ionic Strength Effects
  Formation Constant Values
  Computations
  Checking for Reasonableness and Assessing Accuracy
Overview of Computer Programs
MINEQL+
  Example 1. Acid Mine Drainage
  Example 2. Seawater

MINTEQA2
PHREEQC
  Example. Oxidation of Pyrite
EQ3/6
  EQPT
  EQ3NR
  EQ6
Conclusion
Problems

INTRODUCTION

This chapter describes how to automate the manual calculations and spreadsheet calculations presented in this book. Automated computation involves the same basic steps as manual or spreadsheet calculation:

- Model conception;
- Specification of conditions (e.g., pH, pe, temperature, components, equilibrium phases);
- Specification of component concentrations (e.g., Na, K, Cl);
- Specification of reaction stoichiometries and formation constants;
- Selecting a procedure for ionic strength effects;
- Computation;
- Checking for reasonableness.

Automated computation speeds up only one step: computation.

The computer programs described in this chapter can be run on IBM™ personal computers or compatibles with an 80386, 80486 or Pentium® processor under the Microsoft DOS™ operating system. For some programs, data input files are prepared with a word processing program and saved as MS DOS™ text files. Other computer programs for equilibrium calculations may require the Windows™ or Windows 95™ operating systems. There are as yet no generally available programs for the Apple Macintosh™ operating system.

Computer program solution times are generally in the range of 1 second to 1 minute for calculations that might take hours or days to solve manually or with spreadsheet calculations.

The balance of this chapter addresses: historical/computational context, computational accuracy, setting up the computation, overview of computer programs, examples, conclusion, and problems.

Historical/Computational Context

The examples presented in this chapter deal with a limited subset of the possible chemical equilibrium models which benefit from automated computation. Smith and Missen[1] describe applications of chemical equilibrium analysis. They include:

INTRODUCTION

- Isomers of organic substances;
- Analytical chemistry;
- Chemical processes such as methanol synthesis from hydrogen and carbon oxides;
- Fuel cells;
- Rocket propellants;
- Solutions of nonelectrolytes;
- Formation of nitrogen oxides during fuel combustion;
- Ionic speciation;
- Solubility of electrolytes;
- Redox equilibria;
- Sorption.

This chapter deals only with the last four applications.

Bassett and Melchior[2] note that the concept of chemical modeling of natural hydrologic systems was introduced by Garrels and Thompson[3] in their seawater ion-pair model, which had 17 chemical species at 25°C. Following this effort, development of computer program capabilities increased to robust calculation algorithms that can provide reliable solutions to problems involving: hundreds of chemical species, solubility of electrolytes, ion exchange, sorption, and redox processes.

As noted by Smith and Missen, algorithms for the computation of chemical equilibrium prior to the early 1950s were oriented primarily towards hand calculation methods. They revolved around equilibrium-constant expressions written for a set of stoichiometric equations in terms of concentrations (and later, activities). The resulting set of nonlinear algebraic equations is often solved by the Newton–Raphson method, as has been done for several examples in this book. In contrast, "by viewing the problem of equilibrium computation as a nonlinear optimization problem, White et al.[4] developed an algorithm that solved the problem by 'minimizing the free energy directly.' In effect, this meant that they did not use stoichiometric equations or reactions." The mathematical solution of this free-energy-minimization method involves the use of Lagrange multipliers for constrained optimization. As Smith and Missen note, these two approaches are equivalent in their results. Formation constant values are readily converted to free energy differences.

Computer program algorithms are of no practical concern to persons seeking to automate computations unless the program authors have failed to provide a preprocessor to convert between formation constants and free energy differences. All the programs reviewed in this chapter are able to accept inputs in terms of formation constants.

At least during initial use of chemical equilibrium programs, it is important that the program include a default database of formation constants that has been tested on problems documented in the literature. All the programs reviewed here include one or more default databases. Over time, a serious investigator will build up a personal database of thermodynamic data from primary sources.

COMPUTATIONAL ACCURACY

The accuracy of computed concentration values and activity values depends on:

- The accuracy of measured concentration values;
- The extent of disequilibrium of the system being modeled;
- The extent to which the computation algorithm reproduces experimental observations;
- The accuracy of the formation constant values for the concentration range and temperature of the chemical model.

Measured Concentration Values

It is important to match the complexity of one's equilibrium model to the accuracy of measured concentration values. If experimental accuracy is poor, there isn't much point in setting up an equilibrium model with a great many postulated species. It is possible to become hypnotized by the high precision of computer computations; to postulate new species to achieve a good fit between computed and measured parameters. But this is in vain if the measurements are not correspondingly accurate.

A sensitivity analysis can be run to assess the reasonableness of the model complexity one has chosen. For example, if there is reason to believe that measured concentrations are accurate to 20%, one can repeat computations with fewer and fewer species until disagreement between computed and measured values exceed 20%. This sensitivity analysis provides one with a direct assessment of the predictive power associated with experimental measurements of low accuracy (or low precision).

Disequilibrium

Reaction kinetics limit the approach to equilibrium for many systems, for example, organic redox reactions, As(V)/As(III), alumino-silicate minerals, and dehydration of many transition metal hydroxides and hydrated oxides to form metal oxides. For these systems, it is important to consider equilibrium models as possible end points, but they are often not practical realities for many natural systems.

For redox couples such as As(V)/As(III), oxidation of As(III) to As(V) by dissolved oxygen is a slow process, requiring months to approach equilibrium under natural conditions.[5] For this system, one can model short-term speciation by defining separate species for the arsenic oxidation states. For example, As(III) can be defined in the input data file (and default data file) as Ars. With a word processor, this is a simple search-and-replace operation.

"In the past, speciation computations . . . often included . . . saturation indices for minerals that have never displayed reversible solubility behavior either in laboratory studies or in natural waters. . . . If reversibility has never been shown and there is good reason to believe that they do not attain equilibrium solubility, they should be deleted from equilibrium-based modeling computations and from the interpretation of low-temperature equilibrium mineral assemblages.[6] For example, mineral groups such as smectites, illites and micas have never been shown to control water composition. . . . Such demonstrations, however, are plentiful for minerals such as gypsum and calcite."[7]

For models involving precipitation of ferric oxyhydroxides, it is important to understand that the identity and activity of the precipitated phase depends strongly on the composition of the precipitating solution and the age of the precipitate. Macalady et al,[8] citing Murphy et al.,[9] note that the neutralization of ferric salt solutions first yields monomers and dimers that react to form spherical polycations, 15–35 Å in diameter. The next step is formation of hydroxide-deficient polymers with adsorbed or coprecipitated anions such as chloride, nitrate, and perchlorate. Under acid conditions, these polymers are stable for months at ambient temperatures. An x-ray or scanning-electron-microscope identifiable crystalline solid may form in 2 hours at ambient temperature, or develop in 4 months, or take 15 years. The dominant anion is thought to govern which ferric oxyhydroxide ultimately forms. Sulfate favors goethite and minor lepidocrocite (γ-FeOOH), which slowly converts to goethite in solution over about 1 year. Chloride solutions produce akaganeite (β-FeOOH), which gradually converts to goethite over 15 years. Nitrate solutions at 90°C yield an amorphous material that converts to hematite in a few days. At ambient temperatures, nitrate and perchlorate solutions produce goethite and minor, unstable lepidocrocite (γ-FeOOH). Cited log K_{sp} values are: amorphous $Fe(OH)_3$, -37; hematite, -44.0; and goethite, -44.1.

The relevant points are that it is probably appropriate in most cases to use a log K_{sp} value of -37 for short-term equilibrium computations but to realize that, over time, the amorphous $Fe(OH)_3$ will convert to more stable forms for which log K_{sp} = -44 is more appropriate. Also, if one models precipitation of $Fe(OH)_3$ from concentrated electrolytes followed by exposure to dilute solutions, hydroxide-deficient precipitates can be expected to release adsorbed and coprecipitated anions. This process may not be accurately modeled as a default case.

ION INTERACTION VERSUS ION ASSOCIATION APPROACHES

For the sake of modeling convenience, it would be desirable to be able to accurately characterize soluble chemical species in terms of broadly applicable formation constants and a single-parameter correction factor to account for ionic strength effects, which was discussed in Chapter 2. Here some additional details and comparisons are given.

For solutions of salts of univalent ions up to 0.01 molal, formation constants valid at infinite dilution can be used without activity coefficient corrections for semiquantitative work. In this low concentration range, an extended Debye–Hückel model or a Davies equation model provide good estimates of activity coefficient values. At 0.01 molal ionic strength, the Davies equation estimate for $\gamma_\pm$ is 0.9 for salting-out parameter b values of 0 to 0.3. Thus the error without activity corrections is about ten percent. With Davies equation corrections (see discussion below), the error is estimated to be less than 5%.

For solutions of salts of univalent ions up to 0.05 molal, the Davies equation provides good estimates of activity coefficient values. At 0.05 molal ionic strength, the Davies equation estimate for $\gamma_\pm$ is 0.82 for salting-out parameter values of 0.1 to 0.3. Without activity coefficient corrections, the anticipated error from ionic strength effects is about 18%. To this one would add an expected random error component of about 5% (see discussion below for the Davies equation).

Modified, Extended Debye–Hückel Approach

Nordstrom et al. consider the USGS modified, extended Debye–Hückel approach[10] to be valid to concentrations up to 1 molal for the species covered by the U.S. Geological Survey database. The Truesdell and Jones formulation used in PHREEQC and WATEQF4 is:

$$\log \gamma_i = -\frac{A z_i^2 \sqrt{I}}{1 + B a_i^0 \sqrt{I}} + b_i I$$

A and B are constants dependent only on temperature, a_i^0 is the ion-size parameter, a_i^0 and b_i are ion-specific parameters fitted from mean-salt activity-coefficient data, and z_i is the ionic charge of aqueous species i. Nordstrom does not provide a quantitative estimate of potential errors at 1 molal concentrations. The mineral and aqueous species of the USGS database are limited to those applicable to natural water that contain: Na, K, Li, Ca, Mg, Ba, Sr, Ra, Fe, Al, Mn, Si, C, Cl, S (SO_4), and F. Appropriate extended Debye–Hückel parameters are included with the databases for the WATEQF4 and PHREEQC programs.

A PHREEQC run for H, Li, Na, K, OH, F, Cl, Br, and NO_3 over the range of 0 to 1 molal demonstrates the degree to which the modified, extended Debye–Hückel approach reproduces the Robinson and Stokes data.[11] Mean activity coefficients, $\gamma_\pm$, were calculated as geometric means of the calculated single-ion activities for the component ions as listed in the PHREEQC output. A summary of the data for 1 molal solutions (Table 12.1) shows that the ratio of $\gamma_\pm(D\text{-}H)/\gamma_\pm(\text{expt})$ ranges from 74% to 130%; thus this approach is accurate to about ±25% for most electrolytes. Deviations are about half as great at 0.5 molal as at 1.0 molal.

TABLE 12.1. Ratio of $\gamma_\pm(D\text{-}H)/\gamma_\pm$ (expt) at 1 molal

	OH	F	Cl	Br	NO_3
H	no data	no data	82%	74%	89%
Li	124%	no data	82%	76%	82%
Na	96%	113%	100%	92%	115%
K	81%	91%	99%	93%	130%

ION INTERACTION VERSUS ION ASSOCIATION APPROACHES

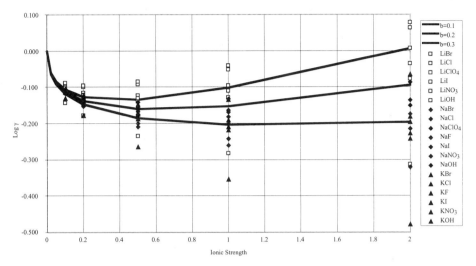

Figure 12.1. Davies equation versus Robinson and Stokes data: Li, Na, K.

Davies Equation

Robinson and Stokes have tabulated activity coefficient values for more than 30 "strong" 1:1 electrolytes, 20 "strong" 1:2 electrolytes, many transition metal electrolytes, and many 2:2 electrolytes. It is a simple matter to plot representative data together with Davies equation predictions.

Figure 12.1 shows Robinson and Stokes data for Li, Na, K, OH, F, Cl, Br, I, ClO_4, and NO_3. Figure 12.2 shows Robinson and Stokes data for H, Rb, Cs, OH, F, Cl,

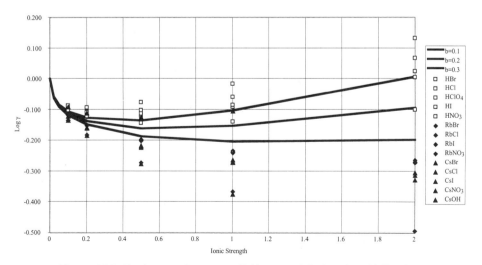

Figure 12.2. Davies equation versus Robinson and Stokes data: H, Rb, Cs.

Br, I, ClO$_4$, and NO$_3$. Davies equation values (calculated with $b = 0.1, 0.2$, and 0.3) for log $\gamma_\pm$ are also shown.

$$\log \gamma_\pm = -A\left(\frac{\sqrt{I}}{1+\sqrt{I}} - bI\right)$$

(A is -0.51 at 25°C, I is the ionic strength, and b is the "salting-out" parameter.) These figures demonstrate that no single value of the salting-out parameter b, can accurately represent log $\gamma_\pm$ values at ionic strengths of 0.1 molal and higher, even for simple 1:1 electrolytes.

At $I = 0.1$ molal, log $\gamma_\pm$ for the Davies equation with $b = 0.1$ to 0.3 varies from -0.117 to -0.107, whereas the Robinson and Stokes observations vary from -0.144 to -0.089. Even allowing for adjustment of the salting-out parameter b, the Davies equation predictions of log $\gamma_\pm$ at an ionic strength of 0.1 molal can be in error by as much as 0.03 log units, a factor of 1.2.

It is possible to estimate the error in $\gamma_\pm$ relative to a best fit for the Davies equation at any particular ionic strength. For the Figures 12.1 and 12.2 data, the standard deviations of observations are given in Table 12.2. The corresponding errors at the 95% confidence limit are approximately $10^{2\sigma}$ (applies for $n = \infty$).

For electrolytes with ions of higher charge, the average errors are thought to be greater.

Pitzer's Ion Interaction Approach[12]

For solutions of salts of univalent, divalent, and trivalent ions up to saturation concentrations, the Pitzer ion interaction approach has been used to fit experimental solution activity coefficient data to within 0.01 log γ units. For spreadsheet calculations of univalent–univalent electrolytes, one first specifies values for 12 parameters:

m = molality
z_M = charge of cation = 1 for univalent–univalent electrolyte
z_X = charge of anion = 1 for univalent–univalent electrolyte

TABLE 12.2. Estimated Error in $\gamma_\pm$

Ionic strength, molal	Standard deviation of log $\gamma_\pm$ (σ)	Corresponding error for $\gamma_\pm$ (95% confidence limit)
0.1	0.014	± 7%
0.2	0.025	± 12%
0.5	0.053	± 28%
1	0.093	± 54%
2	0.162	± 111%

v_M = moles of cation per mole of salt = 1 for univalent–univalent electrolyte
v_X = moles of anion per mole of salt = 1 for univalent–univalent electrolyte
A_ϕ = 0.3915 for 25°C, universal parameter
b = 1.2, universal parameter
α = 2.0, universal parameter
$\beta^{(0)}$ = electrolyte parameter[12]
$\beta^{(1)}$ = electrolyte parameter[12]
C_{MX}^ϕ = electrolyte parameter[12]
I = ionic strength = m for univalent–univalent electrolyte

Three of these parameters are adjustable parameters evaluated through a least-squares regression on data such as the Robinson and Stokes data. The next step involves calculations.

$$x = \alpha\sqrt{I}$$

$$g(x) = \frac{2[1 - (1 + x)\exp(-x)]}{x^2}$$

$$C_{MX}^\gamma = 1.5 C_{MX}^\phi$$

$$B_{MX} = \beta_{MX}^{(0)} + \beta_{MX}^{(1)} g(x)$$

$$B_{MX}^\phi = \beta_{MX}^{(0)} + \beta_{MX}^{(1)} \exp(x)$$

$$B_{MX}^\gamma = B_{MX} + B_{MX}^\phi$$

$$f^\gamma = -A_\phi \left[\frac{\sqrt{I}}{(1 + b\sqrt{I})} + \left(\frac{2}{b}\right) \ln(1 + b\sqrt{I}) \right]$$

$$\ln \gamma_\pm = |z_M z_X| f^\gamma + m\left(\frac{2 v_M v_X}{v_M + v_X}\right) B_{MX}^\gamma + m^2 \left[\frac{2\sqrt[3]{v_M v_X}}{v_M + v_X}\right] C_{MX}^\gamma$$

$$\log \gamma_\pm = \frac{\ln \gamma_\pm}{\ln(10)}$$

The Pitzer approach has been applied principally to a set of electrolytes that do not form strong complexes. Many data are available for 25°C and temperature coefficients are available for a fraction of the 25°C data. Pitzer[12] provides parametric data for calculation of mean activity coefficients (25°C) for:

- 74 univalent–univalent electrolytes, mostly alkali metal hydroxides, halides, chlorates, perchlorates, bromates, nitrites, and nitrates based on the experimental data of Robinson and Stokes;
- 14 carboxylic acid salts of alkali metals (1-1 type) and Tl(I);
- 15 tetraalkylammonium halides;
- 32 sulfonic acids and salts (1-1 type);
- 16 additional organic salts(1-1 type);
- 77 inorganic compounds of (2-1 type);
- 10 sulfonic acids and salts (2-1 type);
- 58 electrolytes (3-1 type);
- 6 electroytes (4-1 type);
- 2 electrolytes (5-1 type);
- 12 electrolytes (2-2 type).

Mixing parameters are provided for 41 binary symmetrical mixtures and 4 binary mixtures without common ions.

The Pitzer approach is applied by investigators with a need to accurately model concentrated solutions and brines within the temperature range of 0 to 350°C.

Formation Constants

Three sources of potential error can be identified for formation constants: supporting electrolyte effects, concentration scale conversions, and extrapolations from zero ionic strength.

Supporting Electrolyte Effects. As the plots of the Robinson and Stokes data illustrate, ion interactions strongly affect activity coefficient values. There are no general rules that allow one to predict whether a particular supporting electrolyte greatly shifts an apparent equilibrium or not. However, analogy and empirical correlation with actual measurements can be helpful. Ionic medium composition is given by Sillen and Martell.[13] For most entries, Smith and Martell[14] give only ionic strength, not composition of the medium.

For most ionic complexes, there are insufficient data to allow an assessment of supporting electrolyte effects. There are data for lead chloride complexes at 25°C in chloride, perchlorate, and nitrate supporting electrolytes over the concentration range of 1 to 4 molar that are summarized in Sillen and Martell. The data for the first three formation constants are plotted in Figs. 12.3–12.5. The data of Tur'yan and Chebotar[16] are not plotted. All other data for the indicated supporting electrolytes have been included.

Data for log β_1 are shown in Fig. 12.3 together with the values selected by Smith and Martell for the NIST compilation and a curve for the Davies equation with $b = 0.14$. The plotted points fall in two series: The chloride and perchlorate data fall close to the plotted curves, and the nitrate data fall in a group well below the other

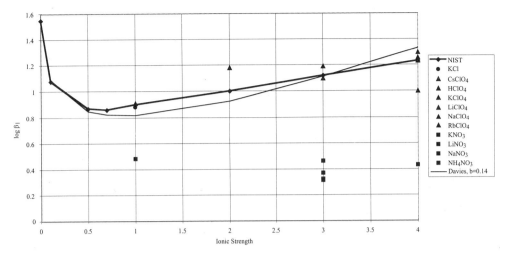

Figure 12.3. PbCl⁺ complex.

data. More than half of the nitrate data points were published by Mironov in a series of papers from 1961 through 1965.[16] Mironov replicated many of the perchlorate data and this is taken as a reason to accept the nitrate data. At 1 molar ionic strength, the formation constant value for nitrate supporting electrolyte is 0.4 log units lower than in chloride/perchlorate electrolyte. At 3 and 4 molar ionic strength, the formation constant values for nitrate supporting electrolyte are 0.8 log units lower than in chloride/perchlorate electrolyte. These are factors of 2.5 and 6.

Data for $\log \beta_2$ are shown in Fig. 12.4 together with the values selected by Smith and Martell for the NIST compilation and a curve for the Davies equation with $b =$

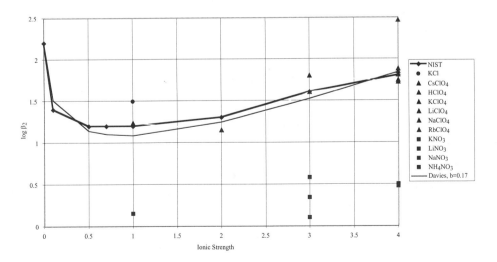

Figure 12.4. PbCl₂ complex.

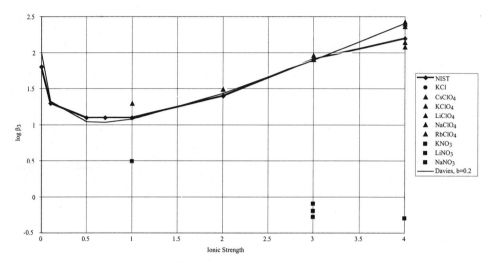

Figure 12.5. PbCl$_3^-$ complex.

0.17. Again, the values for nitrate supporting electrolyte fall well below those in chloride/perchlorate electrolyte. The formation constant values for nitrate supporting electrolyte are about 1 log unit lower than in chloride/perchlorate electrolyte, a factor of 10 lower.

Data for log β_3 are shown in Fig. 12.5 together with the values selected by Smith and Martell for the NIST compilation and a curve for the Davies equation with $b = 0.2$. Again, the values for nitrate supporting electrolyte fall well below those in chloride/perchlorate electrolyte. At 1 molar ionic strength, the formation constant value for nitrate supporting electrolyte is 0.6 log units lower than in chloride/perchlorate electrolyte. At 3 and 4 molar ionic strength, the formation constant values for nitrate supporting electrolyte are about 2 log units lower than in chloride/perchlorate electrolyte, a factor of about 100 lower.

When viewed at ionic strengths of 3 to 4, the stepwise formation constants determined in nitrate supporting electrolyte are about five-fold lower than those determined in chloride/perchlorate supporting electrolyte. From this one example, one might conclude that there is reason to be cautious when applying complex formation constants determined in chloride/perchlorate supporting electrolytes at high ionic strength to other solutions that are not predominantly chloride or perchlorate. The deviations seem to be cumulative for higher ligand numbers.

It is possible to gain some insight into supporting electrolyte effects by plotting log $\gamma_\pm$ for 3 molal solutions of alkali halides, perchlorates, and nitrates. From Fig. 12.6, it is clear that there is a strong effect of anions on mean activity coefficient values. Whatever the mechanism, it is clear that there is a nitrate ion-interaction effect that tends to depress mean activity coefficient values by about 0.3 log units (a factor of 2) for univalent ions with a Kielland ion size (see Chapter 2) of 3. The effect appears to be somewhat less for larger univalent ions. The effect is also less at lower ionic strengths. Judging by the data for lead chloride complexes, the effect is substantially larger for lead.

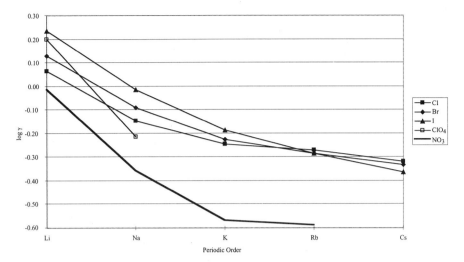

Figure 12.6. log γ at 3 molal.

Conversions between Molal and Molar Scales

Compilations of formation constants may not distinguish between molal and molar concentrations scales. The Smith and Martell compilation (1974–1989) and the NIST Standard Reference Database 46 do not distinguish between molal and molar concentrations scales.[17] This introduces a significant source of error for constants determined in high-concentration background electrolytes. Most, but not all, of the constants determined in high concentration background electrolytes have been determined on the molar concentration scale.[17] For precise work, it is thus necessary to review the primary literature to determine the applicable concentration scale so that conversions to the appropriate scale (for the chosen computation program) can be made.

Conversion from molal to molar units for transition metal chloride complexes determined in concentrated supporting electrolyte is straightforward. For the MX^+ complex,

$$M^{2+} + X^- = MX^+$$

$$\beta_{1,\text{molal}} = \frac{[MX^+]_{\text{molal}}}{[M^{2+}]_{\text{molal}}[X^-]_{\text{molal}}}$$

$$\beta_{1,\text{molar}} = \frac{[MX^+]_{\text{molar}}}{[M^{2+}]_{\text{molar}}[X^-]_{\text{molar}}}$$

$$\frac{[MX^+]_{\text{molar}}}{[MX^+]_{\text{molal}}} = \frac{\rho_{\text{solution}}}{1 + 0.001 m_{\text{elec}}(MW_{\text{elec}})}$$

This same equation applies to all minor components of the solution containing a great excess of the supporting electrolyte (elec). Substituting these concentration factors into the equation for β_{molar} yields the expression

$$\frac{\beta_{1,molar}}{\beta_{1,molal}} = \frac{1 + 0.001 m_{elec}(MW_{elec})}{\rho_{solution}}$$

Similarly,

$$\frac{\beta_{2,molar}}{\beta_{2,molal}} = \left(\frac{1 + 0.001 m_{elec}(MW_{elec})}{\rho_{solution}}\right)^2$$

$$\frac{\beta_{3,molar}}{\beta_{3,molal}} = \left(\frac{1 + 0.001 m_{elec}(MW_{elec})}{\rho_{solution}}\right)^3$$

$$\frac{\beta_{4,molar}}{\beta_{4,molal}} = \left(\frac{1 + 0.001 m_{elec}(MW_{elec})}{\rho_{solution}}\right)^4$$

For 3 molal NaCl, $\rho = 1.113$, $MW_{elec} = 58.5$, and $[1 + 0.001\, m_{elec}(MW_{elec})] = 1.176$. For 3 molal $NaNO_3$, $\rho = 1.159$, $MW_{elec} = 85$, and $[1 + 0.001\, m_{elec}(MW_{elec})] = 1.255$. Results for 3 molal solutions of supporting electrolyte are:

	β_M/β_m in NaCl	$\log \beta_M - \log \beta_m$ in NaCl	β_M/β_m in $NaNO_3$	$\log \beta_M - \log \beta_m$ in $NaNO_3$
MX^+	0.95	−0.024	0.92	−0.035
MX_2	0.90	−0.047	0.85	−0.069
MX_3^-	0.85	−0.071	0.79	−0.104
MX_4^{2-}	0.80	−0.095	0.73	−0.138

For a dense supporting electrolyte solution such as $NaNO_3$, corrections to β values amount to 8% to 27%.

Extrapolation

For ionic equilibrium calculations, much of the variation in activity coefficients demonstrated by the Robinson and Stokes data is incorporated into measured values for formation constants of complex ionic species. This provides a convenient means of removing much of the dependence of activity coefficient values on specific electrolyte ion interactions, as long as one uses formation constants valid for the ionic strength of interest.

For ionic strength values of 0.2 molal and above, formation constant values often vary relatively little with ionic strength, and it is often possible to interpolate between formation constant values from experimental data to match the modeled ionic strength. If the input is for $I=0$, one must then back-calculate $I=0$ formation constant values for the input that will give the correct values for the modeled ionic strength.

An example would be using PHREEQC for modeling the formation of $HgCl_4^{2-}$ concentration in 3 molal sodium chloride. The NIST values for the formation constant (β_4) for $HgCl_4^{2-}$ are log β_4 = 15.6 at zero ionic strength and log β_4 = 16.1 at 3 molal ionic strength. The default activity coefficient correction would be the Truesdell and Jones convention, but this can be overridden by specifying the Davies equation procedure, which assumes a salting-out parameter of 0.3. Since mercury is not included in the default database, it is necessary to enter data for Hg^{2+} $HgCl^+$ $HgCl_2$, $HgCl_3^-$ and $HgCl_4^{2-}$. We consider the options for the formation constant for $HgCl_4^{2-}$. If we specify log β_4 = 15.6 at zero ionic strength and assume $\gamma_{2+} = \gamma_{2-}$, the effective value of log β_4 for a solution of 3 molal ionic strength will be

$$\beta_4 = \beta_4^0 \frac{\gamma_2 \gamma_1^4}{\gamma_2} = \beta_4 \gamma_1^4$$

where

$$\log \gamma_1 = -0.51\left[\left(\frac{\sqrt{i}}{1+\sqrt{i}}\right) - 0.3i\right] = -0.0583$$

so that log β_4 = –0.233 + 15.6 = 15.37. But the correct value of log β_4 is known to be 16.1 at I = 3.0, so we need to calculate the corresponding log β_4^0 from

$$\beta_4^0 = \beta_4 \frac{\gamma_2}{\gamma_2 \gamma_1^4} = \frac{\beta_4}{\gamma_1^4}$$

log β_4^0 = 16.1 + 0.233 = 16.33. This value of log β_4^0 entered into the input data file will produce results corresponding to log β_4 = 16.1 at I = 3.0. Similar calculations need to be made for β_1, β_2, and β_3. Note that for log β_4, this correction of 0.73 log units (16.1-15.37) is much greater than the precision of the formation constant value, which is estimated to be ±0.1.

Calculation of β values for each ionic strength value makes the "automated computation" somewhat less automated, but this may be worthwhile for persons who want to improve the accuracy of results. Computation of formation constants applicable to conditions other than zero-ionic strength has long been the standard approach for seawater and ionic media such as 4M $NaCl_4^0$.

SETTING UP THE COMPUTATION

Each of the examples in this chapter includes the following steps:

1. Model conception;
2. Conditions: pH, pe, temperature, components, equilibrium phases;
3. Component concentrations;
4. Reaction stoichiometries;
5. Formation constants;
6. Correcting for ionic strength effects;
7. Problem solution;
8. Checking for reasonableness and assessing accuracy.

The Model

When reviewing modeling results, it can be helpful to have a complete description of the original intent of each modeling exercise. Examples of model concepts described in this chapter include:

1. Preparing a log concentration diagram for pyrite oxidation in acidic sulfate solutions covering the pH range from about 0 to 3.5 and total iron concentrations from 0.0005 molar to 0.25 molar;
2. Comparing results for spreadsheet and computer solutions to the Garrels and Thompson model of seawater;
3. Preparing a log concentration diagram for neutralization of acid mine drainage with calcium carbonate over the pH range from 1 to 3.5;
4. Modeling the sorption of several multivalent elements to hydrous ferric oxide over the pH range of 5 to 7 with kinetic limitations for the interconversion of arsenic species.

Conditions

Explicit definition of conditions is an aid to efficient setup of computer runs. Solution speciation calculations at a single pH value are straightforward with most computation programs. Solution speciation calculations over a range of pH values may require either many model runs or selection of a special program option.

Though the redox state of a solution may often be ignored, complications sometimes ensue with transition metals, sulfide/sulfate, ammonia/nitrite/nitrate, chloride/perchlorate, etc. For some programs, it may be possible to "turn off" redox equilibrium for certain species. If only one oxidation state of an element is needed for the model, the other oxidation states may be suppressed by redefining the formation constants.

If two or more oxidation states of an element are to exist in disequilibrium, it may be necessary to specify new symbols for specific valence states of ions that you believe are not at redox equilibrium. This is discussed under the heading Computational Accuracy, subheading Disequilibrium.

Component Concentrations

Commonly used concentrations scales are molar, molal, and ppm. As noted in Chapter 1, this text uses the molar concentration scale, moles per liter of solution, except for seawater and as otherwise noted. The molar scale is convenient for laboratory studies wherein solutions are made up with volumetric glassware and temperature ranges are small enough that corrections need not be made for variations in density.

Computation programs intended for calculations over a wide temperature range may be programmed to perform calculations in molal units, moles per kilogram of solvent (e.g., water), to avoid variations due to changes in density. This requires, of course, that formation constants be specified in molal units.

For dilute solutions, the molar and molal scales are essentially equivalent.

Formulas are provided in Chapter 1 and also in the discussion above for conversion of species concentrations between molal and molar units.

Reaction Stoichiometries

Each computation program observes specific conventions for describing reaction stoichiometry. When preparing input data for a particular program, it is necessary to understand these conventions and to specify reaction stoichiometries accordingly.

For buffer calculations, it is important to know whether the model expects dissociation reactions or association reactions.

For programs incorporating Morel tableau notation,[18] it may be necessary to recast all the reactions in terms of a limited set of basis species. For other programs, it may be possible to simply define a complete set of species, using whichever complexes one chooses to define succeeding complexes.

Reaction stoichiometry conventions are described under the summary for each model.

Ionic Strength Effects

As discussed above, manual, spreadsheet, and computer computations all suffer from potential errors due to activity correction procedures that do not correctly predict activity coefficients for the computation at hand. Activity coefficient corrections are reviewed in Chapter 2.

Computation programs commonly invoke any of four activity coefficient correction procedures:

- No activity coefficient corrections;
- An extended Debye–Hückel ion-size-parameter correction, sometimes with Kielland's ion size parameters (Chapter 2);

- Davies equation with a "salting out" parameter;
- Ion interaction approach, such as Pitzer coefficients.

Program input specifications must be checked to be certain of the default procedures including the default database. For example, the default procedure for PHREEQC is the Davies equation with $b = 0.3$. However, the default database, PHREEQC.DAT, contains entries for the Truesdell Jones extended Debye–Hückel parameters, which override the Davies equation.

Formation Constant Values

The starting point for a complete list of formation constants can be the program's default database. One then needs to consider: primary data sources, critical compilations of formation constant data, whether formation constants have been corrected to zero ionic strength, whether all significant complexes were considered in the derivation of the formation constants, and whether the default databases have been checked for errors.

For critical applications, it may be advisable to check both a primary data source and the literature for citations of the primary source. This would be true, for example, for sorption and ion exchange.[19]

The best overall guide to the literature on aqueous complex formation constant data is the series of critical compilations by Smith and Martell.[20] The original six-volume series issued by Plenum (1974–1989) has been continued with a series of CD-ROM volumes[21] available from the National Institute of Standards and Technology (NIST) at www.nist.gov. The Smith and Martell critical compilations focus on selection of the best values and do not include information on composition of the medium, concentration scales (e.g., molal vs. molar), or experimental methods (e.g., isopiestic, potentiometric, etc.). For this information, one must consult primary sources.

An excellent source of formation constants for geochemical modeling is *Revised Chemical Equilibrium Data for Major Water–Mineral Reactions and Their Limitations*.[22] This a critical compilation meant to provide an internally consistent database for species containing fluoride, chloride, oxide, hydroxide, carbonate, silicate, and sulfate species. Enthalpy data are included to allow modeling at temperatures from 0 to 100°C and pressures to 1 bar.

A convenient data source is the tableau summary of stability constants in Morel and Hering's Table 6.3.[23] This summary has been reproduced as Appendix 6.1 in Stumm and Morgan.[24] These tables provide stability constants for formation of complexes and solids from 20 metals and 31 ligands. Constants are given as logarithms of the overall formation constants, β, for complexes and as logarithms of the overall precipitation constants for solids, at zero ionic strength and 25°C. Constants were taken from Smith and Martell,[25] Martell and Smith,[26] Whitfield,[27] Baes and Mesmer,[28] Sunda and Hanson,[29] Sainte Marie et al.,[30] and Morgan.[31] When necessary, constants were extrapolated to $I = 0$ molar using the following values of

SETTING UP THE COMPUTATION

$-\log \gamma_z$ applied to all ions including H⁺ and tri- and tetravalent ions for experimental observations between 0.1 and 4 molar.[23]

I (molar)	$-\log \gamma_z$	Calculated b
0.0	0	0
0.1	$0.11 z^2$	0.246
0.3	$0.13 z^2$	0.330
0.5	$0.15 z^2$	0.240
1	$0.14 z^2$	0.226
2	$0.11 z^2$	0.185
3	$0.07 z^2$	0.166
4	$0.03 z^2$	0.152

For experimental observations less than 0.1 molar, the Davies equation with $b = 0.3$ was applied.[32] However, this table of log γ values is inconsistent with the Davies equation with a salting-out parameter b, value of 0.3. The apparent value of b varies as shown in column 3 of the table. For this reason, calculation results may be expected to vary significantly depending on whether input data files are based on extrapolations from high ionic strength with the tabulated activity coefficient values or by another method selected by experimenters who measured the activity values.

Databases may contain errors. Errors may be major or minor. One form of error is caused by incorrect conversion from the primary data source with one form of the formation constant to the database with another form of the constant. Such errors have been found in the MINTEQA2 database, where conversion to Morel's Tableau format was apparently done incorrectly for a number of ligands (see description of the MINTEQA2 program). Another source of error is incorrect corrections to zero ionic strength.

Computations

Computing the results requires creation of the input data file and running the selected program. These steps are described for each example.

Checking for Reasonableness and Assessing Accuracy

To check for reasonableness, one can compare computed output to similar cases modeled previously, with special care being given to such items as saturation indices for minerals, partial pressures for gases, and pH values.

Assessment of accuracy includes:

- A review of potential disequilibrium (low rate constants);
- Ion interaction effects, formation constant precision;

- Known variability in measured parameters relative to activity coefficient conventions (see, for example, Figs. 12.1 and 12.2 and the accompanying discussion);
- Potential errors from activity coefficient values computed by the program for the modeled ionic strength;
- Errors in interpolating and recalculating formation constant values;
- Failure to convert formation constants to the concentration scale appropriate to the computer program;
- Blatant errors such as incorrect conversion of formation constant values from Morel tableau notation.

Some of this assessment involves simple checking of calculations and some of this assessment involves preparing a quantitative estimate of errors as a percentage of the calculated value.

OVERVIEW OF COMPUTER PROGRAMS

Automated computation programs can be a great convenience for systems with many components, for titration calculations over a wide pH range, for calculations at high ionic strengths, and for computations over a range of temperatures. Another benefit of automatic computation programs is the ease with which a sensitivity analysis can be carried out to identify critical thermodynamic parameters, which then can become the focus for checking original sources of data.

Three generally available computation programs are reviewed and compared in Table 12.3: MINEQL+, PHREEQC, and EQ3/6. A fourth program, MINTEQA2, is briefly summarized, and several others are mentioned. For each of the three reviewed programs, information is provided on: chronology, ordering of software and documentation, language, major program components, processes that can be modeled, ranges for concentration–temperature–pressure, number of chemical components and species, databases available, and example calculations. Example computations are provided for MINEQL+ and PHREEQC.

MINEQL+[33]

The model MINEQL+ is descended from REDEQL,[34] and is a refinement of Westall's model, MINEQL.[35] Refinements from 1989 through 1994 have used Morel's Tableau approach to setting up stoichiometry, equilibria, and mass balances,[36] visualization of stoichiometry, and import or export of data and results to other software. An especially convenient capability is the automatic calculation of log-concentration diagrams via the titration module. The user may specify a range of log-concentration values and total number of data points for a single species.

MINEQL+ computes concentrations of all species, and these may be used to create additional log-concentration diagrams.

Other useful features include import of spreadsheet data, sensitivity analysis, and creation of personal thermodynamic databases. The "Field Data" option allows one to import thousands of data points from spreadsheets created by field monitoring programs, set up Monte Carlo simulations, or assess spatial and temporal variations. With the "Sensitivity Analysis" option, the user may vary every type of input parameter, including thermodynamic data, except the stoichiometric coefficients. This option can be used to model titrations also. With the "Personal Thermodynamic Database" option, the user can enter and save corrections to the default database, as well as enter new data. Changed data are flagged in the tableau displays.

Version 3.0 (1994) of the model runs under the Microsoft DOS operating system on an 80286 Intel-based personal computer or later model. Version 3.0 can also be run on an Apple Power Macintosh with SoftWindows 3.0 PC emulation software. In either case, the AUTOEXEC.BAT and CONFIG.SYS files must contain the bare minimum of commands so that the maximum amount of high RAM is preserved. Neither expanded memory nor extended memory is used.

MINEQL+ can simultaneously model 25 components and 400 species. The temperature range is 5 to 30°C. The intended ionic strength range is 0 to 0.5 molar, consistent with the range over which the Davies equation provides acceptable activity coefficient corrections. Calculations may be performed with or without Davies equation corrections for ionic strength. (Note that MINEQL+ uses a Davies equation with the salting-out parameter $b = 0.24$, not 0.2 as in Chapter 2.) MINEQL+ is intended for ionic equilibrium calculations on water treatment systems and ambient systems including rivers, lakes, and groundwater. However, calculations for seawater are possible with the Davies equation correction turned off by using concentration stability constants valid for seawater. The default database is the U.S. Environmental Protection Agency's MINTEQA2 database.[37,38]

Surface complexation reactions may be modeled with Langmuir adsorption, Freundlich adsorption, triple-layer adsorption, two-layer adsorption, constant capacitance adsorption, or two-layer adsorption on hydrous ferric oxide. The MINEQL+ User's Manual provides examples of each type of adsorption calculation, including three examples contributed by Dr. David Dzombak.

Users of MINEQL+ should be aware of two issues affecting the interpretation of computation results. It is possible to view and export the "concentration of a gaseous or solid phase" expressed as the number of moles the phase dispersed in the solution volume. Thus, if one molar calcium and one molar carbonate is specified for input, and 50% of this precipitates as calcite, the reported concentration value for calcite should be 0.5. At and above saturation, this value can be in error; the program output can list concentration values above the totals present in the system. In the case of a single precipitate, the correct value may be calculated as the totals present in the system minus the sum of all soluble species.

When computations involve Davies equation corrections, the user must ensure that the input data set is charge balanced. Unlike PHREEQC (see below), MINEQL+

TABLE 12.3. Comparison of Three Computation Programs: MINEQL+, PHREEQC, EQ3/6

	MINEQL+	PHREEQC	EQ3/6
Organization	Environmental Research Software	U.S. Geological Survey	U.S. Department of Energy
Primary application	Fresh water and water treatment	Geochemistry	Radwaste disposal
Ordering software and documentation	http://www.agate.net/~ersoftwr/mineql.html Environmental Research Software, 16 Middle Street, Hallowell, ME 04347, (207)622-3340, ersoftwr@agate.net Free software download. Nominal charge for manual.	ftp: brrcrftp.cr.usgs.gov at directory geochem/phreeqc under user name anonymous with the password set equal to your email address U.S. Geological Survey NWIS Program Office 437 National Center, Reston, VA 22092, (703) 648-5695 No charge for download of software or manual.	Technology Transfer Initiatives Program, L-795, Attn: Diana West, Lawrence Livermore National Laboratory, P.O. Box 808, Livermore, CA 94550, (510)423-7678 (tel), (510)422-6416 (fax) Charge for software and manual.
Chronology	1975, MINEQL calculation procedure 1989–1994, MINEQL+ Input/output development	1980, PHREEQE (FORTRAN) 1995, PHREEQC available on Internet	mid-1970s, seawater–basalt model 1992, EQ3/6 Version 7.0
Language	A compiled application running under the Microsoft DOS operating system	A "C" application with source code compiled to run under the Microsoft DOS operating system. Source code can be compiled to run under the UNIX operating system	A FORTRAN 77 application that can be compiled to run on mainframes or under the Microsoft DOS operating system
Computer platforms	Intel 80286-based personal computer or later model; DOS, or DOS shell under Windows, WIN95. Apple Power Macintosh running SoftWindows 3.0 PC emulation software	Intel 80286-based personal computer or later model. Apple Power Macintosh running SoftWindows 3.0 PC emulation software	Mainframe computers. Intel 80286-based personal computer or later model. Not tested on Apple Power Macintosh running PC emulation software

Major program components	A single-component, DOS application with a graphical user interface	A single-component (PHREEQC) computational module operated in batch mode with default data files and other input files prepared with a text editor	EQPT—data file preprocessor EQLIB—software library EQ3NR—speciation solubility code EQ6—reaction path code
Processes that can be modeled	Speciation–solubility, titration, reaction path, adsorption, precipitation disequilibrium (decoupling), redox disequilibrium (decoupling)	Speciation–solubility, titration, reaction path, adsorption precipitation disequilibrium (decoupling), redox disequilibrium (decoupling)	Speciation–solubility, titration, reaction path, solid solutions, precipitation disequilibrium (quantitative), redox disequilibrium (quantitative), kinetics
Concentration scale	Molar (moles/liter of solution)	Molal (moles/kg water)	molal (moles/kg water)
Concentration range	Dilute solutions, optional Davies equation ionic strength, corrections	Dilute solutions, optional extended Debye–Hückel ionic strength corrections	dilute solutions to brines, optional extended Debye–Hückel ionic strength corrections, optional Pitzer model ionic strength corrections
Activity correction options	1 None 2 Davies equation, $b = 0.24$	1 None (manual corrections to input database) 2 Davies equation, $b = 0.3$ 3 Extended Debye–Hückel	1 Davies equation, $b = 0.2$ 2 Extended Debye–Hückel 3 Pitzer ion interaction
Temperature range	5–30°C	0–100°C	0–300°C for COM, SUP & NEA 0–100°C for PIT 25°C for HMW
Pressure range	Atmospheric	Atmospheric	0–86 atm
Chemical components/species/minerals/gases	25/400 model run limit. (Species in MINEQL+ can be composed of soluble complexes, precipitates (minerals) or gases	45/197/54/7 PHREEQC.DAT 66/371/311/7 WATEQ4F.DAT	78/852/886/76 COM 69/315/130/16 SUP 32/158/188/76 NEA 9/17/51/3 HMW 52/68/381/38 PIT

(*continued*)

TABLE 12.3. Continued

	MINEQL+	PHREEQC	EQ3/6
Databases	MINTEQA2 (U.S. Environmental Protection Agency) derived from WATEQF4.DAT	PHREEQC.DAT WATEQ4F.DAT	COM—A composite file specific to the extended Debye–Hückel formalism, a melange of data SUP—A file based on work of Helgeson and others, high internal consistency NEA—A file of the Data Bank of the Nuclear Energy Agency of the European Community HMW—A file based on Harvie, Møller, and Weare for dilute solutions or (seawater) brine calculations, the highest degree of internal consistency PIT—A file based mostly on data summarized by Pitzer, for dilute solutions or brine calculations a melange of data

Example calculations			
1 Fixed pH, CaCO3 system	1 Speciation	1 Seawater major ions
2 pH titration, CaCO3 system	2 Equilibration with pure phases	2 Seawater, Pitzer equations
3 pH calculation	3 Mixing	3 Mineral solubility
4 Langmuir adsorption	4 Evaporation and homogeneous redox reactions	4 pH buffer
5 Freundlich adsorption	5 Irreversible reactions	5 Oxygen fugacity from mineral equilibria
6 Triple-layer adsorption	6 Reaction-path calculations	6 Eh from a redox couple
7 Two-layer adsorption	7 Gas-phase calculations	7 Dead Sea brine
8 Constant capacitance adsorption	8 Surface complexation	8 pHCl as an input
9 Hydrous ferric oxide adsorption	9 Advective transport and cation exchange	9 Seawater precipitates
10 Fixed pe	10 Advective transport, cation exchange, surface complexation, and mineral equilibria	10 High-temperature pH from quench pH
11 Control of pe with oxygen	11 Inverse modeling	11 Microcline dissolution in pH 4 HCl
12 Control of pe with ion ratios	12 Inverse modeling with evaporation	12 Microcline dissolution / fluid-centered flow-through system
		13 Gypsum solubility in brine, Pitzer equations
		14 Alkalinity titration
		15 Kinetics of quartz precipitation

does not require specification of charge balance in the tableau of species. Ionic strength is computed from the species based on the elemental inputs without regard for charge balance. Thus a MINEQL+ model for seawater can be set up without the nonreactive component Cl^-, and will not be charge balanced. Calculations involving Davies equation corrections will be based on an ionic strength of about 0.4 instead of 0.7 unless chloride is included in the elemental inputs.

MINEQL+ Version 3.0 is recommended as a highly usable ionic equilibrium computational program with interactive data input, customizable databases, and excellent display and output capabilities.

Version 4.0 is scheduled for release in 1998. This is to be a Windows program allowing simultaneous on-screen display of data entry screens, spreadsheets, and graphs. New features will include: automatic highlighting of zero values for enthalpy and log K; input wizards to help users with data entry for pH and alkalinity; fixed ionic strength computation options; an improved solids calculation procedure to cope with potential phase rule violations; output of activities; and preparation of predominance area diagrams.

Example computations are provided here for acid mine drainage and for seawater composition.

Example 1. Acid Mine Drainage

For acid mine drainage computations, the first matter to consider is the process being modeled. One may model either the generation of soluble ferric ion, sulfate ion, and acidity by access of water and oxygen to iron sulfide minerals (as in Chapter 7), or the mixing of acid mine drainage with other surrounding waters of known alkalinity. A first step in either model could be to prepare log-concentration diagrams as an aid to understanding the relative predominance of ferric ion complexes.

Pyrite oxidation to ferric ion in moderately acidic sulfate solutions may be expressed as:

$$FeS_2(s) + 3.75\ O_2 + 0.5\ H_2O \rightleftharpoons FeSO_4^+ + SO_4^{2-} + H^+$$

This way of expressing the reaction products emphasizes the large role that $FeSO_4^+$ plays (See Fig. 7.25). Precipitation of ferric iron may be expressed as:

$$FeSO_4^+ + 3\ H_2O \rightleftharpoons Fe(OH)_3 + 3\ H^+ + SO_4^{2-}$$

Approximately one proton is generated by the initial phase of pyrite oxidation. Three protons are generated by ferric ion precipitation. There is no buffering reaction for the pyrite oxidation. On the other hand, ferric iron precipitation buffers pH, especially when acid mine drainage mixes with more alkaline waters.

The acid mine drainage calculations of Chapter 7 were performed using an Excel spreadsheet, but may also be performed using MINEQL+. Formation constant values in Table 7.3 must first be recast into Morel's notation (Table 12.4). The components (Type I in the MINEQL notation) are H_2O, H^+, Fe^{3+}, and SO_4^{2-}. The com-

TABLE 12.4. Speciation Model (Tableau) for Acid Mine Drainage

		Components						Total Conc
Type	Species	H_2O	H^+	Na^+	Fe^{3+}	SO_4^{2-}	log $K°$	(mol/L)
I	H_2O	1					0	
I	H^+		1				0	
I	Na^+			1				0.005
I	Fe^{3+}				1		0	0.005
I	SO_4^{2-}					1	0	0.010
II	OH^-	1	−1				−14.00	
II	$FeOH^{2+}$	1	−1		1		−2.19	
II	$Fe(OH)_2^+$	2	−2		1		−5.70	
II	$Fe_2(OH)_2^{4+}$	2	−2		2		−2.90	
II	$Fe(OH)_{3\,aq}$	3	−3		1		−13.20	
II	$Fe(OH)_4^-$	4	−4		1		−21.60	
II	$Fe_3(OH)_4^{5+}$	4	−4		3		−6.30	
II	HSO_4^-		1			1	1.99	
II	$FeSO_4^+$				1	1	4.04	
II	$Fe(SO_4)_2^-$				1	2	5.38	
V	$Fe(OH)_3$ (a), (Ferrihydrite)	3	−3		1		−3.20	

plexes (Type II) include HSO_4^-, $FeSO_4^+$, $Fe(SO_4)_2^-$, $FeOH^{2+}$, $Fe(OH)_2^+$, and other minor species. Dissolved solids that precipitate, such as $Fe(OH)_3$, are designated Type V. The MINTEQA2 default values for formation constants were replaced by the values listed in Table 12.4, to conform with Chapter 7. The major difference between MINTEQA2 equilibrium constants and those listed in Chapter 7 is in the solubility product of $Fe(OH)_3$ (Ferrihydrite).[39]

On a Power Macintosh™ 7200/75 running the SoftWindows™ 3.0 emulation of an 80486 processor with an 80387 coprocessor, computation with MINEQL+ took 4 minutes for the titration option with 61 data points. Execution on an 80486 or Pentium® processor at a 75 MHz or greater clock speed would be faster. Data were output to Lotus™ files,[40] then transferred to an Excel spreadsheet, and used to create charts. Once this sequence of operations is mastered, preparation of a pH versus Log-C diagram requires only 10–20 minutes.

Figure 12.7 presents results for the pH range 0 to 3.5. Ionic strength is approximately 0.02 but varies with pH for three reasons: To achieve low pH, strong acid must be added; to achieve high pH, alkali must be added; as the species composition changes, so does the distribution of charge. This figure may be compared with Fig. 7.25.

HSO_4^- decreases steadily and SO_4^{2-} increases steadily from pH = 0 to 3.5. Equal concentrations are achieved at pH = 1.7, consistent with Davies equation activity corrections and pK_a = 1.96 at I = 0.

Total soluble iron is constant over the pH range 0.0 to 2.31. Amorphous $Fe(OH)_3$ precipitates at pH = 2.3 and higher, with the result that total soluble iron decreases

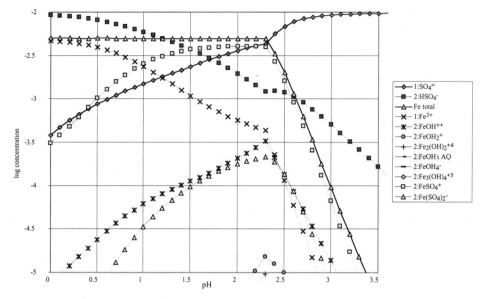

Figure 12.7. Acid mine drainage.

for pH > 2.31. From pH = 0.0 to 0.9, the primary soluble iron species is Fe^{3+}. From pH = 1.0 to > 3.5, the primary soluble iron species is $FeSO_4^+$.

This acid mine drainage computation was repeated for total iron at 0.0005, 0.050, and 0.25 molar. The results are similar to Fig. 12.7. The point of precipitation is at lower pH for higher total Fe, but the change is small.

Fe_{total}	pH
0.0005	2.37
0.005	2.32
0.05	2.16
0.25	2.07

In situations with low alkalinity host rock, once the pH drops below 2, high concentrations of soluble iron species are expected to be present. When acid mine drainage of a pH less than 2 discharges to surrounding waters and mixes to produce a pH greater than 2.4, precipitation of ferrihydrite would be expected to maintain the solution pH around 2.5 until the $FeSO_4^+$ was depleted.

These preliminary calculations of acid mine drainage composition have included ferric iron, sulfate, water, and acid as the components. Only one precipitating phase, $Fe(OH)_3$, has been considered. Other potential precipitating phases have thus been assumed to be kinetically limited. For further detailed modeling, and comparison with experimental or field data, it would be important to carefully review formation constants for $Fe(OH)_3$ (amorphous), as well as ferric complexes with sulfate and hydroxide. There is no apparent thermodynamic limit to the oxidation of pyrite in oxygenated waters to produce ferric ion and sulfate.

Potential sources of error for the Fig. 12.7 results may be assessed. For data points around pH = 2, the concentrations of all species are below 0.005 molar and the ionic strength is approximately 0.01. This is a range where the effect of the value for the "salting-out parameter" b is insignificant. Errors due to activity coefficient corrections are thought to be less than 20%. For calculations involving the point of precipitation at Fe_{total} = 0.25 molar, the potential error is higher.

Likewise, for calculations in the range of pH = 0 to 1, potential errors are expected to be relatively higher. The potential errors could be semiquantitatively assessed by reviewing available formation constant data and by considering the potential effects of MINEQL+'s Davies equation default corrections relative to literature values at known ionic strengths. However, such an assessment would not be expected to materially affect the prediction that Fe^{3+} and $FeSO_4^+$ are the predominant iron species below pH = 0.1; other iron species are present at concentrations at least tenfold lower. Thus the accuracy of computed output is sufficient to confirm overall reaction stoichiometry, to identify major iron species, and to identify the point of precipitation of ferrihydrite at low ionic strength.

In a later section, the sequence of reactions that take place when this acidic solution is neutralized by calcium carbonate will be investigated.

Example 2. Seawater

A classic speciation problem is the Garrels and Thompson[41] model of major inorganic species in seawater using formation constants valid for seawater. A version of this model has been presented in Chapter 10. Stumm and Morgan[42] also provide a review of these calculations.

With MINEQL+, we have attempted to reproduce the Stumm and Morgan results for major species using formation constants valid for seawater. Then seawater results were calculated with the Stumm and Morgan elemental analysis together with three additional formation constant data sets and Davies equation ionic strength corrections. Finally, a sensitivity analysis was performed for the ionic complex with the greatest variability between data sets. Armed with these data, one could determine whether to use a more sophisticated model employing PHREEQC or EQ3/6 (using the Harvie, Møller, and Weare[43] database).

Table 12.5 summarizes the input data. Case 1 uses the elemental concentrations in seawater as presented by Stumm and Morgan (Tableau 6.1, p. 308) and their stability constants. Their model presents concentrations for carbonate and bicarbonate separately, treating bicarbonate as a species not linked to carbonate by a formation constant. For comparison, in the next column are listed constants calculated using $I = 0$ data from Smith and Martell, corrected to $I = 0.7$ by the Davies equation. The Stumm and Morgan value[44] for HCO_3^- formation ($pK_2 = 10.12$) was added to the bicarbonate-based formation constants for $NaHCO_3$, $MgHCO_3^+$, $CaHCO_3^+$, and $H_2CO_3^*$ (all dissolved uncharged CO_2 species) to obtain the carbonate-based con-

TABLE 12.5. Species Composition and Formation Constants for Seawater Models

Species	B(OH)$_3$	Ca^{2+}	Cl$^-$	CO$_3^{2-}$	F$^-$	H$^+$	H$_2$O	K$^+$	Mg^{2+}	Na$^+$	SO$_4^{2-}$	Case 1 log K	Davies log K	Case 2 log K°	Case 3 log K°	Case 4 log K°	Case 5 log K°	Total conc. Mol/L
Ion. str.												SW	0.7	0.648	0.648	0.647	0.652	
Davies b												N/A	0.2	0.24	0.24	0.24	0.20	
B(OH)$_3$	1																	4.76e-04
Ca^{2+}		1																0.0102
Cl$^-$			1															0.544
CO$_3^{2-}$				1														2.42e-03
F$^-$					1													7e-05
H$^+$						1												0.0
H$_2$O							1											0.0
K$^+$								1										0.0102
Mg^{2+}									1									0.0532
Na$^+$										1								0.468
SO$_4^{2-}$											1							0.0282
B(OH)$_4^-$	1					−1	1							−9.24	−9.24	−9.24	−9.24	
CaCO$_{3\,aq}$		1		1								1.89	1.89	3.2	3.15	3.15	3.15	
CaF$^+$		1			1							0.62	0.62	1.1		0.94		
CaHCO$_3^+$		1		1		1						10.41	10.33	11.59	11.59	11.33	11.59	
CaOH$^+$		1				−1	1						−13.01	−12.85	−12.7	−12.6	−12.7	
CaSO$_{4\,aq}$		1									1	1.03	1.05	2.31	2.31	2.31	2.31	

H$_2$CO$_{3aq}$	1				2		16.15	15.68	16.68	16.69	16.68	16.69
HCO$_3^-$	1				1		10.12	9.70	10.33	10.33	10.33	10.33
HF$_{aq}$		1			1				3.2		3.17	
HSO$_4^-$					1	1			1.99	1.99	1.99	1.99
KCO$_3^-$			1					0.07		0.70		0.70
KSO$_4^-$			1			1	0.13	0.22	0.96	0.85	0.85	0.85
MgCO$_{3aq}$	1						2.2	1.66	3.4	2.92	2.98	2.92
MgF$^+$		1		1			1.3		1.8		1.82	
MgHCO$_3^+$	1			1	1		10.33	10.08	11.49	11.34	11.4	11.34
MgOH$^+$				1	−1			−11.51	−11.44	−11.2	−11.79	−11.2
MgSO$_{4aq}$				1		1	1.01	0.97	2.36	2.23	2.25	2.23
NaCO$_3^-$	1						0.97	0.64	1.27	1.27	1.27	1.27
NaF$_{aq}$		1									−0.79	
NaHCO$_3$	1				1		9.56	9.13	10.08	10.08	10.08	10.08
NaOH					−1			−14.06		−13.90		−13.90
NaSO$_4^-$						1		0.07	1.06	0.7	0.7	0.7
OH$^-$					−1			−13.68	−14	−14	−14	−14

stants listed in Table 12.5. The most significant differences are in the constants for formation of HCO_3^- and H_2CO_3. Therefore, differences occur between the Case 1 constant and the Davies constants for all species containing HCO_3^-. Other significant differences include $MgCO_3$ and $NaCO_3^-$. The species $NaSO_4^-$ (major) and $MgOH^+$ (minor) and KCO_3^- (minor) were omitted from the Stumm and Morgan model (see Table 12.6).

Within the rounding errors for data entry, MINEQL+ results for Case 1 agree with those published by Stumm and Morgan, except for H_2CO_3, for which the MINEQL+ value is only 81% of the value listed by Stumm and Morgan.[45]

The computations in Cases 2–5 all refer to free species, not total species. For cases 2, 3, and 4, ionic strength corrections were made by a modified Davies equation with $b = 0.24$ instead of 0.2:

$$\log \gamma_z = -0.51z^2 \left(\frac{\sqrt{I}}{1+\sqrt{I}} - 0.24I \right)$$

to conform to the MINEQL+ default for charged species. For Cases 2, 3, and 4 we used $\log \gamma_o = 0$, again to conform to the MINEQL+ default for uncharged species. For case 5, $b = 0.20$ and $\log \gamma_o = 0.1\ I$ were used to conform with Chapters 9–10. Case 2 was computed using formation constant data for $I = 0$ from the compilation of Morel and Hering.[46]

Case 3a is a MINEQL+ calculation from the formation constants given in Table 11.13. These data are nearly identical to those of Case 2, except for a much lower K for $MgCO_3$ and $NaSO_4^-$. Minor differences include the fluoride species, which are in Case 2 but not Case 3, as well as slightly different values of K for $CaCO_3$, KSO_4^-, and the other magnesium complexes. Case 3b is an Excel spreadsheet calculation using the same formation constants in the method described in Chapter 10, and corresponds to the last column of Table 10.19.

Case 4 was obtained from MINEQL+ using formation constant data of the default database, MINTEQA2. It includes the fluoride complexes but not KCO_3^- or NaOH. K for $CaHCO_3^+$ and $MgOH^+$ are lower than for Cases 2 and 3.

Case 5 was computed by the Excel spreadsheet method described in Chapter 10, with a Davies equation containing $b = 0.2$ and $\log \gamma_o = 0$ for uncharged species. The equilibrium constants are identical to Case 3.

Table 12.6 presents a comparison of major inorganic species for these four cases.

The ratio of each of the concentrations to case 4 (Table 12.7) makes the differences easier to spot. Many of the concentrations agree within a few percent for all cases. Others (boldface) show much larger deviations.

The largest differences are for $[MgOH^+]$: Case 3a gives a result that is a factor of 4 larger than Case 4; Cases 2, 3b, and 5 are a factor of 2.5 higher than Case 4; this is due principally to the lower formation constant used in Case 4 (–11.79 vs. –11.20 in Table 13.5).

Another large difference is for [$MgCO_3$] in Case 2, which is 2.3 times larger than Case 4. Again, the source of this appears to be the difference in formation constants (3.40 vs. 2.92 in Table 13.5).

Other than these, there appears to be no systematic pattern. Differences by a factor of 1.5 to 2 can be seen for:

[SO_4^{2-}]—high in Case 1 because $NaSO_4^-$ was omitted and hence lumped with SO_4^{2-}; low in case 2;

[$NaSO_4^-$]—a significant fraction of total sulfate lumped with [SO_4^{2-}] in case 1; high in Case 2;

[HSO_4^-]—low in case 2; high in cases 3b and 5;

[CO_3^{2-}]—low in Case 1;

[$CaHCO_3^+$]—low for Case 1 and high for Cases 2, 3a, 3b, and 5 compared to Case 4;

[$CaOH^+$]—omitted from Case 1; low in Cases 2, 3b, and 5;

[$MgHCO_3^+$]—low in Case 1.

A comparison, which was made in Chapter 10, between the saturation index calculated from the total and free solubility products, is a test of the consistency of these models. Saturation indices are listed in Table 12.8. Case 4 gives the best agreement: Saturation index calculated from $K_{so}(T)$ – SI from $K_{so}(f)$ is –0.02 log units. Cases 3a, 3b, and 5 are somewhat more negative (0.04–0.09), whereas Cases 1 and 2 are significantly more positive (0.20–0.28).

Table 12.9 provides output for Cases 1–4 of cations and anions and their complexes as percentages of the total mass balance. [SO_4^{2-}] constitutes 32–48% of total sulfate. From this assessment it is clear that errors or data discrepancies involving sulfate will be the most significant. Stumm and Morgan did not include $NaSO_4$ in their model; so their [SO_4^{2-}] is much higher than the others.

Free sulfate concentrations are strongly affected by the formation of sodium, magnesium, and calcium ion pairs, which together account for 52–68% of the total sulfate. From Table 12.6, one can see that [SO_4^{2-}] varies from 9 to 18 mM, [$CaSO_4$] varies from 1 to 1.6 mM, [$MgSO_4$] varies from 6 to 8 mM, and [$NaSO_4^-$] varies from 0 to 12 mM. Variations in [SO_4^{2-}] are mainly attributable to variations in $NaSO_4^-$, which are, in turn, related to the $NaSO_4^-$ formation constants, which vary from absent for Case 1, to 1.06 for Case 2, and 0.70 for Cases 3 and 4. For critical processes dependent on the activity of sulfate in seawater, it would be imperative to examine primary data for the $NaSO_4^-$ formation constant more closely.

MINTEQA2

Like MINEQL+, MINTEQA2[47] has been derived from Westall's model,[48] MINEQL. MINTEQA2 and other EPA exposure assessment models and documentation may

TABLE 12.6. Seawater Major Inorganic Species—Comparisons (all concentrations in millimole/kg)

Species log K from	Case 1 Stumm and Morgan	Case 2 Morel and Hering	Case 3a This work	Case 3b This work	Case 4 MINTEQA2	Case 5 This work
Calc'd by b in Davies	MINEQL+	MINEQL+	MINEQL+	Excel	MINEQL+	Excel Ch. 10
		0.24	0.24	0.24	0.24	0.20
log γ	Seawater	−0.148	−0.148	−0.145	−0.148	−0.1582
log γ_o		0	0	0.11	0	0
$B(OH)_3$		0.403	0.403	0.394	0.403	0.390
Ca^{2+}	8.51	9.04	8.65	8.74	8.69	8.789
Cl^-		544	544	544	544	546
CO_3^{2-}	0.0212	0.0418	0.0474	0.0402	0.0471	0.0446
F^-	0.0351	0.0381			0.0369	
H^+	6.31E-06	6.31E-06	6.31E-06	6.31E-06	6.31E-06	6.31E-06
K^+	9.96	9.99	9.97	9.96	9.97	9.98
Mg^{2+}	44.6	46.5	46.4	46.8	46.1	47.04
Na^+	468	456	461	460.0	460	460.4
SO_4^{2-}	18.1	8.93	12.0	12.81	12.50	13.53
$B(OH)_4^-$		0.0729	0.0729	0.0820	0.0729	0.0861
$CaCO_3$	0.014	0.0389	0.0377	0.0294	0.0375	0.0301
CaF^+	1.24E-03	1.10E-03			7.1E-04	

MINTEQA2

Species						
$CaHCO_3^+$	0.029	0.0599	0.0652	0.0706	0.0357	0.0632
$CaOH^+$		1.02E-04	1.38E-04	1.02E-04	1.75E-04	9.85E-05
$CaSO_4$	1.65	1.06	1.45	1.36	1.44	1.317
H_2CO_3	0.0118	0.0102	0.0116	0.0124	0.0115	0.0117
HCO_3^-	1.75	1.43	1.63	1.69	1.62	1.69
HF		1.92E-07			1.73E-07	
HSO_4^-		1.39E-06	1.97E-06	2.70E-06	1.95E-06	2.83E-06
KCO_3^-			6.03E-04	5.24E-04		5.20E-04
KSO_4^-	0.243	0.207	0.227	0.237	0.225	0.223
$MgCO_3$	0.150	0.315	0.118	0.0929	0.134	0.0948
MgF^+	0.0313	0.0284			0.0286	
$MgHCO_3^+$	0.127	0.245	0.197	0.213	0.223	0.190
$MgOH^+$		0.0135	0.0234	0.0173	0.00598	0.0159
$MgSO_4$	8.25	6.13	6.45	6.05	6.65	5.864
$NaCO_3^-$	0.0924	0.0898	0.103	0.0903	0.102	0.0892
NaF					1.39E-03	
$NaHCO_3$	0.227	0.185	0.212	0.193	0.211	0.2119
$NaOH$			9.18E-04	5.69E-04		6.42E-04
$NaSO_4^+$		11.9	7.42	7.75	7.35	7.271
OH^-	1.59E-03	3.16E-03	3.16E-03	2.23E-03	3.16E-03	2.30E-03

TABLE 12.7. Ratios of calculated concentrations to Case 4

Species	Ratio to Case 4				
	Case 1	Case 2	Case 3a	Case 3b	Case 5
$B(OH)_3$		1.00	1.00	0.98	0.97
Ca^{2+}	0.98	1.04	1.00	1.01	1.01
Cl^-		1.00	1.00	1.00	1.00
CO_3^{2-}	**0.45**	0.89	1.01	0.85	0.95
F^-	0.95	1.03			0.00
H^+	1.00	1.00	1.00	1.00	1.00
K^+	1.00	1.00	1.00	1.00	1.00
Mg^{2+}	0.97	1.01	1.01	1.02	1.02
Na^+	1.02	0.99	1.00	1.00	1.00
SO_4^{2-}	**1.45**	**0.71**	0.96	1.02	1.08
$B(OH)_4^-$		1.00	1.00	1.12	1.18
$CaCO_3$	**0.37**	1.03	1.00	**0.78**	0.80
CaF^+	**1.75**	**1.55**			
$CaHCO_3^+$	0.81	**1.68**	**1.83**	**1.98**	**1.77**
$CaOH^+$		**0.58**	**0.79**	**0.58**	**0.56**
$CaSO_4$	1.15	**0.74**	1.01	0.94	0.91
H_2CO_3	1.03	0.89	1.01	1.08	1.02
HCO_3^-	1.08	0.88	1.01	1.04	1.04
HF		1.11			
HSO_4^-		**0.71**	1.01	**1.38**	**1.45**
KCO_3^-					
KSO_4^-	1.08	0.92	1.01	1.05	0.99
$MgCO_3$	1.12	**2.35**	0.88	**0.69**	0.71
MgF^+	1.09	0.99			
$MgHCO_3^+$	**0.57**	1.10	0.88	0.96	0.85
$MgOH^+$		**2.26**	**3.91**	**2.89**	**2.66**
$MgSO4$	**1.24**	0.92	0.97	0.91	0.88
$NaCO_3$	0.91	0.88	1.01	0.89	0.87
NaF					
$NaHCO_3$	1.08	0.88	1.00	0.91	1.00
$NaOH$					
$NaSO_4^+$		**1.62**	1.01	1.05	0.99
OH^-	**0.50**	1.00	1.00	**0.71**	**0.73**

be downloaded[49] at no charge from the Center for Exposure Assessment Modeling (CEAM), EPA Environmental Research Laboratory in Athens, Georgia. CEAM software products are built using FORTRAN-77, assembler, and operating system interface command languages. Source code, as well as compiled code, may be downloaded. Thus one may modify the code. The compiled version runs on Intel-based 80286 microprocessors or later models under the Microsoft DOS operating system.

Data entry for MINTEQA2 is performed with a standalone preprocessor that automates a batch-mode data entry process. Documentation is excellent, but data entry requires a close reading of the instructions. Data entry involves four edit lev-

TABLE 12.8. Some Derived Quantities

	Case 1	Case 2	Case 3a	Case 3b	Case 4	Case 5
Alk	2.690	2.981	2.818	2.775	2.813	2.779
pH	8.358	8.358	8.358	8.348	8.358	8.358
$[Ca]_T$ (mol/L)	1.020E-02	1.020E-02	1.020E-02	1.020E-02	1.020E-02	1.020E-02
$[CO3]_T$ (mol/L)	2.78E-04	4.86E-04	3.07E-04	2.53E-04	3.21E-04	2.59E-04
$pK_{so\ T}$	6.36	6.36	6.36	6.36	6.36	6.36
SI_T	0.81	1.05	0.86	0.77	0.87	0.78
$[Ca]_f$ (mol/L)	8.51E-03	9.04E-03	8.65E-03	8.74E-03	8.69E-03	8.79E-03
$[CO_3]_f$ (mol/L)	2.12E-05	4.18E-05	4.74E-05	4.02E-05	4.69E-05	4.46E-05
pK_{so_f}	7.28	7.28	7.28	7.28	7.28	7.22
SI_f	0.54	0.86	0.89	0.83	0.89	0.87
$SI_T - SI_f$	0.28	0.20	−0.04	−0.05	−0.02	−0.09

TABLE 12.9. Distribution of selected ion pairs as a percent of total for each case (concentrations are in mole/L)

Case 1	free	CO_3	HCO_3	SO_4	OH	total	% free
free		2.12E-05	1.75E-03	1.81E-02			
Na	4.68E-01	9.24E-05	2.27E-04	incl with free		4.68E-01	100%
K	9.96E-03			2.43E-04		1.02E-02	98%
Mg	4.46E-02	1.50E-04	1.27E-04	8.25E-03		5.31E-02	84%
Ca	8.51E-03	1.40E-05	2.90E-05	1.65E-03		1.02E-02	83%
total		2.78E-04	2.13E-03	2.82E-02	0.00E+00		
% free		8%	82%	< 64%			
Case 2	free	CO_3	HCO_3	SO_4	OH	total	% free
free		4.18E-05	1.43E-03	8.93E-03	3.16E-06		
Na	4.56E-01	8.98E-05	1.85E-04	1.19E-02		4.68E-01	97%
K	9.99E-03			2.07E-04		1.02E-02	98%
Mg	4.65E-02	3.15E-04	2.45E-04	6.13E-03	1.35E-05	5.32E-02	87%
Ca	9.04E-03	3.87E-05	5.99E-05	1.06E-03	1.02E-07	1.02E-02	89%
total		4.85E-04	1.92E-03	2.82E-02	1.68E-05		
% free		9%	74%	32%	19%		
Case 3	free	CO_3	HCO_3	SO_4	OH	total	% free
free		4.74E-05	1.63E-03	1.20E-02	3.16E-06		
Na	4.60E-01	1.03E-04	2.12E-04	7.42E-03	9.18E-07	4.68E-01	98%
K	9.97E-03	6.03E-07		2.27E-04		1.02E-02	98%
Mg	4.64E-02	1.18E-04	1.97E-04	6.45E-03	2.34E-05	5.32E-02	87%
Ca	8.65E-03	3.75E-05	6.52E-05	1.45E-03	1.38E-07	1.02E-02	85%
total		3.07E-04	2.10E-03	2.75E-02	2.76E-05		
% free		15%	77%	44%	11%		
Case 4	free	CO_3	HCO_3	SO_4	OH	total	% free
free		4.71E-05	1.62E-03	1.25E-02	3.16E-06		
Na	4.60E-01	1.02E-04	2.11E-04	7.35E-03		4.68E-01	98%
K	9.97E-03			2.25E-04		1.02E-02	98%
Mg	4.62E-02	1.34E-04	2.23E-04	6.65E-03	5.98E-06	5.32E-02	87%
Ca	8.69E-03	3.74E-05	3.57E-05	1.44E-03	1.75E-07	1.02E-02	85%
total		3.21E-04	2.09E-03	2.82E-02	9.32E-06		
% free		15%	78%	44%	34%		
Case 5	free	CO_3	HCO_3	SO_4	OH	total	% free
free		4.46E-05	1.70E-03	1.35E-02	2.30E-06		
Na	4.60E-01	8.92E-05	2.20E-04	7.27E-03	6.42E-07	4.68E-01	98%
K	9.98E-03	5.20E-07		2.23E-04		1.02E-02	98%
Mg	4.70E-02	9.48E-05	1.90E-04	5.86E-03	1.59E-05	5.32E-02	88%
Ca	8.79E-03	3.01E-05	6.32E-05	1.32E-03	9.85E-08	1.02E-02	86%
total		2.59E-04	2.17E-03	2.82E-02	1.89E-05		
% free		17%	78%	48%	12%		

els, and a multiproblem generator. The database structure has rigid requirements for data formatting, and is daunting to edit.

Serkiz et al. have reviewed the MINTEQA2 database. "The thermodynamic database distributed with the widely used equilibrium speciation model MINTEQA2 contains errors in reactions involving organic ligands. Users of this model should check the equilibrium constants (log K) supplied with the model's database against constants from critically reviewed compilations. The comparison of constants is not straightforward and may involve one or more data reduction steps, such as correcting the log K to zero ionic strength and to a temperature of 25°C. More important, it is necessary to reformulate the reaction from the literature in terms of the same components (reactants) used in MINTEQA2 before making the comparison. . . . Users of other speciation models that have incorporated portions of the MINTEQA2 database such as MINEQL+ . . . should also take note of these database problems."[50] Serkiz et al. provide not only a listing of the sources of errors and corrections to the errors, but also a concise summary of procedures for correcting literature values to 25°C and zero ionic strength.

MINTEQA2 is a well-documented program available at no cost from the U.S. Environmental Protection Agency. As with all computer database files of thermodynamic data, the MINTEQA2 formation constants should be checked against critically reviewed compilations. Relative to some of the other implementations of MINEQL, MINTEQA2 may be more difficult to use.

MINTEQA2 has not been tested by the author on either an Intel-based 80286 (or higher) personal computer or a Macintosh computer.

PHREEQC[51]

Parkhurst and his colleagues at the U.S. Geological Survey have focused on the application of computerized calculation methods to problems in geochemistry, including assessments of acid mine drainage in the western United States. Program capabilities are summarized in the user's guide to PHREEQC.[52] Mole balances for speciation calculations can be defined for any valence state or combination of valence states. Distribution of redox elements among their valence states can be based on a specified pe or on any redox couple for which data are available. The concentration of an element may be adjusted to obtain equilibrium with a specified phase (or a specified saturation index or gas partial pressure). In reaction-path calculations, PHREEQC is oriented more toward system equilibrium than just aqueous equilibrium. Essentially, all the moles of each element in the system are distributed among the aqueous phase, pure solid and gas phases, exchange sites, and surface sites to attain system equilibrium. Mole balances on hydrogen and oxygen allow the calculation of pe and the mass of water in the aqueous phase, which allows water-producing or -consuming reactions to be modeled correctly.

Surface complexation reactions may be modeled with the diffuse double-layer model of Dzombak and Morel[53] and the nonelectrostatic model of Davis and Kent.[54] Surface complexation constants from Dzombak and Morel are included in the default databases for the program. The capability to model ion exchange reactions has been added.

The ability to define multiple solutions and assemblages combined with the capability to determine the stable phase assemblage, leads naturally to one-dimensional, advective transport modeling. PHREEQC provides a simple method for simulating the movement of solutions through a column. The initial composition of the aqueous, gas, and solid phases within the column may be specified, and the changes in composition due to advection of an infilling solution and chemical reaction within the column can be modeled.

A completely new capability added to PHREEQC allows calculation of inverse models. Inverse modeling attempts to account for the chemical changes that occur as a water evolves along a flow path. Assuming two water analyses represent starting and ending water compositions along a flow path, inverse modeling is used to calculate the moles of minerals and gases that must enter or leave solution to account for the differences in composition. PHREEQC allows uncertainties in the analytical data to be defined, such that inverse models are constrained to satisfy mole balance for each element and valence state and charge balance for the solution, but only within specified uncertainties.

A major strength of PHREEQC is inclusion of a five-coefficient temperature correction expression for formation constants. Either this expression or specific enthalpy values may be used for temperature corrections.

For ionic strength activity corrections, PHREEQC provides a default Davies equation correction with the b parameter set to 0.3 (not 0.2) for charged species. For uncharged species the default value is $\log \gamma_o = 0.1I$. If the "gamma option" is used, then the extended Debye–Hückel equation from WATEQ[55] is used (compare Chapter 2)

$$\log \gamma_i = -\frac{Az_i^2 \sqrt{I}}{1 + Ba_i^0 \sqrt{I}} + b_i I$$

where A and B are constants dependent only on temperature, a_i^0 is the ion-size parameter, a_i^0 and b_i are ion-specific parameters fitted from mean salt activity-coefficient data, and z_i is the ionic charge of aqueous species i. (The default database, PHREEQC.DAT, includes the "gamma option" for many species. For cases where one wishes to use only the Davies equation, it is necessary to make appropriate corrections on the input data file.)

For the most accurate results in high-ionic-strength waters, the specific interaction approach to thermodynamic properties of aqueous solutions should be used.[56] Such approaches have been incorporated in EQ3/6 (next section).

Computations with PHREEQC are straightforward on either an 80286 Intel-based PC (or later model) or a Power Macintosh running SoftWindows 3.0 emulation of an 80486 processor with an 80387 coprocessor. Input files are set up with a text editor.

Running PHREEQC involves typing a simple command on the DOS command line, such as, PHREEQC *input output database*. The terms *output* and *database* are optional. If *database* is not included on the DOS command line, PHREEQC uses the default database PHREEQC.DAT. If *output* is not included on the command line, PHREEQC writes the output to *input*.OUT.

The default output report contains a recap of the input followed by initial speciation output (before phase transfer equilibrium is established) and final concentrations. For certain types of calculations, tabular outputs to disk are available. Output concentrations are in molal units.

PHREEQC has no graphical output to the monitor screen.

Due to its application by geochemists of the U.S. Geological Survey, there is a body of thermodynamic data and applications that have been tested with PHREEQC. This makes PHREEQC a particularly good choice for modeling geochemical equilibria, including acid mine drainage. As mentioned above, PHREEQC is well suited to calculations over a range of ambient temperatures applicable to geochemistry. Appropriate corrections for ionic strength effects are easily made.

Example. Oxidation of Pyrite

PHREEQC was used to model the oxidation of pyrite followed by neutralization of the resulting acid mine drainage with calcium carbonate. As mentioned above, the rough stoichiometry of these reactions is:

$$FeS_2(s) + 3.5\ O_2 + H_2O \rightleftharpoons Fe^{2+} + 2\ H^+ + 2\ SO_4^{2-}$$

$$Fe^{2+} + H^+ + 0.25\ O_2 + SO_4^{2-} \rightleftharpoons FeSO_4^+ + 0.5\ H_2O$$

$$FeSO_4^+ + 1.5\ CaCO_3(s) + 1.5\ H_2O \rightleftharpoons Fe(OH)_3(s) + 1.5\ CO_2(g) + SO_4^{2-} + 1.5\ Ca^{2+}$$

$$2\ H^+ + CaCO_3(s) \rightleftharpoons Ca^{2+} + CO_2(g) + H_2O$$

Simulation 1 involved addition of oxygen to water in contact with pyrite. Pyrite activity was held constant at a saturation index of zero. In simulation 2, the pyrite was removed and the solution was equilibrated with atmospheric oxygen. In simulation 3, calcium carbonate was added to the point of saturation with amorphous ferric hydroxide. The partial pressure of carbon dioxide was set to 1 atm, and $CaSO_4$ (gypsum) was allowed to precipitate. In simulation 4, further calcium carbonate was added with CO_2 maintained at 1 atm, and the ferric hydroxide activity and gypsum activity were held constant at a saturation index of zero. The run time for these calculations with 44 reaction steps was 67 seconds on a Power Macintosh™ 7200/75 running SoftWindows™ 80486 emulation software.

The input data file is shown in Table 12.10. The value for the solubility product of amorphous ferric hydroxide in the default database: $\log(Fe^{3+}/[H^+]^3) = 4.89$ was adjusted to the value $\log([Fe^{3+}]/[H^+]^3) = 3.2$ or $pK_{so} = 38.8$, to agree with Table 12.4 and with Table 7.3.[57]

TABLE 12.10. PHREEQC Runtime File for Pyrite Oxidation

```
TITLE Pyrite 6, Simulation 1. Add oxygen, equilibrate with pyrite.
SOLUTION 1 PURE WATER
  pH   7.0
  temp 25.0
PHASES
Fe(OH)3(a)
  Fe(OH)3 + 3 H+ = Fe+3 + 3 H2O
  log_k   3.2
EQUILIBRIUM_PHASES 1
   Pyrite     0.0
REACTION 1
   O2   1.0
   0.0   0.0001  0.0002  0.0005 0.001  0.002  0.005  0.01  0.02
   0.05  0.1  0.2  0.5
SAVE solution 1
END
TITLE Simulation 2. Add oxygen in the absence of pyrite.
USE solution 1
EQUILIBRIUM_PHASES 1
   O2(g)      -0.678
SAVE solution 1
END
TITLE Simulation 3. Add CaCO3.
USE solution 1
EQUILIBRIUM_PHASES 1
   O2(g)      -0.678
REACTION 1
CaCO3    1.0
   0.0  0.01  0.02  0.05  0.055  0.058  0.059  0.060  0.061  0.065
SAVE solution 1
END
TITLE Simulation 4. Add CaCO3 and allow Fe(OH)3 to precipitate.
USE solution 1
EQUILIBRIUM_PHASES 1
   O2(g)      -0.678
Fe(OH)3(a)     0.0
REACTION 1
CaCO3    1.0
   0.059  0.060  0.061  0.065  0.07  0.08  0.10  0.12  0.14  0.16
   0.18  0.20  0.21  0.22  0.23  0.24  0.25  0.30  0.40  0.50
SAVE solution 1
END
```

Other formation constants were found in the default database PHREEQC.DAT (Table 12.11). Where available, parameters for the extended Debye–Hückel activity-coefficient expression given above are supplied. If γ is not supplied, the Davies equation with $b = 0.3$ is used to predict the activity coefficients.

Results are summarized in Table 12.12 and Fig. 12.8. In simulation 1, 0.5 moles of oxygen were added stepwise to 1 liter of water. The proton concentration increased almost in direct proportion to the moles of oxygen added, for example, for $\log[O_2] = -4$, $\log[H^+] = -4.25$, etc. Essentially all the iron from oxidized pyrite was transformed into ferrous iron species. Total ferric iron concentrations were lower by seven orders of magnitude. Amorphous ferric hydroxide was highly unsaturated, with saturation index ranging between -8.9 and -11.9.

In simulation 2, pyrite was removed and the solution was equilibrated with atmospheric oxygen. pH increased slightly to 1.41. Ferrous iron species were essentially completely converted to ferric iron species. The saturation index for amorphous ferric hydroxide rose to -2.34.

In simulation 3, calcium carbonate was added stepwise to the solution, which was equilibrated with oxygen. Gypsum was allowed to precipitate and carbon dioxide partial pressure was set to one atmosphere. At 0.25 moles of calcium carbonate, the pH had increased from 1.41 to 2.205, and the saturation index for amorphous ferric hydroxide rose from -2.34 to -0.06. Essentially all the soluble iron remained in the form of ferric iron species.

In simulation 4, addition of calcium carbonate was continued in the presence of atmospheric oxygen, 1 atm of CO_2,[58] and saturation with gypsum. Amorphous ferric hydroxide was allowed to precipitate. The pH increased about 1 unit, from 2.05 to 3.01, as the amount of calcium carbonate was increased from 0.25 to 0.848 moles. This is the "buffer region" corresponding to the conversion of $FeSO_4^+$ (and other ferric iron species) to amorphous ferric hydroxide. Increasing the calcium carbonate by another 0.002 moles to 0.85 produced a pH change of almost 2 units from 3.01 to 5.04 as the total iron decreased from 140 to 0.039 micromolal.

In summary, these reaction-path calculations have shown that solutions in contact with pyrite and oxygen without a source of alkalinity produce acidic solutions high in ferrous iron species. When these solutions flow away from the pyrite to surficial waters, further aeration converts the ferrous iron species to ferric iron species with a slight increase in pH (0.3 units). This increase in pH is a consequence of the stoichiometry of the conversion ferrous iron to ferric ion:

$$Fe^{2+} + 0.25\ O_2 + H^+ = Fe^{3+} + 0.5\ H_2O$$

Addition of alkalinity raises the pH to the point of precipitation for amorphous ferric hydroxide. This precipitation process buffers the pH in the range of 2 to 3. With the addition of sufficient alkalinity, the pH increases to values close to neutrality.

Results plotted in Fig. 12.8 may be compared directly to Fig. 12.1. In constructing the log C versus $-\log[H^+]$ diagram (Fig. 13.7), the pH was taken as an independent variable, and no thought was given to how it was adjusted—either generating the acid through oxidation of pyrite or neutralizing this acidity with a naturally occurring substance such as $CaCO_3$.

On the other hand, the PHREEQC simulations (Table 12.12, Fig. 12.8) were generated with a specific reaction sequence in mind: Oxidation of sufficient pyrite

TABLE 12.11. Pyrite Oxidation Formation Constants from PHREEQC.DAT

```
SOLUTION_SPECIES
H+ = H+
   log_k    0.000
   -gamma   9.0000  0.0000
Ca+2 = Ca+2
   log_k    0.000
   -gamma   5.0000  0.1650
Fe+2 = Fe+2
   log_k    0.000
   -gamma   6.0000  0.0000
CO3-2 = CO3-2
   log_k    0.000
   -gamma   5.4000  0.0000
SO4-2 = SO4-2
   log_k    0.000
   -gamma   5.0000  -0.0400
H2O = OH- + H+
   log_k    -14.000
   -gamma   3.5000  0.0000
2 H2O = O2 + 4 H+ + 4 e-
   log_k    -86.08
2 H+ + 2 e- = H2
   log_k    -3.15
CO3-2 + H+ = HCO3-
   log_k    10.329
   -gamma   5.4000  0.0000
CO3-2 + 2 H+ = CO2 + H2O
   log_k    16.681
CO3-2 + 10 H+ + 8 e- = CH4 + 3 H2O
   log_k    41.071
SO4-2 + H+ = HSO4-
   log_k    1.988
HS- = S-2 + H+
   log_k    -12.918
   -gamma   5.0000  0.0000
SO4-2 + 9 H+ + 8 e- = HS- + 4 H2O
   log_k    33.65
   -gamma   3.5000  0.0000
HS- + H+ = H2S
   log_k    6.994
Ca+2 + H2O = CaOH+ + H+
   log_k    -12.780
Ca+2 + CO3-2 = CaCO3
   log_k    3.224
Ca+2 + CO3-2 + H+ = CaHCO3+
   log_k    11.435
   -gamma   5.4000  0.0000
Ca+2 + SO4-2 = CaSO4
   log_k    2.300
Fe+2 + H2O = FeOH+ + H+
   log_k    -9.500
Fe+2 + CO3-2 = FeCO3
   log_k    4.380
Fe+2 + HCO3- = FeHCO3+
   log_k    2.0
Fe+2 + SO4-2 = FeSO4
   log_k    2.250
Fe+2 + HSO4- = FeHSO4+
   log_k    1.08
Fe+2 + 2HS- = Fe(HS)2
   log_k    8.95
Fe+2 + 3HS- = Fe(HS)3-
   log_k    10.987
Fe+2 = Fe+3 + e-
   log_k    -13.020
   -gamma   9.0000  0.0000
Fe+3 + H2O = FeOH+2 + H+
   log_k    -2.19
Fe+3 + 2 H2O = Fe(OH)2+ + 2 H+
   log_k    -5.67
Fe+3 + 3 H2O = Fe(OH)3 (aq) + 3 H+
   log_k    -12.56
Fe+3 + 4 H2O = Fe(OH)4- + 4 H+
   log_k    -21.6
2 Fe+3 + 2 H2O = Fe2(OH)2+4 + 2 H+
   log_k    -2.95
3 Fe+3 + 4 H2O = Fe3(OH)4+5 + 4 H+
   log_k    -6.3
Fe+3 + SO4-2 = FeSO4+
   log_k    4.04
Fe+3 + HSO4- = FeHSO4+2
   log_k    2.48
Fe+3 + 2 SO4-2 = Fe(SO4)2-
   log_k    5.38
PHASES
Pyrite
   FeS2 + 2 H+ + 2 e- = Fe+2 + 2 HS-
   log_k    -18.479
O2(g)
   O2 = O2
   log_k    -2.960
END
```

TABLE 12.12. Oxidation of Pyrite and Neutralization of Acid Mine Drainage

Moles of Reagent Added	pH	Fe_t^{2+}(mol/kg)	Fe_t^{3+}(mol/kg)	SI $Fe(OH)_3$(a)
Simulation 1. Oxygen addition to water in contact with pyrite				
0 mole O_2	7.01	5.0e-09	4.4e-17	−8.9
0.0001	4.25	2.9e-05	4.4e-15	−9.4
0.001	3.28	2.9e-04	2.7e-14	−10.1
0.01	2.39	2.9e-03	3.4e-12	−10.8
0.1	1.56	2.9e-02	6.9e-10	−11.4
0.5	1.10	1.4e-01	2.2e-08	−11.9
Simulation 2. Equilibration with oxygen in the absence of pyrite				
	1.41	1.9e-09	1.4e-01	−2.34
Simulation 3. Equilibration with oxygen at 0.21 atm and CO_2 at 1.0 atm in the presence of solid $CaSO_4$ (gypsum)				
0.05 mole $CaCO_3$	1.102	3.0e-09	0.56	−2.98
0.1	1.226	3.9e-09	0.56	−2.60
0.15	1.385	5.4e-09	0.56	−2.10
0.2	1.617	9.0e-09	0.56	−1.38
0.22	1.752	1.2e-08	0.56	−0.97
0.24	1.934	1.8e-08	0.56	−0.42
0.25	2.051	2.4e-08	0.56	−0.06
0.255	2.117	2.7e-08	0.56	+0.14
Simulation 4. Equilibration with oxygen at 0.21 atm and CO_2 at 1.0 atm in the presence of $Fe(OH)_3(s)$ and $CaSO_4(s)$				
0.255 mole $CaCO_3$	2.102	2.0e-08	0.39	0.00
0.3	2.108	1.9e-08	0.36	0.00
0.4	2.124	1.7e-08	0.30	0.00
0.5	2.143	1.4e-08	0.23	0.00
0.6	2.167	1.1e-08	0.17	0.00
0.7	2.204	8.3e-09	0.098	0.00
0.8	2.295	4.1e-09	0.031	0.00
0.83	2.403	2.1e-09	0.011	0.00
0.846	2.720	4.1e-10	0.0010	0.00
0.848	3.010	1.1e-10	1.4e-04	0.00
0.849	3.982	1.3e-12	8.0e-07	0.00
0.850	5.042	1.0e-14	3.9e-08	0.00
0.9	6.660	2.5e-17	1.4e-09	0.00
1.0	7.054	9.4e-18	8.2e-10	0.00

to produce a pH 1 solution, with ferric iron species in equilibrium with atmospheric oxygen, led to a solution much higher in iron and sulfate concentrations than shown in Fig. 12.7. The Fig. 12.7 solution compositions at pH values less than 2 could not be generated through oxidation of pyrite. Since total soluble iron was limited by

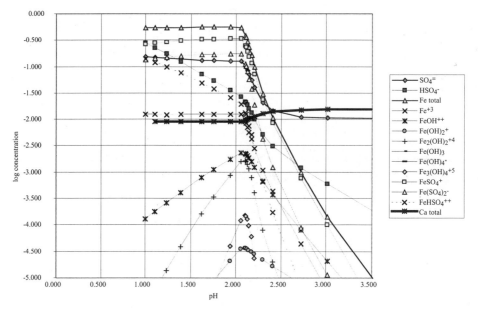

Figure 12.8. Acid mine drainage + $CaCO_3$.

choice, the low-pH solutions could only be created through addition of a strong mineral acid in a laboratory setting.

Addition of calcium carbonate as a neutralizing agent resulted in generation of gaseous carbon dioxide and the precipitation of gypsum. Total iron [principally $FeSO_4^+$ and $Fe(SO_4)_2^-$] is ten times more concentrated in the PHREEQC simulations for pH 1 to 2 than it is in Fig. 12.7. At pH = 2, the PHREEQC simulation (Fig. 12.8) shows 8 iron species present at concentrations above log $C = -5$ whereas Fig. 12.7 shows only 4 species.

Thus log C versus pH diagrams may be useful for assessing the general features of solution composition over a wide pH range, but they are not a substitute for reaction-path modeling of the specific case being examined. Automated computation methods can be of great utility in assessing the constraints inherent in redox, complex formation and precipitation equilibria.

Potential errors in Table 12.12 and Fig. 12.8 output may be assessed. One source of potential error is the use of molar formation constants directly in a computation program programmed in molal units. As noted earlier in this chapter, the potential errors depend on overall solution density, which depends on concentrations, as well as the number of ligands in a complex. For a 3 molal electrolyte, the potential error was shown to range from 8% for β_1 to 27% for β_4. In the present case, the maximum concentration is about 0.5 molal, which is expected to produce an estimated error of less than 20%.

Another potential source of error relates to the deviation of activity coefficient values from the Davies equation prediction with a salting-out parameter value of 0.3 and from the gamma-correction model for those species so modeled. Such nonidealities could change the relative proportions of iron species. At pH 1, the primary iron species are $FeSO_4^+$, $Fe(SO_4)_2^-$, and Fe^{3+}, with $FeSO_4^+$ being the predominant species. Other species are present at much lower concentration. Over the pH range of 1 to about 2.2, the solution ionic strength remains high, and nonideality effects are potentially important. Over the pH range of about 2.2 and higher, the solution ionic strength decreases rapidly, and nonideality effects become less significant. By pH 2.3, the solution ionic strength drops to the vicinity of about 0.03, and nonideality effects are expected to be less important. From Table 12.2, one can see that the 95% confidence limit on log $\gamma_\pm$ is about ±7% or 0.03 log units. At an ionic strength of about 0.03, the expect limit would be about 0.01 log units. Thus the PHREEQC modeling results are sufficiently accurate to identify the important iron species from pH 1 and higher, to identify the buffer region, and to identify the precipitation region within about 0.01 pH units.

EQ3/6

The EQ3/6 software package originated in the mid-1970s. It was originally developed by T. J. Wolery at Northwestern University to model rock–water interactions in hydrothermal and geothermal systems, particularly seawater–basalt interactions in mid-ocean ridge hydrothermal systems.[59] It was brought to the Department of Energy's Lawrence Livermore Laboratory in 1978 by the original author,[60] and there underwent extensive development for modeling geologic disposal of high-level nuclear waste. Technical description of the EQ3/6 model and its specialized data sets will be found in Table 12.3 and the text below.

A primary use for the model and data sets has been the evaluation of disposal options for high-level radioactive waste. It has been used to model the leaching of radionuclides from spent fuel and glass wasteforms, and the rock–water interactions that could take place in the local hydrothermal environment that might be created in the vicinity of an underground waste repository. As a result, the model provides for the modeling of reactions over a temperature range from 0 to 300°C, a pressure range of 1 to approximately 100 atm, and an ionic strength range from 0 to greater than 10 molal.

Many of the refinements that have been incorporated into the code for use in nuclear waste applications are readily adaptable to applications in other environmental areas, such as the evaluation of acid mine waters, low-level radioactive waste, and chemical waste. Some of the more general processes that can be modeled using EQ3/6 include mineral dissolution, mineral precipitation, waste leaching, and incorporation of heavy metals and other inorganic toxic components into secondary minerals. It can be applied to waste treatment, assessment of contaminated sites, assessment of the effects of natural remediation processes, and the design and assessment of engineered remediation processes.

The software allows the user to create and evaluate models that include the effects of chemical equilibrium, disequilibrium, and kinetics. It can handle both dilute waters and high-ionic-strength brines. The database is the most comprehensive of its kind, and includes data for both inorganic and organic species.

The most recent version of this computation program is a set of FORTRAN 77 computer code and associated databases described in a set of four reports.[31] The major components of the computation program are EQPT, EQ3NR, and EQ6.

EQPT

EQPT is the data file preprocessor used to prepare input files. Data input is addressed in *The EQPT User's* Guide. There are five distinct data files, denoted by the suffices **com**, **sup**, **nea**, **hmw**, and **pit**. These are formatted ASCII files are called data0 files. EQPT converts these ASCII files to primitive files, called data1 files, which are used by EQ3NR and EQ6. The user of EQ3NR or EQ6 must select which of the five data files is most appropriate to a given problem. The com, sup, and nea data files are specific to a general extended Debye–Hückel formalism and can be used with either the Davies equation[61] or the B-dot equation.[62] These equations are only valid in relatively dilute solutions. The **hmw** and **pit** data files are specific to the formalism proposed by Pitzer[63] and can be used to model solutions extending to high concentrations. However, the scope of chemical components is smaller. The temperature limits on the data files also vary, from 25°C only to 0–300°C.

The **com** (composite) data file encompasses a broad range of chemical elements and species. It includes the data found on the sup and nea data files, as well as some data found in the hmw data files and other data that do not appear in any of the other data files. Some of these data are estimate based on correlations or extrapolations (as to higher temperature), and are not tied directly to experimental measurements. This melange of data offers less assurance of internal consistency. However, it does allow for modeling aqueous solutions with a high degree of compositional complexity, such as the fluids expected to be found in and about a facility for the geologic disposal of industrial or nuclear waste.

The **sup** data file has a high level of internal consistency among the standard-state thermodynamic data. This data file covers a wide range of chemical elements and species of interest in the study of rock–water interactions (e.g., components that make up the major rock-forming and ore-forming minerals). It also includes a large number of organic species, mostly of small carbon number.

The **nea** data file is something of a specialty item. Its strongest point is a thorough representation of the thermodynamics of uranium species.

The **hmw** data file has the highest degree of internal consistency of any of the five data files, including mutual consistency of activity coefficient data and standard-state thermodynamic data. It can be applied to dilute waters or concentrated brines. However, it only treats the set of major cations and anions present in seawater, including carbonate and bicarbonate. The geochemically important components aluminum and silica are not included. Also, this data file is limited to a temperature of 25°C.

The **pit** data file can also be applied to concentrated brine. It covers a larger set of components, but these mostly involve other cations and anions of strong electrolytes (e.g., lithium and bromide). This data file nominally covers the temperature range of 0–100°C. However, it represents a melange of data, not a carefully crafted internally consistent set.

EQPT in its present form has no input file and no user options. The purpose of the EQPT user's manual is to provide information concerning the data file structure and its processing that might be useful to users who modify the original data files or make up data files of their own. The manual provides a 16-page section that describes the Davies equation, the B-dot equation, Pitzer's equations, and activity coefficients of solid solution components.

EQ3NR

EQ3NR is the speciation solubility code described in *The EQ3NR Theoretical Manual and User's Guide*. Given total inputs for basis species, EQ3NR calculates the distribution of free ions and complexes. It calculates saturation indices, but there is no provision for precipitation. Output from an EQ3NR run is required to initiate an E6 calculation series.

EQ6

EQ6 is the reaction path code described in *The EQ6 Theoretical Manual and User's Guide*. It can be used to perform titration calculations. However, its primary strength is computation of reactions paths involving multiple components and multiple potential precipitating phases of either pure minerals or of solid solutions.

To define a reaction path problem to run on EQ6, the user supplies a thermodynamic model of an aqueous solution, obtained by running EQ3NR, chooses a set of irreversible reactions, provides parameters to define the rates of these processes, and chooses from among the various model options that are available. The code does the rest.

A good example of the reaction path capabilities of EQ6 is the finding of precipitates from multiply supersaturated seawater described in section 6.2 of the EQ6 manual. This example is based on the seawater test case of Nordstrom et al.,[64] and is similar to the models we have presented in Chapter 10 and in Tables 12.3–12.7. The model includes several trace elements, the most significant of which for the present example are iron, aluminum, and silica. The supporting data file is the com file.

Basis species concentrations are input for H_2O, Al^{3+}, $B(OH)_3(aq)$, Br^-, Ca^{2+}, Cl^-, F^-, Fe^{2+}, H^+, HCO_3^-, HPO_4^{2-}, I^-, K^+, Mg^{2+}, Na^+, $SiO_2(aq)$, SO_4^{2-}, Sr^{2+}, and $O_2(g)$. The initial EQ6 calculation for the aqueous phase finds the solution to be supersaturated with 27 pure minerals and 7 solid solutions. EQ6 then begins precipitating minerals one at a time, choosing them according to an algorithm based on weighted values of saturation indices (with the weighting designed to remove bias associated with phases with large molecular formulas such as clays). In order, it first chooses ordered dolomite, then hematite, fluorapatite, muscovite, and the solid solution dioctahedral smectite (a complex iron aluminosilicate). EQ6 determines that hematite should be dropped from the phase assemblage. When the smectite is added to the phase assemblage, it in effect takes the available iron away from the hematite, forcing it to disappear from the system. The final phase assemblage consists of the aqueous solution and four minerals.

The results obtained do not reflect everyday experience on the scale of hours or days. Input fluxes and reaction rates are such that surface seawater is always supersaturated with respect to many minerals. If data on reaction kinetics were available for these reactions, a different final phase assemblage might be obtained. In addition, one might be able to assess whether these reactions do proceed on a geologic time scale and whether modeling results are consistent with sea floor sediment composition.

EQ3/6 is the computer program of choice for modeling complex geological (and waste disposal) phenomena with large numbers of components and potential solid phases, especially over a range of temperatures and ionic strengths. EQ3/6 has the uncommon capability to model precipitation and dissolution of solid solutions. It is a serious computational tool that has been applied to complex problems such as: the origin of the copper ore body at Butte, Montana, seawater–basalt interaction at mid-ocean ridges at both high and low temperature, geochemistry of the Archean hydrosphere and the formation of banded iron formations, exploration of the role of chemical kinetics in geochemical processes, the dissolution of spent fuel nuclear waste form, the kinetics of leaching of borosilicate glass nuclear waste forms, and alteration of cement-based grouts for nuclear waste repositories.

EQ3/6 is probably not the computer program of choice for dilute solutions with few components and few potential solid phases. In addition, EQ3/6 (Version 7.0) has not been programmed to model sorption phenomena such as sorption of trace metals to ferric hydroxide precipitate.

EQ3/6 has not been tested by the author on either an Intel-based 80286 personal computer or a Macintosh computer, but it was used by workers at the Sandia National Laboratories[66] to obtain the activity of HCl in brines. Those results were compared with experimental Harned Cell values of pHCl at $I = 4$, 5, and 6 in Table 11.2 (p. 469). They agreed to within 0.004 to 0.025 log units.

CONCLUSION

Automated computation programs make it possible to predict the fate of chemical elements in the environment and in treatment systems. The convenience of use for these models is greatly affected by the program design and by the availability of data appropriate to the program. As the programs become more capable, and thus more complex, it is important to assess the accuracy of program output by comparison with experimentally determined parameters such as chemical analyses of natural systems, compilations of mean activity coefficients, solubility determinations, etc.

As little as a decade ago, one would not have predicted that it would be possible to: perform an Internet search for computer programs capable of geochemical modeling, download the computation program, download the model, download a default database, order the NIST database,[14] and perform complex computations on a home computer. It is likely that there will be a similar degree of change in succeeding decades.

One can suggest means for obtaining updates on computer programs and databases. Internet searches are likely to be fruitful if one searches for a computer program by name. Science Citation Index searches for authors of computer programs and of technical articles on geochemical or chemical equilibrium modeling, and compilers of databases (e.g., Smith and Martell) will almost certainly lead the way to new developments. Queries to U.S. government agencies such as the U.S. Geological Survey, the Department of Energy, and the Environmental Protection Agency may be helpful if one can identify the appropriate office. These searches can be followed up by direct contacts with appropriate authors.

PROBLEMS

1. Calculate the solubility of mercuric oxide in 0, 0.01, 0.1, 1, and 5 molal NaCl. Use data from Smith and Martell[14] for mercuric chloride complexes. Compare the results of a Davies equation approach and zero-ionic strength formation constants with the results using formation constant values appropriate to the ionic strength.

2. Review the activity coefficient data for Na_2SO_4 determined by Robinson and Stokes,[11] which are:

m	$\gamma_\pm$
0.1	0.452
0.2	0.371
0.3	0.325
0.4	0.294
0.5	0.270
0.6	0.252
0.7	0.237
0.8	0.225
0.9	0.213
1.0	0.204

Using a Davies equation approach, model the Na_2SO_4 solutions assuming no sodium sulfate complexes, then assuming complex formation constant values from Smith and Martell[14] using the zero-ionic strength formation constant, and finally, assuming complex formation constant values from Smith and Martell[14] with extrapolated complex formation constants appropriate to each ionic strength listed above. The formation constants for 25°C are:

I (molar)	$\log \beta_1$	±
0.0	0.73	(±9)
0.1	(0.40)	(±20)
0.5	(0.40)	(±0)
0.7	(0.30)	(±4)

Compare the model output with Robinson and Stokes[11] data. Assume that ion-interaction effects account for the deviations from the model output based on the use of formation constants appropriate to each ionic strength. For models of heavy metals in high sulfate brackish waters, assess the potential error from "ion interaction effects" and from deviations from a simple Davies equation approach with a fixed salting-out parameter.

NOTES

1. William R. Smith, and Ronald W. Missen, *Chemical Reaction Equilibrium Analysis: Theory and Algorithms*, Krieger Publishing Company, Malabar, Florida, 1991.
2. R. L. Bassett, and Daniel C. Melchior, "Chemical modeling of aqueous systems: An overview", in *Chemical Modeling of Aqueous Systems II*, ACS Symposium Series 416, Daniel C. Melchior and R. L. Bassett, Eds., American Chemical Society, Washington, DC, 1990.
3. R. M. Garrels, and M. E. Thompson, *Am. J. Sci.*, **260**:57–66, 1962.
4. W. B. White, S. M. Johnson, and G. B. Dantzig, *J. Chem. Phys.* **28**:751, 1958.
5. L. Edmond Eary and Janet A. Schramke, "Rates of inorganic oxidation reactions involving dissolved oxygen," in *Chemical Modeling of Aqueous Systems II, ACS Symposium Series 416*, Daniel C. Melchior and R. L. Bassett, Eds., American Chemical Society, Washington, DC, 1990.
6. H. M. May, D. G. Kinniburgh, P. A. Helmke, and M. L. Jackson, *Geochim. Cosmochim. Acta*, **50**:1667–77, 1986.
7. Darrell Kirk Nordstrom, L. Niel Plummer, Donald Langmuir, Eurybiades Busenberg, Howard M. May, Blair F. Jones, and David L. Parkhurst, "Revised chemical equilibrium data for major water-mineral reactions and their limitations," in ACS Symposium Series 416, *Chemical Modeling of Aqueous Systems II*, Daniel C. Melchior and R. L. Bassett, Eds., 1990.
8. Donald L. Macalady, Donald Langmuir, Timothy Grundl, and Ala Elzerman, "Use of model-generated Fe^{3+} ion activities to compute E_h and ferric oxyhydroxide solubilities in anaerobic systems," in *Chemical Modeling of Aqueous Systems II*, ACS Symposium Series 416, Daniel C. Melchior and R. L. Bassett, Eds., American Chemical Society, Washington, DC, 1990.
9. P. J. Murphy, A. M. Posner, and J. P. Quirk, *J. Colloid Interface Sci.* **56**:312–19, 1976.
10. A. H. Truesdell and B. F. Jones, *J. Res. U.S. Geol. Survey*, **2**:233–48, 1974.
11. R. A. Robinson and R. H. Stokes, *Electrolyte Solutions, the Measurement and Interpretation of Conductance, Chemical Potential and Diffusion in Solutions of Simple Electrolytes*, 2nd ed. (revised), Butterworths, London, 1968.

12. Kenneth S. Pitzer, "Ion interaction approach: theory and data correlation," in *Activity Coefficients in Electrolyte Solutions*, 2nd ed., Kenneth S. Pitzer, CRC Press, Boca Raton FL, 1991. Table 2 (pp. 100–101) gives parameter values for 76 binary electrolytes.

13. Lars Gunnar Sillen and Arthur E. Martell, *Stability Constants of Metal–Ion Complexes*, Special Publication No. 17, The Chemical Society, Burlington House, London, England, 1964; Lars Gunnar Sillen and Arthur E. Martell, *Stability Constants of Metal–Ion Complexes, Supplement No. 1*, Special Publication No. 25, The Chemical Society, Burlington House, London, England, 1971.

14. R. M. Smith and Arthur E. Martell, *NIST Critically Selected Stability Constants of Metal Complexes Database*, Version 3.0, National Institute of Standards and Technology, Gaithersburg, Maryland, 1997.

15. Ya. I. Turyan and N. G. Chebotar, *Russ. J. Inorg. Chem.* **4**:273, 599, 1959.

16. V. E. Mironov, *Russ. J. Inorg. Chem.* **6**:205, 405, 1961; V. E. Mironov, V. A. Fedorov, and V. A. Nazarov, *Russ. J. Inorg. Chem.* **8**:1102, 2109, 1963; V. E. Mironov, F. Ya. Kulba, V. A. Fedorov, and O.B. Tikhomirov, *Russ. J. Inorg. Chem.* **8**:1328, 2536, 1963; V. E. Mironov, F. Ya. Kulba, and V. A. Fedorov, *Russ. J. Inorg. Chem.* **9**:888, 1641, 1964; V. E. Mironov, F. Ya. Kulba, V. A. Fedorov, and A. V. Federova, *Russ. J. Inorg. Chem.* **9**:1155, 2138, 1964; V. E. Mironov, F. Ya. Kulba, and V. A. Fedorov, *Russ. J. Inorg. Chem.* **10**:495, 914, 1965.

17. Rasmunas J. Motekaitis, personal communication, Texas A&M University, College Station, Texas, 1997.

18. F. M. M. Morel and J. G. Hering, *Principles and Applications of Aquatic Chemistry*, Wiley-Interscience, New York, 1993.

19. J. A. Davis and D. B. Kent, "Surface complexation modeling in aqueous geochemistry," in M. F. Hochella and A. F. White, Eds., *Mineral–Water Interface Geochemistry*, Mineralogical Society of America, *Reviews in Mineralogy*, **23**, Chapter 5, pp. 177–260, 1990; D. A. Dzombak and F. M. M. Morel, *Surface Complexation Modeling—Hydrous Ferric Oxide*, John Wiley, New York, 1990; G. L. Gaines and H. C. Thomas, "Adsorption studies on clay minerals. II. A formulation of the thermodynamics of exchange adsorption," *J. Chem., Phys.* **21**:714–18, 1953.

20. A. E. Martell and R. M. Smith, *Critical Stability Constants*, Vol. 1, *Amino Acids*, Plenum, New York, 1974; R. M. Smith and A. E. Martell, Vol. 2, *Amines*, Plenum, New York, 1975; A. E. Martell and R. M. Smith, Vol. 3, *Other Organic Ligands*, Plenum, New York, 1977; R. M. Smith and A.E . Martell, Vol. 4, *Inorganic Ligands*, Plenum, New York, 1976; A. E. Martell and R. M. Smith, Vol. 5, *First Supplement*, 1982; R. M. Smith and A. E. Martell, Vol. 6, *Second Supplement*, 1989.

21. R. M. Smith and Arthur E. Martell, *NIST Critically Selected Stability Constants of Metal Complexes Database*, Version 3.0, National Institute of Standards and Technology, Gaithersburg, Maryland, 1997.

22. Darrell Kirk Nordstrom, L. Niel Plummer, Donald Langmuir, Eurybiades Busenberg, Howard M. May, Blair F. Jones, and David L. Parkhurst, *Revised Chemical Equilibrium Data for Major Water–Mineral Reactions and Their Limitations*, in ACS Symposium Series 416, Chemical Modeling of Aqueous Systems II, Daniel C. Melchior and R. L. Bassett, Eds., 1990.

23. F. M. M. Morel and J. G. Hering, *Principles and Applications of Aquatic Chemistry*, Wiley-Interscience, New York, 1993.

24. Werner Stumm and James J. Morgan, *Aquatic Chemistry—Chemical Equilibria and Rates in Natural Waters*, 3rd ed., Wiley-Interscience, New York, 1996.
25. R. M. Smith. and A. E. Martell, *Critical Stability Constants*, Vol. 2, *Amines*, Plenum, New York, 1975; R. M. Smith and A. E. Martell, Vol. 4, *Inorganic Ligands*, Plenum, New York, 1976.
26. A. E. Martell and R. M. Smith, *Critical Stability Constants*, Vol. 1, *Amino Acids*, Plenum, New York, 1974; A. E. Martell and R. M. Smith, Vol. 3, *Other Organic Ligands*, Plenum, New York, 1977.
27. M. Whitfield, *Limnol. Oceanogr.*, **19**:235, 1974.
28. C. F. Baes, Jr., and R. E. Mesmer, *The Hydrolysis of Cations*, Wiley, New York, 1976.
29. W. G. Sunda and P. J. Hanson, in *Chemical Modeling in Aqueous Systems*, E. A. Jenne, Ed., ACS Symposium Series 93, American Chemical Society, Washington, DC, 1979.
30. J. Sainte Marie, A. E. Torma, and A. O. Gubeli, *Can J. Chem.*, **42**:662, 1964.
31. J. J. Morgan, in *Principles and Application in Water Chemistry*, S. D. Faust and J. V. Hunter, Eds., Wiley, New York, 1967.
32. Francois M. M. Morel, personal communication to David Cogley, 1997.
33. Produced by Environmental Research Software, Hallowell, ME 04347, (207)622-3340, ersoftwr@agate.net, http://www.agate.net/~ersoftwr/mineql.html.
34. F. M. M. Morel. and J. J. Morgan, *Environ. Sci. Technol.* **6**:58–67, (1972).
35. J. C. Westall, J. L. Zachary, and F. M. M. Morel, *MINEQL, a Computer Program for the Calculation of Chemical Equilibrium Composition of Aqueous Systems*, Tech. Note 18, Dept. Civil Eng., Mass. Inst. Tech. Cambridge, MA, 1976, 91 pp.
36. F. M. M. Morel, *Principles of Aquatic Chemistry*, John Wiley & Sons, Inc., New York, 1982; F. M. M Morel and J. Hering, *Principles and Applications of Aquatic Chemistry*, John Wiley & Sons, Inc., New York, 1993.
37. D. S. Brown and J. D. Allison, MINTEQA1, An Equilibrium Metal Speciation Model: User's Manual, U.S. Environmental Protection Agency, Athens, GA, EPA/600/3-87/012, 1987.
38. S. M. Serkiz, J. D. Allison, E. M. Perdue, H. E. Allen, and D. S. Brown, "Correcting errors in the thermodynamic database for the equilibrium speciation model MINTEQA2", *Water Resources* **30**:1930–33, 1996, is summarized under the description of the MINTEQA2 program.
39. The MINEQL+ default database (MINTEQA2) constant for the reaction

$$Fe^{3+} + 3H_2O = Fe(OH)_3 + 3\,H^+$$

is log K = –4.91, corresponding to pK_{so} = 3 pK_w + log K = 37.1. This is 1.7 log units lower (50 times more soluble) than pK_{so} = 38.8 ± 0.2 quoted in Table 7.3, which was obtained from Smith and Martell [*Critical Stability Constants*, Vol. 4, 1976; Vols. 5 and 6 do not give a solubility product for Fe(OH)$_3$]. Crystalline precipitates have much higher pK_{so} (lower solubility): 41.5 for FeOOH(α), 42.7 for Fe$_2$O$_3$(α). For amorphous Fe(OH)$_3$, 37.1 instead of 38.8 is not unreasonable: other values at I = 0 are 36.35 (Y. Oka, *J. Chem. Soc. Japan* **59**:971, 1938; 36.85 (N. N. Mironov and A. I. Odnosevtsev, *Zhur. neorg. Khim.* **2**:2202, 1957, and 38.6 for "fresh" (R. F. Platford, *Can. J. Chem.* **42**:181, 1964).
40. Using the MINEQL+ output module, Lotus files were created, and stored on the emulated C: drive. Transfer to a volume shared with the Macintosh operating system required

a reconfiguration of the AUTOEXEC.BAT and CONFIG.SYS files and a reboot. Under the Macintosh operating system, these files were then opened with Microsoft Excel.

41. R. M. Garrels and M.E. Thompson, *Am. J. Sci.* **260**:57–66, 1962. See also M. Whitfield, *Mar. Chem.* **1**:251–266, 1973 and D. R. Turner, M. Whitfield, and A. G. Dickson *Geochim et Cosmochim Acta* **45**:855–881, 1981.

42. W. Stumm and J. J. Morgan, *Aquatic Chemistry—Chemical Equilibria and Rates in Natural Waters*, 3rd Ed., Wiley, NY, 1996.

43. C. E. Harvie, N. Moller, and J. H. Weare, The prediction of mineral solubilities in natural waters: The Na–K–Mg–CaH–Cl–SO_4–OH–HCO_3–CO_3–CO_2–H_2O system to high ionic strengths at 25°C, *Geochim. Cosmochim. Acta* **48**:723–51, 1984.

44. log K for formation of bicarbonate from H^+ and carbonate, computed from Stumm and Morgan tableau 6.1, was $\log[HCO_3^-] - \log[H^+] - \log[CO_3^{2-}] = 10.117$, which apparently is intended to apply at $I = 0.7$, but does not agree with any of the usual values, and appears to reflect an inconsistency between their assumed $[HCO_3^-]_f$, $[CO_3^{2-}]_f$, and pH values. Compare with:

$pK_{a2}^\circ(f) = 10.329$ at $I = 0$ (Table 9.1)

$pK_{a2}(f, [H^+]) = 9.70$ at $I = 0.7$ (Davies equation)

$pK_{a2}(f, pH) = 9.86$ at $I = 0.7$ (Table 11.13)

pK_{a2} (SWS, $[H^+]_T$) $= 8.945$ at $I = 0.7$ with ion pairs included (Table 11.9)

Cases 2, 3, and 4, presented in Table 12.6 all use $pK_{a2}^\circ = 10.33$ and apply corrections both for activity coefficients (Davies eq.) and ion pairs.

45. The source of this discrepancy appears to be in the carbonate constants, as mentioned above; in particular that the H_2CO_3 formation constant listed in Tableau 6.6 applies to total bicarbonate, whereas all other formation constants listed in Tableau 6.6 apply to free ligands.

46. Appendix 6.1, of Stumm and Morgan, compiled by F. M. M. Morel and J. G. Hering, *Principles and Applications of Aquatic Chemistry*, Wiley, New York, 1993.

47. J. D. Allison, D. S. Brown, and K. J. Novo-Gradac, *MINTEQA2/PRODEFA2, a Geochemical Assessment Model for Environmental Systems: Version 3.0 User's Manual*, EPA/600/3-91/021, March, 1991.

48. J. C. Westall, J. L. Zachary, and F. M. M. Morel, *MINEQL, A Computer Program for the Calculation of Chemical Equilibrium Composition of Aqueous Systems*, Technical Note 18, Ralph M. Parsons Laboratory for Water Resources and Environmental Engineering, Department of Civil Engineering, Massachusettts Institute of Technology, Cambridge, MA, 1976.

49. Via File Transfer Protocol (FTP) from ftp://ftp.epa.gov/epa_ceam/wwwhtml/software.htm. This FTP site is maintained under the direction of the Center for Exposure Assessment Modeling (CEAM), U.S. Environmental Protection Agency, Environmental Research Laboratory, 960 College Station Road, Athens, Georgia 30605-2720.

50. S. M. Serkiz, J. D. Allison, E. M. Perdue, H. E. Allen, and D. S. Brown, *Correcting Errors in the Thermodynamic Database for the Equilibrium Speciation Model MINTEQA2*, Water Resources, Vol. 30, No. 8, pp. 1930–33, 1996.

51. PHREEQE stands for PH-REdox-EQuilibrium-Equations and is pronouonced "freak."

PHREEQC is written in the C language; PHRQPITZ uses the Pitzer equations to calculate activity coefficients. See L. N. Plummer, D. L. Parkhurst, G. W. Fleming, and S. A. Dunkle, ,"A computer program incorporating Pitzer's equations for calculation of geochemical reactions in brines," U.S. Geological Survey Water-Resources Investigations Report 88–4153, 1988.

52. David L. Parkhurst, The User's Guide to PHREEQC—A Computer Program for Speciation, Reaction-Path, Advective-Transport, and Inverse Geochemical Calculations, U.S. Geological Survey Water-Resources Investigations Report 95-4227, 1992. This guide, the computer program, and data files may be downloaded via the Internet by file transfer protocol (FTP) from brrcrftp.cr.usgs.gov/geochem/phreeqc under user name *anonymous* with the password set equal to one's email address. There is no charge for downloading documentation or software. Alternatively, the documentation and DOS or Unix versions of the software may be ordered from: U.S. Geological Survey NWIS Program Office 437, National Center, Reston, VA 22092, (703) 648-5695.

53. D. A. Dzombak and F. M. M. Morel, *Surface Complexation Modeling—Hydrous Ferric Oxide*, John Wiley, New York, 1990.

54. J. A. Davis and D. B. Kent, "Surface complexation modeling in aqueous geochemistry," in M. F. Hochella and A. F. White, Eds., *Mineral–Water Interface Geochemistry*, Mineralogical Society of America, Reviews in Mineralogy, Vol. 23, Chap. 5, pp. 177–260, 1990.

55. A. H. Truesdell and B.F. Jones, WATEQ, A computer program for calculating chemical equilibria of natural waters, Journal of Research, U.S. Geological Survey, Vol. 2, pp. 233–74, 1974.

56. K. S. Pitzer, "Ionic interaction approach: Theory and data correlation" Chapter 3, K. S. Pitzer, Ed. *Activity Coefficients in Electrolyte Solutions*, 2nd ed. CRC Press, 1991. See also K. S .Pitzer, "Theory—Ion interaction approach," in R. M. Pytkowicz, ed., *Activity Coefficients in Electrolyte Solutions*, 1st ed., Vol. 1, CRC Press, Boca Raton, Florida, 1979, pp. 157–208. C. E. Harvie and J. H. Weare, "The prediction of mineral solubilities in natural waters: The Na–K–Mg–Ca–Cl–SO_4–H_2O system from zero to high concentration at 25°C," *Geochim. Cosmochim. Acta* **44**:981–97, 1980. C. E. Harvie, N. Moller and J. H. Weare, "The prediction of mineral solubilities in natural waters: The Na–K–Mg–Ca–H–Cl–SO_4–OH–HCO_3–CO_3–CO_2–H_2O system to high ionic strengths at 25°C," *Geochim. Cosmochim. Acta*, **48**:723–51, 1984. L. N. Plummer, D. L. Parkhurst, G. W. Fleming, and S. A. Dunkle, "A computer program incorporating Pitzer's equations for calculation of geochemical reactions in brines," U.S. Geological Survey Water-Resources Investigations Report 88–4153, 1988.

57. See discussion in footnote under MINEQL+ example of Acid Mine Drainage. Smith and Martell, Vol. 4, Plenum Press, 1976. The MINEQL+ default database (MINTEQA2) constant corresponds to pK_{so} = 37.1. This is 1.7 log units lower (50 times more soluble) than pK_{so} = 38.8 ± 0.2 from Smith and Martell . For amorphous $Fe(OH)_3$, 37.1 instead of 38.8 is not unreasonable: other values at I = 0 are 36.35 (Y. Oka, *J. Chem. Soc. Japan* **59**:971, 1938; 36.85 (N. N. Mironov and A. I. Odnosevtsev, *Zhur. neorg. Khim.* **2**:2202, 1957, and 38.6 for "fresh" (R. F. Platford, *Can. J. Chem.* **42**:181, 1964). See also H. C. Helgeson, J. M. Delany, H. W. Nesbitt, and D. K. Bird, *Am. J. Sci.*, 278, 1978.

58. If the atmospheric oxygen was present as air, 1 atm of 21% O_2 and 79% N_2, the total pressure would be 2 atm. If the gas phase consisted of 0.21 atm O_2 and 1.0 atm CO_2, the total pressure would be 1.2 atm. The total pressure and the presence or absence of nitrogen has little effect on the results.

59. T. J. Wolery, *Some Chemical Aspects of Hydrothermal Processes at Mid-Ocean Ridges—A Theoretical Study*, Ph.D. thesis, Northwestern University, Evanston, Illinois, 1978.
60. T. J. Wolery, *Calculation of Chemical Equilibrium between Aqueous Solution and Minerals: The EQ3/6 Software Package*, Livermore, CA: Lawrence Livermore National Laboratory, Report UCRL-52658, 1979.
61. C. W. Davies, *Ion Association*, Butterworths, London, 1962. See discussion at the beginning of this chapter.
62. H. C. Helgeson, Thermodynamics of hydrothermal systems at elevated temperatures and pressures, *Am. J. Sci.* **267**:729–804, 1969.
63. K. S. Pitzer, Thermodynamics of electrolytes—*I*. Theoretical basis and general equations, *J. Phys. Chem.* **77**:268–77, 1973; K. S. Pitzer, Thermodynamics of electrolytes—V. Effects of higher-order electrostatic terms, *J. Sol. Chem.* **4**:249–65, 1975.
64. D. K. Nordstrom, et al., "A comparison of computerized chemical models for equilibrium calculations in aqueous systems," in Jenne, E. A., Ed., *Chemical Modeling in Aqueous Systems*, ACS Symposium Series, Vol. 93, American Chemical Society, Washington, D.C., pp. 857–92, 1979.
65. G. Barker, L. H. Brush, S. Free, L. R. Montano, and W. G. Yelton, unpublished data. Sandia National Laboratories, Albuquerque, NM, 1991–1995.

INDEX

A_L for EDTA, 298
A_M for magnesium-EDTA, 297
Abbott Laboratories, 456
accuracy, 503
accuracy, computational, 488
acetate, 4, 7, 410–411
acetic acid, 4, 7, 99–107
acetic acid dissociation constant, extrapolation, 59
acetic acid dissociation equilibrium, 10, 11
acetic acid, glacial as solvent, 148
acetic acid, ionic strength dependence of pKa, 98
acetic acid, temperature dependence of pKa, 98
acetic acid-acetate buffer, 131–133, 135–136
acetic acid-ammonia diagram, 122
acetic acid-formic acid diagram, 118
acetic acid-hydrochloric acid diagram, 120
acid mine drainage, 270–280, 510–513
acid mine water, 270–280, 285–286
acid rain, 149–150
acid, strong, 65–89
acid-base equilibria, effect on potential, 334–339
acidemia, 404
acidosis, 404
acids involving gas phase, 161
acids, monoprotic, 94–156
acids, polyprotic, 157–193
Ackermann, T., 91, 454
Acree, S.F., 200

activity coefficient conventions, 504
activity coefficient for HCl from Harned Cell, 56
activity coefficient, bicarbonate ion, 401
activity coefficient, carbon dioxide, 401
activity coefficient, carbonate ion, 401
activity coefficient, single ion, 462
Activity coefficients, 41 ff
activity coefficients, ratio Debye-Hückel to experiment, table, 490
activity coefficients, Davies equation, 491–497
activity coefficients, single ion, 45
activity coefficients, uncharged molecules, 49, 50
activity in equilibrium constant, 12
activity of water, 477
activity vresus potential at constant pH, 328–329
adipic acid, 158, 193
Aditya, S., 363
age of precipitate, 489
alkalemia, 404
alkalinity in natural waters, 195
alkalinity of seawater, comparison, 521
alkalinity titration curve, experimental, 422–429
alkalinity titration, seawater, 419–427
alkalinity, definition, 375, 378–379
alkalinity, experimental methods, 450
alkalinity, species contributing, 378–379
alkalosis, 404
Allen, H.E., 538

Allison, J.D., 538
α, degree of dissociation, electrolyte, 479
α, degree of dissociation, 102
$α_{12}$, Harned rule coefficient, 468–470
aluminum fluoride complexes, 285
aluminum hydroxide complexes, 260–261, 264–268
amines, equilibria, 159
amines, pH, 178–181
amino acid classical constants, 194
amino acid complexes, 289
amino acids, equilibria, 160–161
amino acids, pH, 182–183
Ammann, D., 316
ammonia, 7, 8
ammonia buffer index, 297
ammonia pH, 113–114
ammonia titrated with acetic acid, 154
ammonia titration with chloroacetic acid, 154
ammonia-acetic acid diagram, 122
ammonia-ammonium equilibria, 161
ammonia-ammonium equilibrium diagram, 111
ammonium, 7, 8
ammonium acetate diagram, 122
ammonium chloride pH, 111–113
ammonium ion, 494
Amyot, M., 287
Anderegg, G., 363
Andersen, N.R., 458
anodic current at platinum electrode, 355–358
Antelman, M.S., 361
Apple Macintosh computer, see Macintosh
Apple Macintosh operating systems, 486
arginine, 160
Arrhenius, S., 150
arsenic acid, 158
arsenic V/III, 488
arterial blood pH, 404
aspartic acid, 160
Astra Pharaeuticals, 456
atmospheric carbon dioxide increase, effect on oceans, 427–434
automated computation methods, 485–541
Avery, G.B., 156
Avogadro number, 38

Baes, C.F., 287, 502, 538
Bard, A, 361–363
barium chromate, 226
barium fluoride, 203, 226
barium sulfate, 203–207, 226
Barker, G., 483, 541
Barner, H.E., 361
base, strong, 65–89
bases, monoprotic, 94–156
bases, polyprotic, 157–193
basicity constant, 154
Bassett, R.L., 459, 484, 487, 536
Batelle Pacific Northwest Laboratory, 482
Bates, N.R., 458–459
Bates, R.G., 62–64, 92, 155, 200, 361, 481–484
Bates-Guggenheim convention for activity coefficient, 463
Battelle Pacific Northwest Laboratories, 484
Bertram, M.A., 454
$β_c$, buffer index in closed solution of carbonate, 392–393
$β_d$, buffer index in closed solution with CaCO3, 394–399
$β_p$, buffer index at constant CO2 pressure, 393–394
$β_{p,s}$, buffer index of CaCO3 at constant CO2 pressure, 394
Betts, J., 459
bicarbonate ion, 514–516, 519–520, 528
bicarbonate ion pair distribution, seawater, 422–427, 435–445, 522
bicarbonate ion, activity coefficient, 401
bicarbonate, ionization to carbonate, 370–372
bidentate, 315
Bird, D.K., 39, 540
Bischoff, W.D., 454
Bishop, F.C., 454
Bjerrum, J., 38, 63, 316–317
Bjerrum, N., 201
blood plasma, 404–407
Blowes, D.W., 287
Boecker, W.S., 459
Boink, A.B.T.J., 457, 460
Bolin, B., 458
borate buffer in NaCl, pH, 470–471

INDEX   **545**

borate ion, 514–515, 518, 520
boric acid, 514–515, 518, 520
boric acid ionization constant, 481
Bower, V.E., 64, 484
Bradshaw, A.L., 422, 427, 458
Breland, J.A., 155
Brewer, L., 361
Brewer, P.G., 458
brines, 461–484
brom thymol blue, 151
bromide ion, 490–491, 497
Brown, D.S., 538
brucite, 203
Brush, L.H., 481, 541
buffer capacity, carbon dioxide and carbonate, 391–400
buffer factor, Revelle, 428–434
buffer index, 133–136
buffer index, acetic acid-acetate, 137
buffer index, carbon dioxide and carbonate, 391–400
buffer index, polyprotic acid, 190–193
buffer index, summary of results, 399–400
buffer solutions, 130–136
buffers, NBS, 463
Buse, D., 457–458
Busenberg, E., 536–537
Butler, J.N., 92, 200, 286–288, 316, 362–364, 422, 455–456, 458, 459, 481–482, 484
Byé, J., 483
Byrne, R.H., 92, 155, 482

cadmium chloride complexes, 240–256
cadmium chloride complexes, distribution diagrams, 242–245
cadmium chloride, mass balance, 244–248
calcite, 489
calcium bicarbonate ion pair, 422–427, 435–445, 513–517, 519–520, 528
calcium carbonate, 230, 235, 381–400, 525–526
calcium carbonate ion pair, 422–427, 435–445, 514–516, 518, 520, 528
calcium carbonate ion pairs, temperature dependence, 451
calcium carbonate precipitation from lactated Ringer's solution, 413–416
calcium carbonate reaction with acid mine drainage, 525–531
calcium carbonate saturation index seawater comparison, 521
calcium carbonate saturation index, seawater model, 439–442
calcium carbonate saturation index, seawater-river water mixture, 446–448
calcium carbonate saturation, constant partial pressure, 380–388
calcium carbonate solubility in pure water, 388–391
calcium carbonate, solubility product, 372–374
calcium carbonate, solubility product, seawater, 420–421
calcium carbonate-carbon dioxide equilibria, table, 381
calcium fluoride, 203, 226, 234, 282
calcium fluoride ion pair, 514–515, 518, 520
calcium hydroxide, 203
calcium hydroxide ion pair, 422–427, 435–445, 514–517, 519–520
calcium in seawater, comparion, 521
calcium ion, 514–515, 518, 520, 525, 528
calcium ion pair distribution, seawater, 422–427, 435–445, 522
calcium phosphate, 230, 235
calcium sulfate, 203, 282
calcium sulfate (gypsum), 529–530
calcium sulfate ion pair, 528, 514–517, 519–520
calcium-NTA complex, 315
calcium hydroxide ion pair, 422–427, 435–445, 528
Cammann, K., 316
carbon dioxide, 158–159, 161, 365–460, 513–515, 525–530
carbon dioxide equilibria for physiological solutions, table, 408
carbon dioxide equilibrium constants, seawater, table, 420
carbon dioxide ionization constant, six forms, 401–404
carbon dioxide partial pressure ($10^{-3.5}$ atm), 384–386

carbon dioxide partial pressure (10^{-6} atm), 385–388
carbon dioxide partial pressure 1 atm, 382–383
carbon dioxide partial pressure 100 atm, 382–384
carbon dioxide partial pressure, function of pH, 409–416
carbon dioxide partial pressure, models for physiological solutions, 407–416
carbon dioxide partial pressure, ocean, as a function of total carbonate, 434
carbon dioxide, activity coefficient, 401
carbon dioxide, calcium carbonate saturation, 380–388
carbon dioxide, Henry's law, 367
carbon dioxide, ionization equilibria, 367–372
carbon dioxide, removal from atmosphere, 427–428
carbon dioxide, source of error in acid-base titration, 449
carbon dioxide-calcium carbonate equilibria, table, 381
carbonate equilibrium constants, seawater, table, 420
carbonate in seawater, comparison, 521
carbonate ion, 514–518, 520, 528
carbonate ion pair distribution, seawater, 522
carbonate ion, activity coefficient, 401
carbonate ionization equilibria, 367–372
carbonate, source of error in acid-base titration, 449
carbonate/bicarbonate buffer, 60
carbonic acid, 513–516, 518–520
case studies on carbon dioxide equilibria, 400–460
cathodic current at platinum electrode, 355–358
CEAM program, 520, 538
ceric-ferrous titration, 349–351
cerium iodate, 203
cesium ion, 491
cesium ion, 494–496
charge balance, 19
charge balance in blood plasma, 406
Charlot equation, 132
Chebotar, N.G., 537

chelates, 291–306
Chen, C.T.A., 155, 482
chloride ion, 490–491, 494–497, 514–515, 518, 520
chloride ion activity coefficient, conventions, table, 473–474
chloride ion selective electrode, 468–473
chlorine redox potential, 327
Choppin, G.R., 155, 482
Christ, C.L., 362, 454, 460
citric acid, 158
Clark formula for pH of weak acid-weak base, 150
Clegg, S., 459–460
closed system, mass balances on carbon dioxide, 379–380
cobaltic-cobaltous redox potential, 327
Cogley, D.R., 485, 538
Coleman, R.L., 457
common ion effect, 204–205
complex formation, 238–317
complex formation constants, table, 240
complex formation, effect on ferric-ferrous potential, 342–348
complex formation, effect on potential, 332–339
complex formation, effect on solubility, 231, 256–259
complexes, formation from solid, 15
complexes, overall formation constants, 14, 15
complexes, stepwise formation, 14, 238–245
complexometric titrations, 300–306
computation, setting up, 500–504
computational accuracy, 488
computer methods, automated, 485–541
computer programs, overview, 504
computers, personal, 1, 485–531
concentration dependence of potential, 323
concentration scales, 9, 10
concentration values, 488
concentrations of components, 501
conditional formation constant for MgEDTA, 299
conditional stability constants, 295–300
conditions, definition, 500
conductimetric end point, 79–81

constant ionic medium, 43
constant ionic medium, 463–464
constant partial pressure ($10^{-3.5}$ atm), 384–386
constant partial pressure (10^{-6} atm), 385–388
constant partial pressure 1 atm, 377–378, 382–383
constant partial pressure 100 atm, 382–384
constant partial pressure, pH and alkalinity, 377–378
copper ammine complexes, 7
copper aquo complexes, 8
copper chromate, 226
copper hydroxide and glycine, 315
copper iodate, 203
copper redox potential, 327
copper-zinc cell, 321–325
Covington, A.K, 316
Crane, F.E., 63
cresol red, 465–466
Culberson, C.H., 63, 155, 456, 482
cumulative distribution diagram, 242–245
cuprous-cupric equilibria, table, 360
current-potential curves, 322
curve-crawling method, 25
Cuta, 454
cysteine, 160

Daniell cell, 321–325, 361
Daniell, J.F., 361
Dantzig, G.B., 536
database of equilibrium constants, 487
Davies equation, 49, 50
Davies equation for 37 °C, 457
Davies equation used in MINEQL+, 505
Davies equation versus experimental activity coefficients, 491–497
Davies model, 489–492
Davies, C.W., 63, 154, 316, 541
Davies–Guggenheim parameter b, table, 503
Davis, A., 288
Davis, J.A., 523, 537, 540
Davis, R., 453–454
Dawson, H.J., 286
Debye Hückel equation, 524
Debye, P., 62
Debye–Hückel activity coefficients, ratio to experiment, table, 490

Debye–Hückel model, 489–490
Debye–Hückel theory, 44
default database, 487
degree of dissociation, 102
degree of formation, 102
Delany, J.M., 39, 540
diammonium hydrogen phosphate, distribution diagram, 184
diammonium hydrogen phosphate, pH, 183–185
dichromate-ferrous titration, 351–354
Dickson, A.G., 63, 155, 286, 316, 455–456, 458, 481–483, 538
dimethylamine titration with chloroacetic acid, 154
dimethylglyoxime, nickel complex, 313–314
Ding, K., 484
(1, 5) diphenylthiocarbazone, see dithizone, 307–312
diphosphoric acid, 158
diprotic acid distribution, approximations, 197–199
diprotic acid, general form, 176
disequilibrium, 488, 503
disodium hydrogen phosphate, pH, 173–174
dissociation constant, weak acid, 10
dissociation constant, weak base, 13, 14
dissociation curve, polyprotic acid, 185–188
dissociation, electrolyte, 479
dissociation, weak acid, 95–110
distribution diagram, complexes, 241–245
distribution diagrams, polyprotic acids, 166–170
distribution ratio, 306
dithizone, acid-base equilibria, 307–312
dithizone, copper complex, 307–312
dithizone, extraction equilibria, 307–312
dolomite, 452
Drenan, J.W., formula for pH in buffer, 152
Driscoll, C.T., 287
Druga, T.R., 200
Dunkle, S.A., 540
Durst, R.A., 316, 482
Dyrssen, D., 316
Dzombak, D.A., 505, 523, 537, 540

Eary, L.E., 536
Edmond, J.M., 459–460
EDTA, 158, 293–295
EDTA acid-base distribution diagram, 294
EDTA complex constants, 295
EDTA, effect on ferric-ferrous potential, 343–344
EDTA, effect on ferric-ferrous potential, 346–347
EDTA, ferrous and ferric complexes, table, 343
EDTA-ferric ion titration, 314
EDTA-Mg buffer, 299–300
EDTA-Mg titration, 302–306
E_H, redox potential in natural systems, 354–358
Ehlers, R.W., 64, 481
Eigen, M., 37
electrochemical cells, 320–328
electrochemical potential, 321
electrolyte effects on formation constants, 494–497
electron balance, 329–331
electronegativity, 319
Ellis, A.J., 38, 63, 453–454
Elving, P.J., 155, 236, 316, 361
Elzerman, A., 536
end point, titration, 78–89
Ender, F., 200
enthalpy, 16, 17
Enting, I.G., 458
entropy, 16
Environmental Research Software, 506, 538
EQ3 program, 469–470
EQ3/6 program, 513, 531–534
EQ3/6 program specifications, 506–508
EQ3NR program, 533, 482–483
EQ6 program, 533
EQPT program, 532
equilibria, simultaneous, 6, 7, 8
equilibrium constant, 10–17
equilibrium constant measurement, 58–61
equilibrium constant sources, 16
equilibrium constant, six forms, 401–404
equilibrium constants for pyrite oxidation model, 528
equilibrium, approach to, 5, 6, 36, 37
equilibrium, definition, 2

equivalence point, 76–78
equivalents versus moles, 92
Eriochrome Black T, 303–306
errors, interpolation, 504
η, sharpness index, 141–147
ethanol as solvent, 89
ethylene diamine, 159–160
ethylene diamine tetra-acetic acid (EDTA), 158, 293–295
ethylene diamine, distribution diagram, 180–181
ethylene diamine, pH, 179–180
ethylene dinitrilo tetra-acetic acid (EDTA), 158, 293–295
ethylenediamine nickel, 292–293
Excel spreadsheeet program, 2, 511
extraction coefficient, 306
extrapolation of formation constants, 498–499

Fedorov, V.A., 537
Federova, A.V., 537
Felmy, A.R., 482
feric-ferrous redox potential, 327
ferrous sulfate ion pair, 528
ferric hydroxide (amorphous solid), 525–526, 529–530
ferric hydroxide complexes, 261, 267–279, 511–513, 528
ferric hydroxide complexes, table, 272
ferric hydroxide solubilities, 261, 272, 285
ferric hydroxo complexes, 511–512
ferric ion, 510–513, 527–528
ferric ion-EDTA titration, 314
ferric oxyhydroxides, 489
ferric sulfate complexes, 511–513, 528
ferric sulfate complexes, table, 272
ferric-bisulfate complex, 286
ferric-ferrous equilibrium, 329–330
ferric-ferrous potential in sulfate, 333
ferric-ferrous potential, effect of pH, 334–337
ferrihydrite, 511–513
ferrous hydrogen sulfate ion pair, 272, 288, 528
ferrous bicarbonate, 528
ferrous carbonate, 528
ferrous ion, 525–528
ferrous iron redox potential, 327

ferrous sulfate ion pair, 525
ferrous sulfide complex, 528
ferrous-ceric titration, 349–351
ferrous-dichromate titration, 351–354
ferrous-ferric equilibrium, 329–330
ferrous-ferric potential in sulfate, 333
ferrous-ferric potential, effect of pH, 334–337
ferrous-hydroxide complex, 528
ferrous-mercuric equilibrium, 329–330
Fischer, R., 483
Fishman, M.C., 456
Fleming, G.W., 540
Flood's diagram, 101–102
Flood, H., 155
fluoride ion, 490–491, 514–515, 518, 520
Fogh-Andersen, N., 457, 460
formal potential, 332
formation constant errors, 494–499
formation constant values, 502
formation constants for pyrite oxidation model, 528
formation constants, ionic strength dependence, 249–256
formation constants, table, 240
formation curve, polyprotic acid, 185–187
formic acid, Gran titration, 153
formic acid, ionization, 148
formic acid-acetic acid diagram, 118
FORTRAN 77 program, 532
fraction titrated, 138
Franks, Felix, 91
Fraser, G.T., 91
Frazier, A.W., 237
free energy, 321, 323
Free, S., 483, 541
Freiser, H., 316
French, D.M., 453
Fridman, Ya. D., 317
fugacity coefficient, carbon dioxide, 12, 38, 453
FZERO program, 39

Gabrielson, C.O., 155
Gaines, G.L., 537
Gamble, J.L., 457
Gamblegram, 406
Garrells, R.M., 362, 459–460, 481, 487, 513, 536, 538

gas phase of acids, 161
Geiger, R.W., 317
GEOSECS (geochemical ocean section) data, 432
geothermal, 531
Getz, C.A., 363
Gieskes, J.M.T.M., 459–460
Giner, J., 362
glass pH electrode, 51
gluamic acid, 160
gluconate, 410–411
glutaric acid, 158, 193
glycine, 160
glycine, acidity constants, 289
glycine, acidity diagram, 291
glycine, copper complex, 289–292
glycine, distribution diagram, 182–184
goal-seek method, 33–35
goethite, 489
Golding, R.M., 38, 63, 453
Goyet, C., 456, 458
Gran titration, 84–87, 425, 450
Gran, G., 92
Greely, R.S., 484
Griffith formula for pH of weak acid-weak base, 150
Gripenberg. S., 455
Grundl, T., 536
Grunwald, E., 156
Grzybowski, A.K., 200
Gubeli, A.O., 538
Guggenheim, E.A., 38, 63–64, 481, 483, 287
Guldberg, C.M, 38
Guntelberg, E., 63
Gurney, E.L., 237
gypsum, 489

H_2SO_4, see sulfuric acid
half-cell potential, 325–328
HALTAFALL program, 39
Hamer, W.J., 91–92, 484
Hanson, P.J., 502, 538
Hansson, I., 427, 456, 458
Harned cell, 53–55, 465
Harned cell, equilibrium constant measurement, 58–59
Harned rule, 468–470
Harned, H.S., 62, 64, 91–92, 453–454, 481–483

Harris, F.J., 361,
Harvie, C.E., 481, 538, 540
Hasselbalch, K.A., 155
Havel, J., 154
Hawley, J. E., 456
HCN equilibrium diagram, 109–110
Heimann, M., 458
Helgeson, H.C., 39, 453–454, 540–541
Helgeson, H.D, 540
Helmke, P.A., 536
hematite, 489
Henderson, L.J., 155
Henderson-Hasselbalch equation, 132
Henry's law, carbon dioxide, 367
Hering, J.G., 155, 502, 537–538
Herlem, M., 92
hexammine nickel, 292–293
Heyrovska, R., 63, 459, 480–481, 484
HCl, see hydrochloric acid
HF, see hydrogen fluoride, hydrofluoric acid
Hill, A.E., 237
Hindman, J.C., 459–460
Historical context of computation, 486–487
Hochella, M.F., 537
HOCl titration with NaOH, 153
Hoffman, A.R., 456
Högfeldt, E., 154
Holland, H.D., 458
Holmes, H.F., 483
homogeneous buffer index, 428–434
Hostetler, P.B., 454
Houghton, R.A., 458
Hückel, E., 62
Hutchinson T.C., 287
hydration number, 478
hydration theory of activity coefficients, 463, 477–480
hydrazine, 160
hydrochloric acid activity in NaCl, pH, 468–474
hydrochloric acid in glacial acetic acid, 148
hydrochloric acid-acetic acid diagram, 120
hydrocyanic acid equilibria, 109–110
hydrofluoric acid equilibria, 107–109, 148
hydrofluoric acid, ionization, hydrogen, 53, 528

Hydrogen electrode, 53, 465
hydrogen fluoride, 514–515, 519–520
hydrogen ion, 4–14, 50–58, 65–78, 95–130, 325–328, 490–494, 514–515, 518, 520, 528
hydrogen ion, aqueous structure, 66
hydrogen ion, standard free energy, 361
hydrogen redox potential, 327
hydrogen selenide, 158–159
hydrogen sulfate ion, 511, 514–517, 519–520, 528
hydrogen sulfate ion, pK_a, table, 475–476
hydrogen sulfide, 158–159, 528
hydrogen sulfide equilibria, 162
hydrolysis of metal ions, 259–270
hydrothermal, 531
hydroxide complexes, 259–270
hydroxide ion pair distribution, seawater, 422–427, 435–445, 522
hydroxide ion, 66–74 490–491, 511, 514–515, 519–520, 528
hydroxyl ion, see hydroxide ion
hydroxyquinoline, 160
hypochlorous acid, titration with NaOH, 153

IBM personal computer, 486
illites, 489
imidazole, 151
indicator cresol red, 465–466
indicator, choice, 152
indicators as weak acids, 123
indicators, acid-base titration, 79
indicators, pH measurement, 126–130, 465–467
Ingri, N., 39
insoluble materials, 488–489
internal resistance, 322
interstitial fluid, 404–405
intracellular fluid, 404–405
intravenous solutions, 404–416
iodic acid, 151
iodide ion, 491–497
ion association, 489
ion interaction, 489
ion pair constants for Normosol, 410
ion pair constants for seawater, table, 435
ion pair model for seawater, equilibria, 423–425

ion pair model for seawater, mass balances, 422
ion pair model, comparison with experiment, 413
ion pair models for physiological solutions, 409–416
ion product of water, 66–70, 372
ion-pair model for plasma, 452
ion-size parameter, 46–49
ionic interactions, 42
ionic stength corrections, 501–502
ionic strength, 45
ionic strength depencence, silver chloride complexes, 257–258
ionic strength dependence of Ka, polyprotic acids, 162–164
ionic strength dependence of pKa, 98
ionic strength dependence, cadmium chloride constants, 249–256
ionic strength dependence, complex formation constants, 249–259
ionic strength, effect on solubility, 210–213
ionization constant table, polyprotic acids, 158
ionization constant, six forms, 401–404
ionization constants, amines and amino acids, 160
ionization of carbon dioxide, 367–372
ionization, weak acid, 95–110
iron hydroxide complexes, 261, 267–279
iron sulfide, 510, 525
iron-ferrous-ferric equilibrium, 329–330
isohydric solutions, 150
IUPAC sign conventions, 321–323
Ives, D.J.G., 482
Izmailov, N.A., 91

Jackson, K.J., 63, 459, 482, 484
Jackson, M.L., 536
Jacobson, R.L., 460
Jagner, D., 63, 316, 482
Jain, D.V.S., 200
Jambor, J.L., 287
Janz, G.J., 482
Jaques, J., 236
Jenne, E.A., 538, 541
Johansson, O, 456
Johnson, K.S., 460
Johnson, S.M, 536
Jones, B.F., 490, 536–537, 540
Jordan, J., 361–363
Jorgensen, C.K., 286

Kahaner, D., 40
Kajander, K., 200
Karl, D.M., 458
Kauffman, G.B., 286
Kent, D.B., 523, 537, 540
Kielland's table of ion sizes, 47
Kielland, J., 63, 287
Kinniburgh, D.G., 536
Klausner, R.D., 456
Kleiber, A., 483
Klotz, I.M., 483
Knap, A.H., 458–459
Knauss, K.G., 63, 482
Kolthoff, I.M., 92, 155, 236, 316, 361–364
Koskinen, M., 200
Kso, 202–237
Ksusalik, P.G., 91
Kugelmass, I.N., 200
Kulba, F.Ya., 537
Kw, 67–70

lactated Ringer's solution, 413–416
lactated Ringer's solution, composition, table, 414
lactated Ringer's solution, ion-pair constants, table, 415
lactated Ringer's solution, mass balances, table, 415
Lagrange multipliers, 487
Lahiri, S.C., 363
Laitinen, H.A., 288, 316–317, 363–364
Lamb, A.B., 286
Lange, P.W., 155
Langmuir surface adsorption, 505
Langmuir, D., 454, 460, 536–537
Lanier, R.D., 155, 484
lanthanum iodate, 203
Larson, A.T., 286
Lassey, K.R., 458
Latimer, W.M., 361
Lawrence Livermore National Laboratory, 482, 531
lead chloride complexes, 494–497
lead formate, 234

lead iodate, 203
lead sulfate, 203, 226
lead-NTA complex, 315
leadl chromate, 226
Lee, T.S., 363
lepidocrocite, 489
Leussing, D.L., 363
Lewis, G.N., 38, 361
Licht, S., 200
Liebig titration, 283
Liebig, J., 288
Lietzke, M.H., 483–484
Lindberg, R.D., 358, 364
Lindsay, W.T., 200
Lingane, J.J., 364
Linke, W.F., 236–237
liquid junction potential, 53, 463–465
Lister, M.W., 288
lithium ion, 490–491, 494–496
logarithmic concentration diagram, 103–110
Logarithmic concentration diagrams, precipitation, 207–208
Longhi, P., 483
Lotus spreadsheet program, 511
Lyons, W.B., 288

Macalady, D.L., 536
Macaskill, J.B., 482–484
MacDougall, F.H., 236
MacInnes convention for single ion activity coefficient, 462
MacInnes, D.A., 481, 483
Macintosh computer, 505, 511, 524–525
Macintosh operating systems, 486
MacIntye, F., 38
MacIntyre, W.G., 460
Mackenzie, F.T., 454, 458
magnesium ammonium phosphate, 230, 235–236
magnesium bicarbonate ion pair, 422–427, 435–445, 513–517, 519–520
magnesium carbonate, 230
magnesium carbonate ion pair, 422–427, 435–445, 514–517, 519–520
magnesium chloride-sodium hydroxide, pH, 483–484
magnesium fluoride, 203, 226, 234
magnesium fluoride ion pair, 514–515, 519–520

magnesium hydroxide, 203, 226
magnesium hydroxide ion pair, 422–427, 435–445, 514–516, 519–520
magnesium hydroxide precipitation, 474–476
magnesium hydroxide saturation diagram, 298
magnesium ion pair distribution, seawater, 522
magnesium sulfate ion pair, 422–427, 435–445, 514–517, 519–520
magnesium sulfate, activity coefficient, 44
magnesium ion, 514–515, 519–520
Malahoff, A., 458
Malhotra, H.C., 200
Malinin, S.D., 63
malonic acid, 193
Manassen, J., 200
manganese dioxide redox potential, 327
manganese equilibria, 331
manganese redox reactions, 338–342
manganese, pH-pe diagrams, 338–342
Margerum, D.W., 40
Marozeau, J., 288
Martell, A.E., 38–39, 91, 154, 199, 236–237, 286–287, 316–317, 361, 363–364, 453–454, 457, 459, 483, 494–497, 502, 537–538
mass and charge balances, carbon dioxide, 374
mass balance on H+, 116–117
mass balances, 17
mass balances for Normosol, 411
mass balances on carbon dioxide, closed system, 379–380
Mattigod, S.V., 288
maximum slope of titration curve, 218–225
May, H.M., 536–537
McBryde, W.A.E., 363
McClendon, J.F., 155
McGaw, Inc., 456
Meema, K.M., 287
Mehrbach, C, 456, 458
Melchior, D.C., 459, 484, 487, 487
Meloun, M., 154
mercuric chloride complexes, 240, 499
mercuric chloride complexes, distribution diagram, 248

mercuric chloride titration, 281
mercuric chloride, mass balance, 248–249
mercuric iodide, 282
mercuric ion as acid, 159
mercuric oxide solubility, 535
mercuric-ferrous equilibrium, 329–330
mercurous chloride, 203
mercurous sulfate, 203, 226
mercuruous-mercuric redox potential, 327
mercury hydroxide complexes, 259–265
Merlainen, P., 200
Mesmer, R.E., 287, 482–483, 502, 538
metal buffers, 299–300
metal indicators, 303–306
methane, 528
methyl orange, 123–124, 150–151
Mg-EDTA buffer, 299–300
Mg-EDTA titration, 302–306
micas, 489
Michaels, A.F., 458–459
Microsoft DOS computer, 505
Microsoft DOS operating system, 486
Miller, G.C., 288
Millero, F.J., 454, 458–460, 484
Milne, P.J., Thurmand, V.L., 454
MINEQL program, 63, 504, 538
MINEQL+ program, 504–517, 538, 540
MINEQL+ program specifications, 506–510
MINTEQA1 program, 538
MINTEQA2 database, 503, 505, 508, 538
MINTEQA2 program, 516–523, 538
MINTEQA2/PRODEFA2, 538
Mironov, N.N., 538, 540
Mironov, V.E., 537
Missen, R.W., 487, 536
Mitchell, R., 287
Mitra, R.P., 200
mixed layer, ocean, carbon dioxide, 427, 431
mixed potential, 356–358
model definition, 500
Mohr titration, 233, 218
molal concentration scale, conversion from molar, 497–498
molar concentration scale, conversion from molal, 497–498
Moler, C., 40
Møller, N., 481, 538, 540

Monoi, H., 457
mononuclear complexes, 240–245
mononuclear equilibria, 240–245
monoprotic acids and bases, 94–156
monoprotic acids, table of pK_a, 96
monoprotic bases, table of pK_b, 97
Montano, L.R., 483, 541
Monte Carlo simulations, 505
Morel, F.M.M., 155, 502, 523, 537, 538, 540
Morgan J.J., 287, 360, 362, 364, 502, 513, 538
Morse, W., 458–460
Motekaitis, R.J., 154, 286, 537
Moyer, H.V., 363
Mucci, A., 458–460
Murphy, G.M., 64
Murphy, P.J., 536
Mussini, P.R., 483
Mussini, T., 483

$\bar{n}$, degree of formation, 185
$NaHSO_4$ titration with NaOH, 153
Nasanen, R., 200
Nash, S., 40
National Bureau of Standards (NBS), 58, 64, 402, 456
National Institute of Standards and Technology (NIST), 64, 456
Nazarov, V.A., 537
NBS (National Bureau of Standards) pH scale, 58, 64, 402, 456
Nernst equation, 323
Nernst, W., 202, 236, 361
Nesbitt, H.W., 540
net current at platinum electrode, 355–358
neutralization of acid mine drainage, 525–531
Newbitt, H.S., 39
Newton's approximation method, 27–32
NH_3, see ammonia,
nickel ammonia complex constants, 292–293
nickel ethylenediamine complex constants, 292–293
Nicholson, R.V., 287
NIST (US National Institute of Standards and Technology), 64, 456
NIST database, 494–496, 502, 535
nitrate ion, 490–491, 494–497

nitrilotriacetic acid, see NTA, 315
nitroprusside indicator, 281
nonqueous solutions, 70
Nordstrom, D.K., 536–537, 541
Normosol-R, 405–456
Novo-Gradac, K.J., 538
NTA (nitrilotriacetic acid) acidity constants, 315
NTA complex formation constants, 315
NTA-calcium complex, 315
NTA-lead complex, 315
nuclear waste disposal, 531

Odnosevtsev, A.I., 538, 540
Oka, Y., 540
open circuit potential, 324–325
Orgel, L.E., 286
osmotic coefficient, 477
overvoltage, 322
Owen, B.B, 62, 64
oxalic acid, 158, 193
oxidation, 319
oxidation number, 319
oxidation of pyrite, 525–531
oxidation-reduction equilibria, 318–364
oxidation-reduction versus complex formation, 320
oxygen, 342, 525–530
oxygen redox potential, 327

$p\varepsilon$ redox potential, 328
paH, 56–58, 63
paHγ_{Cl}, 56–58, 63
Palmer, D.A., 483
Parkhurst, D.L., 523, 536, 537, 540
Parsons, R., 361–363
partition coefficient, 306
Patterson, A., 453
Pauling, L., 360
pcH, 63
Pednekar, S.P., 484
Peng, T.H., 459
Pentium processor, 486
perchlorate ion, 491, 494–497
Perdue, E.M., 538
permanaganate auto-oxidation, 320
permanganate redox potential, 327
permanganate reduction potential, 326–328
pH from Harned cell, 56

pH in brines, 461–484
pH measurement with indicators, 126–130
pH measurement, limitations, 462
pH models for physiological solutions, 407–416
pH of phosphoric acid and salts, 169–175
pH of pure water, 69–70
pH of seawater, comparison, 521
pH of strong acid, 70–72
pH of strong base, 72–73
pH of weak acid, 99–102
pH scale, 50–58
pH scale, extrapolation to I = 0, 57
pH, blood, 404
pH, carbonate/bicarbonate buffer, 60
pH, comparison of spectrophotometric and potentiometric, 128–130, 467
pH, constant carbon dioxide partial pressure, 375–378
pH, definition, 462
pH, magnesium chloride-sodium hydroxide, 483–484
pH, ocean, as a function of total carbonate, 433
pH-pe diagrams, 335–342
pHCl, conversion to pH, table, 473
phenanthroline, effect on ferric-ferrous potential, 344–348
phenanthroline, ferrous and ferric complexes, table, 348
phenolphthalein, 124–125
ϕ, fraction titrated, 138
phosphoric acid, 158–159
phosphoric acid dissociation equilibrium, 13
phosphoric acid, distribution diagrams, 166–170
phosphoric acid, ionic strength and temperature dependence, 162–166
phosphoric acid, pH, 169–172
PHREEQC database, 506
PHREEQC formation constants, 528
PHREEQC output, 529–531
PHREEQC program, 63, 490, 499, 502, 513, 523–531, 540
PHREEQC program specifications, 506–508
PHREEQC runtime file for pyrite oxidation, 526
PHREEQE (pronounciation "freak"), 538

INDEX 555

PHRQPITZ program, 540
phthalic acid, 158
physiological fluids, 404–416
Pitzer activity coefficient approach, 492–494
Pitzer activity coefficient approach, list of parameters, 494
Pitzer calculations for HCl activity, 469
Pitzer K.S., 62–63, 200, 316, 361, 454, 459, 481–484, 492–494, 537, 540, 541
pK_a, boric acid ionization constant, 481
pK_a, hydrogen sulfate ion, table, 475–476
pK_a, ionization constant for indicator cresol red, 465–466
pK_{a1}, carbon dioxide ionization constant, six forms, 401–404
pK_{a1}, ionization of carbon dioxide, 367–370
pK_{a1}, ionization of carbon dioxide, ionic strength dependence, 369
pK_{a1}, ionization of carbon dioxide, table, 368
pK_{a1}, ionization of carbon dioxide, temperature dependence, 370
pK_{a2}, ionization of bicarbonate, 370–372
pK_{a2}, ionization of bicarbonate, ionic strength dependence, 371
pK_{a2}, ionization of bicarbonate, table, 368
pK_H, Henry's law, carbon dioxide, 367
pK_H, Henry's law, carbon dioxide, table, 368
pK_{so}, solubility product of calcium carbonate, 372–374
pK_{so}, solubility product of calcium carbonate, ionic strength dependence, 372
pK_{so}, solubility product of calcium carbonate, temperature dependence, 373
pK_w, ion product of water, 372
pK_w, ion product of water, ionic strength dependence, 374
pK_w, ion product of water, table, 368
Platford, R.F., 538, 540
Plath, D.C., 460
Plumlee, G.S., 288
Plummer, L.N., 459–460, 536–540
pmH, rational pH, 482

Poisson, A., 456, 458
Polya, G., 39
polynomial equations, 24
polyprotic acids and bases, 157–193
polyprotic acids, mixtures, 183–185
Posner, A.M., 536
potassium carbonate ion pair, 422–427, 435–445, 514–516, 519–520
potassium ion, 490–491, 494–496, 514–515, 518, 520
potassium ion pair distribution, seawater, 522
potassium sulfate ion pair, 422–427, 435–445, 514–516, 519–520
potential, 321, 323
potential, effect of OH complexes, 334–339
potential, effect of other equilibria, 332–339
potential, effect of pH, 334–339
potentiometric versus spectrophotometric pH methods, 467, 477
Pourbaix, M., 362
precipitation titrations, 215–225
precipitation, mixing two solutions, 205–207
problem solving procedure, 20, 24
programs, overview, 504
proton condition, 19, 20
proton condition summary, 114–115
Publiano, N., 91
pure water equilibrium, 69
pyrite, 272–280, 510, 525, 528–530
pyrite oxidation products, 270–280, 510–513, 525–531
pyrophosphoric acid, 158
pyrophosphoric acid, pH, 176–178
Pytkowicz, R.M., 456, 460, 540

quadratic equations, 25–27
Quirk, J.P., 536

Ramette, R.W., 155
Randall, M., 38, 361
reaction stoichiometry, 501
reactions, opposing, 4
reasonableness, 503
REDEQL, 504
redox couples, 488

redox equilibria, 318–364
redox potential, 325–328
redox potential, analogy to pH, 328
redox potential, table, 327
redox potential, unique, 354
redox titrations, 348–354
reduction, 319
Revelle buffer factor, 428–434
Revelle, R., 459
Ricci, J.E., 39, 156
Rice, S.A., 484
Riley, J.P., 455–458
Ringbøm, A., 316
Ringer's solution, 405, 456
river water-seawater mixture, 444–448
Rivington, D.E., 288
Robert-Baldo, G., 155. 482
Robinson, R.A., 38, 62–63, 92, 482–484, 491–492, 536, 457
Rondinini, S., 483
Ross, J.W., 482
Rossotti, F.J.C., 154, 200, 286, 316
Rossotti, H., 154, 200, 286, 316
Roy, L.N., 484
Roy, R.N., 64, 316, 454, 456, 481–484, 458, 460
rubidium ion, 491, 494–496
Runnells, D.W., 358, 364

Sainte Marie, J., 502, 538
Salam, M.A., 362
saline physiological solutions, 404–416
salt of weak acid and weak base, 120–123
Salt of weak base and strong acid, 110–113
salting-out parameter, 490
Salutsky, M.L., 236
Sandell, E.B., 155, 236, 316–317
Sandia National Laboratories, 481–483, 533, 541
saturation index of calcium carbonate in lactated Ringer's solution, table, 415
saturation index of calcium carbonate, seawater ion pair model, 439–442
saturation index of calcium carbonate, seawater-river water mixture, 446–448
saturation index, calcium carbonate, seawater comparison, 521

Savage, L., 457–458
Saykally, R.J., 91
Schäfer, K., 200
Scheuerman, R.V., 361
Scholes, S.R., 454
Schramke, J.A., 536
Schroeder, A.H., 316
Schwarzenbach, G, 39, 315
Science Citation Index, 535
seawater equilibrium constants, table, 425
seawater pH scale (SWS), 464, 403–404
seawater speciation, 435–445, 513–523
seawater, alkalinity titration, 419–427
seawater, bicarbonate ionization constant, 419–420
seawater, carbon dioxide equilibria, 416–460
seawater, carbon dioxide equilibrium constants, table, 420
seawater, carbon dioxide ionization constant, 418–420
seawater, composition, 417
seawater, first acidity constant for carbon dioxide, 418–420
seawater, ion pair model, 434–448
seawater, second acidity constant for carbon dioxide, 419–420
seawater-basalt interactions, 531
seawater-river water mixture, 444–448
sebacic acid, 193
Secant method, 32, 33
Seidell, A., 236–237
sensitivity analysis, 441, 505
separation by precipitation, 208–210
serine, 160
Serkiz, S.M., 538
Seyfried, W.E., 484
Shafer, D.K., 458
sharpness index, 141–147
short circuit potential, 324–325
SID, strong ion difference, 406
Siebert, R.M., 454
Siggaard-Andersen, O., 457, 460
Sillén's diagram, 103–110
Sillén, L.G., 38–39, 91, 154–155, 199, 236–237, 288, 316–317, 361–364, 453, 483, 494–497, 537
silver acetate, 226–229, 234–235
silver acetate, solubility, 213–215

silver ammonia complexes, 284
silver bromate, 203
silver bromide, 203
silver chloride, 203
silver chloride complexes, 240
silver chloride electrode, 465
silver chloride redox potential, 327
silver chloride, solubility, 211–213
silver chloride, solubility in chloride, 256–259
silver chromate, 226
silver cyanide titration, 283
silver iodate, 203
silver iodide, 203
silver sulfate, 203
silver sulfate, 226
silver thiocyanate, 203
silver thiosulfate, 284
Silver, L.F., 483
Simmons, J.D., 237
Simms, H.S., 200–201
Simpson, H.J., 459
Singer, P.C., 287
Singleterry, C.R., 483
Skirrow, J., 455
smectites, 489
Smith, G.F., 363
Smith, R.M., 39, 91, 199, 236–237, 287–288, 362–363, 453–454, 457, 459, 494–497, 502, 537, 538
Smith, W.R., 487, 536
Smith, W.T., 484
sodium bicarbonate ion pair, 422–427, 435–445, 514–515, 519–520
sodium bisulfate titration with NaOH, 153
sodium carbonate ion pair, 422–427, 435–445, 514–516, 519–520
sodium chloride supporting electrolyte, 469, 497–498
sodium dihydrogen phosphate, pH, 172–173
sodium fluoride ion pair, 514–515, 519–520
sodium hydroxide in NaCl, effect of magnesium on pH, 474–476
sodium hydroxide in NaCl, pH, 470–471
sodium hydroxide ion pair, 422–427, 435–445, 514–516, 519–520
sodium ion, 47–49, 73–80, 490–491, 494–496, 511, 514–515, 518, 520

sodium ion pair distribution, seawater, 522
sodium nitrate supporting electrolyte, 497–498
sodium redox potential, 327
sodium sulfate activity coefficient, 535–536
sodium sulfate ion pair, 422–427, 435–445, 514–517, 519–520
SoftWindows program, 505, 511, 524–525
Solache-Rios, M., 155, 482
solubility diagrams, logarithmic, 208–209
solubility in pure water, 203–204
solubility product, 15, 202–237
solubility product of calcium carbonate, 372–374
solubility product, calcium carbonate, seawater, 420–421
solubility product, evaluated from data, 213–215
solubility product, table, 203, 226, 230
solubility, calcium carbonate in carbon dioxide solutions, 380–388
solubility, effect of complex formation, 231, 256–259
solubility, effects of acid and base on, 225
solubilty, salt of a weak acid, 225–230
solvent extraction, 306–313
solver program, 1, 33–35
Sørensen, S.P.L., 63
speciation model (tableau) for acid mine drainage, 511
speciation model (tableau) for seawater, 514–515
species fraction versus ligand concentration, 241–242
spectrophotometric pH measurements, 126–130, 463–467
spectrophotometric versus potentiometric pH methods, 129, 467, 477
Sposito, G., 288
spreadsheet, 1
standard free energy, 321, 323
standard potential, 321, 323
Staples, B.R., 62, 64
Stark, J., 362
Stenger, V.A., 92
stepwise formation of complexes, 238–245
Stewart, P.A., 456–457
Stock, D.I., 316
stoichiometry of reactions, 501

Stokes, R. H., 62–63, 38, 92, 457, 482, 484, 491–492, 536
Stoughton, R.W., 483–484
Stráfelda, F., 454
strong acid, pH, 70–72
strong acid-strong base titration, 73–89
strong acids and bases, 65–89
strong and weak acids and bases, mixtures, 117–123
strong base, pH, 72–73
strontium fluoride, 203
strontium sulfate, 203
Stumm, W., 287, 360, 362, 364, 513, 538
succinic acid, 158, 193
Suess, H.E., 459
sulfate in HCl, pH, 119–120, 474–475
sulfate ion, 514–518, 520, 525, 528
sulfate ion, 518
sulfate ion pair distribution, seawater, 422–427, 435–445, 522
sulfate ion, from pyrite oxidation, 510–512
sulfate/hydrogen sulfate equilibrium, 61–62
sulfide, 528
sulfide ion selective electrode, 468
sulfuric acid titrated with ammonia, 154
sulfuric acid, ionization, 148
Sullivan, J.C., 363
Sunda, W. G., 502, 538
Sundquist, E.T., 459
superacid H_4O^{2+}, 91
supporting electrolyte effects on formation constants, 494–497
Svischev, I.M., 91
SWS, seawater scale for pH, 403–404, 464
Sykes, K.W., 288

tableau (speciation model) for acid mine drainage, 511
tableau (speciation model) for seawater, 514–515
tableau summary, 502
Takahashi, T., 458–459
tartaric acid, 158
Taylor, A.W., 237
Teltschik, W., 200
temperature dependence of equilibrium constant, 16
temperature dependence of pK_a, polyprotic acids, 164–166

temperature dependence of pK_a, 98
Thaler, M.S., 456
thallium chloride, 203
thiosulfuric acid, 158
Thomas, H.C., 537
Thompson, M.E., 459, 481, 487, 513, 536, 538
Thompson, S.W., 155, 482
thymol blue, absorbance, 126–127
Tikhomirov, O.B., 537
titration curve fitting, nonlinear, 87–89
titration curve, polyprotic acid, 185–187
titration curve, slope, 218–225
titration end point, 78–89
titration equivalence point, 76–78
titration error, 81–84, 90, 139–147
titration error from carbon dioxide, 196
titration error, acetic acid, 144–146
titration error, polyprotic acid, 188–190
titration of chloride, bromide, iodide, 233
titration, precipitation, 215–225
titration, silver and chloride, 217–218
titration, strong acid-strong base, 74–89
titration, symmetrical precipitation, 218–220
titration, unsymmetrical precipitation, 220–225
titration, weak acid-strong base, 136–139
titration, weak base-strong acid, 136–139
Torma, A.E., 538
triaminopropane, 160
tridentate, 315
tris-hydroxymethylamino methane pH buffer, 466
trisodium phosphate, pH, 174–175
Truesdell, A.H., 490, 536, 540
Turner, D.R., 286, 538
Turyan, Ya. I., 537

US Department of Energy, 506
US Geological Survey, 506, 540
US Geological Survey database, 490
US National Bureau of Standards (NBS, NIST), 456
US National Institute of Standards and Technology (NIST, NBS), 456

van't Hoff factor, 477
Van't Hoff, J.H., 38
vanadium redox potential, 327

Vanderdeelen, J., 454
Vanderzee, C.E., 286
VanSlyke buffer index, 133
VanSlyke, D.D., 155
Vasil'ev, V.P., 317
Vasil'eva, V.N., 317
Vetter, K.J., 363
Via Medical Corp., 457–458
Vogel, K.M., 484
Volhard method, 233

Waage, P., 38
Wagman, D.D., 39, 361
Walton, A.G., 236
WATEQ program, 524, 540
WATEQF4 database, 508
WATEQF4 program, 490
water dissociation equilibrium, 62
water ionization, 65–89
water structure, liquid, 91
water, ion product, 12
weak acid and weak base, salt, 120–123
weak acid dissociation, 95–110
weak acid dissociation, calculation method, 21
weak acid-strong base titration, 136–139
weak and strong acids and bases, mixtures, 117–123
weak base and weak acid, salt, 120–123
weak base pH, 113–114
weak base-strong acid titration, 136–139
weak bases, effect on pH in brines, 474
Weare, J.H., 481, .538, 540
Wedborg, M., 456
Wells, C.F., 362
Wengelen, F., 316

Werner, A, 286
Wesolowski, D.J., 483
Westall, J.C., 504, 538
White, A.F., 537
White, W.B., 487, 536
Whitfield, M., 63, 286, 459–460, 482, 502, 538
Wigley, T.M.L., 459–460
Willey, J.D., 156
Williams, R.D., 458
Wilson, C.A., 156
Wimberley, P.D., 457, 460
Windows operating system, 486
Winn, C., 458
Winter, P.K., 363
WIPP (Waste Isolation Pilot Plant, Carlsbad, NM), 480
Wissbrun, K.F., 453
Wolery, T.J., 63, 459, 482–484, 531, 541
Wong, D., 457–458
Wright, J.M., 200
Wu, Y.C., 484

Yelton, W.G., 483, 541
Young, C.C., 457
Young, T.F., 483
YSZ, yttria-stabilized zirconia membrane for pH, 475

Zachary, J.L., 538
Zeltner, W., 92
Zielen, A.J., 363
zinc redox potential, 327
zinc-copper cell, 321–325
zwitterion, 194